D0204342

THE LIFE AND TIMES OF MODERN PHYSICS

HISTORY OF PHYSICS II

THE LIFE AND TIMES OF MODERN PHYSICS

HISTORY OF PHYSICS II

Edited by
MELBA PHILLIPS
Professor Emeritus of Physics
University of Chicago

Readings from *Physics Today*

Number Five

AIP

American Institute of Physics

New York, New York

1992

Readings from *Physics Today*

PHYSICS TODAY, a publication of the American Institute of Physics, provides news coverage of national and international research activities in physics as well as government and institutional activities that affect physics. Both technical and nontechnical developments are covered by scientific articles, news stories, book reviews, letters to the editor, calendars of meetings, and editorial opinion.

Articles in PHYSICS TODAY are intended to be of interest to—and understandable by—a broad audience of professionals from all subfields of physics as well as people with a general interest in physical science.

The Life and Times of Modern Physics: History of Physics II is the fifth book in a series of volumes that contains reprinted articles and news material from PHYSICS TODAY. The series is produced with the cooperation of Gloria B. Lubkin, editor of PHYSICS TODAY.

Library of Congress Cataloging-in-Publication Data

The Life and times of modern physics : history of physics II / edited by Melba Phillips.
     p. cm.—(Readings from Physics Today : no. 5)
    Includes bibliographical references
    ISBN 0-88318-846-5
    1. Physics—History.   2. Physics—History Sources.   I. Phillips, Melba, 1907–   . II. Series.
QC7.L47   1992
530'.09—dc20
                                                                92-12669
                                                                 CIP

Table of Contents

___ Introduction _____

In 1984 Spencer Weart and I put together a book of historical readings from *Physics Today*, chosen from the entire 35 years of the magazine's prior existence. *Physics Today* is a news magazine, and its first obligation is to report recent advances in physics or its immediate context, but today's physics is rooted in that of yesterday and indeed subsumes it. Just how this has taken place has been of growing interest, as indicated by the more and more frequent appearance of historical articles after a few initial years. The rise in interest among physicists and the emergence of attention from professional historians reflect a recognition of the ever closer interdependence of society and science today. The growth continues, and the accumulation of historical articles within the last decade is more than any single book of readings can accommodate. Various other modes of publication have been used or are under way. Pieces on the history of nuclear weapons have been included in another reprint volume, *Physics and Nuclear Arms Today*, in one example. A special issue on lasers was an outgrowth of the Laser History Project (founded in 1982 by four societies, including the American Physical Society and the Optical Society of America) which has its own publications program. The articles on Richard Feynman are being collected for another book. But what remains beyond such plans for alternate reprinting is by no means inferior.

Physics Today is not what is called an archival journal, but takes just pride in giving its readers previews of works in progress, preliminary or partial results. This objective applies to historical works as well as to results of purely scientific research. Expanded or supplemented treatments of several subjects included in this present volume have been subsequently published as books, but the original articles are of no less interest in themselves. Examples are *German National Socialism and the Quest for Nuclear Power, 1939–1948*, a biography of *Chan-*

drasekhara Venkata Raman, and *Landau, The Physicist and the Man*. As a similar example we note that Robert Marshak's account of the Rochester Conferences is given more fully in *Pions and Quarks*, a volume on particle physics which derives from a Fermilab symposium (see For Further Reading).

References to sources are given at the end of most articles in this collection. A number of articles have elicited from readers letters of comment, often critical but sometimes corroborative. The two pieces included here which have produced the most copious yield of responses are "Heisenberg and Goudsmit and the German atomic bomb" and Langmuir's lecture on pathalogical science; the corresponding series of published letters follows each of these articles in this volume. Some of the other articles have also evoked letters that raised interesting or debatable points, but unfortunately it has not been practical to insert them here.

When the articles available and appropriate for a second history reprint book were considered as a group and compared with the earlier volume a shift in emphasis was revealed. While there is as much physics as before, the physicists themselves take on a larger role as human beings, despite the fact that there was considerable biography in the earlier volume as well as a number of personal accounts of important developments. The same trend has been noticed elsewhere: in a review of seven "lives" in *Science* magazine (3 August 1990) we find "the number of such works [autobiographies] continues to swell, the more so if one includes another category of books similar in character, recollections of the living and recently dead by their friends and associates. Some 20 such personal books have been reviewed in *Science* in the past two years, for instance ..." (This number is not restricted to physicists, to be sure, but two of the seven "lives" are those of Raman and Landau mentioned above.)

The personal character of the present collection is stengthened by the inclusion of several obituaries. A great deal of history is included in obituary notices, although *Physics Today* has not pursued a firm policy on printing them. The first to appear were for John T. Tate and K. T. Compton, and featured principally their roles as statesmen of physics. The magazine was not given nor did it assume responsibility for disseminating such information, but gradually began to do so. Although brief, some of these notices are very perceptive summaries of both the professional and the nonprofessional characteristics of their subjects, and are excellent history.

Three of the four sections of this reprint volume exhibit the trend in history of science toward recording the personal aspects of scien-

tists as well as the professional. The first section is an eclectic compilation of pieces having to do with physics and public concerns. These concerns vary from the technological to the psychological. Some will appear more timely or more controversial than others. With such variety, however, it seems more than likely that most readers will find some pieces of concern to themselves!

It is a pleasure to acknowledge the advice and encouragement of Spencer Weart, Director of the American Institute of Physics Center for History of Physics, and Gloria Lubkin, Editor of Physics Today, during the course of my putting this book together.

Melba Phillips

Author Affiliations

Finn Aaserud is director of the Niels Bohr Archive in Copenhagen. He was formerly associate historian at the American Institute of Physics, Center for History of Physics.

Alexei A. Abrikosov is now at the Argonne National Laboratory.

John F. Ahearne is executive director of Sigma Xi, The Scientific Research Society, Research Triangle Park, NC, a former chairman of the Nuclear Regulatory Commission and head of the National Research Council's Committee on Risk Perception and Communication.

Henry A. Barton (1889–1983) was director of the American Institute of Physics from its beginning (1931) until his retirement in 1957.

Leo L. Beranek is director of Bolt, Beranek National Laboratory. He formerly did research in acoustics at the Jefferson Physical Laboratory at Harvard.

Laurie M. Brown is professor of physics and astronomy at Northwestern University in Evanston, Illinois.

Karl K. Darrow (1891–1982) was secretary of the American Physical Society from 1941 to 1966.

David H. DeVorkin is curator for the history of astronomy and the space sciences in the department of space history at the Smithsonian's National Air and Space Museum, in Washington D.C.

Paul P. Ewald (1888–1985) achieved renown for his pioneer work in crystallography. He taught and did research in several European universities before coming to the United States, where he was head of physics at the Polytechnic Institute of Brooklyn.

Murray Gell-Mann is Robert A. Millikan Professor of Theoretical Physics at the California Institute of Technology.

Vitaly L. Ginzburg is in the theoretical physics department of the P. N. Lebedev Physical Institute of the Russian Academy of Sciences in Moscow.

Gerald Holton is Mallinckrodt Professor of Physics and professor of history of science at Harvard University.

J. David Jackson is professor of physics at the University of California, Berkeley.

Aiyasami Jayaraman is a Distinguished Member of the technical staff at AT&T Bell Laboratories in Murray Hill, New Jersey.

Karen E. Johnson is professor in the department of physics at Saint Lawrence University, Canton, New York.

Martin J. Klein is the Eugene Higgins Professor of history of physics and professor of physics at Yale University. He is now the senior editor of the collected papers of Albert Einstein.

Irving Langmuir (1881–1957), winner of the 1932 Nobel Prize for chemistry, worked for many years in the General Electric Company research laboratories.

John S. Laughlin is chairman of the department of medical physics at Memorial Sloan-Kettering Cancer Center in New York.

Giorgio Margaritondo is with the École Polytechnic Federale, Lausanne, Switzerland.

Robert E. Marshak is University Distinguished Professor of Physics, Emeritus, at the Virginia Polytechnic Institute and State University in Blacksburg, Virginia. Much of his career was spent at the University of Rochester.

Edwin M. McMillan (1907–1991) was professor emeritus of physics at the University of California, Berkeley. He was awarded the Nobel Prize in chemistry in 1951.

Arthur I. Miller is professor of history and philosophy of science at University College London, in London, England.

Albert E. Moyer is an associate professor of the history of science at Virginia Polytechnic Institute and State University, in Blacksburg, Virginia.

Joseph S. Mulligan is professor of physics at the University of Maryland, Baltimore County, in Catonsville, Maryland.

Janet Oppenheim is professor of history at the American University, Washington, D.C.

Abraham Pais is Detlev Bronk Professor Emeritus at Rockefeller University, in New York.

Melba Phillips is professor emeritus, University of Chicago, and now lives in New York.

John R. Pellam was professor of physics at the California Institute of Technology, in Pasadena, California.

Anant Krishna Ramdas is a professor of physics at Purdue University.

William R. Sears is professor of aerospace and mechanical engineering at the University of Arizona, Tucson.

Emilio Segrè (1905–1989) was emeritus professor of physics at the University of California at Berkeley. He was awarded the Nobel Prize in physics in 1959.

Robert W. Smith is a historian at the National Air and Space Museum of the Smithsonian Institution, and also teaches the history of science at The Johns Hopkins University.

John Stachel is professor of physics at Boston University and was editor of the first two volumes of the Collected Papers of Albert Einstein.

Fritz Stern is Seth Low Professor of History at Columbia University.

Lloyd S. Swenson, Jr. is professor of history at the University of Houston, in Houston, Texas.

John H. Van Vleck (1899–1980) was on the faculty of Harvard University for most of his very distinguished career. He was awarded the Nobel Prize in 1977.

Mark Walker is a member of the history department at Union College, in Schenectady, New York.

Spencer R. Weart is director of the Center for History of Physics at the American Institute of Physics.

Charles Weiner is professor of history of science and technology at Massachusetts Institute of Technology and was the director of the Center for History of Physics at the American Institute of Physics, 1964–1973.

Victor F. Weisskopf is institute professor emeritus at the Massachusetts Institute of Technology.

Catherine L. Westfall is an assistant professor of the history of science and technology at the Lyman Briggs School, Michigan State University.

George Wise is a historian in the Communications Operation of the General Electric Research and Development Center at Schenectady, New York.

Chapter 1
____ Physics and Public Concerns____

In a sense all science, even the most abstract, is of public concern, but increasingly science and technology are very directly involved in public policy. In this section we consider various aspects of this involvement. As emphasized in the lead article, mistrust of technical experts is a growing problem in science policy. Scientists and engineers can do much to improve both public and official understanding of policy issues, but thus far the scientific community has not been of much help. Here we have recommendations from John Ahearne, who is well experienced in such matters, on how to do better.

Among the most pressing policy questions in the past half century have been those related to nuclear arms. As noted in the general introduction, much history of these problems has appeared in *Physics Today,* and is reprinted in *Physics and Nuclear Arms Today.* We include here two more recent articles which add to this history: one by Mark Walker on the still controversial question of the German atomic bomb during the war, the other by Robert Marshak's piece on the growth of particle physics during the Khrushchev détente, a period which was interrupted by the U-2 incident in 1960. The lack of communication between the scientists of the superpowers over many years was clearly attributable to both countries.

Another subject of very general concern is pseudoscience. Very few physicists today take psychic research seriously, unlike some of the greatest of the late 19th century figures, including Lord Rayleigh, J. J. Thomson, and William Crookes, whose attitudes are analyzed in a fascinating article by Janet Oppenheim. Casting his net more widely, Irving Langmuir discussed "pathological science" in a 1953 talk printed in 1989. While some of Langmuir's cases are primarily examples of scientific error or self-deception, others are controversial in principle, as were the psychic phenomena and spiritualism in Victorian England. That they remain so is evidenced by current response in letters to the editor.

Other developments do not seem controversial. The changes in scale and style undergone by research during World War II were reflected in the great industrial laboratories, as George Wise has detailed for the General Electric Research Laboratory at Schenectady. That the spurt in growth of medical physics since the 40s is even more striking is indicated in an article by John Laughlin (1983) which could be vastly extended only a few years later. But "Big Physics" involves controversy of a kind quite different from that surrounding pseudoscience. As noted by Catherine Westfall, the recent contest for a site for the Superconducting Super Collider is reminiscent of that for Fermilab a quarter century earlier. The stakes are even higher this time, but the similarities are as interesting as the contrasts.

Were the physicists working in the 40s and 50s aware of the pace of rapid development, and of the problems inherent in the expansion into what became "Big Science"? Not fully, of course, but Charles Weiner's look backward in 1973 on the occasion of the 25th anniversary of the founding of *Physics Today* has echoes of misgivings as well as optimism. There were no signs of complacency.

—— Chapter 1: Physics and Public Concerns ——

ADDRESSING PUBLIC CONCERNS IN SCIENCE

If the public doesn't understand us, perhaps it's because we aren't listening. Answering the right questions in science policy depends on a sustained dialogue with the concerned public.

John F. Ahearne

PHYSICS TODAY/SEPTEMBER 1988

Mistrust of technical experts by the general public is a growing problem in science policy. The National Science Board reports that whereas 80 percent of the public believes that scientists work for the good of humanity, 55 percent believes that their knowledge gives scientists a power that makes them dangerous. The image of the crazed scientist plotting to master and destroy remains a staple of both children's and adult fiction, as Spencer Weart explained in an article in this magazine (June 1988, page 28).

Americans generally recognize that technological advances are changing their lives and the world. The public expects science to accomplish wonders, and our political system responds to public attitudes and provides funding for science and technology. Yet the public's perceptions of science and of policy decisions involving science often differ from prevailing attitudes in the scientific community. Scientists and engineers have not been much help in developing rational technology policies and bear significant responsibility for the public's confusion. We can do better.

In a survey conducted by the Harris poll for the Office of Technology Assessment in late 1986 (see the table, next page), about 70 percent of the responding adult Americans described themselves as very interested or somewhat interested in science and technology, and the same proportion said their understanding of science was very good or adequate. More than 80 percent of the respondents said they were very concerned or somewhat concerned with science policy. Yet in a poll conducted for the National

Science Board (see the table), nearly half of the responding adults disagreed with the statement that human beings evolved from earlier forms of animal life, and about the same proportion agreed that rocket launches affect the weather.

Describing one aspect of science illiteracy, William Clark of Harvard University commented, "Society's attitudes toward risks such as cancer and nuclear reactors are not readily distinguishable from its earlier fears of the evil eye."[1]

Technological factors are significant in many current questions of public policy, including:

▷ Should the Federal government fund the Superconducting Super Collider?
▷ Should the United States build a space station?
▷ Where and how can high-level nuclear wastes be disposed of safely?
▷ Can herbicides and pesticides be used safely?
▷ What can be done about AIDS?
▷ Should scientists have the unrestricted right to manipulate genetic material, and should altered material be patentable?
▷ How serious a health hazard is radon in homes?
▷ What kind of program, if any, should the United States have to develop space-based defenses against ballistic missiles?
▷ Can the United States regain (or maintain) a competitive position in world markets?
▷ How can the United States become a leader in emerging technologies such as those associated with high-temperature superconductors?
▷ Will the US education system, including research universities, meet the country's future needs for scientists and engineers?

The final decisions about such matters generally are made by politicans, who take a wide variety of political, economic and international factors into account. Scientists and organizations representing scientists have become quite adept at bringing their concerns directly to politicians. But political decisions are made in a broader context of interactions among scientists and engineers, the

John F. Ahearne is Executive Director of Sigma Xi, The Scientific Research Society, Research Triangle Park, NC, a former chairman of the Nuclear Regulatory Committee and head of the National Research Council's Committee on Risk Perception and Communication. Thus article is based on a keynote address Ahearne delivered to the Georgetown Symposium on Nuclear Radiation and Public Health Practices in the Post-Chernobyl World on 18 September 1987.

Public understanding and appreciation of science: Poll results

Harris poll How much interest do you have in scientific and technological matters?	Very interested 23%	Somewhat interested 48%	Rather uninterested 11%	Not at all interested 18%	Not sure 1%
How concerned are you about government policy concerning science and technology?	Very concerned 32%	Somewhat concerned 50%	Not very concerned 11%	Not at all concerned 7%	Not sure 1%
How would you rate your basic understanding of science and technology?	Very good 16%	Adequate 54%	Poor 28%	Not sure 1%	

National Science Board poll	Agree	Disagree	Don't know
Human beings as we know them developed from earlier species of animals	45%	47%	7%
Some numbers are especially lucky for some people	43%	53%	4%
Rocket launches and other space activities have caused changes in our weather	44%	44%	12%

managers and operators of technical systems, and concerned citizens, all of whom have considerable influence on the evolution of policy involving technology.

Having been a member of all three groups, I conclude that scientists and engineers have the most to learn about the process of making sound technology policy, but also that they can make a difference. I will therefore concentrate on scientists and engineers and will mention only briefly some of the problems associated with the other two groups.

The failure to communicate

Scientists, engineers and other technologists can be separated into three subgroups: those who know in depth the science and engineering associated with a given policy; those who know a lot of science or engineering but are not experts on the specific issues in dispute; and those who operate high-technology systems but do not truly understand the technology they are using.

Of the true experts, unfortunately, many cannot communicate their knowledge. They are not able to simplify their discussions so that they can be translated by the media or understood by lay people. In some cases this lack of communication is due not to an inability but rather to a belief that the effort is not worthwhile. Some scientists believe that writing for the general public is a waste of their time because it is of little professional benefit and does little good.

Some good scientists do work at communication but do not deal effectively with the media. Many scientists are, quite properly, reluctant to say more than they know. Journalists have a tendency to treat this reluctance as equivocation and to describe it as such to the public. Many lay people believe that if you know something, you should be positive and unconditional about it. Therefore they conclude that when a scientist refuses to be definitive, it is equivocation or at least indicates that the scientist does not know much about the area.

Regrettably, what a scientist can be positive about is often not what the lay person is interested in. This difficulty faces all scientists and engineers who try to deal with the media. Nevertheless my belief is that most representatives of the media will take the time to try to understand if it is obvious that the technologist is making an effort to help them understand.

My hypothesis is that when an expert does not communicate effectively, it usually stems from inability or unwillingness. Failure to communicate well also can be connected, however, with an overestimation or overvaluation of one's own expertise. People who are well informed about science and engineering in general but not about the specific policy questions in dispute should not be called experts, but often they believe they are experts. In contrast to those who are aware of all the complexities bearing on the issues at hand, the less informed often take a paternalistic or maternalistic attitude toward the general public. Sometimes they express the belief that controversy would disappear if only the public were better educated—if only, that is, the public became as well informed as they believe themselves to be. Sometimes they act as though the solution to conflict is simply for the public to trust them and what they claim.

The attitude that education is the answer to everything was apparent in articles in the 1970s that importuned the public to understand "true risks" and to rank those risks relative to one another. This approach has not been abandoned, as was demonstrated by a 1986 book, *Improving Accuracy and Reducing Costs of Environmental Benefit Assessment*,[2] which presented a "health risk ladder" that included being hit by lightning, getting an x ray, being hit by a car, riding a motorcycle and smoking a pack of cigarettes a day.

The attempt to address technological risk management by ranking risks for different hazards has led to an attempt to define acceptable risk. I can sympathize with this approach because I participated in it for many years. When I was on the Nuclear Regulatory Commission, I tried unsuccessfully to get the National Academies to undertake a study of the comparative risks of coal and nuclear power, believing that the development of an objective view by a credible organization would help the debate on the risks of nuclear power. I am now shifting to the position of those who have concluded, with social scientists Harry J. Otway and Detlof von Winterfeldt, that "the acceptable risk formulation has provided increasingly elaborate and precise answers to the wrong question."[3] The questions are wrong because they do not arise from sustained dialogue with the concerned public.

In a recent study of the sources of conflict on environmental issues,[4] the researchers sought to explain the conflicts in terms of the characteristics and views of the participants. They interviewed people who had been involved in developing environmental policy, including lawyers, scientists and government officials. Nearly three-quarters of those polled labeled "public misunderstanding" as a major source of such conflicts. But they did not agree on what the public did not understand. The respondents whose educations had given them "hard expertise [viewed] environmental conflict as scientific rather than political, while those . . . individuals educated in the humanities or social sciences reject[ed] knowledge differentials as a major source of controversy. . . . Physical scientists, as expected, endorse[d] knowledge differentials, and reject[ed] value differences." Thus those who understand technology see the conflict as being between themselves and those who do not understand technology. But those who do not understand technology do not see this understanding as central to conflicts about the environment.

Unfortunately, many people who are untrained in science and technology believe that understanding technology is not important to understanding the risks of technology.

Trusting in trust

The accident at Three Mile Island destroyed a large reactor and nearly bankrupted a major company. The cleanup has been under way for nine years, is still not completed and will cost one billion dollars. But health studies done by the Pennsylvania State Department of Health and the US Department of Health and Human Services indicate that there have been no significant adverse *physical* health effects associated with that accident and that there are unlikely to be many.

The Chernobyl nuclear accident, the worst accident known at a nuclear power plant, led to at least 32 deaths, the hospitalization of several hundred people and a high radiation exposure to many thousands of Soviet citizens. Nevertheless, the immediate deaths in the surrounding region were far fewer than what some previous studies had estimated would result from such a massive release of radiation.

Consequently some members of the nuclear industry have said that TMI showed how well built reactors are and that Chernobyl showed that the worst accident would not be a calamity. While perhaps scientifically correct, this argument should not lead to the conclusion that nuclear power is now acceptable to the public. The flaw in that conclusion is that it avoids addressing the public's concerns and is based on the attitude that "if only the public were educated, they would agree with us."

"Trust me" is still used by US government officials

Managers of Chernobyl plant stand trial in USSR, July 1987. From left: Victor Bryukhanov, the ex-director; Anatoli Dyatlov, the former deputy chief engineer; and Nikolai Fomin, the former chief engineer. The Soviet judge who sentenced Bryukhanov to ten years in labor camp said there had been a lack of responsibility and control at the plant.

responsible for technology as the principal answer to the question, "Why are you doing that?" This attitude characterizes the approach the government has taken to locating nuclear waste sites, starting with the Atomic Energy Commission's efforts in Kansas, continuing with the Energy Research and Development Administration's search in the Middle West, and now seen for many years in the Department of Energy's efforts.

This approach has led to a highly polarized situation, as Clark Bullard observed last year in a seminar at Oak Ridge National Laboratory. "The lines are clearly drawn," said Bullard, a professor at the University of Illinois and the chairman of a committee responsible for advising a group of Middle Western states on the selection of a low-level nuclear waste site. "It is a battle between the technocrats and the public over whose values the technology will ultimately reflect."

Not only the general public is now skeptical about scientists and science officials. Many early advocates of nuclear power convinced executives at electric utilities of its advantages. These executives went ahead—aggressively—with ambitious nuclear programs, much to the chagrin of later executives, who found themselves saddled with adverse rulings on rates by public utility commissions.

The latest obstacles facing such utility executives are

prudency hearings. In these hearings, many years after construction of a plant began, public utility commissions examine whether it was wise for the electric utility to have built the plant. Frequently the commission decides it was not, and rate payers do not get charged for the plant.

Speaking at the annual convention of the Edison Electric Institute in Cincinnati last year, Jerry D. Geist, the retiring president of EEI, said, "Technology was the Siren who beckoned us to nuclear power—power that was supposed to be too cheap to meter but turned out to be too expensive to bill."

Managers and technicians

Those who are engaged in managing or operating high-technology systems often have had substantial technical training, often not. Their chronic weakness is complacency, and their failings lead the public to question the competence and judgment of the scientists and engineers who design such systems.

Complacency can be reflected in many ways: a lack of recognition by management that increased attention needs to be given to technologies whose use has potentially serious consequences; inadequate attention by operators, based upon a belief that the technology is so well developed that monitoring is not really needed; a belief that it is not important to understand the technology; and a lack of attention to mundane matters such as regular maintenance.

Aircraft accidents have been attributed to complacency in the cockpit. In 1978, for example, a DC-8 crashed in Portland, Oregon, when the plane ran out of fuel. The plane had been circling the airport while the crew tried to solve a landing gear problem. The flight engineer mentioned that the plane was running out of fuel, but apparently the message did not register with the captain and cocaptain, and the plane crashed. Last year a plane crashed in Detroit, probably because the crew had neglected to extend the flaps during takeoff.

Underlying the problems that led to the space shuttle Challenger disaster was a disbelief that the technology was hazardous—complacency on the part of people who did not fully understand the shuttle system. "In 1971," the Rogers commission on the Challenger accident found, "NASA went back to the drawing board, aware that development cost rather than system capability would probably be the determining factor in getting the green light for shuttle development."[5] Another study concluded that the shuttle represented "an effort to build one vehicle to serve many roles" and that "the inevitable result [was] a very complex and somewhat fragile vehicle."[6]

The pressures on NASA increased under the current Administration. The same day that the initial orbital tests were concluded, President Reagan announced a national space policy. "The first priority of the [space transportation system] program," he stated, "is to make the system fully operational and cost effective in providing routine access to space." But under the new policy, "resources were strained to the limit," the Rogers commission found. Astronaut Henry Hartsfield told the commis-

sion: "Had we not had the accident, we were going to be up against the wall.... Somebody was going to have to stand up and say we have got to slip the launch because we are not going to have the crew trained." Arnold Aldridge, the shuttle program manager, said, "Intentional decisions were made to defer the heavy build-up of space parts procurement in the program so that the funds could be devoted to other, more pressing activities." The Rogers commission concluded: "Those actions resulted in a critical shortage of serviceable space components. To provide the parts required to support the flight rate, NASA had to resort to cannibalization. Extensive cannibalization of spares...became an essential *modus operandi* in order to maintain flight schedules."

In a revealing visit to the Marshall Space Flight Center, the late Richard Feynman found a drastic difference in the estimates of accident risk made by shuttle engineers and by a middle-level manager. Taking a secret ballot, Feynman found that the engineers estimated the risk of an accident as about 1 in 200, the manager as 1 in 100 000 (see PHYSICS TODAY, February 1988, page 26).

Many reports on the Three Mile Island accident showed that operator misunderstanding initiated the accident. Similarly, reviews of the Chernobyl accident have highlighted the complacency that had afflicted the crew of that plant.

In my published examination of the two reactor accidents, I suggested that the more serious similarities between them stemmed from general complacency: "The operators of both plants took a series of steps that were deliberate and defeated safety systems. The operators at Chernobyl had no simulator training for the accident sequence that occurred. Similarly, TMI operators were never trained for the sequence of the stuck-open [valve], and instructions on how to handle such an event were not written in their emergency procedures. Both the TMI and Chernobyl accident reviews found weaknesses in the approval of operating procedures. Another common feature was that the operators did not understand their plants."[7]

The Soviet judge, in sentencing the director of the Chernobyl plant to ten years in a labor camp, said, "There was an atmosphere of lack of control and lack of responsibility at the plant," adding that "workers on duty played cards and dominoes or wrote letters."[8]

The concerned citizen

Some people are against all technology; some are afraid of technologies they do not understand; and some are opposed only to the technologies that affect the local environment. Many concerned citizens, however, try sincerely to understand confusing and complex issues.

People who are against technology or against anything new tend to be the most dedicated opponents of projects. They sincerely believe that technology is wrecking our culture. The attitude is not new. Consider how contemporaries of Galileo commented on his discovery of the moons of Jupiter: "Jupiter's moons are invisible to the naked eye, and therefore can have no influence on the

Earth, and therefore would be useless, and therefore do not exist."[9]

Members of this group can be quite sophisticated; in many cases they subscribe to E. F. Schumacher's view that "small is beautiful." These concerned citizens perceive that large-scale technologies such as nuclear power plants have been sold by deceit and tend to change society in undesirable directions.

Certain members of this group, those opposed to nuclear power, have been characterized by a political scientist as seeing themselves "standing with nature against the insensitive engineer. . . . Nuclear power is seen as an ultimate transmutation of matter. . . . It is man's manipulation of nature, expressive of an arrogance unknown in earlier technologies. It is viewed as an unnatural act. . . . Just as Prometheus took fire from the gods and was punished for all eternity, the nuclear opponent believes nature will make us suffer for a parallel transgression."[10]

Although the people who oppose technology may not be many, they are sincere and strong in their beliefs. They participate, often effectively, in debates, public meetings, letter writing and other activities that number among the advantages of living in a democracy.

The members of the second group, people who are afraid of technology essentially because they do not understand it, tend not to trust anyone who argues the citizen should not be worried, whether or not they understand the argument. The corollary is that they tend to believe anyone who says things are worse than they seem. Organizations that lobby against local waste dumps or nuclear power plants include many individuals who belong to this group. They believe they are being asked to accept on faith the safety of a technology.

Some concerned citizens have a special agenda that they prefer not to state explicitly. This agenda is based sometimes on opposition to big government, other times on protecting the local environment. Often these citizens are affluent and prefer to have costs imposed on others while they reap the benefits. "Not in my backyard" is the way political scientists and sociologists characterize their customary attitude. Something of this attitude was evident when California opposed siting a nuclear plant inside the state but was quite willing to let Arizona build the Palo Verde stations and export power to California, or when it opposed siting coal plants in-state but endorsed the idea of the Rocky Mountain states building large coal plants that would supply electricity to California.

Of course, affluence does not necessarily coincide with selfishness. Sometimes affluence enables one to take a more objective view of costs and benefits. The situation is similar to the debates between less developed countries and the highly industrialized countries concerning the costs and benefits of environmental regulation versus economic growth.

Perhaps the largest group of concerned citizens is the last—the people who do not fully understand the technologies at issue and are skeptical about strong claims by participants on either side of the debate. They do not believe that technology is automatically bad, nor do they believe government is automatically wise. These people will enter a proceeding or hearing with reasonably open minds. They will listen to arguments. They will value substance more than the appearance of sincerity. They will focus more on rationality than on rhetoric. In the end, they will be forced to reach a decision based on incomplete information. And they will decide. This subgroup of concerned citizens has not been well served by many technologists.

Lies, lawyers and leanings

Government officials often are responsible for implementing policies made elsewhere, and even though these officials can have a very important impact on science and technology policy, they often serve only fleetingly in decision-making positions. Problems associated with technology may not figure importantly in evaluations of their performance in office, and they may not be well trained for their jobs in the first place.

Like the scientists and engineers who consider themselves more expert than they really are, government officials have a tendency to adopt a "trust me" posture. These people do not understand democracy. They do not understand the concept of the consent of the governed, and they do not accept that the people have a right to be wrong. Thus, in a recent proceeding on AIDS, one author suggested: "In presenting information to the public, it is at least an open question as to whether or not honesty really is the best policy. The opposite of honesty is, of course, dishonesty or lying. No one can advocate lying as a public policy. . . . But there are at least two other options: exaggeration, underplaying."[11]

Law schools promote two beliefs that adversely affect technology policy: that the adjudicatory process is the best and possibly the only way to establish facts; and that a lawyer can learn enough about anything in a short time—a few days, or weeks at most. Although many lawyers are committed to improving the lot of society, these attitudes can lead to significant problems in developing sound technology policy.

The attitude of the adjudicatory process is abhorrent to science, since it seems to be based on the premise that people will not tell the truth unless pressured under oath. George Bernard Shaw is quoted as having said, "The theory of the adversary system is that if you have set two liars to exposing each other, eventually the truth will come out."[12] Informed discussion is not possible in that climate. Consequently, many scientists and engineers shy away from involvement in such conflict resolution approaches, leaving the public to be ill served by less than the best people.

The second legal belief stems from the need for lawyers to master a case quickly. Their high intellects often make this possible in nontechnical areas. But where science or technology is critical, lawyers tend to make bad policy. Lawyer–managers learn the issues at a superficial level, and often they want people around them who talk at that same level. Therefore lawyer–managers populate their staffs with bright young people who are articulate and industrious, but know little and unfortunately do not realize it. These lawyer–managers usually get along very well with members of the media, who also often learn an issue only at the same superficial level. This syndrome is captured in something Kierkegaard said in *Fear and Trembling*: "He had in addition a most unusual gift for explaining what he himself had understood. There he

SIDNEY HARRIS

When scientists get into really heated arguments, people prick up their ears, sensing that material interests and not merely scientific truths are at stake. As disputes have proliferated between "expert witnesses" representing opposing sides in court cases, the public has begun to be skeptical about the objectivity of science and technology.

stopped. Nowadays, one goes further and explains more than one has understood."

Worse still are the ideologues of the right and the left—those who know what the right decision is before analysis has been done. Beginning with the 1972 reelection of President Nixon, administrations have stressed ideology before competence to a disturbing and increasing degree.

From 1969 to 1983 I worked for Cabinet officers, in the White House, and on a Federal commission. After Nixon took office, resignations were requested from many assistant secretaries, who were replaced with choices more in sympathy with the White House ideology. Competence was still desired. When the Carter Administration arrived, belief in the position of the left became a critical test for White House and assistant secretary positions. The Administration did prefer bright people, but substantive knowledge was less important than ideological conviction. The Reagan Administration swung back to requiring right-wing views and placed even less emphasis on competence and much more on true belief.

Recommendations

Scientists and engineers can help improve public understanding of policy issues and policy itself by doing the following:

▷ Be responsible for understanding the technology you deal with, and be alert for surprises. Understanding that many scientific discoveries resulted from intelligent observation of experimental accidents should alert technologists to watch for the unexpected.

▷ Do not tolerate complacency.

▷ Listen to and discuss issues with the public. The public's resources will be used and their lives will be affected by your technologies. This listening should be a true dialogue. A public hearing should be a *hearing*, not, as a recent New York City Board of Estimates meeting was described in a news account, only a "public talking."

▷ Do not accept incompetence in government. The government often is derided by academic technical experts. But academic technologists understand technology and should not allow superficial treatments of technology to pass for understanding on the part of government officials.

Scientists and engineers often have not demanded competence from officials. Worse, they themselves have sometimes demonstrated a willingness to depart from their normal standards of professional behavior when policy is at stake. Harvey Brooks, the former dean of applied science at Harvard University, once pointed out: "Scientists inexperienced in the political arena, and flattered by the unaccustomed attentions of men of power, are often inveigled into stating their conclusions with a confidence not warranted by the evidence and . . . not subject to the same sort of prompt corrective processes that they would be if confined within the scientific community."[13]

▷ Provide impartial "friends of the court" to give expert testimony in proceedings involving controversial technological issues. In the United States, all controversial issues seem to end up in court. "Expert" witnesses proliferate— and often they strongly disagree. The public senses that these are often conflicts between contending interests, not competing objective views. As a result the public has become skeptical of the objectivity of science and technology. Recently a controversy regarding whether the large funds spent on cancer research have accomplished much heated up the pages of *Science* and the general press. Observing this controversy, Daniel Greenberg wrote: "When scientists become abusive, pay attention. The departure from professional decorum means something important is at stake."[14]

Heinz Pagels, one of our most prominent writers on science and science policy until his tragic death in July, suggested last year the establishment of a bureau similar to the Office of Technology Assessment to serve as a "friend of the court" when needed.[15] The American Physical Society has provided advice on major public policy issues via reports by special panels such as the 1976 study group on light-water reactor safety, the 1978 study group on nuclear fuel cycles and waste management, and the 1987 study group on the science and technology of

Nuclear Regulatory Commission meets (below) with the public in Harrisburg, Pennsylvania, on 9 November 1982 to consider the proposed restart of the Three Mile Island Unit 1 nuclear plant, which had been shut down since the TMI accident in March 1979. From left: NRC commissioners Thomas M. Roberts, John F. Ahearne, Nunzio J. Palladino, Victor Gilinsky and James K. Asseltine. Inset: Ahearne confers with Palladino, NRC chairman at that time.

directed-energy weapons.

Professional societies such as APS, the American Chemical Society and IEEE should identify and assemble experts who are able and willing to do *pro bono* service for the courts. A panel from the appropriate society could use consensus agreement to address issues that a court needs addressed. These friends of the court should be paid through the court system. If a scientist is needed to address a scientific issue, or an engineer to address an engineering issue, one of these friends of the court would appear. If a scientist or an engineer wanted to appear for one of the sides in the case, he or she could appear not as an "expert" witness but rather as an "advocate" witness.

References

1. W. C. Clark, "Witches, Floods, and Wonder Drugs," R-22, Institute of Resource Ecology, University of British Columbia, Vancouver, B. C. (January 1980).

2. W. Schultz, G. McClelland, B. Hurd, J. Smith, *Improving Accuracy and Reducing Costs of Environmental Benefits Assessment*, vol. IV, Center for Economic Analysis, University of Colorado, Boulder, Colo. (1986).

3. H. J. Otway, D. von Winterfeldt, Policy Sci. **14**, 255 (1982).

4. T. Dietz, P.C. Stern, R.W. Rycroft, "Definitions of Conflict and the Legitimization of Resources: The Case of Environmental Risk," Sociological Forum (1988).

5. Report of the Presidential Commission on the Space Shuttle Challenger Accident, Washington, D. C. (6 June 1986).

6. National Commission on Space, *Pioneering the Space Frontier*, Bantam, New York (1986).

7. J. F. Ahearne, Science **236**, 677 (1987).

8. "Chernobyl Officials Are Sentenced to Labor Camp," The New York Times, 30 July 1987, p. A5.

9. A. Williams-Ellis, *Men Who Found Out*, Coward–McCann, New York (1930), p. 43; quoted in Congressional Research Service Report CB-150 (29 May 1969), p. 32.

10. A. Hacker, Electric Perspectives, Summer 1980 (Edison Electric Institute publ. no. 07-80-22), p. 11. See also S. Weart, *Nuclear Fear: A History of Images*, Harvard U. P., Cambridge, Mass. (1988).

11. H. M. Sapolsky, in *AIDS: Public Policy Dimensions*, United Hospital Fund of New York (1987), p. 108.

12. G. B. Shaw, quoted in M. J. Saks, Technol. Rev. **90**, 43 (August/September 1987).

13. H. Brooks, Proc. Am. Philos. Soc. **119**, 259 (1975).

14. D. Greenberg, quoted in The Public Interest **88**, 151 (Summer 1987).

15. H. R. Pagels, Science Focus **2** (1), 2 (Summer 1987). ∎

HEISENBERG, GOUDSMIT AND THE GERMAN ATOMIC BOMB

Contrary to accounts based on Heisenberg's claims, the German fission research effort in World War II was indeed a nuclear weapons program, and contrary to Goudsmit's interpretations, the German team knew what it was doing.

Mark Walker

PHYSICS TODAY/JANUARY 1990

The question of whether German scientists would have been willing to make atomic bombs for Adolf Hitler has excited persistent interest. Just why this is so is a topic I have explored elsewhere.[1] Here, I contend that the roots of the controversy about the role of the German scientists are to be found mainly in the period immediately after the war, not in the war itself.

When I set out several years ago to write a doctoral dissertation on Germany's wartime nuclear program, I found that the secondary literature was confusing, contradictory and largely undocumented. Almost all the literature turned out to derive either from a polemic interpretation of the German war work set forth by Samuel Goudsmit or from an apologetic interpretation proffered by Werner Heisenberg. In particular, the well-known book by Robert Jungk, *Brighter than a Thousand Suns* (Harcourt, Brace, New York, 1958), did much to amplify Heisenberg's claim that the German scientists had conspired to deny Hitler a bomb. While this conspiracy theory was rebutted in David Irving's *German Atomic Bomb* (Simon and Schuster, New York, 1968), Irving perpetuated Heisenberg's false claim that an erroneous calculation by Walter Bothe had seriously retarded the German effort. Irving's book has influenced books on related but broader topics, and even when such books have met high scholarly standards, they have tended to follow in the footsteps of the previous secondary literature.[2]

This article focuses on the illuminating and troubling dialogue that took place after the war between two colleagues and former friends, Goudsmit and Heisenberg—the debate that did so much to shape subsequent literature.

Goudsmit, who had discovered electron spin with George Uhlenbeck (see PHYSICS TODAY, December 1988, page 34) and who for many years was editor in chief of The American Physical Society, was one of the most influential physicists in America after World War II. But he is perhaps best known outside of the physics community for his work in scientific intelligence. During the last few years of the war, Goudsmit served in Europe as a reserve officer and the ranking scientific member of the Alsos Mission, an extraordinary intelligence-gathering unit of the American nuclear weapons project. Goudsmit and his colleagues hunted down the German scientists who had been involved in applied nuclear fission and isotope separation research, in the process seizing scientific reports and materials, destroying experimental apparatus and arresting physicists and chemists. Goudsmit's heroic account of his adventures was presented in his book *Alsos* in 1947 (Tomash, Los Angeles; second edition, American Institute of Physics, 1988).

In many respects, by the time Goudsmit received his discharge from the army, he was an embittered man. His mother and father had died at Auschwitz. The loss of his parents and the horrific legacy of National Socialist Germany were blows that he would feel keenly the rest of his life. Understandably, Goudsmit no longer was completely objective when it came to Germany, German science or German scientists. Immediately after the war, he advocated a sink-or-swim policy for German science. But he eventually came to soften and qualify his stance toward German science, and he did so partly as a result of his emotional public and private debate with Heisenberg.

Heisenberg, already a world-famous physicist when the war began, was an ambitious man and a German patriot who, after being badly bruised in the 1930s by attacks by the physicsts Johannes Stark, an early supporter of Hilter, had had to solicit the assistance of the SS to defend his position. After the war Heisenberg was eager to be recognized as a leader of Germany's physics community.[3] In his exchanges with Goudsmit, he sought to portray

Mark Walker is a member of the history department at Union College, in Schenectady, New York. This article is based on his book *German National Socialism and the Quest for Nuclear Power* (Cambridge U. P., New York, 1989).

Samuel Goudsmit (left), a member of a military intelligence force during the war and a leader of the physics community in postwar America, clung tenaciously to the mistaken idea that the Germans thought an atomic bomb would consist of a nuclear reactor in which thermal neutron reactions went out of control.

Werner Heisenberg, an ardent German nationalist but not a National Socialist, sought to dissociate himself and other leading physicists in postwar Germany from Nazi science by portraying physics administrators in the Third Reich as incompetent.

his wartime responsibilities as greater than they actually were without incurring the moral onus of having tried to build a superbomb for Hilter. At the same time, he tried to dissociate himself and most German physicists from the wartime leaders of German fission research by insinuating, without coming right out and saying so, that the administrators had been contaminated with Nazism, so that they would be lumped together with the proponents of *deutsche Physik*—the antirelativistic, supposedly more Germanic theories expounded by physicists like Philipp Lenard and Stark (see the article by Fritz Stern in this volume).

A common attribute of Heisenberg's apologetic account and Goudsmit's polemic was, ironically, a distinct exaggeration of Heisenberg's importance. Heisenberg appeared as though he had controlled and dominated the entire German nuclear fission project. This oversimplified version of the project's history sat easily with a stereotype that was and is widely held by scientists: the view that scientific progress results mainly from a few "great" scientists having profound ideas.

Another commonality between Goudsmit and Heisenberg, also deeply based in scientific stereotype, had to do with scientific objectivity and the politicization of science. Goudsmit and Heisenberg tended to classify project scientists as objective and apolitical, even though they plainly were doing their work in furtherance of well-defined and well-understood political objectives. Both men tended to classify the physicists administering the project, equally inappropriately, as political hacks.

The wartime project

In April 1939, a few months after the discovery of fission, the German physicist Wilhelm Hanle delivered a lecture on possible applications of nuclear energy making use of a uranium–graphite pile. Georg Joos, a colleague at Göttingen, transmitted a report on Hanle's lecture to the Ministry of Education, which in turn passed it on to Abraham Esau, who was in charge of physics at Ger-

The German Decision for Heavy Water

The initial German experiments on carbon as a moderator in nuclear reactors were done by Paul Harteck in Hamburg, who used dry ice, and Walther Bothe in Heidelberg, who used graphite. Bothe concluded on the basis of experiments with graphite he had obtained from Siemens that a uranium machine meeting the requirements of Heisenberg's theoretical model would not work because graphite would absorb too many thermal neutrons. Subsequent experiments by Wilhelm Hanle showed, however, that Bothe's conclusion was erroneous: Hanle showed that even the very pure Siemens graphite contained boron and cadmium, strong absorbers of thermal neutrons, and that these impurities had been lost to the air when Bothe reduced the graphite to ash. Hanle reported his results to Army Ordnance, which took them fully into account in opting for heavy water as a reactor moderator.

Heisenberg already had demonstrated that a uranium machine relying on graphite would require much more uranium and much more moderator than a machine relying on heavy water. The cost of producing large quantities of graphite that was free of cadmium and boron seemed prohibitively high to Army Ordnance. Supplies of heavy water could be much more readily assured, and several

Walther Bothe, winner with Max Born of the 1954 Nobel Prize in Physics, was a leading experimental physicist on the German fission research project.

German scientists already were highly expert in dealing with heavy water. It was a German physical chemist, Karl-Friedrich Bonhoeffer, who persuaded Norsk Hydro, a Norwegian company, to begin commercial production of heavy water after American researchers discovered in 1932 that heavy water could be separated from light

water by means of electrolysis. Karl Wirtz, an assistant to Bonhoeffer, had done a lot of work with heavy water, and so had Harteck and Klaus Clusius, a physical chemist in Munich who was a key figure in Germany's uranium isotope separation effort. In August 1940, Robert Döpel demonstrated experimentally in Leipzig that heavy water would make an excellent moderator.

By this time Germany had occupied Norway, and it was soon decided that the quickest, most efficient and cheapest means of assuring a large heavy-water supply would be to boost Norwegian production. Norsk Hydro was ordered to increase its annual production rate from 20 liters to 1 metric ton. In 1941 Norsk Hydro was ordered to install a novel catalytic conversion technology designed by Harteck and his assistant Hans Suess.

It was recognized all along that ordinary water might also be a satisfactory moderator but would require the use of enriched uranium. But an isotope separation technique developed by Clusius and Gerhard Dickel ran into unforeseen problems, and centrifuge technology could not be brought to fruition within a meaningful time. Thus the Germans came to focus on developing a reactor fueled by natural uranium and moderated by heavy water.   **FIGURE 1.**

---

many's Reich Research Council.

Meanwhile, Nikolaus Riehl, a former student of Otto Hahn and Lise Meitner who was working as an industrial physicist at the Auer Company in Berlin, brought nuclear fission to the attention of the Army Ordnance Office. Independently of Hanle and Riehl, the Hamburg physical chemists Paul Harteck and Wilhelm Groth wrote to the Army drawing attention to the possibility of making nuclear explosives. Their communication was reviewed at Army Ordnance, probably by Erich Schumann, the head of research, and Kurt Diebner, the resident expert on nuclear physics. Esau and Diebner were PhD physicists, and Schumann was qualified to teach physics at the university level.

Two uranium workshops, in April and September 1939 led to the convening of two conferences, in September and October of that year, dedicated among other things to the feasibility of using nuclear energy and to the theory of chain reactions. As a result of the conferences it was decided by the Army Ordnance Office to distribute work on nuclear fission among several institutes, rather than concentrate it at the Kaiser Wilhelm Institute at Berlin–Dahlem. Important assignments were given to the University of Hamburg (Harteck), Leipzig (Robert Döpel, Heisenberg), the Army Research Center at Gottow (Diebner), the Kaiser Wilhelm Institute for Medical Research in Heidelberg (Bothe) and the Kaiser Wilhelm

Institute for Physics (Fritz Bopp, Diebner, Carl-Friedrich von Weizsäcker, Karl Wirtz).

In the fall of 1939 the Dutch physicist Peter Debye was asked to take German citizenship as a condition of his managing military research as director of the Kaiser Wilhelm Institute of Physics. When he declined and left to take a position at Cornell, Diebner took over the institute. In July 1942, after the institute was returned to the Kaiser Wilhelm Society, Heisenberg was made director. Esau was the senior official in charge of nuclear physics at the Reich Research Council from November 1942 to December 1943, when he was replaced by Walther Gerlach. Throughout the war Gerlach, Bothe, Heisenberg, Harteck, Hahn and Klaus Clusius—all scientists of the first rank—were involved in the scientific work and the administration of the German fission project.

The first phase of the project was devoted to establishing the feasibility of building a "uranium machine," that is, a prototype of a reactor capable of producing energy and explosive material for nuclear weapons. This was accomplished by the end of 1941, almost exactly in parallel with work in the United States and Great Britain. At this juncture the project was evaluated, just when Germany's fortunes in the war were changing dramatically as a result of Germany's failure to subdue Britain, the first Russian counterattacks and Pearl Harbor.

A 150-page report to Army Ordnance, prepared

Paul Harteck, a physicist at the University of Hamburg, was perhaps the most dynamic and effective member of the German nuclear fission project. Exploiting excellent connections with German industry, he made important contributions to work on moderators, uranium enrichment and reactor design.

probably by Diebner and younger project scientists, strongly recommended an industrial-level effort to build a working reactor and to produce fissionable materials, despite the admittedly rather distant prospects for actual construction of a bomb. In light of Germany's urgent military situation, competing demands on resources, and scarcity of certain critical materials—notably materials suitable for use as reactor moderators—Army Ordnance decided against industrialization of the project. This was the final verdict, which never was reassessed.

The project was transferred to the Kaiser Wilhelm Society, in keeping with its designation as basic research (albeit *kriegswichtig*—important to the war), but then was shifted back again to the Reich Research Council. By now the council was part of the sprawling jurisdiction of Reichsmarschall Hermann Göring, the Air Force chief, but Albert Speer, the head of armaments and later war production, also took an interest in the fission project. While bureaucratic battles associated with the project's loose organization occasioned some unhappiness among project scientists, the disaffection appeared to have everything to do with personal rivalries and wounded vanity and nothing to do with the fundamental objectives of the project, which were well understood and accepted on all sides.

Goudsmit attacks

During the course of 1946, Goudsmit published several popular articles on physics under Nazism.[5] These essays had both a pedagogical function and a political aim: to illustrate the debilitating effect of fascism on science, and to argue against tight military control of postwar American research. Although Goudsmit presented his account as authoritative, he grossly misrepresented the German scientific achievement. Goudsmit's initial distortion of the record was unintentional, the result of sloppy research, but once he had staked out positions, he refused to recant. Among other erroneous statements, Goudsmit claimed that the Germans conceived of an atomic bomb as a nuclear reactor gone out of control and that the Germans did not seriously consider using plutonium for atomic bombs.

Heisenberg and his close friend and younger colleague von Weizsäcker, who often spoke with one voice, had tried to publish an account of the German nuclear fission research as soon as the imprisoned German scientists had been released from confinement in England and returned to Germany, but they were stopped by the British occupation officials. But the following year Heisenberg received permission to proceed with publication of an article on the German fission project, which appeared in *Die Naturwissenschaften* in November 1946.[5] Heisenberg composed a preliminary draft—which fortunately has survived—and sent copies to selected colleagues for criticism. By this time Heisenberg had seen a copy of the "Smyth report" and thus knew a considerable amount about the successful American nuclear weapons project.

When Heisenberg's article is compared with sources documenting the history of the German nuclear fission project, several important discrepancies emerge. First of all, Heisenberg slighted the project scientists and administrators connected to Army Ordnance and those who held other high positions in the Third Reich's science policy bureaucracy—particularly Diebner, Esau and Schumann. For example, Heisenberg gave Gerlach, Esau's successor as head of the German nuclear fission effort, credit for several of Esau's innovations and accomplishments.

Scientists like Esau and Schumann were in many respects an embarrassment to the rest of their colleagues in postwar Germany, for they had held highly visible positions under National Socialism. Even after the end of the war and the revelations about German atrocities, these professionally respectable, if not world-class, physicists often admitted their wartime support of Nazi Germany unrepentantly. In other words, they represented exactly that from which Heisenberg and others were trying to distance themselves.

Graphite and plutonium questions

The second discrepancy that stands out in Heisenberg's draft article of 1946, and one that surprised and annoyed Bothe, was Heisenberg's assertion that Bothe's measurement of the diffusion length of neutrons in carbon had been a mistake that had hindered the further progress of the entire nuclear fission project. Bothe protested, and Heisenberg accepted the rewording that his Heidelberg colleague suggested. Nevertheless, this theme, "if only we had tried graphite...," continued to circulate within Heisenberg's circle and beyond. Eventually Bothe became a scapegoat—the scientist who had made "the mistake" that had kept the Germans from achieving a chain reaction.

The reasonable and justifiable decision for heavy water and against carbon as a moderator was made on economic grounds by those responsible at Army Ordnance, in full knowledge of the potential of carbon as a moderator (see Figure 1). Why then did Heisenberg claim otherwise? By the time he wrote his draft article, of course, he had studied the Smyth report, and he had noted that the Americans—in contrast to the Germans—had used graphite for their nuclear piles. The facts that the Germans had not and failed, were used to reach the dubious conclusion that the Germans would have succeeded, or certainly would have gone much further, if only Bothe had not made his "error" and they had chosen graphite as their moderator.

A third discrepancy in Heisenberg's printed article— but not in his draft!—concerned von Weizsäcker's theoretical discovery of the explosive properties of plutonium. In the draft article, Heisenberg mentions von Weizsäcker's discovery that an operating nuclear reactor produces uranium-239, whose transuranic daughter products have the same properties as uranium-235. Heisenberg thus wrote in the draft that an energy-producing nuclear reactor could be used to produce materials for nuclear explosives.

Apparently one of Heisenberg's colleagues thought that this passage was too explicit, for the published version was much more circumspect. Instead of stating flatly that nuclear explosives can be produced using materials from an operating nuclear reactor, the sentence now allowed that von Weizsäcker's work made it more probable that an energy-producing nuclear reactor could manufacture fissionable materials, but immediately added that the "practical execution" of this process had not been discussed at the time. The form the final version took could be taken to imply that this lack of discussion was intentional.

A fourth important discrepancy was Heisenberg's description of how the Germans came to conclude that nuclear weapons could not influence the further course of the war. Here Heisenberg grossly misrepresented one important aspect of this question by attributing falsely the decision not to shift the research up to the industrial level of production to a meeting of a small group of Kaiser Wilhelm Society scientists with Speer in June 1942. Heisenberg implied that it was Speer who decided that the scientists should focus on the design and construction of a "peaceful" nuclear reactor rather than on military applications.

In fact, the decision to leave the research at the laboratory scale was made by Army Ordnance alone, more than half a year before the meeting with Speer. But apparently Heisenberg preferred to attribute this crucial judgment to a meeting in which he personally played a considerable role rather than to Army Ordnance. This was consistent with his apparent strategy of implying that project scientists had actively refrained from working on a bomb project for Hitler and of dissociating himself and other project scientists from Army Ordnance, whose

scientists and science policy makers were discredited in the postwar era.

The alleged conspiracy

In the second-to-last paragraph of the published version of his 1946 article, Heisenberg presented an excellent summary of why the Germans had not attempted the production of nuclear weapons on an industrial scale: Until 1942, the Germans anticipated an early end to the war and were uninterested in weapons that could not be used in the immediate future; after 1942, it was evident that a bomb could not be built before the war was over, and Germany's steadily deteriorating military situation hampered any further progress.

Had Heisenberg left it at that, his ultimate conclusion about Germany's fission effort would have been unimpeachable. But both the draft version and the published version ended with a paragraph implying that the German scientists—especially the circle around Heisenberg—had held themselves back from producing nuclear weapons for the Nazi state because of moral scruples. Indeed this passive resistance was supposed to have stopped the German production of such weapons. In the published article, Heisenberg wrote that the German physicists had striven from the very beginning to keep the control of the project in their hands, and that they had used the influence they had as experts to steer the research away from the manufacture of nuclear weapons.

Of course it is possible that some of the German scientists who worked on nuclear energy and weapons, including Heisenberg, neither intended nor desired that German nuclear weapons be created and used. But Heisenberg's claim that the scientists willfully hindered the creation of nuclear weapons for Hitler's government is not supported by the documentary evidence, and it is intrinsically implausible. Why should they have feared and tried to prevent that which they knew could not and would not be done before the end of the war?

Goudsmit's second offensive

Goudsmit was angered after reading a translation of Heisenberg's article in the British journal *Nature*. According to Goudsmit, Heisenberg's account had all the earmarks of being meant for consumption in Germany: It was meant to appeal to national sentiment that German science was good and pure and could not fail.

In Goudsmit's opinion (which was poorly based in fact), Heisenberg had not owned up to the failures of the German nuclear fission project; indeed he had told a "tale of success." But what really enraged Goudsmit was Heisenberg's attempt to seduce the casual reader into believing that the German scientists had made a "deliberate decision" to refrain from making nuclear weapons. Goudsmit argued, correctly, that the Germans themselves had thought that they were progressing satisfactorily in that direction.

Heisenberg's article was probably one reason why Goudsmit decided to take his case to a broader audience and so wrote the popular book *Alsos* in 1947. In order that *Alsos* not go unnoticed, Goudsmit arranged for it to be previewed in *Life* magazine on 20 October 1947.

Alsos posed and purported to answer a question that was of keen interest to Goudsmit and of great topical importance at a time when the organization of big science was at issue in the United States: Why did German science fail where the Americans and British succeeded? Goudsmit's answer was that science under fascism was not, and probably could never be, the equal of science in a

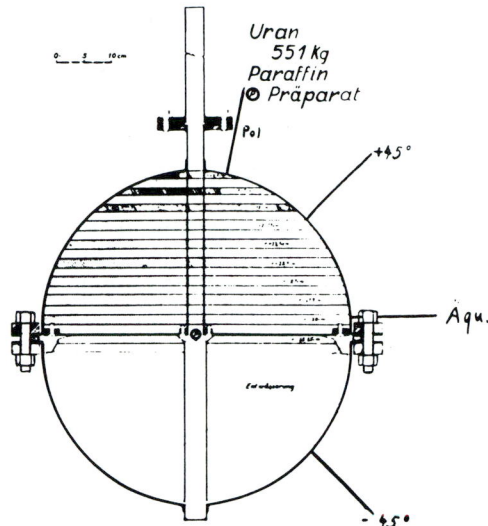

Uran
551 kg
Paraffin
Präparat

BROOKHAVEN NATIONAL LABORATORY

Alternating horizontal layers of uranium powder and paraffin (diagram, above) were characteristic of the early German reactor experiments. Heisenberg's insistence on completing this kind of experiment late in the war, when other scientists such as Karl-Heinz Höcker and Kurt Diebner of Army Ordnance wanted to proceed rapidly with construction of reactors consisting of uranium lattices suspended in heavy water, may have held back German progress toward a self-sustaining nuclear reaction. The spherical design of the early German reactors (right) apparently fortified Goudsmit in his misconception that the reactors themselves were intended to be bombs. In his book *Alsos* he reproduced drawings of such reactors and labeled them "Germany's atom bomb."

FIGURE 2.

democracy. In Goudsmit's opinion, the "totalitarian climate" of Nazi Germany led to complacency, politics in science and hero worship, all of which adversely affected the German research.

Goudsmit's account of the Alsos Mission and the German nuclear fission research program was basically the same as that set out in his earlier articles. Heisenberg was portrayed as a tragic figure, an extreme nationalist led astray by the Nazis and made to appear foolish by the revelations of Hiroshima. But Goudsmit was concerned with issues larger than the German nuclear energy and weapons project. Using Gestapo records that he himself considered suspect, Goudsmit unfairly dismissed Schumann and other National Socialist science policy administrators as incompetent Nazis and drew an arbitrary line of demarcation between the "good scientists"—good in both the professional and moral senses—and the "Nazi scientists." Thus Goudsmit did exactly what Heisenberg had done, although his motives were quite different.

Goudsmit's concluding chapter sharply criticized what he saw as American complacency about its scientific and military superiority over the Soviet Union, and he attacked those who wanted to continue wartime restrictions on nuclear science in the United States. He used the example of Heisenberg to argue that isolation, secrecy and governmental control ruin science.

Heisenberg–Goudsmit correspondence

The publication of *Alsos* touched off a fascinating correspondence between Goudsmit and Heisenberg. The exchange brought out a central issue in postwar psychology—the issue of whether the past should simply be buried, so that one could move on unencumbered, or whether it

needed to be addressed and worked through.

Without having read *Alsos*—indeed there is no record that he ever read it—Heisenberg wrote to Goudsmit in the fall of 1947 and enclosed a copy of his article from *Die Naturwissenschaften*. Using polite language, Heisenberg remarked that he had seen several of Goudsmit's articles and had gotten the impression that Goudsmit was unaware both of the details of the German fission effort and the psychological situation in Germany during the war.

After reminding Goudsmit that he had misled Heisenberg in the spring of 1945 (when Goudsmit still was subject to the secrecy requirements of the Manhattan Project) by telling him that the Americans were not working on nuclear weapons, Heisenberg went on to describe the wartime situation in Germany in a manner difficult to reconcile with his conduct during the war. On one hand, he wrote, it was clear to the scientists what "heinous consequences" the victory of Nazism in Europe would have, but on the other hand, in view of the hatred toward Germans with which the war had saturated Europe, they could hardly look forward to the country's utter defeat. The situation, Heisenberg argued, led them to assume a more "passive and humble manner."

After some delay, Goudsmit finally replied to Heisenberg on 1 December 1947 by criticizing Heisenberg's justification of this "passive and humble manner." Goudsmit had been deeply disappointed to learn of Heisenberg's attempts at a compromise with the Nazis, he said. What surprised him most was that Heisenberg did not see that such a compromise was impossible. Attempts to convince the Nazis of the soundness of relativity and quantum theory—an effort Heisenberg emphasized in his opening letter—seemed to Goudsmit so out of place.

How, Goudsmit asked, could Heisenberg have hoped to be successful, or have thought that these were important issues? In Goudsmit's opinion, not the period under Hitler but the present was the right time for the "more humble manner" Heisenberg described.

Heisenberg replied in early 1948, and this time his tone was colder and less polite. First of all, he made it clear that the question he now considered most important was whether the Germans had known how an atomic bomb would have worked. He then went on to refute Goudsmit's claims on this score by showing that the Germans had understood both fast-neutron chain reactions and the potential of plutonium. Letting a little sarcasm leak into his tone, Heisenberg remarked that obviously Goudsmit had accidentally overlooked the reports that would have given him the correct picture. Only after they had agreed on the "facts" of the German scientific achievement would Heisenberg be willing to discuss the political motives behind the work.

However, Heisenberg wanted to comment on a few points raised by Goudsmit's letter. First of all, he pointed out that he had always believed that German science had suffered under Nazism, especially because of the expulsion of many capable scholars from Germany and the advancement of nonsensical theories such as *deutsche Physik*. Moreover, Heisenberg had made critical comments about *deutsche Physik* in public at a time when such action had been dangerous. Also, it never would have occurred to Heisenberg to think that the German physicists were any different from their Allied counterparts. But how could Goudsmit continue to overlook the fact that the German physicists also found themselves in a different psychological situation from their colleagues in England and America?

Heisenberg also commented on what Goudsmit had described as compromises with Nazism by denying that he had been so naive as to believe that there was much chance of winning over SS leader Heinrich Himmler, and he bluntly stated that he would have "criminally" neglected his duty if he had not, at least in his small circle, tried to shatter the "delusion" of the dictatorship. In particular, Heisenberg had never had the slightest sympathy for the people who withdrew from all responsibility during the Third Reich but then in a safe dinner conversation would tell someone that National Socialism would ruin Germany and Europe, just wait and see.

At the end of his letter, Heisenberg turned to the present situation in Germany and his views on dealing with the past. It was difficult, he wrote, to win the hearts and minds of people through the force of arms, especially because of the "indescribable misery" in Germany. What the Germans needed was not a hateful settling of accounts with the past, but instead a quiet reconstruction of a life worthy of a human being. In any case, Heisenberg assured Goudsmit that he could be certain that the German physicists would gladly participate in any effort that would contribute to a "better world understanding."

Intervention by van der Waerden

At this point the Dutch mathematician Bartel L. van der Waerden, a friend of Heisenberg, former countryman of Goudsmit and a very perceptive observer, briefly entered the debate. Van der Waerden had spent the war years at the University of Leipzig, where for many years he was Heisenberg's colleague. After the war he went to teach at Johns Hopkins, where he read *Alsos* in March 1948 and experienced the American debate over the book firsthand.

Subsequently he wrote both Goudsmit and Heisenberg. Van der Waerden told Goudsmit that there was one point in *Alsos* that he did not understand. Did Goudsmit mean, he asked, that the German physicists, knowing who Hitler was, had planned "the horrible crime" of putting an atomic bomb into his hands? Goudsmit replied (obscurely) that such an act would not have been a crime; rather it would have been something that could not be stopped. (Perhaps he had in mind the "inevitability" that the US weapons scientists often talked about, once the feasibility of an atomic bomb was established in the Manhattan Project.)

According to Goudsmit, the Germans thought that making an atomic bomb was much more difficult than it actually was, and thus the question of conscience was not so urgent for them. This argument pervaded all of Goudsmit's publications on the German nuclear fission project and was an attempt to reconcile the evidence that he had gathered with his profoundly ahistorical and noncontextual preconceptions of how science and technology work. In his view, since the Americans had succeeded in building an atomic bomb, the Germans should somehow have known that it was feasible. Since the German scientists and authorities had decided that this task was *not* feasible, Goudsmit concluded that this decision must have been a mistake, and searched until he thought he had found the German error in the concept of a nuclear pile as a bomb—an error they never made.

(In November 1944 the Alsos Mission heard from a secondhand source that Hahn's institute had estimated the minimum mass of an atomic bomb at eight tons. In hindsight it is evident that the source was confusing the mass of a reactor with the mass of a bomb, but Goudsmit attributed the confusion to the whole project; a subsequent misreading of a letter by Gerlach fortified Goudsmit in his conviction that the Germans thought the reactor itself would be the bomb.)

Van der Waerden wrote Heisenberg immediately after his conversation with Goudsmit, noting that documents in Goudsmit's office verified Heisenberg's claims about what the Germans had known about plutonium and nuclear weapons, but adding that in his opinion questions such as the "complacency" of the German physicists or what Heisenberg and other scientists had understood or overlooked were insignificant.

The two Dutch-born scientists also had discussed the "question of guilt," van der Waerden told Heisenberg. In the end they still disagreed over the psychological question—what Heisenberg's group would have done if they had made greater progress—but agreed someone should not be condemned for what he might have done if the situation had been different.

Although van der Waerden was defending Heisenberg, he was also critical of his friend and former colleague. In a second letter, sent the very next day to Germany, van der Waerden asked for more information. Heisenberg had informed high-ranking German authorities about the potential of nuclear explosives during the war; had he also considered the question of responsibility at that time? Had Heisenberg's statements touting the military potential of the project been a mere ruse to get money for physics?

As Heisenberg's "lawyer" (van der Waerden's term), van der Waerden said he had enough evidence to defend him, but as Heisenberg's friend he also wanted to believe that under any circumstances Heisenberg's decency would have been stronger than the combination of his nationalism and ambition.

Van der Waerden made one last attempt to reconcile Goudsmit and Heisenberg before he returned to Holland. Naturally Heisenberg was right to claim, van der Waerden argued in his second letter, that his efforts on behalf of

Von Laue and Morrison on the Arming of 'Himmler and Auschwitz'

Goudsmit's book *Alsos* touched off an emotional debate in the *Bulletin of the Atomic Scientists*. Philip Morrison, a Manhattan Project scientist who had also been involved with scientific intelligence, reviewed the book and attacked Heisenberg's suggestion that the Germans had not wanted to create nuclear weapons. Like their counterparts in America and Britain, Morrison stated, the German scientists had worked for the military as best as their circumstances allowed. The difference, which Morrison found unforgivable, was that they worked for the "cause of Himmler and Auschwitz," for the burners of books and the takers of hostages.

With justification, Morrison and others in America believed that the Germans were implying that German scientists had been morally superior to their American and emigré counterparts because they had not built a bomb—and because in particular, they had not built bombs that were actually used to destroy cities.

Morrison's attack provoked a response from Max von Laue, perhaps the only German physicist residing in Germany who still commanded respect in America at that time. Von Laue was a more outspoken critic of National Socialism than Heisenberg or von Weizsäcker, but as acting director of the Kaiser Wilhelm Institute for Physics, he had helped oversee the work on applied nuclear fission and was aware of its military potential. Von Laue's reply to Morrison was published in German in *Physikalische Blätter*, so that many German readers saw only his forceful comments and not the original critiques by Goudsmit and Morrison.

Von Laue attacked Morrison for the "monstrous suggestion" that German scientists as a body had worked for Himmler and for Auschwitz. He recognized that Goudsmit had lost many of his closest relatives in concentration camps and what unutterable pain the mere word "Auschwitz" must evoke in the Jewish physicist. But von Laue then went on to draw the conclusion that neither Goudsmit nor his reviewer Morrison was capable of an unbiased judgment.

Just because a few German scientists had managed to avoid being drawn into the maelstrom, von Laue wrote, this did not mean that all could have. Von Laue argued that whereas open refusal to participate in war work would have led to catastrophic personal consequences, a fictitious compliance often allowed German physicists to shield younger researchers from the war. Sometimes, von Laue continued, the physicists protected "political suspects" from concentration camps by assigning them work labeled "of military importance." Von Laue asked ironically whether such scientists and internal refugees should be labeled "armorers of Himmler and Auschwitz." Von Laue said that articles such as Morrison's merely kept "alive hate."

Regretting that he had to take issue with von Laue's "moving statement," Morrison pointed out in the same issue that he had accused the German scientists of working not for Himmler, but for Himmler's cause, the victory of Nazi Germany.

For Morrison, Goudsmit and many people who had been in or had fled from countries dominated or threatened by Germany, working for a German victory in World War II was seen—after the fact—as equivalent to working for the cause of Himmler and Auschwitz. But, during the Third Reich few Germans had distinguished clearly between the known military and political goals of the National Socialist government and those of Germany or the German people. Once the full extent of Nazi crimes was revealed and undeniable, Germans such as von Laue and Heisenberg advanced a retrospective argument implying that efforts toward a German victory and for the good of the German people had been separable from allegiance or service rendered to Nazism. **FIGURE 3.**

German physics represented a significant success. But on the other hand, van der Waerden could understand the negative reaction of Goudsmit and others toward those same efforts. This negative reaction was admittedly illogical. But emotionally it was comprehensible. Could Heisenberg associate with the SS and leading National Socialists, even exert influence over them, van der Waerden asked, without compromising himself? Van der Waerden believed that he could, but could understand if others did not feel that way.

In a letter written on 28 April 1948, Heisenberg replied to van der Waerden by addressing the moral question once again. When he knew near the end of 1941 that a nuclear reactor would work and that nuclear weapons probably could be built, Heisenberg explained, he had been "shocked" by the thought of such weapons in the hands of some ruler, and not only Hitler. In any case, Heisenberg flatly stated, he would have considered it a crime to make atomic bombs for Hitler. But Heisenberg also considered it unfortunate that these weapons were given to other rulers and were used by them. During the past few years, Heisenberg said, he had learned something that his friends in the West did not want to understand: During times like those in Germany during the war, hardly anyone can avoid committing crimes or supporting them by doing nothing. Heisenberg hastened to add that he did not mean he would have been prepared to commit any sort of crime for Hitler.

A failure to communicate

Goudsmit wrote Heisenberg in late September of 1948 and admitted that he had been wrong about certain details, but now he grasped mistakenly at another perceived German deficiency by claiming that the Germans had not understood that an atomic bomb would rely on a fast-neutron chain reaction. However, Goudsmit's main aim in this letter was again to stress how science had suffered under Nazism and how political interference ruins science. As Goudsmit himself realized, the contents of this letter differed little from the previous one he had sent Heisenberg.

After spending several letters discussing the moral aspects of the German nuclear fission project, only to have Goudsmit stubbornly continue to make false statements about Germany's scientific achievements, Heisenberg dealt almost exclusively with the question of scientific competence in replying to Goudsmit's latest missive. The German physicist noted with pleasure Goudsmit's grudging partial admission of error. Now that Goudsmit had agreed that the Germans had known about plutonium, Heisenberg wanted him to make this admission public, in the *Bulletin of the Atomic Scientists.*

But Heisenberg did more than gloat over Goudsmit's discomfiture. Naturally, he wrote, he agreed that a totalitarian system greatly damages science, but in *Alsos* this conclusion was based on false arguments, which Heisenberg found very unfortunate.

Goudsmit found Heisenberg's answer "impertinent" in turn and could not understand why the recognition of the value of Heisenberg's scientific work meant so much to his German colleague. Heisenberg may well have been asking himself why Goudsmit could not recognize the admittedly modest German achievement.

Before Goudsmit could write his German colleague or publish anything in the *Bulletin*, Heisenberg took his case directly to the American people by means of an interview published in *The New York Times*. He devoted most of the interview to refuting Goudsmit's claims about the German scientific work, but he also addressed the moral question carefully: Because of their sense of decency, most of the leading German scientists had disliked the totalitarian system, he claimed; yet as patriots, when called upon to work for the government, they could not refuse.

Goudsmit responded in a letter to the *Times* by admitting that his portrayal in *Alsos* was an oversimplification, but he still insisted stubbornly that the Germans had had only a very vague notion of how an atomic bomb works.

On 11 February 1949 Goudsmit wrote Heisenberg again, advancing the same arguments, and in an exasperated tone asked whether further correspondence made any sense. Heisenberg responded with his most curt letter yet, stating bluntly that he would have preferred no public discussion of the German nuclear energy and weapons research, that through his articles and book Goudsmit repeatedly had spread false information about the German work, that it was time that a correct description of the German research was presented in the newspapers for a change, and that he was disappointed that Goudsmit did not recognize Heisenberg's right to take his case to the public as Goudsmit had done so often. It is clear that in the end Goudsmit's unfair criticism of Heisenberg's scientific abilities and achievement exasperated and embittered the German physicist. The damage done to his

reputation as a physicist may have come to bother him more than the criticism he received for serving the Nazism. The debate between Goudsmit and Heisenberg left outsiders with the impression that Heisenberg was morally insensitive or even obtuse, but that impression is not altogether warranted. Private wartime correspondence shows that Heisenberg was indeed troubled by his compromises with National Socialism. He also was troubled by the fate of individuals. In 1943, at the request of a Dutch physicist, Heisenberg wrote a letter to authorities on behalf of Goudsmit's parents, who were about to be sent to a concentration camp. Even for a person in Heisenberg's position, this was a very dangerous move at that time. Yet throughout the long and painful correspondence with Goudsmit after the war, Heisenberg never mentioned his intervention on behalf of Goudsmit's parents.

Last letters and visit

In the course of their protracted debate, both Goudsmit and Heisenberg had come very close to losing their tempers, and they now realized that it was pointless to continue. Goudsmit wrote one last letter in which he repeated his litany of arguments, told Heisenberg that he would not bring up the matter again, and expressed the hope that they could continue to correspond about physics. Heisenberg replied immediately and thanked him for his letter, which this time Heisenberg had been very happy to receive. Goudsmit could be certain, Heisenberg promised, that he would not stir up the controversies of the past few years again, and especially not in public. Exactly like Goudsmit, Heisenberg said, he believed that it was more interesting to discuss physics than the unpleasant past.

Both men worked very hard the rest of their lives to restore international cooperation in science. Despite the loss of his parents, Goudsmit went out of his way to assist German scientists. In the same issue of the *Bulletin of the Atomic Scientists* that contained a debate between Philip Morrison and Max von Laue on the German scientists' role during the war (see Figure 3), Goudsmit contributed an essay that showed he had changed his mind about the sink-or-swim policy he once had advocated for German science. Entitled "Our Task in German," Goudsmit's article once again called for international scientific cooperation, and this time he extended his hand to Germans as well. Goudsmit expressed the belief that American scientists should morally support those German colleagues who were worthy of confidence, and he added that there were many of them. Americans did not have to agree with all of their opinions but should make allowances for the disturbing circumstances under which these Germans had lived and were still living. Americans had to communicate with them, Goudsmit maintained, as in the days before Hitler. It was as if Goudsmit had been converted by the disturbing forces that he had helped to unleash.

In 1950 Heisenberg visited Goudsmit at Brookhaven National Laboratory in Long Island, New York, and they discussed only physics.

References

1. M. Walker, *Vierteljahrshefte für Zeitgeschichte* **38**, 1 (1990).
2. A. Beyerchen, *Scientists under Hitler*, Yale U. P., New Haven, Conn. (1977). R. Rhodes, *The Making of the Atomic Bomb*, (Simon and Schuster, New York (1988).
3. J. Radkau, *Aufstieg und Krise der deutschen Atomwirtschaft, 1945–75*, Rowohlt, Reinbeck (1983).
4. S. Goudsmit, *Rev. Sci. Instrum.* **17**, 49 (1946); *Bull. At. Sci.* **1**, 4 (1946); *Sci. Illustrated* **1**, 97 (1946).
5. W. Heisenberg, *Naturwissenschaften* **33**, 325 (1946). ∎

LETTERS ON HEISENBERG, GOUDSMIT AND THE GERMAN 'A-BOMB'

PHYSICS TODAY/MAY 1991

In Mark Walker's strange new view ("Heisenberg, Goudsmit and the German Atomic Bomb," this volume, page 10), Germany's uranium project, so feared by the Allies, so surprisingly inconsequential in fact, was not a failure at developing atomic weapons.[1] Nuclear *power* was its animating goal, and progress, under sound National Socialist management, was highly creditable.

But how accurate is Walker's account? Much of it derives from material accessible only in Germany, but much else is based on open sources available in the archives of AIP's history library in New York, and familiar to me from the year and a half I worked with Samuel Goudsmit at the *Physical Review*. Comparing Walker's account with the available records turned up some remarkable differences, of which I can list only a few here:

▷ Walker extravagantly dismisses as the product of Goudsmit's "profoundly ahistorical and noncontextual preconceptions" his conclusion that the Germans decisively overestimated the difficulty of making bombs. In fact this overestimate was the controlling assumption of the uranium project's agenda. This much was later acknowledged by Werner Heisenberg[2] ("we regarded the necessary technical effort as rather greater than, in fact, it was") and by Carl-Friedrich von Weizsäcker[3] ("I must admit that we also overestimated the difficulty of the problem"; "we had thought it would be even more difficult and so this was sufficient reason not to try it").

▷ Though Walker makes much of Goudsmit's refusal to do so, Goudsmit did, after a long, painstaking investigation, acknowledge that the senior German physicists had an accurate theoretical conception of fast-fission uranium bombs and of plutonium breeding.[4]

▷ The Alsos mission did *not* destroy apparatus. Heisenberg did *not* write a letter to German authorities on behalf of Goudsmit's parents; he re-

plied to a plea from the Dutch physicist Dirk Coster. Heisenberg's letters to Goudsmit "never mentioned his intervention on behalf of Goudsmit's parents," as Walker states, presumably because Goudsmit by then had long known about the letter, via Coster and Max von Laue, and perhaps because Heisenberg's oddly vague response arrived too late to be of any use. Goudsmit was *not* a reserve officer (he declined a military commission) and did not, of all things, write a "heroic" account of the mission in his book *Alsos*[5] (see the index entry under "Rabi, I. I.").

▷ Goudsmit never held the stereotype view of science as a series of isolated works of great minds, as Walker has it. He advanced the opposite view (in, for example, "It Might as Well Be Spin"[6] and "Guess Work: The Discovery of the Electron Spin,"[7] whose titles suggest how Goudsmit thought of his own "great" contribution).

▷ The basis of Goudsmit's assessment of the uranium project's bomb plans nowhere includes the idiotic identification of spherical reactors with spherical bombs, as in Walker's caricature. Nor did Goudsmit assert that any existing reactors were meant to be bombs, as Walker suggests by substituting the literal "Germany's atom bomb" for the rhetorical "Germany's 'atom bomb'" in Goudsmit's figure caption.

Goudsmit did err, but for reasons very different from those Walker gives, in concluding that the German physicists missed the concept of fast-fission bombs. Germany's uranium work, as the Alsos mission found it at the end of the war, stood roughly where the Allies' had been in late 1941 to early 1942, when the feasibility of fast-neutron explosives had just been established by Merle Tuve's fast-neutron cross-section measurements on uranium-235. Prior to March 1941, it was anyone's guess whether a reasonably small mass of uranium really could, as Otto Frisch and Rudolf Peierls had suggested the year

before, support an explosive chain reaction, or whether it would prove necessary to take advantage of the larger cross sections at lower neutron energies by using moderators. Finding no indication that the German project had resolved this question, and ample evidence of interest in weapons based on slow-neutron fission,[8] Goudsmit reasonably but incorrectly concluded that the project physicists had not grasped the possibility of fast-fission weapons. This point he eventually set right.

Goudsmit's encounter with the German uranium project impressed him with how badly physics had fared under the Nazis. Where the Manhattan Project had advanced with such conspicuous success under a coordinated leadership involving respected scientists at high levels in government, military and industrial decisions, the Nazi system had slighted science. Where autonomy was allowed, as in the Luftwaffe research organization, success (in rocketry, in jet fighter development) was forthcoming. But, lacking a unified and influential organization like the American National Defense Research Council until late in the war (and then mainly on paper), the uranium project was weakly represented and fragmented. Thus, at the time Enrico Fermi's Chicago pile CP-1 went critical with 6 tons of uranium metal, in late 1942, the German project had some $7\frac{1}{2}$ tons in hand—but divided between competing groups unwilling to share scarce uranium, heavy water and other resources. Eugene Wigner and Hans Bethe, who were well placed to assess the difficulties, had reckoned that Germany could stockpile bombs by the end of 1943. That not even a self-sustaining chain reaction had been achieved by war's end came as a great surprise. Yet CP-1 cost no more than one million dollars, an amount available to the Germans for the asking.[9] Nor did Germany lack the industrial sophistication or the desire: Nuclear weapons are mentioned enthusiastically in the project documents again and again.

Why then did the project get no further? Goudsmit emphasized four factors:

1. That the Germans, mistakenly believing themselves far ahead of the Allies, felt no competitive urgency.

2. That Nazi political control and interference burdened the project in a variety of ways.

3. That the uranium project physicists decisively overestimated the difficulty of the task (because they failed to appreciate fully the plutonium alternative; because, as Goudsmit initially thought, they conceived of bombs as depending on slow-neutron reactions; and because the academically oriented theoreticians lacked aptitude and enthusiasm for industrial undertakings, even on the scale of cyclotrons).

4. That wartime conditions worked against the project.

Walker allows only the last of these, with a passing acknowledgment of the first. The project administrators, as he sees it, chose with admirable correctness to forgo an industrial-scale weapons program as incompatible with wartime priorities. And so, Walker says, a considered decision was taken in 1942 not to press forward. ("This was the final verdict, which never was reassessed.") Walker's principal evidence for this remarkable conclusion, a single, anonymous Army Ordnance Office report, suggests an extraordinary faith in committee organization charts. Certainly, the Allied enterprise followed no such orderly course, as Richard Rhodes's vivid account makes particularly clear.[10] Powerful persuasion from the Allied physicists themselves time and again overturned the cautious positions of government officials formally vested with power. Not least for such reasons, surely, did Heisenberg stress the importance of his meeting with the armaments minister, Albert Speer, whose considerable power—to provide funds, priorities, influence—lay beyond any Ordnance Office committee's "final verdict."

That political interference impeded the project Walker dismisses as so much prejudice (on Goudsmit's side) or self-serving misrepresentation (on Heisenberg's). The uranium project administrators were "professionally respectable physicists," Walker insists, not political hacks. Erich Schumann, the head of army ordnance research, was "qualified to teach physics at the university level," he tells us, but he doesn't mention Schumann's specialty (the physics of piano

strings), his title (professor of military physics) or the mock title that suggests his standing with the project physicists themselves—"the professor of military music." If the senior administration (with the isolated exception of Walther Gerlach, at the end of the war) was distinguished in more than political guile, Walker has not given us evidence for it. The six talented physicists he lists as "involved in the scientific work and the administration" include only one (Gerlach) with authority beyond laboratory level. Goudsmit's opinion that such figures as Schumann, Rudolf Mentzel and Bernhard Rust were scientifically incompetent political men is hardly belied by their credentials. Rust, Hitler's minister of education, was "scientifically illiterate," in Rhodes's words. His subordinate Mentzel, the chief of all research in German universities, has not in any account other than Walker's been described as a capable scientist. Both, however, held high honorary ranks in the SS (*Obergrüppenführer* and *Brigadeführer*, respectively); Schumann was a Wehrmacht general.

One point that no reader of *Alsos* could miss, in some three dozen pages dealing with these men, is Goudsmit's own low opinion of four particular political administrators and his reasons for it (though "nincompoops," the word he used for two of them, was edited out of the book's page proofs by the publisher). *Alsos* makes the important point that the Gestapo, too, held critical reports on several of these men. Walker, however, telescopes the entire picture into a single, completely misleading sentence: "Using Gestapo records that he himself considered suspect, Goudsmit unfairly dismissed Schumann and other National Socialist science policy administrators as incompetent. . . ."

For one to understand the Heisenberg–Goudsmit letters, some of the record omitted by Walker needs to be restored: the letters themselves, and the conversations of the German physicists interned at Farm Hall, secretly taped by British intelligence.[11–13] (Key portions of the transcripts, which are still classified, were reproduced in *Alsos*, though secrecy restrictions forced Goudsmit to do this with a coy device that did not do justice to the certainty of his evidence.[11] Walker ignores the actual conversations in favor of the memorandum prepared by the Germans for public release.) The Farm Hall transcripts confirm three points of interest here: first, that the German physicists conceived the construction of

atomic weapons to be vastly difficult—far beyond even the combined resources of the Allied nations; second, that such basic questions as the critical mass for a uranium bomb had not, by war's end, been settled in Germany; third, that von Weizsäcker and others planned an artful reinterpretation of their embarrassingly slight showing—namely, that the physicists had chosen, on principle, not to pursue atomic weapons.[11,12] Heisenberg, at the time *Alsos* was written, still endorsed this comforting fiction, and it did not escape Goudsmit's notice that the point on which the Heisenberg letters so strongly insisted—that the physicists had well understood how to create fission bombs—was the very premise of von Weizsäcker's invention. Goudsmit thus had reason to question whether the physicists really had understood fission bombs and plutonium, and to view Heisenberg's later protests to the contrary with a certain wariness.

Heisenberg's letter of 5 January 1948 conveys his evidence that the German physicists understood fast-neutron fission weapons and plutonium breeding, but it is evidence of an oddly thin, inferential kind. Among the wartime progress reports, he points out, is a speculation (by Walther Bothe) that protactinium might support an explosive fast-neutron chain reaction. As proof that such reactions had also been contemplated in uranium, Heisenberg reproduces from memory a slide on which he had illustrated the neutron multiplication to be expected in a large mass of pure U-235. The slide itself, from a 1942 lecture to Luftwaffe officials "adapted to the intelligence level of a Reich Minister of that time," he presumes lost. Lastly, as evidence that the plutonium alternative had been appreciated, Heisenberg cites a 1940 report in which von Weizsäcker reasoned that slow-neutron capture by U-238 in a natural-uranium-fueled reactor should produce the transuranic element neptunium (Eka Re-239, as it was then called), with expected fission properties similar to those of U-235. That Heisenberg's evidence consisted of no more than this—an open-ended speculation about an impractically rare element; a theoretical sketch on a single, lost lecture slide; and a plausible conjecture—could only have encouraged Goudsmit's doubts.

Heisenberg, we might surmise, felt entitled to be taken at his word, yet Goudsmit was less concerned with what Heisenberg knew than with the common currency available to the uranium project. Why, for example, to the thousand-plus experiments undertaken by the Allies to investigate plutonium, had Germany done none—not even the vital cross-section and yield measurements needed to confirm or exclude the plutonium option, measurements Emilio Segrè and Glenn Seaborg had done with cyclotron-generated microsamples in May 1941? That Germany had been slow to acquire cyclotrons was, as Goudsmit observed, more a statement of the problem than an explanation of it. Von Weizsäcker's 1940 speculation on plutonium was noteworthy, yet Louis Turner, at Princeton, had outlined the idea (correctly identifying plutonium rather than neptunium as the fissile end product of U-238 neutron capture) at about the same time.[14] For Turner, however, such bare speculation was so far from real knowledge as hardly to warrant his withholding it from publication in the *Physical Review* on secrecy grounds. "It seems as if it was wild enough speculation so that it could do no possible harm," he had written to Leo Szilard.

Alsos, according to Walker, is unreliable because after Goudsmit's parents were killed at Auschwitz he "no longer was completely objective." The insupportable implication here, that the word of the oppressors is intrinsically more, rather than less, reliable than the word of their victims, was addressed in the very *Bulletin of the Atomic Scientists* article (April 1948) Walker examines at length. As Walker renders it, Philip Morrison's "Reply to Dr. von Laue" seems an apologetic qualification of his earlier *Bulletin* statement attacking Heisenberg's claim that the Germans had not tried to create nuclear weapons. Morrison's actual words, however, are quite different:

I am of the opinion that it is not Professor Goudsmit who cannot be unbiased, not he who most surely should feel an unutterable pain when the word Auschwitz is mentioned, but many a famous German physicist in Göttingen today, who could live for a decade in the Third Reich, and never once risk his position of comfort and authority in real opposition to the men who could build that infamous place of death.

Goudsmit's views are "oversimplified," Walker tells us, "deeply based in scientific stereotype" or "the result of sloppy research." To judge by what can actually be checked against the documents, however, the confusion, exaggeration and distortion here are Mark Walker's own contribution to the record.

References

1. See also M. Walker, *German National Socialism and the Quest for Nuclear Power, 1939–1949*, Cambridge U. P., New York (1989).
2. W. Heisenberg, *Der Teil und das Ganze*, R. Piper, Munich (1969), p. 245.
3. C. F. von Weizsäcker, transcript of lecture given at the Harvard University Science Center, 29 March 1974. (This and all other archival material referred to here is in the collection of the Niels Bohr Library of the American Institute of Physics in New York.)
4. S. A. Goudsmit, letter to *The New York Times*, 9 January 1949; *Year Book of the American Philosophical Society*, Am. Philos. Soc., Philadelphia (1976), p. 74. "Samuel Abraham Goudsmit," Harvey Mudd College Oral History Project on the Atomic Age, Claremont Graduate School, Claremont, Calif. (1976), p. 82.
5. S. A. Goudsmit, *Alsos*, Henry Schuman, New York (1947).
6. S. A. Goudsmit, lecture given at The American Physical Society meeting in New York, 2 February 1976; published in PHYSICS TODAY, June 1976, p. 40.
7. S. A. Goudsmit, Delta, Summer 1972, p. 77.
8. S. A. Goudsmit, *Alsos, op. cit.*, pp. 177–181; letter to B. L. van der Waerden, March 1948 (in Dutch).
9. R. Rhodes, *The Making of the Atomic Bomb*, Simon and Schuster, New York (1986), p. 436.
10. R. Rhodes, *op. cit.*
11. S. A. Goudsmit, *Alsos, op. cit.*, pp. 132–139.
12. M. von Laue, letter to P. Rosbaud, 4 April 1959, translated in A. Kramish, *The Griffin*, Houghton Mifflin, Boston (1986), pp. 245–247.
13. R. V. Jones, *The Wizard War*, Coward, McCann and Geoghegan, New York (1978), pp. 473, 483.
14. L. A. Turner, Phys. Rev. **69**, 366 (1946); submitted 29 May 1940 but voluntarily withheld from publication.

JONOTHAN L. LOGAN
11/90   *Westford, Vermont*

Mark Walker states that "contrary to accounts based on Heisenberg's claims, the German fission research effort in World War II was indeed a nuclear weapons program." I contradict this statement on the basis of the *same* documents Walker used: formerly secret reports[1] on the German nuclear fission research of World War II (henceforth called the German uranium project) kept at the Kernforschungszentrum in Karlsruhe, as well as letters and documents in the Werner-Heisenberg-Archiv, Munich.

To decide this question one should first recall very briefly the historical facts. The German uranium project was started—after some preliminaries—in September 1939 by the Heereswaffenamt (Army Ordnance Office). The following problems were presented to the scientific experts assembled at the inaugural meeting:

It is the task of the participants to work out all preparatory steps in order to answer uniquely the question of whether nuclear energy can be produced on a technical scale. Of course, it would be very nice if the answer turned out to be positive and if one succeeded in opening a new source of energy. This would very probably also have military importance. A negative result, however, would be likewise important, since one could then be certain that the enemy has no access to it [that is, nuclear energy] either.[2]

The research carried out up to early 1942 did indeed nearly answer these questions. Theoretical and experimental work made it apparent that a self-sustaining nuclear chain reaction could very likely be established in a machine (reactor) containing natural uranium as fuel and heavy water as moderator; also, the creation of nuclear weapons seemed to be feasible in principle by proceeding along either of two paths: production of a suitable amount (tens of kilograms) of U-235 by isotope separation, or breeding enough fissionable transuranium material in an already functioning uranium machine.[3] The scientists reporting to the German authorities (the Heereswaffenamt and Reichsforschungsrat) in February 1942 and later also declared that a "nuclear explosive" (*Kernsprengstoff*) would not be available without several years of enormous technical, industrial and financial effort. At that time the Heereswaffenamt retired from the uranium project, which continued, however, to be a secret project rated "important for the war" (*kriegswichtig*). Further experiments were carried out to achieve a critical reactor, on the one hand, and isotope separation (of uranium and hydrogen), on the other hand. In spite of some progress, at the end of the war (May 1945) neither a functioning reactor nor larger amounts of U-235 existed in Germany. (It might be mentioned that the isotope separation effort was directed toward getting material for a smaller uranium machine or a machine running with light water as moderator.)

Careful study of the material documenting the above story does not uncover any serious work, theoretical or experimental, on a "nuclear weapon," not even during the time when the German uranium project was supervised by the Heereswaffenamt, a military authority. True, the possibility of such weapons was mentioned occasionally, more or less in passing, in some of the reports. (Only one of the roughly 150 reports submitted through the end of February 1942 dealt explicitly with "the requirement for the utilization of uranium as an explosive"—Paul O. Müller's sketchy report of six pages, dated 31 May 1940. Müller proposed to use a mixture of water and uranium oxide in which the isotope U-235 was enriched—anything but an efficient explosive.) It is also true that in the beginning of 1942, when the project was given up by the Heereswaffenamt and the danger arose that it might be dropped altogether, the scientists involved tried to rescue their research by emphasizing that it was *kriegswichtig*—because otherwise they would not obtain the required funds, nor the necessary materials (uranium, heavy water, steel and rare metals), nor the junior scientists and assistant scientific personnel (freed from military service) needed for the work. Their reports and talks stressed any possible military use of nuclear energy, be it for machines propelling tanks and submarines or for explosives. No action or work followed from these words; nevertheless the state authorities (including Albert Speer's ministry for war production) kept the project alive. Hence I do not see any justification for calling the German uranium project a "nuclear weapons program."

Let me finally mention a further weak point in Walker's argument. In discussing the details of a 1946 review of the German uranium project by Werner Heisenberg—a decent English translation appeared in *Nature*[4]—Walker claims that "when this review is compared with sources documenting the history of the German nuclear fission project, several important discrepancies emerge." Whatever one thinks about these discrepancies—I have some trouble discovering any in the examples mentioned by Walker—one must always keep in mind that any report given some years later on an extended project will strongly reflect the personal recollections and opinions of the writer; it certainly cannot yield a detailed, "document proof" account of the historical events (especially if the documents were, as in this case, not available to the writer). Instead of accusing Heisenberg personally of any inaccuracies, Walker should have scolded those historians who base their reconstructions of the whole story on a report by a single actor in the game.

References

1. For a bibliography, see L. R. David, I. A. Warheit, *German Reports on Atomic Energy: A Bibliography of Unclassified Literature*, TID-3030, Technical Information Service, Oak Ridge, Tenn. (1952).
2. E. Bagge, Fusion **6**(6), 26 (1985).
3. See, for example, the report delivered by Heisenberg on 26 February 1942, "Die theoretischen Grundlagen der Energiegwinnung aus der Uranspaltung," reprinted in *W. Heisenberg: Collected Works*, vol. A II, Springer-Verlag, Berlin (1989), p. 517.
4. W. Heisenberg, Nature **160**, 211 (1947).

HELMUT RECHENBERG
*Max-Planck-Institut für Physik
und Astrophysik
Munich, Germany*

2/90

Mark Walker's conclusions that Samuel Goudsmit "clung tenaciously to mistaken ideas," that his approach was "profoundly ahistorical and noncontextual" and that Werner Heisenberg played a relatively minor role in the German nuclear effort are controversial, misleading and at best only partially correct.

Since the following comments have a distinctly personal flavor, it is proper to explain my own direct involvement in the events of the times. Like Goudsmit, I was born and educated in the Netherlands. I came to the United States to complete my PhD with Enrico Fermi, who had just emigrated to America. Fermi was scheduled to lecture in Ann Arbor the summer of 1939 (on cosmic rays), and I came to Ann Arbor a few months before the start of the summer session. While there, I met Goudsmit, Otto Laporte and of course Fermi, as well as a number of visitors to the summer symposium. All during the summer discussions focused on cosmic rays and nuclear physics, with a comparable amount of time (and intensity) spent on the turbulent and frightening political events of that summer. Heisenberg visited for a week late in July and stayed at Goudsmit's home. He left to return to Germany early in August. I had met Heisenberg before, in Hendrik Kramers's seminar in Leiden, and I saw quite a bit of him during his visit to Ann Arbor. I also saw a lot of Goudsmit while at Michigan. And well after the war, from 1964 on, when I was a professor at the State University of New York at Stony Brook, I saw Goudsmit, who was at Brookhaven, very frequently.

Like Goudsmit's family, my parents, many members of my family and many friends were exterminated in the Holocaust. Goudsmit and I often discussed our fluctuating respective reactions of anger and guilt—and our fear and even terror of the possibility of a renewed wave of barbarism (anywhere in the world). But painful as these reactions may be, to dismiss Goudsmit's conclusions concerning the successes and failures of German and American science on the grounds that he could not be objective because of his personal experiences is totally unjustified and ignores the intellectual integrity of Goudsmit (and myself). This of course doesn't mean Goudsmit's analysis is correct. Emotions do affect attitudes; they influence the assignment of personal guilt and personal responsibility; but they do not invalidate an argument and do not alter the facts. Goudsmit's arguments deserve to be discussed and analyzed on their own merits, not ignored in a cavalier fashion, as Walker does.

Perhaps the most serious objection to Walker's article is that he totally fails to place the Heisenberg–Goudsmit confrontation in the proper scientific and personal context. The relationship between Heisenberg and Goudsmit involved three distinct elements. The background of their earlier personal and scientific interactions, the changing relations between American and European physics, and their widely differing political opinions were all crucial ingredients in their angry exchanges.

In 1925 both Goudsmit and Heisenberg were members of the brilliant new generation of quantum physicists. It was widely expected that both would make major contributions to quantum theory. Both of them surely did. But after about three or four years, atomic spectroscopy, the field of Goudsmit's special expertise, became less central in physics, and his contributions to physics started to diminish and become less basic. Heisenberg, by contrast, continued (at least for some time) his brilliant exploits. Right after Goudsmit discovered the electron spin, Niels Bohr invited him to come to Copenhagen to study the problem of *ortho* and *para* helium. Goudsmit went, but made no progress whatsoever and returned a little disillusioned to Leiden. Heisenberg followed Goudsmit to Copenhagen and solved the helium problem completely. It was that very achievement that was mentioned in his Nobel Prize citation. Goudsmit often mentioned that episode. He stated on numerous occasions, "Heisenberg's solution was way beyond me." It is hard to know the effect of a single incident, but this one must have had a substantial impact. It is certain that by 1939, Goudsmit felt that physics had passed him by. He was disappointed about his contributions to physics, and he had severe doubts that he was capable of understanding, let alone contributing, to the then current physics. (Heisenberg never had such doubts.)

Goudsmit's disappointments made him at times depressed, often angry and always cynical. These cynical attitudes, combined with strong anti-Nazi feelings, caused him to be abrasive. He worried incessantly about the future of Europe. Goudsmit was not particularly interested in politics, but his tendencies were liberal rather than conservative, international rather than national. By contrast, Heisenberg was a strong German patriot, a true believer in Germany's historic destiny. He often said during the war that he hoped that Germany would win. Although hardly surprising, this hope was totally unacceptable to Goudsmit.

Yet another source of tension had to do with Goudsmit's and Heisenberg's shared belief that as leading members of the international physical community, they were expected to meet certain standards of behavior, intellectual integrity, personal compassion and individual accountability. Goudsmit felt this was incompatible with an allegiance to Nazi Germany. He thus felt that Heisenberg had not lived up to these standards. By the same token, Heisenberg argued that those who had not been subject to the insidious pressure of a ruthless totalitarian regime had no right to sit in judgment of those who had suffered through it.

A third area of conflict was the shift of the center of theoretical physics from Europe to the United States. As beautifully analyzed by Silvan S. Schweber,[1] by the middle of the 1930s, American physics, helped by a large influx of foreign physicsts, had evolved into a powerful independent discipline with a style and approach all its own, combining the abstract, theoretical European approach with the more direct, pragmatic American methodology. This American approach was particularly successful during (and after) World War II. The resulting shift, accelerated by the deterioration of European physics, was difficult to accept for Heisenberg and for many others (Wolfgang Pauli, Kramers, Carl-Friedrich von Weizsäcker). In fact they never did fully accept it.

Goudsmit understood better than Heisenberg that the change of the scientific hegemony from Europe to the US was an important element in their personal conflict. That is why Goudsmit was so irritated by the automatic assumption of German superiority and was outraged at Heisenberg's suggestion that he would be willing to lecture on the "uranium problem" to the American physicists (including Fermi, Eugene Wigner, J. Robert Oppenheimer and Hans Bethe—the very people who had built a bomb and constructed a pile).

In early August 1939, while Heisenberg was staying with Goudsmit, Laporte, another old friend of Heisenberg's, gave a party for him. I and a few other graduate students were asked to function as bartenders and waiters. There was actually not much to do, so we could pay close attention to the conversations. There was really only one central topic. Fermi had just left Fascist Italy to come to the US; Heisenberg had decided to return to Nazi Germany. The crucial part of their argument was whether a decent, honest scientist could function and maintain his scientific integrity and personal self-respect in a country where all standards of decency and humanity had been suspended. Heisenberg believed that with his prestige, reputation and known loyalty to Germany, he could influence and perhaps even guide the government in more rational channels. Fermi believed no such thing. He kept on saying: "These people [the Fascists] have no principles; they will kill anybody who might be a threat—and they won't think twice about it. You have only the influence they grant you." Heisenberg didn't believe the situation was that bad. I believe it was Laporte who asked what Heisenberg would do in case of a Nazi–Soviet pact. Heisenberg was totally unwilling to entertain that possibility: "No patriotic German would ever consider that option." The discussion continued for a long time without resolution. Heisenberg felt Germany needed him, that it was his obligation to go back. Fermi did not think there was anything anyone could do in Italy (or Europe); he was afraid for the life of his wife (her father was later killed); and so he felt it was better to make a fresh start in the US. But none of the decisions had come easy. The role of physics and physicists was mentioned off and on.

After the party was over everybody left in a state of apprehension and depression. Although there was no clear anticipation of the turbulent events to come, it was evident that theoretical physicists would no longer be a happy, unconcerned group of brilliant young men matching their intelligence against the secrets of the universe.

Reference

1. S. S. Schweber, Hist. Stud. Phys. Sci. **17**(1), 55 (1986).

MAX DRESDEN
Stanford Linear Accelerator Center
8/90   *Stanford, California*

Mark Walker's interesting article brought me back to September 1944, when Samuel Goudsmit's intelligence team visited the physics lab of the N. V. Philips Gloeilampen Fabrieken in Eindhoven, the Netherlands, right after Eindhoven was liberated.

I had joined the Philips labs in Eindhoven as a research physicist, working on vacuum tube electronics and noise problems, after obtaining my PhD degree in 1934. Before that I had studied experimental physics at the University of Groningen, the Netherlands, from 1928 to 1934. My physics professors were Dirk Coster (experimental physics), Frits Zernike (theoretical physics) and Ralph de Laer Kronig (wave mechanics), and my mathematics professors were J. G. van der Corput (analysis) and B. L. van der Waerden (linear algebra).

Coster was very active in helping the Jews. In 1939 he had traveled to Berlin and led Lise Meitner to safety. He also had pleaded in vain with the German authorities to release Goudsmit's elderly parents from the concentration camp and spare their lives.

Goudsmit's intelligence team consisted of several groups. In hindsight the most important was the nuclear physics group. It had to inquire about German atomic bomb development without revealing Allied progress in the field. I showed the electronics group some German crystal diode detectors with 5-cm half-wave antennas. They were much surprised, for it was their first concrete evidence that the Germans were working on 10-cm radar. A third group worked with the commercial department to detect large shipments of vacuum tubes to German locations that were not on the Allied intelligence list. Some time earlier an alert commercial administrator had noted huge shipments of vacuum tubes to an obscure village called Peenemünde and alerted Allied intelligence via the underground. Allied bombers bombed Peenemünde heavily, retarding the German rocket program substantially.

During one of the breaks Goudsmit took me aside and asked me what I knew about his parents. I told him of Coster's efforts—that he had not succeeded in having Goudsmit's parents released and that therefore the worst had to be feared. Later it turned out that they had died in Auschwitz. It was difficult for me to be the bearer of such sad tidings.

Walker mentions that Heisenberg also intervened with the German authorities on behalf of Goudsmit's parents. The question is now whether Coster's and Heisenberg's intercessions were isolated events or part of a larger effort. Knowing Coster, and bearing in mind that he knew practically all the important German physicists, I believe the latter.

Van der Waerden played a very active role in the early years of wave mechanics. He must therefore have known Heisenberg since 1925. After 1930 they were colleagues in Leipzig, though it is not known how much they interacted during that time. Either fact makes it understandable why van der Waerden came to Heisenberg's defense after the war.

Our research lab did not suffer from German interference, but after 1 January 1944 we had to work on German research contracts. Slowdown tactics were successfully employed at first, but it was soon clear that this was not a long-term solution. Much stricter rules were announced in early September 1944, with the threat of death penalties. But on the same day Antwerp fell, and two weeks later Eindhoven was liberated. Our ordeal was over.

There is not enough understanding about what it means to work under an unfriendly totalitarian regime. Your options are quite limited. Outright refusal leads to prison or execution. Slowdown techniques are possible but cannot become obvious. I therefore have considerable sympathy for the positions taken by van der Waerden, Heisenberg and especially Max von Laue.

A. van der Ziel
University of Minnesota
Minneapolis, Minnesota

2/90

WALKER REPLIES: I am grateful to the editors of PHYSICS TODAY for an opportunity to respond to my critics. First of all, I would like to direct all interested persons to my book[1] and to my supplementary article on the postwar controversy surrounding the "German atomic bomb."[2] My article in PHYSICS TODAY, like all articles that appear there, was restricted in regard to length and the number of footnote references. The issues under discussion are complex, and I recognize that "Heisenberg, Goudsmit and the German Atomic Bomb" does not and could not have done them complete justice. Taken together, my book and supplementary article not only provide extensive references for the evidence I present but also place the postwar controversy between Samuel Goudsmit and Werner Heisenberg in the proper context. I will respond to my critics by referring to my book and that article, not merely to my publication in PHYSICS TODAY.

I consider Jonothan Logan's criticism of me to be unfair. Rather than my work being "strange," "extravagant," a "caricature," "misleading," "confused," "exaggerated" and "distorted," I have provided a fair and objective account of what happened and why. For the sake of brevity, I will respond to the most important of Logan's assertions in the order they appear in his letter.

Goudsmit's conclusion that the Germans overestimated the difficulty of making atomic bombs rested on the assumption that the Germans believed that such a bomb would require tons of uranium. This assumption is false, as I have demonstrated in my book, especially on page 48. The postwar statements by Heisenberg and Carl-Friedrich von Weizsäcker cited by Logan refer to the great costs of the industrial production of nuclear weapons and were written at a time when German physicists were concerned with explaining why they had not made more progress toward their goals.

Logan writes that Goudsmit did acknowledge that these German scientists had known about fast-fission uranium bombs and plutonium breeding, and he refers the reader to a letter Goudsmit published in *The New York Times* in 1949 and to two publications from 1976. Goudsmit did recant at the very end of his life, but that in no way contradicts my argument that he had not in the period covered by my book. When I reread Goudsmit's 1949 letter, however, I saw the following passage, which can hardly be reconciled with Logan's account: "Finally, the German physicists also missed the crucial point that a bomb is a reaction produced by fast neutrons in plutonium or in U-235. Fast neutrons are mentioned by them only in the hope that they might perhaps produce a chain reaction in the abundant isotope U-238." This assertion by Goudsmit is false.

I do not understand Logan's distinction between writing a letter and answering a plea. As I understand the English language, it is possible both to respond to a plea and to write a letter. I stand by my characterization of Goudsmit's account of the Alsos mission as "heroic."

Goudsmit repeatedly portrayed the German conception of an atomic bomb incorrectly, for example, in *The Bulletin of the Atomic Scientists*: "The German line of thought was as follows.... An atomic bomb is an uranium engine which gets out of control.... To make a bomb of pure plutonium never entered their minds, or at least was not considered feasible and not taken seriously. The idea of using a pile to produce plutonium and to make a bomb out of that material came to them only slowly, after the detailed radio descriptions of our bomb in August 1945."[3]

I do not rely on a "single, anonymous Army Ordnance Office report," as Logan claims. This report contains a bibliography of the 134 scientific and technical reports on which it was based. I not only read through each and every one of those 134 reports; I also researched thoroughly the dozens of letters exchanged between scientists and Army Ordnance in regard to this research.

Arguably it is Logan's criticism of my portrayal of Erich Schumann that is most typical and most revealing of his approach to history. There is no reason to doubt that Goudsmit told Logan that Schumann was incompetent, and I am willing to believe that Goudsmit believed this himself. But it is not true. I have seen Schumann's personnel file at the archives of the Humboldt University in Berlin, and these documents provide a very different picture. Schumann received his *Habilitation* (the right to teach) in 1929 at the University of Berlin— arguably the most prestigious teaching institution for physics in the world at the time—in "Systematic Musical Science," which included acoustics, an area of physics. His command of acoustics was examined and approved at that time by (among others) Max Planck, Max von Laue and Walther Nernst. Schumann received the *Venia Legendi* (the right to teach an additional subject) for the entire discipline of experimental and theoretical physics in 1931. His command of physics was examined and approved at that time by (among others) Nernst, von Laue and Erwin Schrödinger. But Adolf Hitler did not come to power until January of 1933. Schumann taught physics in Berlin from 1929 until the Second World War forced him to devote all his time to administration. Moreover, all the courses he taught in the Third Reich, including classes with titles like "Military Science," he had already taught in 1931. Does the fact that Schumann was interested in musicology necessarily mean that he was a bad physicist? Schumann was a ruthless and unscrupulous administrator, he was as convinced a National Socialist as they came, and he had no qualms about using science to create weapons with previously unknown destructive capacities, as proven by his sponsorship as an administrator in Army Ordnance of the rocket research in Peenemünde. But he was not incompetent.

Logan's reference to the so-called Farm Hall tapes is irrelevant. First of all, these recorded conversations, if they exist, have not been released. Thus the brief excerpts that some have claimed are genuine cannot be checked for accuracy. Second, and what is far more important, the portrayal these excerpts suggest, that these German scientists did not understand how an atomic bomb would work, is contradicted by evidence that they understood very well by 1942 how to make such a bomb in principle, and it is unlikely that they would have forgotten this knowledge in the meantime.

Logan argues that Heisenberg's postwar letters to Goudsmit are speculative, thereby implying that Goudsmit was correct to be skeptical of his German colleague's claims. But Logan has unaccountably refrained from mentioning the other evidence I present in my book on page 218. (Since Logan cites my book in his first footnote, I assume that he took the time to read it thoroughly and carefully before criticizing me.) Bartel van der Waerden visited Goudsmit and subsequently wrote Heisenberg and told him that he had seen documents in Goudsmit's office that verified Heisenberg's claims about what the Germans had known about plutonium and nuclear weapons.

Finally, it is unfortunate that Logan has accused me of putting words into Philip Morrison's mouth, for my account of what Morrison published is

fair. But what I consider to be especially unjust is for both Logan and Max Dresden to accuse me of arguing that Goudsmit's claims should be rejected *because* he had suffered at the hands of Germans and therefore was no longer completely objective. I have never written or said any such thing. Goudsmit's claims about the quality of the German research on nuclear power during World War II are *objectively* false, as anyone who examines the reports the German scientists composed during the war can see. I mentioned Goudsmit's loss of objectivity to suggest an explanation for why Goudsmit not only made false claims about the German work but also refused to correct those claims in public even after evidence had been presented to him that demonstrated that he had been incorrect. My book deals with a controversial topic, and I expect to receive criticism, but no other critics have so gravely misconstrued my words and intent.

I also believe that Dresden's criticism of me is unfair. My conclusions are not "misleading" or partially incorrect, and my work is not "cavalier," as I believe Dresden might see if he were to take the time to read my book thoroughly and carefully. Similarly, I believe that a reader who examined the 204 pages in my book that precede my discussion of the controversy between Goudsmit and Heisenberg would see that I have tried very hard to place this debate in the proper context. The history and anecdotes that Dresden relates in his letter are interesting but not relevant for the issues raised in my article.

I am grateful to Helmut Rechenberg for his tacit willingness to "agree to disagree." As far as I can see, our difference of opinion can be summed up in the following question: Did the German scientists *try* to make nuclear weapons during the Second

World War? But this question has no one answer. It depends on what one means by "try." If trying to make nuclear weapons means making the massive industrial efforts, spending the billions of marks, employing the thousands of scientists and engineers, and building the factories that were all obviously needed to manufacture nuclear weapons, then the Germans did not try. However, if trying to make nuclear weapons means making efforts to produce known nuclear explosives—plutonium and uranium-235—in steadily increasing amounts as quickly as possible without interfering with the war effort, then the Germans did try. In my book, I tried to leave this question open, so that each reader could decide for him- or herself which interpretation is justified. In my condensed article in PHYSICS TODAY, this discussion unavoidably was simplified.

Finally, I would like to say that I agree completely with A. van der Ziel. I have considerable sympathy for individuals who have to work and live under any totalitarian regime, and I have tried very hard to express this sympathy in my work.

References

1. M. Walker, *German National Socialism and the Quest for Nuclear Power, 1939–1949*, Cambridge U. P., New York (1989); German edition, *Die Uranmaschine: Mythos und Wirklichkeit der deutschen Atombombe*, Siedler-Verlag, Berlin (1990).

2. M. Walker, Vierteljahrshefte für Zeitgeschichte **38**, 45 (1990); English version in T. Meade, M. Walker, eds., *Science, Medicine, and Cultural Imperialism*, St. Martin's, New York (1991), p. 178.

3. S. Goudsmit, Bull. Atom. Scientists **1**, 4 (1946).

MARK WALKER
Union College
Schenectady, New York

3/91

THE KHRUSHCHEV DÉTENTE AND EMERGING INTERNATIONALISM IN PARTICLE PHYSICS

Nikita Khrushchev initiated a period of growing Soviet–American cooperation in particle physics, rudely interrupted by the 1960 spy-plane incident.

Robert E. Marshak

PHYSICS TODAY/JANUARY 1990

Mikhail Gorbachev's program of *glasnost* and *perestroika*—now in its fifth year—has wrought striking changes inside the Soviet Union, and even more spectacular metamorphoses outside its borders. Among the positive repercussions of the Gorbachev program on the international scene, we have seen major improvements in international scientific cooperation. In particle physics the improvement during the years of the Gorbachev détente appears less dramatic than it does in space science, for example. But that's largely because the opportunity to solidify East–West cooperation in particle physics had been seized much earlier—during the Khrushchev détente of 1956–60, a time when this field, with its large and expensive accelerators, stood alone as the epitome of "big science."

The idea that Nikita Khrushchev was the first reformer of the Stalinist system was only given official Soviet sanction nine months ago, with the belated Soviet publication of his famous February 1956 "de-Stalinization" speech. However, Khrushchev's secret speech[1] was known to many Russian scientists when I visited the USSR in May 1956. Moreover, Andrei Sakharov refers to the speech in his tract *Progress, Coexistence and Intellectual Freedom* (published in the West in 1968, but not in the Soviet Union):

"This bold speech, which came as a surprise to Stalin's accomplices in crime, and a number of associated measures—the release of hundreds of thousands of political prisoners and their rehabilitation, steps toward a revival of the principles of peaceful coexistence and toward a revival of democracy—oblige us to value highly the historic role of Khrushchev. . . ."[2]

The recent awakening of interest in Khrushchev's role as a spiritual forerunner of Gorbachev has induced me to take another look at my papers from the first decade (1950–60) of the "Rochester" conferences on high-energy physics,[3] and to record here the surprisingly close correlation that I have found between the evolution of the

Robert Marshak is University Distinguished Professor of Physics, Emeritus, at the Virginia Polytechnic Institute and State University in Blacksburg, Virginia.

Lev D. Landau (center) has Murray Gell-Mann (left) and the author to lunch at his home during the Moscow conference on the physics of high-energy particles in May 1956, when the Khrushchev détente was in its first flower.

Khrushchev détente and the development of Soviet–American cooperation in particle physics during the latter half of that decade. I was at the time chairman of the physics department at the University of Rochester, where these (then annual) international conferences began. I hope that this exercise will fill some lacunae in the postwar history of science, and that it will also help us to deal constructively with the challenges and opportunities of the present Gorbachev détente.

Before discussing the impact of the Khrushchev détente on international cooperation in particle physics, I will sketch the scientific and sociological milieu before 1950 and then comment briefly on the emerging internationalism in particle physics, as seen through the prism of the Rochester conferences during the early 1950s.

Particle physics before 1950

The foundations of particle physics were laid in the prewar decade of the 1930s through advances in nuclear physics, cosmic-ray physics and quantum field theory. International cooperation in those days consisted primarily of western and central European cooperation, and the major forum for the exchange of ideas among the leading European physicists was the triennial Solvay Congress,[4] which met in Brussels starting in 1911. The 1933 Solvay Congress, the last one before the war, was attended by only one American—Ernest Lawrence. The subject was: The Structure and Properties of Atomic Nuclei. Werner Heisenberg and Wolfgang Pauli, among others, correctly identified many properties of the strong and weak nuclear interactions in terms of the newly discovered neutron and the newly postulated neutrino.

Several other important prewar contributions to the maturing discipline of particle physics should be noted: Enrico Fermi's theory of beta decay (1934); Hideki Yukawa's meson theory of nuclear forces (1935); and the discovery of the muon in cosmic rays by Carl Anderson and Seth Neddermeyer (1937). The muon, wrongly

thought to be the meson predicted by Yukawa, lent premature support to his theory. By the end of the war, the experiments of Marcello Conversi, Ettore Pancini and Oreste Piccioni in German-occupied Rome had cast serious doubt on the mesonic character of the muon; the Lamb shift of the hydrogen spectrum had been measured; and Edwin McMillan at Berkeley and Vladimir Veksler at the Lebedev Institute in Moscow had formulated the principle of phase stability in high energy particle accelerators, independently of one another.

The end of hostilities in 1945 left central Europe, the Soviet Union and Japan in shambles. The World War was quickly followed by the Cold War. A grateful American nation expressed its appreciation for the special contributions of physicists to victory in the "hot war" by investing heavily in basic science facilities (including high-energy accelerators) at universities and national laboratories. Consequently, by 1950, no less than seven high-energy accelerators were in operation in the United States. In those days, high energy meants anything exceeding 100 MeV—just enough to make pi mesons. Unbeknownst to the West, the Soviets had exploited Veksler's phase-stability discovery to construct several high-energy machines of their own.

Apart from the burgeoning interest in accelerator experiments, cosmic-ray experimentalists throughout the world, with improved nuclear-emission and cloud-chamber techniques, were making important contributions to particle physics. To cap all this, theoretical interest in particle physics was intense and worldwide, whetted by the triumphs in quantum electrodynamics and meson theory flowing from the three Shelter Island conferences held during 1947–49 under the leadership of Robert Oppenheimer.[5] These conferences, held on a small island of the eastern end of New York's Long Island, had a profound impact on the conceptual restructuring of particle physics immediately after the war. But they played only a minor role in furthering international cooperation in the field because they were essentilly small, American theoretical meetings. The Shelter Island conferences having served theirpurpose, Oppenheimer decided to discontinue them after 1949.

The early Rochester conferences

As the new chairman of the physics department at the University of Rochester in the spring of 1950, I thought that the growth of global interest and experimental capabilities in particle physics justified a new American initiative to replace the discontinued Shelter Island meetings by a series of conferences that would give equal weight to the participation of accelerator experimentalists, cosmic-ray experimentalists and theorists. I also believed that this new series of gatherings could serve to help reestablish the international community of science— at least in particle physics—that had been sundered by World War II and its aftermath.

I would not have predicted that the first decade of these Rochester conferences would play such a significant role in furthering international cooperation in our field— particularly Soviet–American cooperation. But that is what happened, chiefly because of Khrushchev's accession to power in the Soviet Union. I should like to retell this part of the story[3] because of its obvious parallels with the present situation.

But first I will recall the humble beginnings of the Rochester conference, and how its scope and international stature grew during the first half of the decade. To fine-tune our objectives, we deliberately kept the first two Rochester conferences modest in duration, in the number of participants and in their funding. The first Rochester conference lasted only one day (16 December 1950), the total attendance was 50 (with a few foreign guests who were already in the country) and financial support was provided by local industry. The physics included accelerator results on nucleon interactions and meson production and decay, as well as relevant cosmic-ray results. The British cosmic-ray results were communicated by one returning American! And there were spirited exchanges with the theorists. The following year, more spectacular results were obtained from accelerators and cosmic rays, and the second Rochester conference, now lasting two days (11–12 January 1952), bubbled with excitement. Experimenters were unfolding the world of "strange particles," theorists were responding with bold hypotheses and particle physics was entering a period of tremendous vitality.

The success of these first two conferences led to the raising of sights in all respects. The next three Rochester conferences (1953–55) acquired the pretentious title: Annual Conference on High Energy Nuclear Physics. They each had 150–175 participants, lasted three days and received support from several US government agencies in addition to industry. These three conferences were also more consciously international, with representatives from Western Europe, Asia, Australia and Latin America.

Cosmic-ray physicists still dominated the subject of strange particles, although by 1955 its center of gravity was shifting sharply to accelerator data. There was an abundance of good accelerator results, including confirmation of the Gell-Mann–Nakano–Nishijima strangeness scheme and the first electron-proton scattering data.

Visa problems

The fifth conference achieved the distinction of ensuring American visas for all foreign invitees by overcoming the restrictions of the McCarran–Walter Immigration Act, one of the evil consequences of the McCarthy era. Because success in neutralizing the American-imposed political

At the third annual Conference on High Energy Nuclear Physics, at the University of Rochester in December 1952. Left to right are Hideki Yukawa, Edwin McMillan, Carl Anderson and Enrico Fermi.

Under the watchful eyes of Nikita Khrushchev (left portrait) and Nikolai Bulganin, the author lectures to the 1956 Moscow high-energy-physics conference. About a dozen American physicists were invited to this conference, organized by the Soviets shortly after Khrushchev's secret February 1956 speech denouncing Stalin.

obstacle to free scientific communication was essential to progress in international cooperation, a few details may be in order. Under the provisions of the McCarran–Walter Act passed in 1952, the entry of foreigners into the US for short cultural and scientific visits was treated on a par with the admission of regular immigrants. The practical consequence was that a scientist could be denied a visa to attend an international conference if "derogatory" information (for example, belonging to a "fellow-traveler" organization) was in the hands of the local American consul. The only loophole was to secure a waiver from the US Attorney General, and no American consul took the trouble or incurred the risk of making such a recommendation.

Shortly before the start of the 1955 Rochester conference, I discovered to my chagrin that about ten foreign invitees had been denied visas. I immediately departed for Washington, where I persuaded the State Department's Security Division to request the necessary waivers directly from the Attorney General instead of waiting for local consuls to do so. As a result, all the affected foreign invitees were able to attend our conference. This procedure worked so well during the remainder of the decade that our foreign guests at succeeding Rochester conferences were unaware that a few of them had required special waiver treatment.

Thus was the McCarran–Walter barrier to international scientific communication overcome. In connection with the tenth Rochester conference (1960), a *New York Times* editorial moralized: "The High Energy Conference at Rochester now is one whose importance is so obvious that it has been possible to get the official waivers needed to admit foreign scientists who would otherwise be barred by the McCarran–Walter Act. But, in other cases, the difficulties have tended to divert international meetings from our territory.... It is time we removed those artificial barriers which impede international scientific contacts and hinder the progress of our own research." There was a feeling at the time that the *Times* editorial had a positive effect, but it wasn't until 1987 that McCarran–Walter was modified in this regard!

The sixth Rochester conference, in April 1956, marked the transition from little to big science in particle physics. Until the sixth conference, the decision to give equal treatment to accelerator and cosmic-ray results had guaranteed lively discussion. But 1956 was the conference at which the discovery of the antiproton at the new Berkeley Bevatron was reported. It was also the year in which careful accelerator measurements that could not be done with cosmic rays established the full magnitude of the tau–theta dilemma—the apparent existence of two heavy mesons with equal masses but opposite parities. This puzzle led to the correct Lee–Yang hypothesis of parity violation in all weak interactions, so critical for the development of modern particle physics. This was also the year accelerator results on strange particles were coming so fast that Bob Leighton was led to remark: "Next year those people still studying strange particles using cosmic rays had better hold a rump session of the Rochester conference somewhere else."

I will now turn to the initiation of Soviet–American cooperation in particle physics with the sixth Rochester conference, and I will attempt to trace the close connection over the next four years between the flowering and wilting of this cooperation and the growing warmth and ultimate collapse of the Khrushchev détente.

The Khrushchev détente

Stalin died in March 1953. Lavrenti Beria, Stalin's NKVD chief, was executed inb 1954 and Georgi Malenkov was ousted as Soviet premier in February 1955 by Khrushchev, who finally took over as Party General Secretary. In July 1955 Khrushchev and Premier Nikolai Bulganin met with President Dwight Eisenhower and other Western leaders in Geneva, initiating the so-called spirit of Geneva. This meeting was followed five months later by the very amiable Atoms for Peace conference in Geneva, at which small numbers of Soviet and American physicists were able to exchange views and hopes for future contact.

I prefer to date the true commencement of the Khrushchev détente from his "de-Stalinization" speech to the 20th Congress of the Soviet Communist Party in

February 1956. This wide-ranging attack on the evils of Stalinism signaled a more open and tolerant approach to the conduct of Soviet society, both internally and externally. Although Khrushchev spoke in secret session, positive repercussions of this speech, and of the Atoms for Peace conference, were soon felt in the particle-physics communities of the USSR and the US. A small Soviet delegation attended the sixth Rochester conference in April 1956, and a month later a larger American delegation attended the quickly arranged Physics of High Energy Particles conference in Moscow.

The great impetus given to Soviet–American cooperation in particle physics by the exchange of these visits was slowed only slightly when the Soviet army crushed the Hungarian rebellion in November 1956. The severity of the American reaction did persuade Soviet authorities to prohibit Soviet invitees from attending the seventh Rochester conference, in April 1957, but this setback was only temporary.

I do not know how many obstacles had to be overcome in the Soviet bureaucracy before permission was granted to invite Americans to the May 1956 Moscow conference. But I can testify that the American particle-physics community had to engage in intensive lobbying before permission was granted to invite our Soviet colleagues to the previous month's Rochester conference. The major obstacle on our side was the State Department's nervousness, or so it appears, about pushing the Cold War thaw too fast after the Eisenhower–Khrushchev summit meeting in Geneva in July 1955.

The State Department dodged the issue by insisting on formal approval by the Atomic Energy Commission, whose chairman was Lewis Strauss. This turned out to be a formidable task.[3] Two AEC members, John von Neumann and Willard Libby, were persuaded at an early stage, by myself and Herbert Anderson respectively, to support the Soviet invitations. However, the vote of Chairman Strauss was crucial, and my colleague Arthur Roberts, the director of the sixth Rochester conference, had to mobilize a superstar "hit" team that included Victor Weisskopf, Eugene Wigner and Edward Teller to persuade Strauss to support inviting the Soviets. The successful outcome of this effort was a landmark in the restoration of East–West exchanges in many scientific fields. In particular, it opened the door to the rapid growth of Soviet–American cooperation in particle physics during the next several years.

The beginning of a warm relationship between the American and Soviet particle-physics communities was not an end in itself. We considered it a step, albeit a major one, towards the full reestablishment of international cooperation in particle physics. Fortunately the opportunity to consolidate international cooperation arose the following year (September 1957), on the occasion of the General Assembly meeting of the International Union of Pure and Applied Physics in Rome. In the spring of 1957 the US national commission of IUPAP, of which I was a member, had agreed to support the creation of a high-energy physics commission under the aegis of IUPAP. It was also necessary for the Soviet Union to join IUPAP. So,

In Moscow at the May 1956 high-energy-physics conference organized by the Soviets, the author (right) stands with Yakov Zel'dovich (left) and Hungarian physicist George Marx.

with the encouragement of Robert Brode, the chairman of our national commission, I contacted Igor Tamm about the possibility of such a move. The suggestion was timely, and it was quickly implemented. A distinguished Soviet delegation, under the chairmanship of Abram Ioffe, showed up at the Rome meeting. In Rome, I was able to approach Ioffe directly about the American proposal and, with his concurrence, the General Assembly approved the creation of a new IUPAP Commission on High-Energy Physics. (Its name was eventually changed to the present one: IUPAP Commission on Particles and Fields.)

The membership of the IUPAP commission and its mandate were quickly established.[3] It was decided that the six seats on the first IUPAP Commission would be divided equally among the United States, the Soviet Union and Western Europe. (A Japanese member, Yukawa, was soon added.) The first appointees were: C. J. Bakker (chairman) and Rudolph Peierls from Western Europe; myself (secretary) and Wolfgang Panofsky from the US, and Tamm and Veksler from the Soviet Union. It was also decided that the venue of future "Rochester" conferences would rotate among the three regions, starting with CERN

In May 1960 (front row, right to left) A. I. Akhiezer of the Kharkov Physical–Technical Institute and his son escort Caltech physicist Robert Bacher into the Institute. Bacher was part of a US delegation sent to implement the McCone–Emelyanov Atomic Energy Agreement resulting from the 1959 Eisenhower–Khrushchev summit. This visit of American physicists to the Soviet Union was effectively derailed by the U-2 spy-plane incident, which happened a few weeks earlier.

(Geneva) in 1958, Kiev in 1959 and back to Rochester in 1960. The 1958 conference at CERN became the Eighth Annual International Conference on High Energy Physics, although it was the first outside of Rochester. Except for the change from "annual" to "biennial" after 1960, the present system of Rochester conferences was in place in 1957, so that the upcoming 25th conference will take place in Singapore this August—forty years after the first one.[6]

The mandate of the IUPAP Commission at the start was broader than simply organizing "an appropriate number of international conferences each year" and of encouraging "rapid exchange of the latest scientific results"—the usual charges assigned to IUPAP commissions. A third charge was added that went considerably further and made possible the introduction of new modes of international cooperation: "to encourage international collaboration among the various high energy laboratories to ensure the best use of the facilities of these large and expensive installations."[3] This would become a central concern of IUPAP Commission activities.

The IUPAP Commission

I will now trace the close correlation between the actions of the commission and the unfolding Khrushchev détente (and American government response to it) during the period 1957–60. As the détente warmed up during the spring of 1959, President Eisenhower released a statement on 17 May concerning American–Soviet cooperation in particle physics:

"As a first step in the direction of international collaboration looking toward the development of new high-energy accelerators, representative scientific groups from other countries, including the USSR, should be encouraged to meet with us in order to lay plans for cooperative research on new accelerator concepts. The National Academy of Sciences should be requested to advise on the best means for accomplishing this objective."[3]

In consultation with Brode and Detlev Bronk, president of our National Academy of Sciences, I arranged—in my capacity as secretary of the IUPAP Commission—for a

discussion of Eisenhower's proposal at the Commission's forthcoming Kiev meeting in July. The conference would be asked "to consider the possibility of international collaboration and cooperation in the planning for and design of future large accelerators, and the creation of favorable conditions for their use."[3] There was unanimous agreement at the Kiev meeting that this proposal "merited further intensive study," and an IUPAP study committee was set up, consisting of Edward Lofgren, Leland Haworth and Panofsky from the US, P. T. Dzhelepov and Veksler from the Soviet Union and Bakker, T. G. Pickavance and Giorgio Salvini from Western Europe.

The study committee convened in Geneva in September and agreed unanimously that international cooperation in particle physics should comprise:
▷ the exchange of technical information on all particle accelerators above an energy of 100 MeV, in use or under construction
▷ exchange of information on the functional characteristics of accelerators in the planning stage
▷ cooperation in the exploration of new ideas for future accelerators designed for purely scientific purposes
▷ cooperation in scientific research using high-energy accelerators.
The Russians suggested that a formal proposal along these lines could come from CERN or the US National Academy of Sciences to the Soviet Academy of Sciences. This was most auspicious, considering that the Iron Curtain in particle physics had only been lifted three years earlier.[7]

The momentum of increasing warmth in Soviet–American particle-physics cooperation did not stop with the IUPAP study committee's recommendations. The second Eisenhower–Khrushchev meeting in Washington in September 1959 produced the McCone–Emelyanov Agreement between the United States and the Soviet Union. (John McCone was chairman of the US AEC and V. S. Emelyanov was head of the USSR Main Administration for Utilization of Atomic Energy.) The agreement made provision for:

"reciprocal exchanges of unclassified information and visits of scientists to unclassified areas, and for the exploration of the desirability of unclassified projects jointly sponsored by the two countries... including high energy physics."

The McCone–Emelyanov agreement thus offered an opportunity for the American and Soviet particle-physics communities to implement the agreements worked out at the IUPAP study committee's Geneva meeting. Our AEC promptly appointed an American delegation, consisting of Robert Bacher, George Kolstad, Lofgren, Robert Wilson and myself, with instructions to visit half a dozen high-energy physics laboratories in the Soviet Union during May 1960 and then negotiate with our Soviet colleagues detailed recommendations for bringing the study committee's proposals to fruition. It was always understood that "CERN must be brought into activities fostered under the McCone–Emelyanov Agreement."[8]

The spy-plane episode

Our planned visit to the Soviet Union was soon in trouble. As our delegation was preparing for its 11 May departure for Moscow, an American U-2 spy plane was shot down by the Russians on 1 May 1960 in the course of a reconnaissance flight over Soviet territory. Khrushchev was furious and insisted on an American apology and repudiation of our overflight policy. At the same time, he offered to meet with President Eisenhower in Paris as scheduled, *if* these conditions were met. In the midst of crisis and uncertainty, a quick decision had to be taken as to whether our delegation should proceed to the Soviet Union and attempt to pursue the objectives of the McCone–Emelyanov agreement. The US State Department decided in favor of the trip, apparently hoping that the summit meeting would take place, and we left for the Soviet Union on 11 May, as scheduled.

When we arrived in the USSR we were greeted cordially by our Soviet hosts, and our scheduled tour began with a visit to the Dubna laboratory on 13 May. The next few days were also spent pleasantly with director A. I. Alikhanov and his colleagues at the Institute for Theoretical and Experimental Physics in Moscow. But the political situation was deteriorating rapidly. When we arrived at the Lebedev Institute on 18 May, the worst had happened in Paris several hours earlier. At a press conference Khrushchev had called off the summit meeting. We were read the riot act by gentle Veksler, who had worked so hard within the IUPAP framework to broaden and deepen Soviet–American cooperation in particle physics.

The Lebedev visit was the turning point in our visit. I should like to quote from Kolstad's recollection of Veksler's reaction: "The latest events will hinder cooperation, the position of the Soviet scientists is now very difficult, and he personally, and the rest of the Soviet scientists, stand firmly behind their Government in the present difficulty... The Soviet Academy of Sciences had decided to cooperate and send a delegation of scientists to the tenth Rochester conference. Now [Veksler] is not sure that this position will not be changed. He said that they are now in a very difficult position—the summit meeting is all off."[9]

Needless to say, the American delegation was very upset. But we argued that, despite political mistakes by both sides, we scientists must surmount political grievances and continue friendly relations. This plea struck a responsive cord, and Veksler became quite conciliatory. He responded (as Kolstad recalls) that "it is very important that the scientists themselves must do nothing to worsen this relationship between the governments. He

At Kharkov in May 1960, Physical–Technical Institute director Sinelnikov (2nd from right) and his wife welcome American visitors (left to right) Edward Lofgren, George Kolstad (in front of an unidentified Russian), Robert Bacher and Robert Wilson (far right), members of the delegation sent to implement the McCone–Emelyanov Atomic Energy Agreement.

At the tenth Rochester conference, in August 1960. Arrayed in front of a University of Rochester fraternity house are (left to right) Emilio Segrè, C. N. Yang, Owen Chamberlain, T. D. Lee, Edwin McMillan, Carl Anderson, I. I. Rabi and Werner Heisenberg.

was very glad that we had come at this time, in spite of the difficulties. 'Now, let's get on with the visit.' "

We did continue with our schedule, visiting laboratories in Leningrad, Sukhumi, Yerevan and finally Kharkov. This was the first American visit to the Kharkov laboratory since the Soviet purges in the 1930s. But the facts were that Khrushchev had been humiliated by the American government, the détente was destroyed and our McCone–Emelyanov visit was destined to end in abysmal failure. On the final day in Moscow, our two-week visit was supposed to culminate in a meeting with a large group of Soviet colleagues, to draft implementing recommendations to our governments with regard to the IUPAP study committee's proposals. Instead, we Americans were ushered into a solitary audience with Emelyanov, who lectured us on the perfidy of our government and dismissed us forthwith.

There was no question that the U-2 incident and the breakdown of the Eisenhower–Khrushchev summit in Paris soured American–Soviet relations in particle physics, and of course on the wider stage, where they were soon followed by the Berlin Wall (1961) and the Cuban missile crisis (1962). When we left Moscow, we were dubious whether a Soviet delegation would attend the tenth Rochester conference, coming up in three months. But a Soviet delegation did show up in considerable strength in Rochester in August, partly, I think, because the confer-

ence was under the aegis of the IUPAP Commission, and partly because the Soviet particle-physics leadership was keenly aware of American efforts to further international cooperation in our field.

On a fine Sunday morning during the conference, Bob Wilson spontaneously organized an informal exchange of views among a small group of American, Soviet and Western European delegates on the reactivation of the IUPAP study committee's recommendations. At the conference banquet, AEC chairman McCone, the after-dinner speaker, gave noble support to the goal of increasing cooperation in particle physics between the United States and the Soviet Union. But the damage to American–Soviet relations could not be undone so quickly. Cooperation in particle physics sank to minimal levels for many years to come.

CERN and ICFA

I should not like to conclude this account on such a grim note. Three decades after the U-2 incident, the Rochester conference is still very much alive, as are numerous other IUPAP Commission-sponsored conferences and workshops. Fortunately, CERN was not tarnished by the spy-plane fiasco in Soviet–American particle-physics cooperation. To quote John Adams, who was acting director of CERN in 1960–61: "During the past year, CERN and the USSR have taken some small steps towards further cooperation

Gersh Budker, director of the Novosibirsk Institute of Nuclear Physics, follows the author's argument about kaon decay at the tenth Rochester conference, in August 1960. About 20 Soviet physicists showed up at Rochester, despite the rancorous aftermath of the U-2 spy plane incident.

(for example, exchanging scientific personnel for several months at a time between the two laboratories)."[10] The CERN–Dubna and CERN–Serpukhov collaborations to which Adams was alluding apparently led to informal meetings between Western and Eastern European particle physicists, to discuss questions of international collaboration and future planning in particle physics (under the third charge to the IUPAP Commission), and then to a formal meeting in 1967, to which Americans were invited.

The record[11] indicates that these tripartite meetings continued at irregular intervals until the International Committee for Future Accelerators (ICFA)was formally established as a subcommittee of the IUPAP Commission in 1977, with roughly the same objectives as the defunct 1959 IUPAP study committee. (The study committee ceased to exist after the failure of the McCone–Emelyanov mission in 1960.) Initially, the mandate of ICFA went beyond that of the IUPAP study committee by adding the charge that "international collaboration should provide for studies leading towards the realization of a next generation of superhigh-energy facilities... It is expected that these facilities will be so large that their realization will be possible only by pooling the resources of all regions concerned into common international projects." At its 1985 meeting, ICFA lowered its sights to something more realistic under present circumstances: "to organize workshops to study problems related to superhigh-energy accelerator complexes and their international exploitation, and to foster research and development of the necessary technology."[12]

This more modest ICFA mandate pretty much returns the goals of international cooperation in particle physics to the 1960 level—before the Khrushchev détente ended. Perhaps the worldwide particle-physics community will be able to reassert its more ambitious goals of international collaboration under the conditions of the new Gorbachev détente, and then proceed to the creation of joint international projects in which Soviet, Eastern European and Chinese particle physicists can become equal partners with their Western colleagues in these undertakings. This would be the ultimate stage of international cooperation in particle physics in a peaceful and democratic world.

References

1. English translation given in *Khrushchev Speaks*, U. Michigan P., Ann Arbor (1963), p. 207.

2. A. Sakharov, *Progress, Coexistence and Intellectual Freedom*, Deutsch, London (1968).

3. Robert E. Marshak papers, library archives of Virginia Polytechnic Inst. and State U.; microfiche copies available at the AIP Niels Bohr Library in New York, U. of Rochester library and CERN library.

4. J. Mehra, *The Solvay Conferences on Physics*, Reidel, Boston (1975).

5. S. S. Schweber, in *Shelter Island II*, R. Jackiw, N. Khuri, S. Weinberg, E. Witten, eds., MIT P., Cambridge (1985), p. 301.

6. R. E. Marshak, in *Proc. Int. Conf. on Restructuring of Physical Sciences in Europe and United States 1945–60*, M. DeMaria, M. Grilli, eds., U. of Rome (1988).

7. R. E. Marshak, Bull. At. Sci., June 1970, p. 92.

8. E. B. Skolnikoff, memo to R. E. Marshak, 4 May 1960, in ref. 3.

9. G. A. Kolstad, "Visit of US High Energy Physics Team to USSR, May 1960," report to the Atomic Energy Commission (1960), p. 22, in ref. 3.

10. From notes of R. E. Marshak on "Informal meeting organized by J. Adams at his home on 5 June 1961," attended by E. Amaldi, D. Blokhintsev and R. E. Marshak, in ref. 3.

11. E. L. Goldwasser, in *Proc. 19th Int. Conf. on High Energy Physics*, S. Hommay, M. Kawaguchi, H. Miyazawa, eds., Physical Society of Japan, Tokyo (1979), p. 961. Y. Yamaguchi, in *Proc. 1985 Int. Symp. on Lepton and Photon Interactions at High Energies at High Energies*, M. Konuma, K. Takahashi, eds., Nissha, Kyoto (1986), p.826.

12. W. O. Lock, Europhys. News, April 1989, p. 54.

PATHOLOGICAL SCIENCE

Certain symptoms seen in studies of 'N rays' and other elusive phenomena characterize 'the science of things that aren't so.'

Irving Langmuir

Transcribed and edited by Robert N. Hall

PHYSICS TODAY/OCTOBER 1989

Irving Langmuir spent many productive years pursuing Nobel-caliber research (see the photo on next page). Over the year, he also explored the subject of what he called "pathological science." Although he never published his investigations in this area, on 18 December 1953 at General Electric's Knolls Atomic Power Laboratory, he gave a colloquium on the subject that will long be remembered by those in his audience. This talk was a colorful account of a particular kind of pitfall into which scientists may stumble.

The tape recording that was made of Langmuir's colloquium has been lost or erased. However, in 1966, a microgroove disk transcription that was made of this tape was found among the Langmuir papers in the Library of Congress. The disk recording is of poor quality, but most of what Langmuir said can be understood with a little practice. Robert N. Hall, a former colleague of Langmuir's at General Electric, transcribed the disk and edited it to make an internal report for the company. At that time, a small amount of editing was felt to be desirable. Some abortive or repetitious sentences were eliminated. Hall wrote the epilogue. Figures from corresponding publications were used to represent Langmuir's blackboard sketches. These agree in essence, if not in every detail, with Langmuir's descriptions. Some references were added for the benefit of anyone wishing to undertake a further investigation of this subject.

The disk recording has been transcribed back onto tape, and a copy is on file in the Whitney Library at General Electric. A slightly abbreviated version of the talk was published in Speculations in Science and Technology **8,** *77 (1985).*

Physics today *has added a few more illustrations and has done some additional light editing to improve the readability while still maintaining the spontaneous flavor of this talk. Bracketed phrases in roman type are substitutions for the original words made for clarity or precision. Italicized bracketed phrases are editorial insertions. Deletions from the original text are marked with ellipses.*

Irving Langmuir earned the 1932 Nobel Prize in Chemistry for work dealing with the adsorption of monolayers of molecules on surfaces. He spent his career at the General Electric Company in Schenectady, New York, working there from 1909 until his retirement in 1950. His research included such phenomena as thermionic emission and the properties of liquid surfaces. Over the years, Langmuir also explored the subject of "pathological science," although he never published his investigations on this topic. He died in 1957.

The thing started in this way. On 23 April 1929, Professor Bergen Davis from Columbia University came up and gave a colloquium in this Laboratory, in the old building, and it was very interesting. He told Dr. Whitney and myself and a few others something about his talk beforehand. He was very enthusiastic about it and he got us interested in it. [*Langmuir may have remembered this date incorrectly. The date on a letter Langmuir wrote just a few days after Davis's talk is 23 April 1930. Willis R. Whitney was director of the General Electric Research Laboratory.*]

Davis and Barnes experiment

I'll show you right on this diagram what kind of thing happened. [*See Figure 1*]. Davis [*and his colleague Arthur Barnes*] produced a beam of alpha rays from polonium in a vacuum tube. [They] had a parabolic hot cathode electron emitter with a hole in the middle, and the alpha rays came through it and could be counted by scintillations on a zinc sulfide screen with a microscope. The electrons were focused on [a] plate so that for a distance there was a stream of electrons moving along with the alpha particles.

Now you could accelerate the electrons and get them up to the velocity of the alpha particles. To get an electron to move with that velocity takes about 590 volts, so if you put 590 volts [on the grid], accelerating the electrons, the electrons would travel along with the alpha particles. The idea of the experiment was that if they moved along together at the same velocity they might [combine]. Thus the alpha particle would lose one of its charges. It would pick up one electron so that, instead of being a helium [ion] with two positive charges, it would only have one charge. Well, if an alpha particle with a double charge has one electron, [its energy levels are] just like [those of] the Bohr theory of the hydrogen atom . . . , with a Balmer series, and you can calculate the energy necessary to knock off the electron and so on.

Well, Davis and Barnes found that if [the electron's] velocity was made to be the same as that of the alpha particle, there was a loss in the number of deflected particles. If there were no electrons, for example, and no magnetic field, all the alpha particles would be collected [at the upper screen]. They counted something of the order of 50 counts per minute there. If you put on a magnetic field you could deflect the alpha particles so they would go down [to the lower screen]. But if they picked up an electron, then they would only have half the charge, and therefore they would only be deflected half as much and they would not strike the screen.

Now the results that they got, or said that they got at the time, were very extraordinary. They found that not only did these electrons combine with the alpha particles when the electron velocity [corresponded to] 590 volts, but also at a series of discrete differences of voltage. When the velocity of the electrons was less or more than that velocity by perfectly discrete amounts, then they could also combine. All the results seemed to show that about 80% of them combined. In other words, there was about an 80% change in the current when the conditions were right. Then they found that the velocity differences had to be exactly the velocities that you can calculate from the Bohr theory. In other words, if the electron happened to be going with a velocity equal to the velocity that it would have if it were in a Bohr orbit, then it would be captured.

Of course that makes a difficulty right away, because in the Bohr theory when there is an electron coming in from infinity it has to give up half its energy to settle into the Bohr orbit. Since it must conserve energy, it has to radiate out, and it radiates out an amount equal to the energy that it has left in the orbit. So if the electron comes in with an amount of energy equal to the amount it is going to end up with, then it has to radiate an amount of energy equal to twice that. Nobody had any evidence for that. So there was a little difficulty, which never was quite resolved, although there were two or three people including some in Germany who worked up theories to account for how that might be. Sommerfeld, for example, in Germany, worked up a theory to account for how the electron could be captured if it had a velocity equal to what

it was going to have after it settled down into the orbit.

Well, there were these discrete peaks, each one corresponding to one of the energy levels in the Bohr theory of the helium atom, and nothing else. Those were the only things they recorded. So you had these discrete peaks. Well, how wide were they? They were *one-hundredth of a volt wide*. In other words, you had to have 590 volts. That would give you equal velocities. But there were other peaks, and I think the next velocity would be at about 325.01 volts. If you had that voltage, then you got beautiful capture. If you didn't, if you changed it by one hundredth of a volt— nothing. It was sharp. They were only able to measure to a hundredth of a volt so it was an all-or-nothing effect. Well, besides the peak [at 590 volts], there were 10 or 12 different lines in the Balmer series, all of which could be detected, and all of which had an 80% efficiency. [*See Figure 2.*] [The lapha particles] almost completely captured all the electrons when they got exactly on the peak.

Well, in the discussion [following Davis's talk], we questioned how, experimentally, you could examine the whole spectrum, because each count, you see, took a long time. There was a long series of alpha particle counts that took two minutes at a time, and you had to do [them] 10 or 15 times and you had to adjust the voltage to a hundredth of a volt. If you had to go through steps of a hundredth of a volt each and cover all the range from 330 up to 900 volts, you'd have quite a job. Well, they said they didn't do it quite that way. They had found by some preliminary work that [the peak voltages] did check with the Bohr orbit velocities, so they knew where to look for them. They sometimes found them not exactly where they expected them, but they explored around in that neighborhood and the result was that they got [the peaks] with extraordinary precision—so high, in fact, that they were sure they'd be able to check the Rydberg constant more accurately than it can be done by studying the hydrogen spectrum, which is something like one in 10^8. At any rate, they had no inhibitions at all as to the accuracy which could be obtained by this method, especially since they were measuring these voltages within a hundredth of a volt.

Anybody who looks at the setup would be a little doubtful about whether the electrons had velocities that were fixed and definite within a hundredth of a volt, because this is not exactly a homogeneous field. The distance over which the alpha particles and electrons were moving along together was only [about] 5 [cm].

Well, in [Davis's] talk, a few other things came out that were very interesting. One was that the percentage of capture was always around 80%. The curve would come

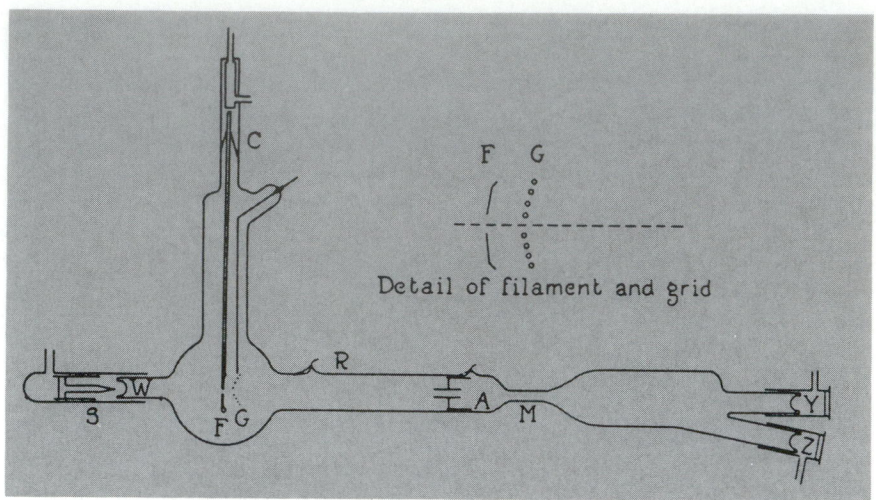

Apparatus used by Arthur Barnes and Bergen Davis in a 1929 experiment at Columbia University. Alpha particles from source S entered the tube through the window W and would either travel straight through to the window Y or be deflected by a magnetic field at M and leave through window Z. Scintillations from zinc sulfide screens beyond both exit windows signaled which path the alpha particles took. Electrons were produced at the filament F, accelerated by the anode grid G and focused toward the anode A. Thus electrons traveled along with alpha particles for a short distance. Barnes and Davis claimed the electrons were captured by the alpha particles when their velocities were equal. If so, the alpha particle would be deflected somewhere between Y and Z, and the scintillation count would decrease. The Columbia scientists later retracted their claims. (From ref. 1.)   **Figure 1.**

along at about 80% and there would be a sharp peak up here and another sharp peak here and, well, all the peaks were about the same height.

Well, we asked him, how did this depend upon current density? "That's very interesting," he said. "It doesn't depend at all upon current density."

We asked, "How much could you change the temperature of the cathode?"

"Well," he said, "that's the queer thing about it. You can change it all the way down to room temperature."

"Well," I said, "then you wouldn't have any electrons."

"Oh yes," he said. "If you check the Richardson equation and calculate, you'll find that you get electrons even at room temperature and those are the ones that are captured."

"Well," I said, "there wouldn't be enough to combine with all the alpha particles, and besides that, the alpha particles are only there for a short time as they pass through, and the electrons are a long way apart at such low current densities, at 10^{-20} amperes or so."

[Davis] said, "That seemed like quite a great difficulty. But," he said, "you see it isn't so bad because we now know that the electrons are waves. So the electron doesn't have to be there at all in order to combine with something. Only the waves have to be there and they can be of low intensity and the quantum theory causes all the electrons to pile in at just the right place where they are needed." So he saw no difficulty. And so it went.

On-site inspection

Well, Dr. Whitney likes the experimental method, and these were experiments—very careful experiments, described in great detail. The results seemed to be very

interesting from a theoretical point of view. So Dr. Whitney suggested that he would like to see these experiments repeated with a Geiger counter instead of [relying on] counting scintillations, and C. W. [Clarence] Hewlett, who was here working on Geiger counters, had a setup and it was proposed that we would give [Davis] one of these, maybe at a cost of several thousand dollars or so for the whole equipment, so that he could get better data. But I was a little more cautious. I said to Dr. Whitney that before we actually give [the counter] to [Davis] and just turn it over to him, it would be well to go down and take a look at these experiments and see what they really mean. Well, Hewlett was very much interested, and I was interested so only about two days after [Davis's] colloquium, we went down to New York. We went to Davis's laboratory at Columbia University, and we found that they were very glad to see us, very proud to show us all their results. So we started in early in the morning.

We sat in the dark room for half an hour to get our eyes adapted to the darkness so that we could count scintillations. I said I would first like to see these scintillations with the field on and with the field off. So I looked in and I counted about 50 or 60. Hewlett counted 70, and I counted somewhat lower. On the other hand, we both agreed substantially. What we found was this: These scintillations were quite bright with your eyes adapted, and there was no trouble at all counting when these alpha particles struck the screen. They came along at a rate of about one per second. When you put on a magnetic field and deflected them out, the count came down to about 17, which was a pretty high percentage—about 25% background. Barnes was sitting with us, and he said that's probably radioactive contamination of the screen. Then Barnes counted and he got 230 on the first count and about

200 on the next, and when he put on the field, [the count] went down to about 25. Well, Hewlett and I didn't know what that meant but we couldn't see 230. Later, we understood the reason. . . .

Well, I don't want to spend too much time on this experiment. I wrote a 22-page letter about these things and I have a lot of notes. The gist of it was this. There was a long table at which Barnes was sitting, and he had another table where an assistant named Hull sat looking at a big scale voltmeter, or potentiometer really. It had a scale that went from one to a thousand volts and, on that scale that went from one to a thousand, he read hundredths of a volt. He thought he might be able to do a little better than that. At any rate, you could interpolate and put down figures, you know. Now the room was dark except for a little light on which you could read the scale on that meter. And it was dark except for the dial of a clock.

[Barnes] counted scintillations for 2 minutes. He said he always counted for 2 minutes. Actually, I had a stopwatch and I checked up on him. [The intervals] were sometimes as low as 1 minute and 10 seconds and sometimes 1 minute and 55 seconds, but he counted them all as 2 minutes, and yet the results were of high accuracy!

Well, we made various suggestions. One was to turn off the voltage entirely. Then Barnes got some low values around 20 or 30, or sometimes as high as 50. Then to get the conditions on a peak he adjusted the voltage to 200 and—well, some of those readings are interesting. [One figure I put down was] 325.01. There he got only a reading of 52, whereas before, when he was on the peak, he got about 230. He didn't like that very much so he tried changing this to 325.02—a change of one-hundredth of a volt. And there he got 48. Then he went in between. [The counts] fell off, you see, so he tried 325.015, and then he got 107 [counts]. So that was a peak.

Well, a little later, I whispered to Hull, who was adjusting the voltage, holding it constant, and I suggested to him to make it one tenth of a volt different. Barnes didn't know this and he got 96. Well, when I suggested this change to Hull, you could see immediately that he was amazed. He said, "Why, that's too big a change. That will put it way off the peak." That was almost one tenth of a volt, you see. Later I suggested taking a whole volt.

Then we had lunch. We sat for half an hour in the dark room so as not to spoil our eyes and then we had some readings at zero volts and then we went back to 325.03 [volts]. We changed by one-hundredth of a volt and there [Barnes] got 110 counts. And now he got two or three readings at 110.

The denouement

And then I played a dirty trick. I wrote out on a card of paper ten different sequences of V and 0. I meant to put on a

certain voltage and then take it off again. Later I realized that that [trick wouldn't quite work] because when Hull took off the voltage, he sat back in his chair—there was nothing to regulate at zero so he didn't. Well, of course, Barnes saw him whenever he sat back in his chair. Although the light wasn't very bright, he could see whether [Hull] was sitting back in his chair or not, so he knew the voltage wasn't on, and the result was that he got a corresponding result. So later I whispered, "Don't let him know that you're not reading," and I asked him to change the voltage from 325 down to 320 so he'd have something to regulate. I said, "Regulate it just as carefully as if you were sitting on a peak." So he played the part from that time on, and from that time on Barnes's readings had nothing whatever to do with the voltages that were applied. Whether the voltage was at one value or another didn't make the slightest difference. After that he took 12 readings, of which about half were right and the other half were wrong, which was about what you would expect out of two sets of values.

I said: "You're through. You're not measuring anything at all. You never *have* measured anything at all."

"Well," he said, "the tube was gassy. The temperature has changed and therefore the nickel plates must have deformed themselves so that the electrodes are no longer lined up properly."

"Well," I said, "isn't this the tube in which Davis said he got the same results when the filament was turned off completely?"

"Oh, yes," he said, "but we always made blanks to check ourselves, with and without the voltage on."

He immediately—without giving any thought to it—he immediately had an excuse. He had a reason for not paying any attention to any wrong results. It just was built into him. He just had worked that way all along and always would. There is no question but [that] he is honest: He *believed* these things, absolutely.

Hewlett stayed there and continued to work with [Barnes] for quite a while, and I went in and talked it over with Davis and he was simply dumbfounded. He couldn't believe a word of it. "It absolutely can't be," he said. "Look at the way we found those peaks before we knew anything about the Bohr theory. We took those values and calculated them and they checked exactly. Later on, after we got confirmation, in order to save time, to see whether the peaks were there, we would calculate ahead of time." He was so sure from the whole history of the thing that it was utterly impossible that there never had been any measurements at all that he just wouldn't believe it.

Well, [Davis] had just read a paper before the research laboratory at Schenectady, and he was going to read the paper the following Saturday before the National Academy of Sciences. . . . And he wrote me that he was going to

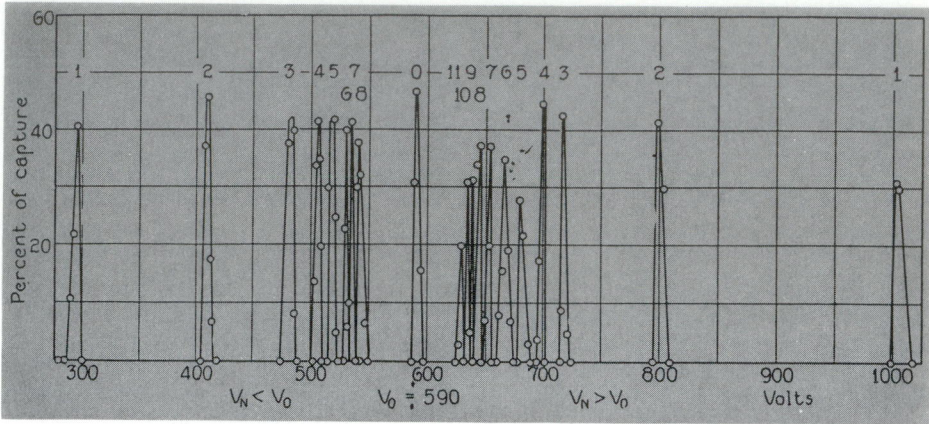

Peaks signaled electron capture, according to Barnes and Davis (see the figure on page 39). Plotted here is the decrease in the number of scintillation counts when a magnetic field was turned on, as a function of the accelerating voltage applied to electrons. Fewer scintillations meant that fewer alpha particles were arriving at the screen—because, the researchers claimed, the alpha particles were capturing electrons. Peaks appeared only at certain discrete voltages, whose energies corresponded to the energy levels of electrons in the Bohr theory. Numbers above each peak correspond to the principal quantum number n. The central peak, labeled 0, is at the voltage where the electrons and alpha particles have the same speed. (From ref. 1.)   **Figure 2.**

do so on the 24th [of April 1930]. I wrote to him on the day after I got back. Our letters crossed in the mails and he said that he had been thinking over the various things that I had told him, and his confidence wasn't shaken. So he went ahead and presented the paper before the National Academy of Sciences.

Then I wrote him a 22-page letter giving all our data and showing really that the whole approach to the thing was wrong; [Barnes] was counting hallucinations, which I find is common among people who work with scintillations if they count for too long. Barnes counted for six hours a day and it never fatigued him. Of course it didn't fatigue him, because it was all made up out of his head. He told us that you mustn't count the bright particles. He had a beautiful reason for why you mustn't pay any attention to the bright flashes. When Hewlett tried to check his data [Barnes] said: "Why, you must be counting those bright flashes. Those things are only due to radioactive contamination or something else." He had a reason for rejecting the very essence of the thing that was important. So I wrote all this down in my letter and I got no response, no encouragement. For a long time Davis wouldn't have anything to do with it. He went to Europe for a six months' leave of absence, came back later and I took up the matter with him again.[1]

In the meantime, I sent Bohr a copy of the letter that I had written to Davis, asking [Bohr] to hold it confidential but to pass it on to various people who would be trying to repeat these experiments—to Professor Sommerfeld and other people. It headed off a lot of experimental work that would have gone on. And from that time on, nobody ever made another experiment except one man in England who didn't know about the letter that I had written to Bohr.[2] And he was not able to confirm any of it. Well, a year and a half later, in 1931, there was just a short little article in the *Physical Review* in which [Davis and Barnes] said that they hadn't been able to reproduce the effect.[3] "The results reported in the earlier paper depended upon observations made by counting scintillations visually. The scintillations produced by alpha particles on a zinc

sulfide screen are a threshold phenomenon. It is possible that the number of counts may be influenced by external suggestion or autosuggestion to the observer." And later in that paper they said that they had not been able to check any of the older data. And they didn't even say that the tube was gassy.

To me, [it's] extremely interesting that men, perfectly honest, enthusiastic over their work, can so completely fool themselves. Now what was it about that work that made it so easy for them to do that? Well, I began thinking of other things. I had seen R. W. Wood and told him about this phenomenon because he's a good experimenter and doesn't make such mistakes himself very often—if at all. [*Wood was a physicist from Johns Hopkins University.*] And he told me about the N rays that he had an experience with back in 1904. So I looked up the data on the N rays.[4]

N rays

In 1903, Blondlot, who was a well-thought-of French scientist and a member of the Academy of Sciences, was experimenting with x rays, as almost everybody was in those days. [*René-Prosper Blondlot, shown in the photo on next page, was a physicist at the University of Nancy.*] The effect that he observed was something of this sort. I won't give the whole of it; I'll just give a few outstanding points. He found that if you have a hot wire, a platinum wire or a Nernst filament, or anything that's heated very hot inside an iron tube, and you have a window cut in it and you have a piece of aluminum about $\frac{1}{8}$ of an inch thick on it, some rays come out through that aluminum window. Oh, [the window] can be as much as 2 or 3 inches thick and [these rays can still] go through aluminum but not through iron. The rays that come out of this little window fall on an. . .object [that is barely illuminated by a light source], so that you can just barely see it. You must sit in a dark room for a long time. [Blondlot] used a calcium sulfide screen which could be illuminated with light and gave out a very faint glow which could be seen in a dark room. Or he used a source of light from a lamp shining through a pinhole and maybe through another pinhole so as to get a faint light on a white surface that was just barely visible.

Now he found that if you turn this lamp on so that these rays that came out of the little aluminum slit fell on the piece of paper that you were looking at, you could see [the paper] much better. Oh, *much better!* And therefore you could tell whether the rays would go through or not. [Blondlot] said later that a great deal of skill is needed. He said you must'nt ever look at the source. You shouldn't look directly at it. He said that would tire your eyes. Look away from it and, he said, pretty soon you'll see [the piece of paper], or you won't see it, depending on whether the N rays are shining on [it]. In that way, you can detect whether or not the N rays are acting. [*See Figure 3.*]

Well, he found that N rays could be stored up in things. For example, you could take a brick. He found that N rays would go through black paper and would go through aluminum. So he took some black paper, wrapped a brick up in it and put it out in the street and let the sun shine through the black paper into the brick. Then he found that the brick would store N rays and give off the N rays even with the black paper on it. [You could] bring [the brick] into the laboratory and hold it near the piece of paper that you were looking at—faintly illuminated—and you could see [the paper] much more accurately. Much better, if the N rays are there, but not if [the brick was] too far away. . . .

Well, you'd think he'd make such experiments as this—to see if with ten bricks he got a stronger effect than he did with one. No, not at all. He didn't get any stronger effect. It didn't do any good to increase the intensity of the light. You had to depend upon whether you could see it or whether you couldn't see it. And there, the N rays were very important.

Now, a little later, [Blondlot] found that many kinds of things gave off N rays. A human being gave off N rays, for example. If someone else came into the room then you probably could see [the faintly illuminated paper]. He also found that if someone made a loud noise that would spoil the effect. You had to be silent. Heat, however, increased the effect—radiant heat. Yet heat itself wasn't the same thing as N rays. N rays were not heat because heat wouldn't go through aluminum. Now he found a very interesting thing was that if you took the brick that was giving off N rays and held it close to your head, it went through your skull and it allowed you to see the paper better. Or you could hold the brick near the paper; that was all right too.

Now he found that there were some other things that were like negative N rays. He called them N' rays. The effect of the N' rays was to decrease the visibility of a faintly illuminated slit. That worked too, but only if the angle of incidence was right. If you look at [the paper] tangentially you found that the [N' rays] increased the intensity. [The intensity] decreased if you looked at [the paper] normally and it increased if you looked at it tangentially. All of which was very interesting. And he published many papers on it—one right after the other. Other people did too, confirming Blondlot's results. There were lots of papers published, and at one time about half of them were

confirming the results of Blondlot. You see, N rays ought to be important because x rays were known to be important, and alpha rays were, and N rays were somewhere in between, so N rays must be very important.

Enter R. W. Wood

Well, R. W. Wood heard about these experiments—everybody did, more or less. So R. W. Wood went over there. At that time Blondlot had a prism, quite a large prism of aluminum, with a 60° angle. [He also] had a Nernst filament with a little slit about 2 mm wide. There were two slits, 2 mm wide each. The beam [from the filament] fell on the prism and was refracted and he measured the refractive index. . . . He found that it wasn't monochromatic, that there were several different components to the N rays. . . . He could measure three or four different refractive indices, each to two or three significant figures, and he was repeating some of these and showing how accurately repeatable they were, showing it to R. W. Wood in this dark room.

Well, after this had gone on for quite a while and Wood found that he was checking these results very

René Blondlot in the robes of the French Academy of Sciences. The French physicist believed he had found a new form of radiation, which he named N rays to honor the University of Nancy, where he worked. (Photo courtesy of Irving Klotz, Northwestern University; and the French Academy of Sciences.)

accurately, measuring the position [of the beam on] the little piece of paper within a tenth of a millimeter, although the slits were 2 mm wide. Wood asked him about that. He said: "How? How can you, from just the optics of the thing, with slits 2 mm wide, how can you get a beam so fine that you can detect its position to within a tenth of a millimeter?"

Blondlot said: "That's one of the fascinating things about the N rays. They don't follow the ordinary laws of science that you ordinarily think of." He said: "You have to consider these things all by themselves. They are very interesting but you have to discover the laws that govern them."

Well, Wood asked him to repeat some of these measurements, which he was only too glad to do. But in the meantime, the room being very dark, R. W. Wood put the prism in his pocket and the results checked perfectly with what [Blondlot] had before. Well, Wood rather cruelly published that.[5] And that was the end of Blondlot.

Nobody accounts for the methods by which [Blondlot] could reproduce those results to a tenth of a millimeter. Wood said that he seemed to be able to do it but nobody understands that. Nobody understands lots of things. But some of the Germans—Pringsheim was one of them—came out later with an extremely interesting story. They had tried to repeat some of Blondlot's experiments and had found this: One of the experiments was to have a very faint source of light on a screen of paper. To make sure that you are seeing the screen of paper you hold your hand up and move it back and forth. And if you can see your hand move back and forth then you know it is illuminated. One of the experiments that Blondlot made was that the [illumination] was made much better if you had some N rays falling on the piece of paper. Pringsheim was repeating these in Germany and he found that if you didn't know where the paper was, whether it was in front of or behind your hand, [the experiment] worked just as well. That is, you could see your hand just as well if you held it in back of the paper as if you held it in of front of it. Which is the natural thing, because this is a threshold phenomenon. And a threshold phenomenon means that you don't know, *you really don't know*, whether you are seeing it or not. But if you have your hand there, well, of course, you see your hand because you *know* your hand's there and that's just enough to win you over to where you know that you see it. But you know it just as well if the paper happens to be in front of your hand instead of in back of your hand, because you don't know where the paper is but you *do* know where your hand is.

Mitogenetic rays

Well, let's go on. About 1923 there was a whole series of papers by Gurwitsch and others. [*Alexander Gurwitsch was a professor at the First State University of Moscow.*] There were hundreds of them published on mitogenetic rays.[6] There are still a few of them being published. I don't know how many of you have ever heard of mitogenetic rays. They are rays that are given off by growing plants, living things, and they were proved, according to Gurwitsch, to be something that would go

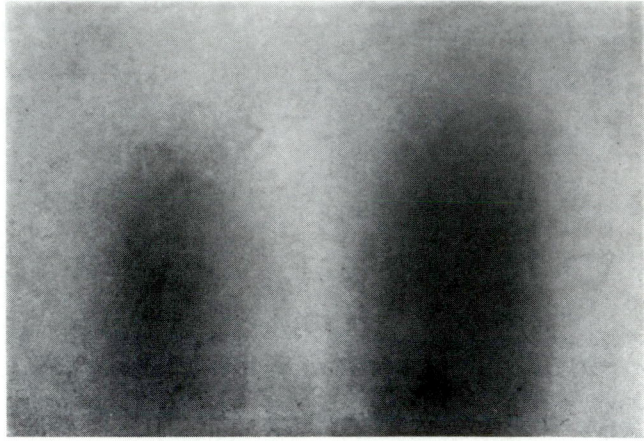

N rays could enhance illumination of a dimly lit object, or so Blondlot claimed. Among other evidence, he published the above photogravure, made by exposing film to a small luminous spark. The darker exposure (right) was made in the presence of N rays, mysterious radiation emanating from a hot wire. (From *Comptes Rendus*, 22 February 1904.) **Figure 3.**

through quartz but not through glass. They seemed to be some sort of ultraviolet light.

The way they studied these [rays] was this: You had some onion roots—onions growing in the dark or in the light—and the roots would grow straight down. Now if you had another onion root nearby, and this [first] onion root was growing down through a tube or something, going straight down, and another onion root came nearby, [the first onion root] would develop so that there were more cells on one side than the other. One of the tests they had made at first was that the root would bend away. And as it grew it would change in direction, which was evidence that something had traveled from one onion root to the other. And if you had a piece of quartz in between [the change would occur], but if you put glass in between it wouldn't. So this radiation would not go through glass but it would go through quartz. [*See Figure 4.*]

Well, it started that way. Then everything gave off mitogenetic rays—anything that remotely had anything to do with living things. And then they started to use photoelectric cells to check it, and whatever they did they practically always found that if you got the conditions just right, you could *just* detect [mitogenetic rays] and prove it. But if you looked over those photographic plates that showed this ultraviolet light you found that the amount of light was not much bigger than the natural particles of the photographic plate so that people could have different opinions as to whether it did or did not show this effect. The result was that less than half of the people who tried to repeat these experiments got any confirmation of it, and so it went. Well, I'll go on. . .

Symptoms of sick science

The Davis–Barnes experiment and the N rays and the mitogenetic rays all have things in common. These are cases where there is no dishonesty involved but where people are tricked into false results by a lack of understanding about what human beings can do to themselves in the way of being led astray by subjective effects, wishful thinking or threshold interactions. These are examples of pathological science. These are things that attracted a

> ## Symptoms of Pathological Science
>
> ▷The maximum effect that is observed is produced by a causative agent of barely detectable intensity, and the magnitude of the effect is substantially independent of the intensity of the cause.
> ▷The effect is of a magnitude that remains close to the limit of detectability or, many measurements are necessary because of the very low statistical significance of the results.
> ▷There are claims of great accuracy.
> ▷Fantastic theories contrary to experience are suggested.
> ▷Criticisms are met by *ad hoc* excuses thought up on the spur of the moment.
> ▷The ratio of supporters to critics rises up to somewhere near 50% and then falls gradually to oblivion.

great deal of attention. Usually hundreds of papers have been published on them. Sometimes they have lasted for 15 or 20 years and then they gradually have died away. Now here are the characteristic rules [*see the box above*]:
▷ The maximum effect that is observed is produced by a causative agent of barely detectable intensity. For example, you might think that if one onion root would affect another due to ultraviolet light then by putting on an ultraviolet source of light you could get it to work better. Oh no! *Oh no!* It had to be just the amount of intensity that's given off by an onion root. Ten onion roots wouldn't do any better than one and it didn't make any difference about the distance of the source. It didn't follow any inverse square law or anything as simple as that. And so on. In other words, the effect is independent of the intensity of the cause. That was true in the mitogenetic rays and it was true in the N rays. Ten bricks didn't have any more effect than one. It had to be of low intensity. We know why it had to be of low intensity: so that you could fool yourself so easily. Otherwise, it wouldn't work. Davis–Barnes worked just as well when the filament was turned off. They counted scintillations.
▷ Another characteristic thing about them all is that these observations are near the threshold of visibility of the eyes. Any other sense, I suppose, would work as well. Or many measurements are necessary—many measurements—because of the very low statistical significance of the results. With the mitogenetic rays particularly, [people] started out by seeing something that was bent. Later on, they would take a hundred onion roots and expose them to something, and they would get the average position of all of them to see whether the average had been affected a little bit... Statistical measurements of a very small effect... were thought to be significant if you took large numbers. Now the trouble with that is this. [Most people have a habit, when taking] measurements of low significance, [of finding] a means of rejecting data. They are right at the threshold value and there are many reasons why [they] can discard data. Davis and Barnes were doing that right along. If things were doubtful at all, why, they would discard them or not discard them depending on whether or not they fit the theory. They didn't know that, but that's the way it worked out.
▷ There are claims of great accuracy. Barnes was going to get the Rydberg constant more accurately than the spectroscopists could. Great sensitivity or great specificity—we'll come across that particularly in the Allison effect.
▷ Fantastic theories contrary to experience. In the Bohr theory, the whole idea of an electron being captured by an alpha particle when the alpha particles aren't there, just because the waves are there, [isn't] a very sensible theory.
▷ Criticisms are met by ad hoc excuses thought up on the spur of the moment. They always had an answer—always.
▷ The ratio of the supporters to the critics rises up somewhere near 50% and then falls gradually to oblivion. The critics couldn't reproduce the effects. Only the supporters could do that. In the end, nothing was salvaged. Why should there be? There isn't anything there. There never was. That's characteristic of the effect. Well, I'll go quickly on to some of the other things.

Allison and isotopes

The Allison effect is one of the most extraordinary of all.[7] It started in 1927. There were hundreds of papers published in [journals such as] the *Physical Review*, the *Journal of the American Chemical Society*—hundreds of papers. Why, they discovered five or six different elements that were listed in the Discoveries of the Year. There were new elements discovered—alabamine, virginium. A whole series of elements and isotopes were discovered by Allison. [*Fred Allison was at the Alabama Polytechnic Institute.*]

The effect was very simple. There is the Faraday effect, by which a beam of polarized light [is rotated when it passes] through a liquid which is a magnetic field. The plane of polarization is rotated by a longitudinal magnetic field. Now that idea has been known for a long time and it has a great deal of importance in connection with light shutters. At any rate, you can let light through [*See Figure 5.*] or not depending upon the magnetic field. Now the experiment of Allison's was this. They had a glass cell and a coil of wire around it and [they had] wires coming up here, a Lecher systems. [There was] a spark gap so that a flash of light came through [the lens] and went through one Nicol prism and then another one. You would adjust [the second Nicols prism] with a liquid like water or carbon disulfide or something like that in the cell so that there was a steady light. If you had a beam of light and you polarized it and then you turned on a magnetic field, why you see that you could rotate the plane or polarization. There would be an increase in the brightness of the light when you put [on] a magnetic field.

Now they wanted to find the time delay, how long it takes [for the Farraday effect to occur]. So they had a spark, and the same field that produced the spark induced a current through the coil. By sliding this wire along the trolley of the Lecher system, they could cause a compensating delay [in the second cell, where the field was

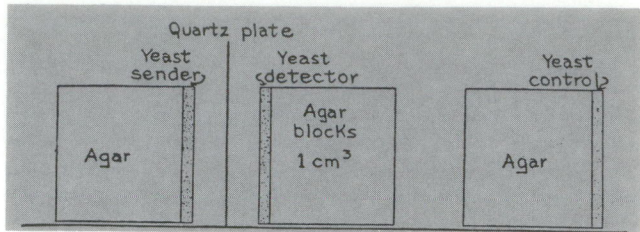

Detection of mitogenetic rays. These rays were believed to be a form of ultraviolet radiation, emitted by biological materials, that influenced the growth of other biological materials. The experiment shown here was one way to test whether mitogenetic rays from the yeast on one block of agar would go through the quartz and affect yeast on the facing block. (From ref. 6.) **Figure 4.**

reversed]. The sensitivity of this thing was so great that they could detect differences of about 3×10^{-10} seconds. By looking in they could see these flashes of light, the light from the sparks, and they tried to decide as they changed the position of this trolley whether it got brighter or dimmer. They set it for a minimum and measured the position of the trolley. They put in this [second] glass tube a water solution and added some salt to it. And they found that the time lag was changed. . . . They got a change in the time lag depending upon the presence of salts.

Now they first found—very quickly—that if you put in a thing like ethyl alcohol you got one characteristic time lag, and with acetic acid another one that was quite different. But if you had ethyl acetate you got the sum of the two. You got two peaks. So that you could analyze ethyl acetate and find the acetic acid and the ethyl alcohol. Then they began to study salt solutions and they found that only the metal elements counted, but they didn't act as ions. That is, all potassium ions weren't the same, but potassium nitrate and potassium chloride and potassium sulfate all had quite characteristic different points, which were a characteristic of the compound. It was only the positive ion that counted and yet the negative ions had a modifying effect. But you couldn't detect the negative ions directly.

Now they began to see how sensitive it was. Well, they found that any [concentration of] more than about a 10^{-8} molar solution would always produce the maximum effect. You'd think that that would be kind of discouraging from the analytical point of view, but no, not at all. And you could make quantitative measurements to about three significant figures by diluting the solutions down to a point where the effect disappeared. Apparently, it disappeared quite sharply when you got down to about 10^{-8} or 3.42×10^{-8} in concentration, or something of that sort. . . Otherwise you would get it so that you could detect the limit within this extraordinary degree of accuracy.

Well, they found that things were entirely different—even in these very dilute solutions—in sodium nitrate from what it was with sodium chloride. Nevertheless, it was a characteristic which depended upon the compound, even though the compound was dissociated into ions at those concentrations. That didn't make any difference, but it was fact that was experimentally proven. They then went on to find that the isotopes all stick right out like sore thumbs with great regularity. In the case of lead, they found 16 isotopes. These isotopes were quite regularly spaced so that you could get 16 different positions and you could assign numbers to those so that you could identify them and tell which they were. Unfortunately, you

couldn't get the concentrations quantitatively; even the dilution method didn't work quite right because [the isotopes] weren't all equally sensitive. You could get them relatively but only approximately. Well, [this effect] became important as a means of detecting elements that hadn't yet been discovered, like alabamine and elements that are now known and filling out the periodic table. All the elements in the periodic table were filled out that way and published.

But a little later, in 1945 or '46, I was at the University of California. Owen Latimer who is now head of the chemistry department there—not Owen Latimer, Wendell Latimer—had had a bet with G. N. Lewis in 1932. [*Gilbert Lewis was also a chemist at Berkeley.*] He said: "There's something funny about this Allison effect, how they can detect isotopes." He had known somebody who had been down with Allison and who had been very much impressed by the effect, and he said to Lewis: "I think I'll go down and see Allison, to Alabama, and see what there is in it. I'd like to use some of these methods."

Now people had begun to talk about spectroscopic evidence that there might be traces of hydrogen of atomic weight 3. It wasn't spoken of as tritium at that time but as hydrogen of atomic weight 3 that might exist in small amounts. There was a little spectroscopic evidence for it, and Latimer said: "Well, this might be a way of finding it. I'd like to be able to find it." So he went and spent three weeks at Alabama with Allison. Before he went he talked over with G. N. Lewis what he thought the prospects were, and Lewis said, "I'll bet you ten dollars you'll find that there's nothing in it." And so they had this bet. Latimer went down there and he came back. He set up the apparatus and made it work so well that G. N. Lewis paid him the ten dollars. He then discovered tritium and he published an article in the *Physical Review*[8]—just a little short note saying that using Allison's method he had detected the isotope of hydrogen of atomic weight 3. And he made some sort of estimate as to its concentration. to its concentration.

Well, nothing more was heard about it. I saw [Latimer] then, 7 or 8 years after that. I had written these things up before, about this Allison effect, and I told him about [my observations concerning pathological science] and how the Allison effect fit all these characteristics. Well, I know that at that time at one of the meetings of the American Chemical Society there was great discussion as to whether to accept papers on the Allison effect. There they decided, no, they would not accept any more papers on the Allison effect, and I guess the *Physical Review* did too. At any rate, the American Chemical Society decided that they would not accept any more manuscripts on the Allison effect. However, after they had adopted that as a firm policy, they did accept one more a year or two later because here was a case where all the people in the faculty had [made up] twenty or thirty different solutions. . . . They had labeled them all secretly and they had taken every precaution to make sure that nobody knew what was in these solutions, and they had given them to Allison. He had used his method on them and he had gotten them all right, although many of them were at concentrations on the order of 10^{-6} molar. That was sufficiently definite—good experimental methods—and it was accepted for publication by the American Chemical Society, but that was the last.[9] You'd think that would be the beginning, not the end.

Anyway, Latimer said: "You know, I don't know what was wrong with me at that time. After I published that paper I never could repeat the experiments again. I haven't the least idea why. But," he said, "those results were wonderful. I showed them to G. N. Lewis and we both agreed that it was all right. They were clean-cut. I checked them myself every way I knew how to. I don't know what else I could have done, but later on I just couldn't ever do it again."

I don't know what it is. That's the kind of thing that happens in all of these [cases]. All the people who had anything to do with these things find that when [they] get through with them, [they can't account for them]. You can't account for Bergen Davis saying that they didn't calculate those things from the Bohr theory, that they were found by empirical methods without any idea of the theory. Barnes made the experiments and brought them in to Davis, and Davis calculated them up and discovered all of a sudden that they fit the Bohr theory. He said Barnes didn't have anything to do with that. Well, take it or leave it, how did he do it? It's up to you to decide. I can't account for it. All I know is that there was nothing salvaged at the end, and therefore none of it was ever right and Barnes never did see a peak. You can't have a thing halfway right.

Extrasensory perception

Well, there's Rhine. [*Joseph B. Rhine was a parapsychologist then at Duke University.*] I spent a day with Rhine at Duke University at the meeting of the American Chemical Society, probably about 1934. Rhine had published a book and I'll just tell you a few things. First of all, I went in and told Rhine . . . the whole story [I have just told you]. I said these [traits of pathological science] are the characteristics of those things that aren't so. They are all characteristics

of your thing too. He said: "I wish you'd publish that. I'd love to have you publish it. That would stir up an awful lot of interest." He said: "I'd have more graduate students. We ought to have more graduate students. This thing is so important that we should have more people realize its importance. This should be one of the biggest departments in the university."

Well, I won't tell you the whole story with Rhine because I talked with him all day. He uses cards which you guess at [before they are turned] over. You have extrasensory perception. You have 25 cards and you deal them out face down, or one person looks at them . . . on the other side of a screen . . . and you read his mind. The other thing is for nobody to know what the cards are, in which case they are turned over without anybody looking at them. You record them and then you look them up and see if they check, and that's telepathy, or clairvoyance rather. Telepathy is when you can read another person's mind.

Now a later form of the thing is for you to decide now and write down what the cards are going to be when they are shuffled tomorrow. That works too.

All of these things are nice examples in which the magnitude of the effect is entirely independent of magnitude of the cause. That is, the experiments work just as well when the shuffling is to be done tomorrow as when it was done some time ago. It doesn't make any difference in the results. There is no appreciable difference between clairvoyance and telepathy, although if you try to think of mechanisms for the two, it should be quite different. [It's rather difficult to think of a mechanism] . . . to get the cards to telegraph you all the information that's in them as to how they are arranged and so on, when they are stacked up on top of each other, and to have it given in the right sequence. On the other hand, it is conceivable that there may be some sort of mechanism in the brain that might send out some sort of unknown messages that could be picked up by some other brain. That's a different order of magnitude—a different order of difficulty. But they were all the same from Rhine's point of view.

Well, now, [I've told you a few of the things that I know about Rhine]. There are many more I could [tell] you. Rhine, being in quite a philosophical mood, said "It's funny how the mind tries to trick you." He said: "People don't like these experiments. I've had millions of these cases where the average is about 7 out of 25." You'd expect 5 out of 25 to come out right by chance and on the grand average they come out, oh, out of millions or hundreds of millions of cases, they average around 7. Well, to get 7 out of 25 would be a common enough occurrence but if you take a large number and you get 7,

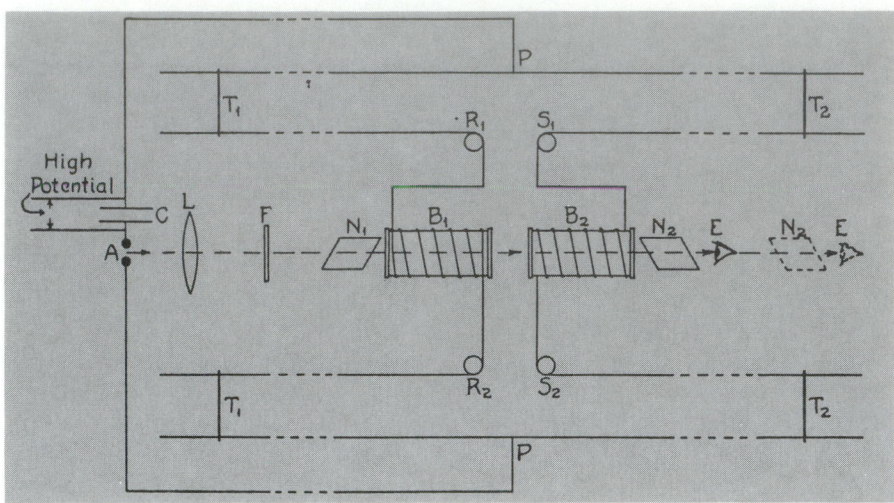

Apparatus for studying the "Allison effect." Cells B1 and B2 were filled with liquid. When a current flowed through the coils of wires surrounding these cells (from R1 to R2 and from S1 to S2), the resulting magnetic field rotated the plane of polarization of any light passing through the cells, because of the Faraday effect. Discharge of the capacitor C produced a flash of light from the spark gap at A and, at the same time, sent current through the coils. The light was focused by lens L and polarized by the Nicol prism N1. Cell B1 rotated the plane of polarization, and cell B2 was moved until it produced an exactly compensating rotation: At that point the time for light to travel between the cells just equaled the time lag in the Faraday effect in the two cells. Allison claimed that the time lag, as measured by the position of cell B2, was a unique characteristic of each isotope present in solution in the cells. This effect became a way of identifying new isotopes and elements—until it was recognized as pathological science. (From ref. 7.)   **Figure 5.**

well you doubt the statistics or the statistical application or, above all, what I think of, and what I want to give you reasons for thinking, is the rejection of a small percentage of the data.

Before I get into what Rhine said, [I'll first] say this [about] David Langmuir, a nephew of mine, who was in the Atomic Energy Commission. When he was with the Radio Corporation of America a few years ago, he and a group of other young men thought they would like to check Rhine's work. So they got some cards and they spent many evenings together finding how these cards turned up and they got well above 5. They began to get quite excited about it and they kept on, and they kept on, and they were right on the point of writing Rhine about the thing. And they kept on a little longer and things began to fall off, and fall off a little more, and they fell off a little more. And after many, many, many days, they fell down to an average of 5—grand average—so they didn't write to Rhine. Now if Rhine had received that information, that this reputable body of men had gone ahead and gotten a value of 8 or 9 or 10 after so many trials, why, he would have put it in his book. How much of that sort of thing [goes on]? When you are fed information of that sort by people who are interested—how are you going to weigh the things that are published in the book?

Now, an illustration of how it works is this. [Rhine] told me: "People don't like me." He said: "I took a lot of cards and sealed them up in envelopes and I put a code number on the outside and I didn't trust anybody to know that code. Nobody!"

[*A section of the speech is missing at this point. It evidently described some tests that gave scores below* 5.] "...the idea of having this thing sealed up in the cards as though I didn't trust them and therefore to spite me they made it purposely low."

"Well," I said, "that's interesting—very interesting, because you said that you'd published a summary of *all* the data that you had. And it comes out to be 7. It is now within your power to take a larger percentage, including those cards that are sealed up in those envelopes, which could bring the whole thing back down to 5. Would you do that?"

"Of course not," he said. "That would be dishonest."

"Why would it be dishonest? The low scores are just as significant as the high ones, aren't they?"

"They proved that there's something there just as much, and therefore it wouldn't be fair."

I said: "Are you going to count them? Are you going to reverse the sign and count them or count them as credits?"

"No, no," he said.

I said: "What have you done with them? Are they in your book?"

"No."

"Why, I thought you said that all your values were in your book. Why haven't you put those in?"

"Well," he said, "I haven't had time to work them up."

"Well, you know all the results. You told me the results."

"Well," he said, "I don't give the results out until I've had time to digest them."

I said: "How many of these things have you?"

He showed me filing cabinets—a whole row of them. Maybe hundreds of thousands of cards. He had a filing cabinet that contained nothing but these things that were done in sealed-up envelopes. And they were the ones that gave the average of 5.

Well, we'll let it stand at that. A year or so later, [Rhine] published a new volume of his book. In that, there's a chapter on the sealed-up cards in the envelopes and they all come up to around 7. And nothing is said about the fact that for a long time they came down below 5. You see, [Rhine] knows, if they come below 5, he knows that it isn't fair to the public to misrepresent this thing by including those things that prove just as much a positive result as though they came out above [5]. It's just a trick of the mind that these people did to try to spite [him] and of course it wouldn't be fair to publish.[10]

Flying saucers

I'm not going to talk about flying saucers very much except just this. A flying saucer is not exactly science, although some scientific people have written things about them. I was a member of General Schwartz's [*This name is uncertain.*] advisory committee after the war, and we held some very secret meetings in Washington in which there was a thing called Project SIGN. I think it was s-i-g-n. Anyway, it was hushed up. It was hardly even talked about. It was the flying saucer stuff, concerned with gathering the evidence and weighing and evaluating the data on flying saucers. And [General Schwartz] said: "You know, it's very serious. It really looks as though there is something there."

Well, afterwards I told him this story here. I said that it seems to me from what I know about flying saucers they look like [pathological science]. Well, anyway, it ended up by two men being brought to Schenectady with a boiled down group of about 20 or 30 best cases from hundreds and hundreds that they knew all about. I didn't want them all. I said to pick out about 30 or 40 of the best cases, and bring them to Schenectady. [I promised that we would] spend a couple of days going over them. . . .

Most of them were Venus seen in the evening through a murky atmosphere. Venus can be seen in the middle of the day if you know where to look for it—almost any clear bright day, especially when Venus is at its brightest—and sometimes it has almost caused panic. [There has been] traffic congestion in New York City when Venus is seen in the evening near some of the buildings around Times Square, and people thought it was a comet about to collide with the Earth, or somebody from Mars, or something of that sort. That was a long time ago. That was 30 or 40 years ago. Venus still causes flying saucers.

Well, they only had one photograph or two photographs, taken by one man, [that puzzled me]. It looked to me like a piece of tar paper when I first saw it, and the two photographs showed the thing in entirely different shapes. I asked for more details about it. What was the weather at the time? Well, they didn't know but they'd look it up. And they got out some papers and there it was. It was taken about 15 or 20 minutes after a violent thunderstorm out in Ohio. Well, what's more natural than some piece of tar paper picked up by a little miniature twister and being carried a few thousand feet up into the clouds? And it was coming down, that's all. So what could it be? "But it was going at an enormous speed." Of course the man who saw it didn't have the vaguest idea how far away it was.

That's the trouble. If you see something that's up in the sky, a light or any kind of an object, you haven't the vaguest idea of how big it is. You can guess anything you like about the speed. You ask people how big the Moon is. Some say it is as big as your fist, or as big as a baseball. Some say as big as a house. Well, how big is it really? You can't tell by looking at it. How can you tell how big a flying saucer is?

Well, anyway, after I went through these [cases], I didn't find a single one that made any sense at all. There was nothing consistent about them. They all suffered from these facts: They were all subjective. They were all near threshold. You don't know what the threshold is exactly in detecting the velocity of an object that you see up in the sky, when you don't know whether it's a thousand feet or ten thousand feet or a hundred thousand feet up. But they all fitted in with this general pattern, namely, that there didn't seem to be any evidence that there was anything in them. Anyway, the men [on the committee] were convinced and they ended Project SIGN. And later the whole [project] was declassified, and it was written up by the *Saturday Evening Post* about 4 or 5 years ago. At any rate, that seemed to be the end of it. But of course the newspapers wouldn't let a thing like that die. It keeps coming up again and again and again, and the old story keeps coming back again. It always has. It's probably hundreds of years old anyway.

Epilogue (R. N. Hall)

Pathological science is by no means a thing of the past. The search for some record of Langmuir's lecture began in 1965 out of curiosity about two phenomena—the photomechanical and electromechanical effects—that were being reported with increasing frequency in papers from a number of laboratories around the world. The experiments that were described in these publications conformed to the first 5 characteristic symptoms of pathological science precisely as Langmuir had outlined them. Further work[11] disclosed the subjective nature of these observations, and helped to bring this field of investigation to its final stage—the decline toward oblivion. Many readers will recall subsequent examples of phenomena that exhibit some of the characteristics of pathological science listed by Langmuir and that may be of a similar nature.

References

1. Eight months after the visit of Langmuir and Hewlett to Columbia and this exchange of letters, Barnes published a paper on the Davis–Barnes effect in Phys. Rev. **35**, 217 (1930).
2. H. C. Webster, Nature **126**, 352 (1930).
3. B. Davis, A. H. Barnes, Phys. Rev. **37**, 1368 (1931).
4. R. Blondlot, *The N-Rays*, Longmans, Green, London (1905). J. G. McKendrick, Nature **72**, 195 (1905).
5. R. W. Wood, Nature **70** (1904); Phys. Z. **5**, 789 (1904). W. Seabrook, *Doctor Wood*, Harcourt Brace, New York (1941), ch. 17.
6. A. Hollaender, W. D. Claus, J. Opt. Soc. Am. **25**, 270 (1935).
7. F. Allison, E. S. Murphy, J. Am. Chem. Soc. **52**, 3796 (1930). F. Allison, Ind. Eng. Chem. **4**, 9 (1932). S. S. Cooper, T. R. Ball, J. Chem. Ed. **13**, 210 (1936), also pp. 278, 326. M. A. Jeppesen, R. M. Bell, Phys. Rev. **47**, 546 (1935). H. F. Mildrum, B. M. Schmidt, Air Force Aero Prop. Lab. report AFAPL-TR-66-52 (May 1966).
8. W. M. Latimer, H. A. Young, Phys. Rev. **44**, 690 (1933).
9. Langmuir may have been referring to the paper by J. L. McGhee, M. Lawrentz, J. Am. Chem. Soc. **54**, 405 (1932), which contains the statement "In December 1930 one of us (McGhee) handed out by number to Prof. Allison twelve (to him) unknowns which were tested by him and checked by two assistants 100 percent correctly in three hours." See also T. R. Ball, Phys. Rev. **47**, 548 (1935), which describes additional tests in which unknowns were identified.
10. See, for example, G. R. Price, Science **122**, 359 (1955) and replies on 6 Jan. 1956. M. Gardner, *Fads and Fallacies in the Name of Science*, Dover (1957).
11. R. E. Hannemann, P. J. Jorgenson, J. Appl. Phys. **38**, 4099 (1967); R. N. Hall, *Proc. 9th Intl. Conf. Physics of Semiconductors*, Moscow (1968), p. 481.

Physics and psychic research in Victorian and Edwardian England

Lord Rayleigh, J. J. Thomson, William Crookes and Oliver Lodge were among the physicists who joined the Society for Psychical Research in the late 19th century, a time of public fascination with spiritualism and psychic phenomena.

Janet Oppenheim

PHYSICS TODAY/MAY 1986

Scientists in industrialized England commanded public respect, even veneration. Their widely publicized discoveries promised to elucidate the mysteries of nature, to harness natural forces for human purposes, to increase productivity in fields and factories and to improve the odds in the struggle against disease. The scientific method itself was celebrated as the surest means of attaining truth.

For all the conviction that science would provide blessings to humanity, however, much of the British public in the Victorian and Edwardian decades also feared its triumphant advances. In those years—roughly from 1840 to the outbreak of World War I in 1914—the dark side of science attracted increasing attention. People realized that technological progress could damage both individual health and age-old patterns of communal society. It could pollute the environment and create potent new weapons for destroying the very civilization that gloried in scientific knowledge. Most important of all to many British men and women, the findings of science could undermine

Janet Oppenheim is professor of history at the American University, Washington, DC. She is the author of *The Other World: Spiritualism and Psychical Research in England, 1850–1914* (Cambridge U.P., New York, 1985) and other works in modern British cultural and intellectual history.

traditional religious beliefs without offering a substitute.

These anxious Victorians shuddered in 1874 when the physicist John Tyndall, in his notorious Belfast Address as president of the British Association for the Advancement of Science, proclaimed the hegemony of science over religion, the adequacy of physical, material forces to explain whatever remained mysterious in nature. They hoped that a compromise could somehow be arranged, that religion could be made more compatible with the teachings of modern science and that science, consequently, would pose less of a threat to the basic tenets of Christianity. In these decades, the attempt to reconcile science and faith preoccupied British spiritualists, who firmly believed in human survival after death and in the possibility of communion with disembodied human spirits. It similarly engaged the attention of psychic researchers, who claimed to approach psychic phenomena with a critical mind, objectively gathering the facts needed for strict scientific evaluation and harboring no preconceived explanatory theories. The British physicists who, rejecting Tyndall's proclamation, sympathized with the goals of psychic research form the subject of this article.

Open-minded investigators

Spiritualists believed that they could effect the union of science and the

Christian religion because they used the empirical methods of the former to attempt to prove a fundamental teaching of the latter—the immortality of the soul. There was nothing that occurred at séances, they asserted, that had not been seen, heard, touched or smelled by witnesses in full possession of their senses. Furthermore, mediums were occasionally subjected to laboratory experiments, whose results, spiritualists insisted, had the same conclusive authority as any findings from a physicist's apparatus. If the possibility of communion with the dead could be thus indisputably affirmed, the promise of human immortality would cease to be metaphysical aspiration and would assume the imposing status of scientific fact. In a period when the age-old Judeo–Christian convictions about man's place in the universe were under assault from a wide range of sources—including evolutionary biology, geology, anthropology and textual studies of the Bible—spiritualism's claim to offer an alternative source of religious reassurance was the principal reason for its appeal.

Given the Victorian admiration for science, however, that claim necessarily depended heavily on the spiritualists' scientific pretensions. Only if they could emulate the practices of reputable scientists could spiritualists command credibility. Thus the spiritualists eagerly sought endorsement

William Crookes, 1832–1919. Crookes vigorously pursued psychic research in his home laboratory in London during the early 1870s. He aroused the wrath of many professional colleagues by his endorsement of the reality of a "psychic force" and by his promotion of mediums specializing in materializations.

the Society for Psychical Research carefully avoided any public commitment to the reality of séance phenomena and was thus able to attract a number of distinguished scientists. Two of the most celebrated were Nobel Prize-winning physicists: John William Strutt (third Baron Rayleigh) and Joseph John Thomson.

Rayleigh and Thomson were not active psychic researchers, but both served for decades on the council of the Society for Psychical Research, and Rayleigh became president of the society in 1919, the year of his death. In part Rayleigh's involvement in psychic research arose from his fondness for Eleanor Sidgwick, his sister-in-law and principal assistant in his redetermination of the ohm, ampere and volt, work he undertook as Cavendish Professor of Experimental Physics at Cambridge in the early 1880s. Sidgwick and her husband, Henry Sidgwick, Knightbridge Professor of Moral Philosophy at Cambridge, were at the center of the activities of the Society for Psychical Research, and Rayleigh's membership may have reflected more a courteous desire to promote their organization than a deep commitment to psychic inquiry.

Nevertheless Rayleigh's interest in the subject was genuine. As early as 1874 he had written to Sidgwick about psychic manifestations:

I am quite amazed at the little interest most people take in the

from leading scientists and were outraged when most of the British scientific profession rejected their credentials as scientific investigators. Yet the issue did not reduce to a simple conflict of spiritualist versus scientist. Just as numerous spiritualists revered the achievements of professional scientists, so were there scientists who would not join in any facile repudiation of spiritualism. They sought to maintain an open mind about psychic phenomena and séance manifestations. Some became committed spiritualists themselves, while others never ventured beyond the more cautious role of psychic researcher. In both cases, they approached the phenomena in the hope that these bizarre occurrences might reveal hitherto unknown scientific

laws or unimagined powers of human beings.

Within Victorian and Edwardian physics, all possible degrees of opposition to spiritualism and psychic research mingled with varying shades of support. Tyndall, Michael Faraday and William Thomson (Baron Kelvin) were outspoken in their contempt for the gullibility and ignorance of spiritualists, but several of their professional colleagues joined the Society for Psychical Research when it was established in London in 1882. The membership of that organization, which is still flourishing today, came to include luminaries of the political, intellectual and social worlds in late 19th- and early 20th-century England. Unlike most earlier British spiritualist associations,

J. J. Thomson, 1856–1940. Thomson's interest in psychic phenomena prompted him to join the Society for Psychical Research. As to the reality of such phenomena, however, he could only reach "the Scottish verdict—not proven."

question. A decision of the existence of mind independent of ordinary matter must be far more important than any scientific discovery could be, or rather would be the most important possible scientific discovery.

Whatever the strength of his Christian beliefs, Rayleigh had no doubts about the inadequacy of materialism—the philosophy that nothing exists in the universe but matter, that all phenomena, even mental activity, are the result of material agency. Rayleigh was far more sympathetic with the tenets of spiritualism, which upholds the reality of nonmaterial forces and their influence in the world at large. As Rayleigh wrote[1] to a correspondent in 1910, "I have never thought the materialist view possible, and I look to a power beyond what we see, and to a life in which we may at least hope to take part."

Thomson, Rayleigh's successor as Cavendish Professor at Cambridge, shared Rayleigh's gift for penetrating beyond the visible. Rayleigh received the Nobel Prize in 1904 for his discovery of a hitherto unnoticed component of the atmosphere, argon; Thomson probed the secrets of the atom and earned the Nobel Prize in 1906 for his discovery of the electron. The riddles of the universe intrigued Thomson, and it was entirely consistent with his open-mindedness that he should devote five pages of his memoirs to dowsing for water, a technique he fully believed[2] to be efficacious. However, Thomson was no more able than Rayleigh to reach definite conclusions about psychic phenomena. Both men remained willing to be convinced, for example, that telepathic communication, independent of all sensory assistance, was

possible, but they agreed that no incontestable proof was forthcoming. Neither allowed his sympathy with the goals of psychic researchers to compromise his judgment as a scientist.

Balfour Stewart, professor of physics at Owens College, Manchester, and Thomson's mentor there, also joined the Society for Psychical Research, and served as its second president. Stewart, who had been director of Kew Observatory, was no narrow specialist—he pursued investigations and published work in physics, mathematics, chemistry, meteorology and astronomy. He never sought to challenge the authority of the experimental method, but, unlike Rayleigh and Thomson, he believed that the methodology of the physical sciences could readily accommodate psychic phenomena. While repudiating the spiritualist explanations of those phenomena, he nonetheless maintained that forces as yet hidden from science were at work in the cosmos. *The Unseen Universe or Physical Speculations on a Future State*, a book that Stewart published anonymously in 1875, expressed his profound desire to portray the natural world in terms that both science and theology could accept.

Accordingly, Stewart sketched an invisible universe, one beyond the competence of scientists to investigate but linked to the visible world by bonds of energy. As he told[3] William F. Barrett, professor of physics at the Royal College of Science, Dublin, and a founding member of the Society for Psychical Research, there are "strong grounds for supposing that our environment is something very different from that to which the atomic materialists would wish to confine us." In his rejection of materialism, Stewart purposefully refused to impose any limits on the realm of natural law. A statement in an article that he wrote for *Nature* in 1871 exemplifies[4] his attitude toward the scientific enterprise:

Let us suppose that a man comes before us as a witness of some strange and unprecedented occurrence. Here it is evident that we are not entitled to reject his testimony on the ground that we cannot explain what he has seen in accordance with our preconceived views of the universe, even although these views are the result of a long experience; for by this means we should never arrive at anything new.

Stewart well understood how often prejudices within the scientific commu-

Lord Rayleigh (John William Strutt), 1842–1919. Like Thomson, Rayleigh suspended his judgment concerning the events he witnessed at séances. He believed psychic manifestations to be puzzles of paramount importance to modern science.

BETTMANN ARCHIVE

nity mold the course of research and dictate which subjects receive professional approval or neglect. His own receptivity to unverified hypotheses provoked criticism from coworkers. Arthur Schuster, successively professor of applied mathematics and professor of physics at Owens College, and himself a member of the Society for Psychical Research, found[5] Stewart gullible; Peter Guthrie Tait, the professor of natural philosophy at Edinburgh who aided Stewart in writing *The Unseen Universe*, subsequently dismissed[5] some of Stewart's ideas as "weird and grotesque." If this was the reaction to a physicist who carefully shunned spiritualism, how much harsher was the professional reaction to colleagues who actually embraced spiritualist beliefs?

Among those British physicists in the Victorian and Edwardian decades who accepted the spiritualist hypothesis as a possible explanation for séance manifestations, William Crookes and Oliver Lodge were preeminent. Both men came to espouse the reality of communion with the dead and to believe that disembodied human spirits provoked many of the events that they witnessed at séances. Lodge's opinions evolved in a straight line from skepticism to whole-hearted belief, but Crookes's views on spiritualism vacillated.

William Crookes

Crookes was a free-lance chemist and physicist of remarkable ability who launched his career in 1861, at the age of 29, with his discovery of the element thallium. Although he failed to secure a permanent academic position, he established himself professionally as an editor of scientific journals and a technical consultant to government organizations. From his home laboratory he pursued experiments on cathode rays, highly rarefied gases and rare-earth elements—experiments that earned him the admiration of the scientific community and a host of professional honors, including knighthood in 1897 and the Order of Merit in 1910. For part of his professional life, however, he also provoked scorn and ridicule as he tried to prove the reality of psychic phenomena with scientific apparatus.

Crookes was initially drawn to psychic research when his brother Philip died of yellow fever in 1867. Crookes's diary and some correspondence from the early 1870s suggest[6] that he briefly, and privately, embraced spiritualist beliefs while publicly retaining the pose of a noncommitted scientific expert. His series of tests with some of the best-known mediums of the day, especially the renowned Daniel Dunglas Home, led him to announce[7] publicly "the existence of a new force, in some unknown manner connected with the human organisation, which for convenience may be called the Psychic Force."

Crookes always insisted that he maintained rigid control over his experimental séances, and in the published summaries of his experiments he emphasized the mechanical contrivances he devised to test a medium's powers, as if the equipment itself eliminated all possibility of human error or deceit. Yet a medium of Home's skill could manipulate Crookes, allowing the scientist to believe that he had control over the conditions prevailing at the séances, which took place between 1871 and 1873. Indeed, in 1889, when Crookes printed[8] some of his actual séance notes in the *Proceedings of the Society for Psychical Research*, it became evident that his sittings with Home had not been as rigorously supervised as he had implied. If spirit commands issued through the medium ordered "Hands off the table," for example, all hands, including Crookes's, were immediately removed. Home may have easily imposed his own conditions while seeming to offer the scientist the utmost cooperation. After all, if Home disliked Crookes's arrangements for a séance, he had only to produce no manifestations whatsoever, blaming the failure on uncongenial circumstances. When phenomena did occur, including Home's customary repertoire of levitation, disembodied hands and the handling of hot coals, there was no guaran-

Oliver Lodge, 1851–1940. Lodge's early caution concerning the possibility of thought transference gave way to an unshakable belief in the reality of communication with the dead. His Christian faith in the immortality of the soul was inextricably connected with his belief in the ether of space.

tee that Crookes really directed the proceedings.

Galvanometer as monitor. Even greater interpretative problems concern Crookes's series of sittings with the winsome young medium Florence Cook in 1874–75. Cook was one of numerous mediums in the 1870s who specialized in the production of full-form materializations—entire spirit bodies that emerged from behind partitions to promenade among séance participants while the allegedly entranced medium, usually tied to a chair, slumbered out of sight. Cook's materialized spirit "control"—the personality directing her actions as a medium—was known as "Katie King." This spirit became a familiar sight in Crookes's London laboratory, and Crookes sought to record her appearance with an elaborate barrage of cameras. His equipment in these séances also included a galvanometer, by means of which the medium became part of an electric circuit. When Cook was attached to this contraption, Crookes assumed that she could not impersonate a materialized spirit without producing telltale fluctuations in the galvanometer readings. Nonetheless Katie emerged undaunted, and Crookes abandoned his earlier caution to proclaim[9] the unequivocal reality of Katie King as a materialized spirit entirely separate from the medium. Understandably, not a few scientific eyebrows were raised, and a controversy began, which continues today, over Crookes's relationship with Cook. Some commentators insist[10] that Cook was Crookes's mistress and that the scientist conspired with her to dupe the British public.

Far greater is the likelihood that Crookes himself was duped. He would not be the only scientist whose self-confidence led to self-deception and whose technical expertise gave him a false sense of security. Proud of his own astute powers of observation, he may well have failed to grasp how readily he could be misled. Tests in 1875 with another medium, Annie Eva Fay, suggest, for example, how a clever performer might have tricked the galvanometer. Although Fay sat behind a curtain, with her hands attached to the apparatus in such a way that she could scarcely move without noticeably altering the galvanometer in front of the curtain, a bell nonetheless rang on her side of the barrier, a hand emerged to offer a book, and a box of cigarettes was thrown through the curtains. All the while the galvanometer reading held steady, and Crookes believed that he had forestalled all opportunities for fraud. Subsequent psychic researchers, however, have demonstrated[11] ways in which Fay could have used other parts of her body, or even a resistance coil, to maintain the electric circuit while freeing her hands for other purposes.

After 1875 Crookes retreated to his earlier public position of suspended judgment. He had always known of the

Florence Cook (above) and "Katie King," Cook's materialized spirit "control." Cook was still an adolescent when she became a celebrity in London's spiritualist circles. King, her spirit control, looked remarkably like her. (Cook photograph courtesy Harry Price Library, University of London.)

trickery that abounded in spiritualist circles, and the discovery that Mary Showers, another materialization medium and a friend of Cook, had consistently produced fraudulent manifestations during experiments with him in 1874 and 1875 apparently persuaded him to cease his investigations. (In 1880 Cook herself was caught impersonating a materialized spirit.) Certainly Crookes was tired of allegations concerning his conduct toward attractive female mediums. "For myself," he confided[12] to Lodge many years later, "I have been so troubled by hints and rumors in connection with Miss Cook, that I shrink from laying stress on what I tried with her mediumship and rely on phenomena connected with Dan. Home's mediumship when saying anything in public."

A mental duality? Crookes said very little on the subject for the next 40 years, although he did serve on the council of the Society for Psychical Research and, from 1896 to 1899, as its president. Then in 1916, when Crookes was in his eighties, he was devastated by his wife's death and proved susceptible to the lure of spiritualist reassurance. William Hope, a fraudulent "spirit photographer," persuaded Crookes to accept a photograph of his wife's spirit presence. Despite clear marks of double exposure, Crookes regarded the picture as undeniable proof of the self's survival of bodily death. When Lodge tried to warn him of Hope's chicanery, Crookes dismissed[13] the need for precaution, explaining, "I went into the question of photographic trickery many years ago, and from confessions and admissions I had from tricksters I am acquainted with all the dodges possible." Ever the expert, Crookes refused to believe that anyone could fool him.

It was just such an attitude that had prompted a sustained attack on Crookes during the 1870s by the distinguished physiologist William B. Carpenter, who was registrar of the University of London. Carpenter waged an unremitting campaign against endeavors that he perceived to be misapplica-

tions of modern science: phrenology, mesmerism and spiritualism, for example. Crookes particularly aroused his wrath, for he saw Crookes betraying extraordinary talents to dabble in pseudoscience. Carpenter deplored the fact that so gifted a scientist should waste his time and stifle his critical faculties in an effort to prove the reality of a psychic force, and he could only suppose[14] that a "curious 'duality' of Mr. Crookes's mental constitution" allowed him to turn a blind eye on quackery and deception.

In fact, however, Crookes's work was all of a piece, without clear division between his physical and psychic experiments. Although the former were far more successful than the latter, they were prompted by the same curiosities and aspirations. Throughout his career he was willing to take risks, proposing startling hypotheses before he could demonstrate their validity. In 1879, for example, he addressed the British Association for the Advancement of Science on the subject of radiant matter and informed[15] his audience that highly rarefied gases approach a fourth state of matter, "as far removed from the state of gas as a gas is from a liquid." Crookes was pursuing an important physical problem when he investigated the properties of gas molecules in high vacua, but his commitment to the concept of matter in a fourth state barred him from reaching J. J. Thomson's momentous conclusion of the late 1890s: his proof that cathode rays, which had prompted Crookes's speculations on radiant matter, were in fact electrons.

One of the reasons the fourth state of matter intrigued Crookes was his belief that some such unseen phenomenon might explain psychic manifestations. Indeed, from 1870 until his death in 1919 he balanced two approaches to spiritualism: At times he believed in real spirit visitations, but more frequently he sought solutions within the prevailing creed of scientific naturalism. At different times in his life, and under varying circumstances, he inclined more toward one stance than the

other, but neither ever disappeared completely from the Crookesian perspective. Each was consonant with Crookes's lifelong attraction to "the shadowy realm between Known and Unknown," "the border land where Matter and Force seem to merge into one another," where he assumed[16] that "the greatest scientific problems of the future will find their solutions."

Oliver Lodge

After the complexity and ambiguity of Crookes's opinions, Lodge's approach to psychic phenomena appears comparatively simple. In the course of his long affiliation with the Society for Psychical Research, from 1884 until his death in 1940, Lodge made the transition from cautious psychic researcher to outspoken spiritualist, with none of the apparent backtrackings and detours that characterized the development of Crookes's attitudes. Lodge's conversion to spiritualism came through long study of the mental phenomena of mediumship, rather than through the more flamboyant and far more suspect physical phenomena, such as full-form materializations, that Crookes studied. Finally, Lodge differed significantly from Crookes in that religion played a major role in molding Lodge's ultimate interpretation of his psychic inquiries. Lodge could not study the phenomena of mediumship in a morally neutral context and brought to his work for the Society for Psychical Research a deeply rooted Christian faith.

Lodge's distinguished career as a physicist took him from London, where he studied at the Royal College of Science and University College, to Liverpool, where he held the chair of physics at the newly established university from 1881 to 1900, to Birmingham, where he served as the first principal of the university from 1900 to 1919; he was knighted in 1902. He contributed substantially to the development of wireless telegraphy, while his experiments on the relative motion of matter and ether made him one of the grandfathers of the special theory of relativity. Yet Lodge is known

Capture of a 'spirit.' The sequence of events depicted here chronicles the exposure of a medium masquerading as a materialized spirit. The particular drama recorded in this illustration occurred in 1879; the captor was Sir George Sitwell.

above all as an apologist for spiritualism and is particularly associated with the spiritualist revival that occurred during World War I. His publicized conviction[17] that his son Raymond Lodge, killed in Flanders in 1915, was sending communications to his grieving family provided comfort to thousands and brought Lodge a wider audience than his scientific writings had ever attracted.

It is clear that Lodge attained his spiritualist faith not in response to Raymond's death, but as a consequence of his long involvement in the activities of the Society for Psychical Research. From the start his attention focused on thought transference, or "telepathy," a word coined in 1882 by Frederic W. H. Myers, a leading member of the society. Lodge's introduction to the subject came in Liverpool when a prominent city merchant and governor of University College asked the young physics professor to test two employees of his drapery firm for thought-reading powers. In a report that he submitted to the *Proceedings of the Society for Psychical Research* in 1884, Lodge came to the significant conclusion[18] that "one person may, under favourable conditions, receive a faint impression of a thing which is strongly present in the mind, or thought, or sight, or sensorium of another person not in contact." At the time Lodge had no ideas concerning the way thoughts might be transferred without sensory contact, but he was definitely interested in pursuing the question further.

Leonora Piper. Throughout the 1880s Lodge remained content to believe in the possibility of telepathy between the living without proceeding to argue for potential communication with the spirits of the dead. He took that momentous step only after 1890, following séances with Leonora Piper, a trance medium from Boston whom the Society for Psychical Research had invited to England. Piper had acquired a reputation for astonishing her sitters with detailed information about their lives—information that the sitters were convinced she could not have

gained through normal means. While in England Piper allowed investigators from the Society for Psychical Research to search her luggage and to scrutinize her mail, but they found nothing to suggest that she turned to outside sources, human or literary, for the contents of her trance conversations.

Lodge's contribution to the lengthy report the investigators wrote about Piper in 1890 reveals the perplexity he felt at that time. He expressed unequivocal confidence that Piper's trance condition was genuine, but he argued that her spirit control—a self-proclaimed French doctor named Phinuit—was "probably a mere name for Mrs. Piper's secondary consciousness." Lodge conceded that Phinuit occasionally "fished" for information from his sitters, but he held that well-developed thought transference between the living was the best way to explain Piper's command of obscure, intimate knowledge about total strangers. Lodge had become convinced by 1890 that conscious, purposeful thought transference between persons not in contact, but in close proximity, had been indubitably established, and thus he used telepathy between the living as an explanation rather than as a hypothesis to be proved. He acknowledged, however, that the kind of telepathy that would explain Piper's mediumship—where the agents were not conscious of thinking about the ideas seemingly transmitted to her—was nowhere near to being experimentally proved.

Even further from satisfactory explanation was a class of telepathic revelations that Lodge allowed himself to consider only at the end of the report. These revelations, which emerged only rarely from Piper's trances, contained facts unknown to any sitter present. Telepathy from distant persons might be involved, Lodge suggested, although he was not yet sure that thought transference could operate between minds separated geographically. Only as "a last resort" did he suggest[19] "telepathy from deceased persons."

In subsequent years that last resort came to occupy an increasingly central place in Lodge's thought. He was deeply impressed by the improvement in the quality of Piper's telepathic communications: A new, more reliable spirit control gradually replaced Phinuit, and Piper turned more to "automatic scripts" than to the direct voice utterances that dominated her séances before 1890. The automatic scripts, apparently written without conscious effort by the entranced medium, left a record of Piper's mediumship that could be examined carefully by the Society for Psychical Research. Lodge was similarly impressed by another series of automatic scripts—the so-called cross correspondences—produced for the Society for Psychical Research by a diverse group of trance writers over a period of about 30 years commencing in 1901. Although the "automatists" were geographically separated and not in all cases personally acquainted, they produced some 3000 scripts that when studied minutely seemed to relate to each other through a similar and elaborate symbolism. By the early years of the 20th century Lodge was convinced of human immortality and did not hesitate to publicize his beliefs in such books as *The Immortality of the Soul* (1908) and *The Survival of Man* (1909).

Will to believe. Inevitably, Lodge was mercilessly ridiculed for his opinions. An article in the first volume of the scientific quarterly *Bedrock* grouped Lodge with his fellow physicist and psychic researcher Barrett and dismissed[20] them both:

It is not necessary either to regard the phenomena of so-called telepathy as inexplicable or to regard the mental condition of Sir W. F. Barrett and Sir Oliver Lodge as indistinguishable from idiocy. There is a third possibility. *The will to believe* has made them ready to accept evidence obtained under conditions which they would recognise to be unsound if they had been trained in experimental psychology.

Certainly there were other psychic researchers within the Society for Psychical Research who remained skeptical about Piper's mediumship and sought explanations in terms of secondary personalities or the heightened powers of perception possible in the trance state. Lodge, however, was adamant in his conviction that the survival of the personality beyond physical death had been demonstrated empirically.

The reasons for Lodge's beliefs—or will to believe—involved both his scientific and his religious views of the universe. Trained in the heyday of Victorian physics, he was deeply committed to the vision of universal continuity to which his illustrious predecessors had so handsomely contributed. As the contours of post-Newtonian physics emerged, he shrank from their implications and deplored "the modern tendency . . . to emphasise the discontinuous or atomic character of every-

thing." At a time when the ether of space was ceasing to be a meaningful concept for modern physics, Lodge insisted on its existence as a sort of cosmic glue without which "there could hardly be a material universe at all." It is significant that he gave the title "Continuity" to his 1913 presidential address[21] to the British Association for the Advancement of Science, in which he made these remarks.

Personal immortality was as essential to Lodge's concept of continuity as was the ether, for just as he sought to prove that no empty space divides the physical fabric of nature, so he insisted that no gap severs life from death. In this second, related assertion, the voice of Christian faith spoke out distinctly. In his 1910 book *Reason and Belief*, Lodge "proved" the persistence of the soul distinct from the body by reminding[22] his readers that "Christ did not spring into existence as the man Jesus of Nazareth. The Christ spirit existed

through all eternity. At birth he became incarnate." Lodge saw nothing incongruous in offering the Christ spirit as part of what he claimed to be a scientific demonstration.

The critics' philosophy

Lodge's strategies for reconciling science and faith won him little applause from fellow scientists in the early 20th century. While they honored his contributions to physics, they refused to agree that his spiritualist speculations formed an integral part of his professional work. Their objections to the intrusion of psychic phenomena into physical science centered on the fundamental question of evidence. Psychic phenomena were incapable of proof or disproof by the acknowledged methods of observation and experimentation under controlled circumstances, with all variable factors identified. By contrast, they were hopelessly erratic and utterly subjective. Critics who refused to credit stories of nonmaterial agents at work in the universe understood the difference between a hypothesis whose invalidity could be demonstrated and one incapable of such demonstration. Theories concerning the existence of mind apart from body, like arguments for the reality of spirit forces, belonged to the second category. For these reasons the majority of British scientists in this period found that psychic phenomena merited neither their investigation nor their belief. The British Association for the Advancement of Science and the Royal Society never inquired into alleged psychic occurrences; they were not attempting to ostracize wayward scientists, but to avoid wasting time on a futile task, one that could lead to no reasonably certain knowledge.

Those few physicists who did join the Society for Psychical Research of course disagreed with the majority opinion. Yet they also disagreed significantly among themselves. Those who, like Rayleigh and Thomson, remained psychic researchers only, withholding judgment on human immortality, could separate their rejection of materialism

Daniel Dunglas Home, 1833–86. Home was the most renowned medium of the 19th century. He entertained European royalty and British aristocrats and cooperated with scientists who sought to investigate his powers.

from their scientific research. For those who, like Lodge, endorsed spiritualism, the repudiation of materialism became an underlying assumption of all their professional endeavors.

At the turn of the century the British scientific profession as a whole held no *a priori* commitment to a philosophy of materialism. In fact, many scientists were religious men for whom Christianity meant more than a few appearances in church. In rejecting spiritualism and psychic research, they were turning their backs on speculation that was not conducive to further scientific discovery and that risked linking modern science with the medieval magic from which it had comparatively recently emerged. They were not, however, ruling out metaphysical questions as inappropriate to science. As Thomson explained[23] in 1907:

> There is indeed one part of Physical Science where the problems are very analogous to those dealt with by the metaphysician. . . . To some men this side of Physics is peculiarly attractive, they find in the physical universe with its myriad phenomena and apparent complexity a problem of inexhaustible and irresistible fascination. Their minds chafe under the diversity and complexity they see around them, and they are driven to seek a point of view from which phenomena as diverse as those of light, heat, electricity, and chemical action appear as different manifestations of a few general principles.

In the attempt to enunciate those principles and to find the hidden pattern, or unifying framework, of the universe, the physicists of the Society for Psychical Research shared with many of their critics a common goal.

* * *

I am grateful to the Society for Psychical Research for permission to quote from material in its archives.

References

1. Both letters are quoted in R. J. Strutt (fourth Baron Rayleigh), *Life of John William Strutt, Third Baron Rayleigh, O.M., F.R.S.* [1924], Univ. Wisconsin Press, Madison (1968), pp. 67, 361.

2. J. J. Thomson, *Recollections and Reflections*, G. Bell, London (1936), p. 158.

3. Barrett Papers, box 2, A2, no. 19, 26 December 1881, Archives of the Society for Psychical Research, London.

4. B. Stewart, Nature **4**, 237 (27 July 1871).

5. A. Schuster, *Biographical Fragments*, Macmillan, London (1932), p. 215; P. G. Tait, Proc. R. Soc. London **46**, xi (1889).

6. See the diary extracts and correspondence in E. E. Fournier D'Albe, *The Life of Sir William Crookes, O.M., F.R.S.*, T. Fisher Unwin, London (1923), pp. 141, 170, 198.

7. W. Crookes, Q. J. Sci., July 1871, reprinted in M. R. Barrington, K. M. Goldney, R. G. Medhurst, eds., *Crookes and the Spirit World*, Souvenir, London (1972), p. 22.

8. W. Crookes, Proc. Soc. Psychical Res. **6**, 114 (1889).

9. See Crookes's three lengthy letters to the Spiritualist Newspaper in 1874, reprinted in M. R. Barrington, K. M. Goldney, R. G. Medhurst, eds., *Crookes and the Spirit World*, Souvenir, London (1972), p. 130.

10. See T. H. Hall, *The Spiritualists: The Story of Florence Cook and William Crookes*, Helix, New York (1963).

11. C. Brookes-Smith, J. Soc. Psychical Res. **43**, 26 (March 1965).

12. Lodge Collection, box 2, no. 357, 5 July 1909, Archives of the Society for Psychical Research, London.

13. Three letters between Crookes and Lodge, in Lodge Collection, box 2, nos. 363–5, December 1916, Archives of the Society for Psychical Research, London.

14. W. B. Carpenter, Nineteenth Century **1**, 256 (April 1877).

15. Quoted in E. E. Fournier D'Albe, *The Life of Sir William Crookes, O.M., F.R.S.*, T. Fisher Unwin, London (1923), p. 284. Also see R. K. DeKosky, Isis **67**, 36 (March 1976).

16. Quoted in E. E. Fournier D'Albe, *The Life of Sir William Crookes, O.M., F.R.S.*, T. Fisher Unwin, London (1923), p. 290.

17. O. J. Lodge, *Raymond, or Life and Death*, Methuen, London (1916).

18. O. J. Lodge, Proc. Soc. Psychical Res. **2**, 190 (1884).

19. F. W. H. Myers, O. J. Lodge, W. Leaf, W. James, Proc. Soc. Psychical Res. **6**, 446, 648 (1889–90).

20. I. Tuckett, Bedrock **1**, 204 (July 1912).

21. O. Lodge, *Continuity. The Presidential Address to the British Association for 1913*, Putnam, New York (1914), pp. 21, 29.

22. O. Lodge, *Reason and Belief*, Methuen, London (1910), p. 47.

23. J. J. Thomson, *On the Light Thrown by Recent Investigations on Electricity on the Relation Between Matter and Ether. The Adamson Lecture Delivered at the University on November 4, 1907*, Manchester U.P., Manchester (1908), p. 5. □

SECOND OPINIONS ON 'PATHOLOGICAL SCIENCE'

PHYSICS TODAY/MARCH 1990

If Irving Langmuir's posthumous article (page 35, this volume) is interpreted as providing some characteristics of those experiments that *may* be "pathological" and therefore should be scrutinized with more than the usual care, it will give the scientific community a valuable tool.

The language and tone of the article, unfortunately, encourage quite another interpretation of those characteristics: one in which the characteristics are in and of themselves sufficient to brand an experiment as "pathological," thereby sidestepping any need for "bothersome" critical analysis of the methods and claims of the experiment. Since the characteristics constitute acceptable practice, trivial, incidental errors or features wholly external to the experiment, this interpretation does science a considerable disservice. Already I have seen the label "pathological science" applied to experiments without any attempt at real analysis.

The underlying pathology that Langmuir identifies is subjectively biased measurement or data selection. His second characteristic—that the effect is close to the perceptual threshold, or a small per-trial effect size is made statistically significant by combining many observations—boils down to there being an opportunity for this underlying pathology to enter. But the existence of the opportunity does not establish the existence of the fault. The necessary precautions (blind judging or including unambiguous criteria in the initial design) are now routinely taught as part of undergraduate laboratory practice and are generally applied. Indeed, if this characteristic constitutes a fault, then a very large proportion of all contemporary experimental science would have to be discarded.

The fourth characteristic—"Fantastic theories contrary to experience are suggested"—could be rephrased in a more positive way as "Exciting new theories invalidating previous thinking are suggested." Unless one can point to a logical flaw or to an actual experimental result contradicted by the new theory, one has no objective basis for deciding whether a theory is pathological or "merely" revolutionary. In any case, truly revolutionary experiments are frequently accompanied by flawed attempts to explain them. It is ironic that Langmuir's primary example of a theory that isn't "very sensible"—"the whole idea of an electron being captured by an alpha particle when the alpha particles aren't there, just because the waves are there"—appears to be a simple statement of quantum tunneling as expressed in the language of the then current de Broglie pilot-wave formalism.

In my experience, new ideas, especially seemingly revolutionary new ideas, are generally only universally accepted if supported by overwhelming experimental evidence. If the experimental evidence is merely "exceptional" it is met with a mixed reception. Unless the experimental technology is quickly developed to produce overwhelming evidence, enthusiasm wanes. This loss of enthusiasm occurs even with the production of independent good evidence. Langmuir's sixth characteristic—"The ratio of supporters to critics rises up to somewhere near 50% and then falls gradually to oblivion"—may therefore simply be rephrased "The experiment is controversial"—hardly damning to any but the most conservative.

Langmuir's third characteristic is that "there are claims of great accuracy." There are many reasons why such claims or apparent claims might be made. The most obvious is that the claims are justified. Another is that the authors might be presenting their raw measurements without intending to imply that all the digits are significant. Still another reason is for rhetorical effect (as in, for example, Langmuir's unrealistic claims of great accuracy implicit in his verbatim quotes of conversations occurring decades previously). Unless it can be demonstrated that a claim of great accuracy is unjustified, no flaw has been demonstrated. Un-

less it can be shown that the inflated accuracy is crucial to supporting the theoretical claims, the flaw is minor. In fact this flaw is commonplace and generally is simply ignored in uncontroversial work.

The fifth characteristic—"Criticisms are met by *ad hoc* excuses thought up on the spur of the moment"—is of no diagnostic use, since it is simply characteristic of any scientist defending a pet theory, conventional or unconventional. Langmuir gives instantly apparent, plausible explanations for experimental results seemingly in contradiction to established theory. Those he criticizes give "*ad hoc* excuses thought up on the spur of the moment." Both are, in fact, doing exactly the same thing: demonstrating that a plausible explanation for the apparent anomaly does exist, so that it does not constitute a falsification of the defended theory. Neither, in all likelihood, feels any great commitment to the proposed explanation's being the correct one. It is sufficient merely to demonstrate that *some* plausible explanation exists. It is then up to the critic (either of conventional or of new theory) to show that, in fact, no plausible and adequate explanation exists. This is the essence of orderly scientific dispute.

The first characteristic—"The magnitude of the effect is substantially independent of the intensity of the cause," and the minimum causal intensity is at the threshold of detectability—is the only one with real, independent diagnostic use. One has good reason to be suspicious of any experiment that shows this characteristic. But reason for suspicion is not equivalent to reason for rejection. We certainly would not wish to categorize Galileo's experiments with inclined planes, George Biddell Airy's experiments with a water-filled telescope, Albert A. Michelson and Edward W. Morley's interferometer experiments or Heike Kamerlingh Onnes's discovery of superconductivity as pathological.

I do rather strongly suspect that

Langmuir's characteristics can be useful in alerting us to the need for deeper analysis of an experiment. While, as I attempted to show above, each of his characteristics is individually innocent, their appearance together in a single experiment should raise serious doubts. They must be applied strictly, carefully and as objectively as possible, however, and they should not be taken as a basis for rejecting an experiment. The need for rigor here is acute: I was unable to think of a single controversial experiment, however eventually resolved, to which its critics could not have applied the label "pathological science," taking no more liberties with these characteristics than Langmuir did in some of his examples. If we are not careful, his message simply becomes "Reject all controversial claims."

CHRISTOPHER COOPER
1/90   *Arlington, Massachusetts*

I was delighted to see that Irving Langmuir's informative and enjoyable lecture on "pathological science" has finally been made available to the general scientific community. Teachers at universities tend to spend all of their time informing their students of the success stories of science. However, if students are to develop into creative and discriminating scientists, they would do well to have some passing familiarity also with some of the failures. It may help them sort out which current new ideas are going to last and which are going to disappear.

The last section of Langmuir's talk, dealing with "flying saucers" (now referred to as UFOs), needs some clarification.[1] The Air Force did not close down and declassify its entire research project with the ending of Project Sign in 1949. Sign was replaced by Project Grudge, which in 1952 was in turn replaced by Project Blue Book. All three projects were classified, but not at a high level. The Air Force finally closed Project Blue Book in 1969, on the recommendation of Edward Condon, who directed a study on behalf of the Air Force at the University of Colorado from 1966 to 1968. (It is unfortunate that the Condon report,[2] the basis of the recommendation, is marred and weakened by serious inconsistencies.[3])

The article in *The Saturday Evening Post* that Langmuir referred to was probably that of Sidney Shallett.[4] The general mentioned by Langmuir, with a name that sounds like "Schwartz," was probably General Carl Spaatz, Air Force Chief of Staff from 1947 to 1948.

A more general commment is the following: Very few young scientists have heard of the Davis and Barnes experiment, N rays, mitogenetic rays or the Allison effect. By contrast, all young scientists have heard of extrasensory perception (a part of "parapsychology") and UFOs. It is worth inquiring whether this difference is a reflection of a basic difference between the two groups of topics.

When we look at Langmuir's list of "symptoms of pathological science," we find that the parapsychology and UFO problems do not now present all those symptoms, nor did they present them at the time Langmuir gave his talk. For instance, in neither case have there ever been "claims of great accuracy," and in neither case has the ratio of supporters to critics (among the scientific community) risen to anything approaching 50%.

I suggest that the parapsychology and UFO problems are, in fact, *not* examples of pathological science. I have suggested elsewhere[5] that they may better be understood as examples of "heretical science." I have in mind the interpretation that in addition to being based only on shaky and shifting evidence, a topic of pathological science arises from *within* a scientific discipline (presenting mainly an intellectual challenge, but also to some extent an internal political challenge), whereas, by contrast, a topic of heretical science arises or has strong support *outside* the scientific community (presenting the scientific community with both an intellectual challenge and an external political challenge).

It seems likely that problems of pathological science are resolved more rapidly than those of heretical science partly because they are more sharply defined and partly because it is possible (though not easy) to cover them in scientific meetings and publications. Problems of heretical science would no doubt be resolved more rapidly if they were discussed freely, fully and objectively in scientific journals. Most scientific journals are not open to such discussion; an exception is the *Journal of Scientific Exploration*, published jointly by the Society for Scientific Exploration and Pergamon Press.

References

1. See, for instance, D. M. Jacobs, *The UFO Controversy in America*, Indiana U. P., Bloomington (1975).
2. E. U. Condon, D. S. Gillmor, *Scientific Study of Unidentified Flying Objects*, Bantam, New York (1969).
3. P. A. Sturrock, J. Sci. Exploration 1, 75 (1987).
4. S. Shallett, Saturday Evening Post, 30 April 1949, p. 20; 7 May 1949, p. 36.
5. P. A. Sturrock, New Scientist, 24/31 December 1988, p. 1644.

PETER A. STURROCK
Stanford University
11/89   *Stanford, California*

Irving Langmuir's lecture on pathological science, transcribed and edited by Robert N. Hall, contains numerous errors in its section on extrasensory perception that need to be corrected. Much of what Langmuir says about his meeting with J. B. Rhine, who founded the institution of which I am director, is misleading and even false.

First, as to the statements of fact, Langmuir says that he met Rhine "probably about 1934," that Rhine "published a new volume of his book" that contained "a chapter on the sealed-up cards in the envelopes," that the ESP scores obtained using these cards "all come up to around [an average of] 7" per 25 trials and that "nothing is said about the fact that for a long time they came down below 5."

It is fortunate that the records of Rhine's experiments and his extensive correspondence are kept in the archives at Duke University Library, so that we can check on the accuracy of some of the statements made by Langmuir.

▷ Langmuir did not recall the date of his meeting with Rhine correctly. The meeting was not in 1934, but rather in 1937, when Langmuir was in the Durham area in connection with a meeting of the American Chemical Society.

▷ Rhine published *New Frontiers of the Mind* in 1937. It contains no chapter on sealed-up cards. In fact, there is no such *chapter* in any of Rhine's books.

▷ It is false to state that Rhine ever wrote in any of his books that the mean run score obtained with sealed-up ESP cards was around 7.

▷ Contrary to what Langmuir says, Rhine and his associates reported in table 7 of their 1940 publication *Extra-sensory Perception After Sixty Years* six series of experiments in which subjects were tested with ESP cards enclosed in sealed opaque envelopes. One of these series, conducted by Rhine, gave an average of 4.89. This was the series Langmuir was referring to. Langmuir was therefore clearly wrong in implying that Rhine did not report his negative results. The average run score of the six series together is 5.21, which is significantly different from the mean chance expectation of 5.00 ($z = 7.74$).

The implication of Langmuir's statements on parapsychological research is unambiguous. The reports of successful ESP experiments, it is suggested, are due to selection of data. He implies that parapsychologists report only significant results and find an excuse to discard and reject results contrary to their expectations. Even the most vociferous critics of parapsychology today do not subscribe to this view because there are numerous published studies that simply cannot be explained away by such an assumption.

Langmuir refers to the declining scores in the ESP experiments of his nephew David Langmuir. If his account of these experiments is accurate, they seem to fit very well with the general pattern of results in this field. Such declines are quite common and are to be expected unless the experimenter is careful to introduce novelty and other incentives to keep up proper motivation of the subjects. In fact, even Rhine's star subjects eventually failed to keep up their levels of high scoring. David Langmuir wrote to Rhine on 14 April 1937 acknowledging the latter's sending him test materials, including cards, and promising to keep Rhine informed of the test results. If David Langmuir had not been scared away from continuing his contact with Rhine by such skeptics as Irving Langmuir, perhaps the outcome of those experiments would have been different.

It would seem that Irving Langmuir was a man with a mind of his own, a mind that was tricked by his own preconceptions. For a man of his scientific stature and intellectual capabilities, Langmuir did surprisingly little to ensure that his statements about ESP research were backed up by facts, and he was unwilling to face up to those facts that he did find. For example, in May 1938 he gave a lecture at Union College in which, among other things, he criticized Rhine's research. In the *Schenectady Gazette* dated 5 May 1938, Langmuir is quoted as saying, "Although I cannot as yet find anything wrong with the experiments of Professor Rhine, after having spent two days in studying the methods and experiments, I can't help feeling that it has all the indications of being wrong scientifically."

On learning about the Union College talk, Rhine wrote to Langmuir for clarification, but as far as we know he never heard from Langmuir. It is clear that Langmuir's critique of Rhine's work was based not on anything he found wrong with it, but on his perception that it fit his "symptoms of sick science," which themselves are quite arguable, to say the least. I do not believe any serious

student of philosophy of science today would agree with his criteria for pathological science. Clearly, his diagnosis relative to ESP research was wrong.

K. RAMAKRISHNA RAO
*Foundation for Research
on the Nature of Man*
11/89   *Durham, North Carolina*

The Langmuir lecture on "pathological science," originally given to a small group at General Electric, has now reached a much larger audience through its publication in PHYSICS TODAY. I first came across a copy of this lecture in 1972 and was especially struck by the vigor with which Langmuir attacked J. B. Rhine, on the basis of an interview that had taken place almost 20 years before the lecture was given. Since Rhine was still alive and active I decided to ask his opinion of Langmuir's accusations, guessing (correctly) that neither Langmuir nor anyone else had thought to send him a copy. I received a prompt reply. Since this letter was definitely not intended for a wide audience I will limit myself to two short quotations from it: "[Langmuir] has not hesitated to make up things to suit his pattern," and "He quoted me as saying things I could easily convince a fair-minded, honest third person I would never have said." Langmuir evidently felt that he could dispose of Rhine (and, presumably, the whole field of parapsychology) on the basis of his interpretation of this one interview; Rhine, however, was prepared to defend himself if necessary, and confident that he could do so effectively.

In view of this profound disagreement between the two men I would urge readers not to base their own assessment of parapsychology on Langmuir's account but to look for more information. It is certainly not true, for example, that parapsychology has come to an end, as Langmuir's portrayal of it would have led us to expect. But the opposition between the minority of scientists who support parapsychological findings and the much more powerful group that actively opposes them is as strong as it was in Langmuir's day, and apparently no closer to being resolved. Just as Langmuir's lecture is incomplete without Rhine's reply, so the recent report by the National Research Council[1] should only be read in conjunction with the reply from the Parapsychological Association.[2]

References

1. D. Druckman, J. A. Swets, eds., *Enhancing Human Performance*, Natl. Acad. P.,

Washington (1988).
2. J. A. Palmer, C. Honorton, J. Utts, *Reply to the National Research Council Study on Parapsychology*, Parapsychological Association, Research Triangle Park, N. C. (1988).

PETER R. PHILLIPS
Washington University
11/89   *St. Louis, Missouri*

Robert Hall's transcription of Irving Langmuir's talk on pathological science was entertaining; however, there is now reason to doubt Langmuir's belief that selective reporting can sufficiently explain the statistical significance of (presumably) all types of parapsychological experiments.

For example, an article by Roger D. Nelson and myself addresses the problems of selective reporting and methodological quality in a large body of experiments studying the possibility of consciousness- or observer-related effects on physical systems.[1] Several other recent articles in the physics literature lend further support to the likelihood of such effects.[2]

One of Langmuir's symptoms of pathological science is that support for it "falls gradually to oblivion." Four years after Langmuir's talk, J. B. Rhine and his colleagues formed the Parapsychological Association. Twelve years later, the Parapsychological Association was elected an affiliate of the American Association for the Advancement of Science, and it remains an active scientific society today. Within the past two years, the National Research Council and the Office of Technology Assessment have both examined the parapsychological literature in some detail. That doesn't sound like declining interest to me.

References

1. R. D. Nelson, D. Radin, Found. Phys. **19**, 1499 (1989).
2. See, for example, R. G. Jahn, B. J. Dunne, Found. Phys. **16**, 721 (1986); E. J. Squires, Eur. J. Phys. **8**, 171 (1987).

DEAN RADIN
Contel Technology Center
10/89   *Chantilly, Virginia*

Irving Langmuir's description of cases of "pathological science" is interesting as history not usually available to anyone not close to the events described. It is especially interesting because at about the same time that Langmuir gave his recorded talk he was the central figure in an illusory interpretation of the results of a series of cloud-seeding experiments. As part of Project Cirrus, conducted jointly by the Signal Corps, the Office of Naval Research and General Electric from 1947 to 1953, silver iodide was introduced into clouds in New Mexico at seven-day intervals. Statistical analysis of rainfall and temperature records covering much of the US convinced Langmuir that periodic seeding was capable of producing marked and very widespread effects on atmospheric circulation and rainfall.[1] He also was convinced that seeding with dry ice had resulted in a drastic shift in the path followed by an Atlantic hurricane. Langmuir's claims were met with disbelief or skepticism by nearly all meteorologists, and a panel appointed by the American Meteorological Society found that Langmuir's claims had not been demonstrated "if the available evidence is interpreted by any acceptable scientific standards."[2] An extended and illuminating account of these events has been published by Horace Byers.[3]

The Weather Bureau, in response to the extravagant claims Langmuir made for Project Cirrus, launched the Cloud Physics Project in 1947. The consistently unspectacular results stimulated challenges by Cirrus adherents. I recall an AMS meeting in the early 1950s at which Langmuir got into a shouting match with Ross Gunn, a Bureau of Standards physicist who represented the Cloud Physics Project.

There is no evidence that Langmuir ever was convinced by the criticism of his interpretation of the Cirrus results. His final paper on the subject was published as a GE Research Laboratory Report in 1955. Langmuir had made fundamental contributions to the understanding of cloud physics as well as to innumerable topics in basic chemistry and physics.

Yet in this case he was unable to see around his own idiosyncratic interpretation. There are lessons here for all of us.

References

1. I. Langmuir, Bull. Am. Meteorol. Soc. **31**, 386 (1950).
2. B. Haurwitz, E. Emmons, G. Wadsworth, H. C. Willett, Bull. Am. Meteorol. Soc. **31**, 346 (1950).
3. H. Byers, in *Weather and Climate Modification*, W. N. Hess, ed., Wiley, New York (1974), p. 3.

ROBERT G. FLEAGLE
University of Washington
10/89   Seattle, Washington

Near the close of Irving Langmuir's interesting account, reference is made to the planet Venus's being mistaken for "a comet about to collide with the Earth, or somebody from Mars."

Venus caused some concern at Los Alamos in late 1944 or early 1945. We had heard of the balloon bombs that the Japanese had launched, which were supposed to drift eastward across the Pacific and then land and do damage in the United States when they exploded. We learned later that one such bomb had landed in Oregon and that several people were killed when it exploded. The military people, who were responsible for safety and security, were concerned that such a bomb might arrive at Los Alamos.

One bright sunny day as I walked back from lunch to the Technical Area I noticed many people looking up at a bright point in the sky. There, drifting overhead, was one of the dreaded bombs. Some people on the street were using binoculars and thought they could see the payload attached to the spherical balloon. Rumors abounded about triangulation measurements that gave numbers for the elevation and direction of motion of the balloon. Interest remained high through much of the afternoon. But at sunset the object had drifted well to the west, and the next morning the object rose in the eastern sky. We were all amused that a planet had caused so much concern.

ALBERT A. BARTLETT
10/89   University of Colorado at Boulder

I took exception to the article "Pathological Science." While the article is informative and interesting, I could not help but feel from its timing and tone that it was a thinly veiled indictment of the researchers who have reported observing cold fusion. It is still premature to dismiss those findings. Perhaps when and if the cold fusion results are retracted, this article will provide an interesting perspective on the phenomenon. Until that time it should be noted that cold fusion is not unlike many other new phenomena that seem at first to be counterintuitive, controversial and irreproducible. One need only think back to the appearance of high-temperature superconductivity for one example. Perhaps PHYSICS TODAY might review the initial resistance to acceptance that other landmark discoveries have faced, rather than hopping on the bandwagon of those who find it rewarding to engage in a cold fusion witch-hunt.

MICHAEL WIXOM
11/89   Ann Arbor, Michigan

Science at General Electric

Research underwent profound changes in scale and style during the Second World War; that this transformation also stirred industry is illustrated by the changes at one large laboratory in the postwar era.

George Wise

PHYSICS TODAY/DECEMBER 1984

The decade after 1945 saw substantial changes at the General Electric Research Laboratory in Schenectady both in the way in which research was done and in the people doing the research. The passing of the generations at a research laboratory in a small city in Upstate New York reflected a broader change undergone after World War II by all of American science. Science and technology had helped win the war. Scientists and engineers now appeared to be the bulwarks of national defense and the preservers of national prosperity. In the heady postwar climate, no one worried much about where science ended and where technology began. There was enough for both. Inventions and innovations launched before or during the war—televisions, plastics, atomic energy, electronics, computers—now loomed as huge and expanding opportunities. A person who had embarked on a career in science now had a wide choice of well-paying jobs in government or industry.

To the young scientist in 1946, the prospect of employment at General Electric was indeed an attractive one. Some of the scientists at the laboratory, such as Saul Dushman, Albert Hull, William D. Coolidge and Irving Langmuir, were widely known by reputation. Dushman had written a widely used book about the theory and practice of vacuum techniques. Hull had invented several important electronic devices, and had recently completed a term as president of the American Physical Society. Coolidge had been GE's research director from 1932 until 1945, and had earlier improved filaments and x-ray tubes used around the world. Langmuir was probably the first Nobel laureate the job candidate had ever met.

If he got offered a job by GE, the young recruit would not be asked to work with these giants. He would be asked to help replace them. GE's long-running experiment in industrial research was changing its guard.

The candidate for employment at General Electric may have come east from Berkeley, Chicago or Los Alamos, north from Oak Ridge, or west from Cambridge to explore opportunities in industry. His trip to the banks of New York State's Mohawk River took him back in time as well: from the scientific revolution of the mid-20th century to a revolution in electrical technology that had begun more than half a century earlier. In 1886, Thomas Edison had moved his machinery works to Schenectady. In 1946, its row on row of red and brown brick buildings still suggested more of Edison than of Einstein. Factory workers poured castings, hand-crafted precision fittings, shaped turbine buckets, assembled motors, and

George Wise is a historian in the Communications Operation of the General Electric Research and Development Center at Schenectady, New York.

Volney C. Wilson developed high-power vacuum tubes for the study of the direct conversion of heat into electricity. (General Electric photo.)

Katharine B. Blodgett, the first woman PhD physicist in American industry, demonstrates experiments in surface chemistry at the opening of General Electric's new Research Laboratory building at the Knolls site in 1950.

GENERAL ELECTRIC

turned out the heavy apparatus that was the major product of the world's largest electrical company. Near the main gate of the works stood buildings 5 and 37, the home of the GE Research Laboratory.

The laboratory had been born half a mile and half a century away from its 1946 site. On the banks of the Erie Canal in December 1900, GE's chief consulting engineer, Charles Proteus Steinmetz, had led out of his boarding house and into a barn in the backyard of his house an instructor in chemistry from MIT named Willis R. Whitney. The barn became GE's first laboratory, and Steinmetz installed Whitney there as GE's first director of research. Physics soon came to play a major role alongside chemistry. The laboratory's most important achievement in the prewar era was the invention by Coolidge, a physicist, of a process for turning the brittle metal tungsten into flexible wire suitable for use as light bulb filament. Coolidge succeeded Whitney in 1932 as the laboratory's director. By 1945, the laboratory had 630 employees, of whom 160 were trained scientists or engineers. About half of them worked on physics and its applications, about 30% on chemistry, 15% on metallurgy and 5% on mechanical investigations.[1]

The laboratory marked a milestone for the world of science and technology, as well as for GE. "The interdisciplinary research laboratory has been the most significant institution for modern technological development," historian Edwin Layton has stated.[2] He added: 'The rise of interdisciplinary research is sometimes associated with the founding of the General Electric Research Laboratory."

Another historian, Kendall Birr, concluded[3] in a history of the laboratory's first half century, "One characteristic of the laboratory stands out above all others: its success." He attributed that success to four causes. First, from the beginning GE's top officers granted the lab substantial independence and solid financial backing. Second, the lab recruited outstanding researchers. Third, it did not insist on exclusive dedication to research but carried out development work, troubleshooting, and testing, and put results before research. And, fourth, the lab had the good fortune to be associated with one of the financially strongest and most technically diverse companies in the world.

In describing the conditions that had produced these successes, Birr also noted that those conditions were changing. When the Research Labora-

tory had been created in 1900, General Electric had recently been centralized into a single company ruled from corporate headquarters. In the postwar period, GE was decentralizing into a collection of largely independent operating units. The company had concerned itself exclusively with electricity in 1900. In the 1940s, it was diversifying. Many businesses of the old GE, such as incandescent lamps, x-ray tubes and electronics, rested on strong patents. In the postwar period, some new technologies, such as plastics, would lend themselves to that kind of protection. But other big-growth areas—solid state devices, computers, jet engines, atomic power—would be much less influenced by patents. And, finally, in 1900 GE had been a pioneer in industrial research. By 1950, many rivals in government and universities as well as industry competed for bright young scientists.

The choice before GE was either to stand pat on the tradition that had made it a successful model of industrial research or to change with the times. General Electric chose to change.

A new guard

By 1946, Coolidge had joined Whitney in retirement; and Langmuir, Dushman and Hull were soon to follow.

Francis P. Bundy, one of the inventors of a method for making diamonds, examines the die that forms the heart of the high-pressure and high-temperature system used for this purpose. The photo was taken in 1971.

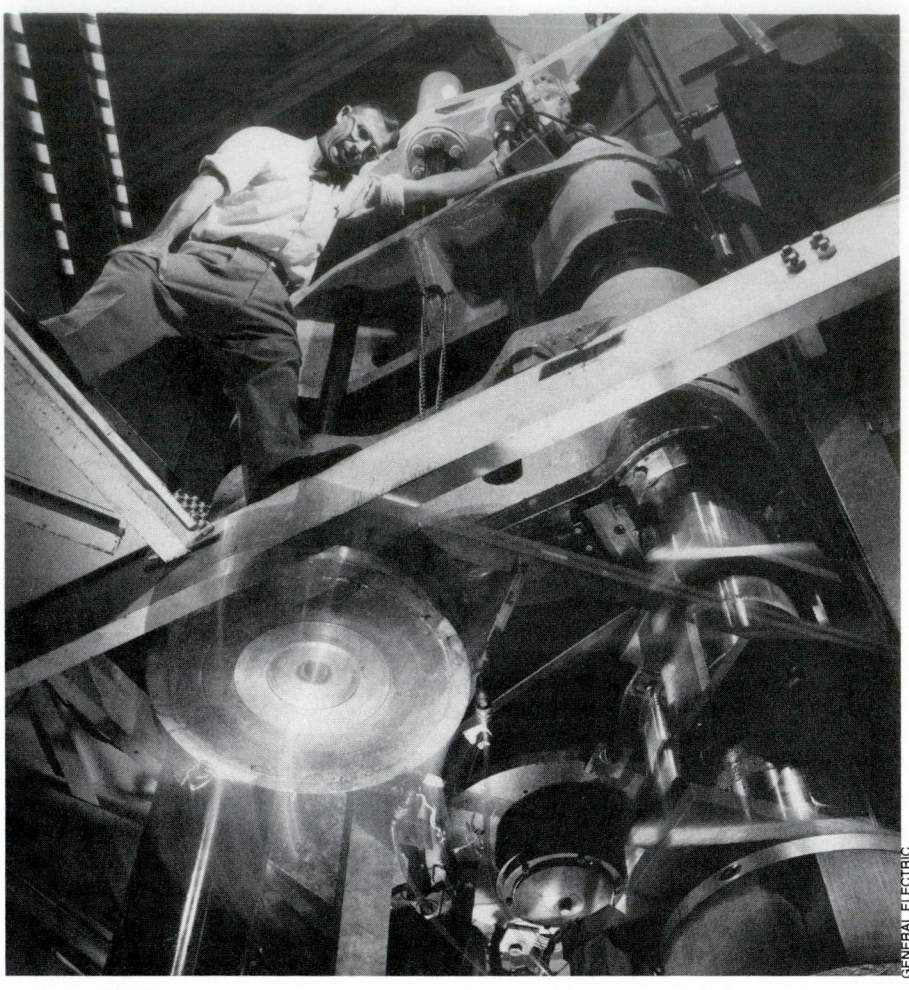

GENERAL ELECTRIC

To replace them, a generation hired in the 1930s should have been coming of age. But few researchers had been hired during the depression decade. GE had to fill the gap left by this lost generation with a new guard. The job fell to one of the last scientists hired in the prosperous 1920s, the man who had become the laboratory's director in 1945, C. Guy Suits.

Suits was a short, trim, balding physicist with a straightforward manner; a pharmacist's son from Wisconsin whose interests leaned more to the shop and the outdoors than to mathematics or philosophy. As an undergraduate student in physics at Wisconsin University, and later, as a PhD candidate at the Swiss Federal Technical Institute, Suits had never seen an industrial scientist and had never talked to a professor who recommended an industrial career. But he had read of the work of Langmuir, Coolidge and Hull. He had little interest in teaching, and his versatility would have been difficult to confine within the classroom. He had played the clarinet professionally to earn his way through college and would later prove just as proficient at turning out furniture in his home workshop, making dresses for his wife, piloting a plane, designing boomerangs, and stalking game in the wilds of Alaska.[4]

He joined the Research Lab in 1929 and immediately began earning a solid reputation for his work in physics and electronics. He conceived an original and effective method for studying the characteristics of electric arcs and designed nonlinear circuit elements, whose uses included the switches for the blinking gigantic GE monogram that identified the Schenectady Works. His superiors, Hull and Coolidge, were impressed with his organization, self-direction and independence. Although he had never managed more than a handful of people, he was chosen in 1940 to become the lab's assistant director and Coolidge's heir apparent.

Before he could move up further, the war broke out and so offered him a new management opportunity. As director of the National Defense Research Council's Division 15, he led some 1000 scientists and engineers in developing countermeasures to enemy radar.[5] The success of the effort catapulted him into the position of a promising junior member of the nation's science establishment.

He also met some of the nation's outstanding young scientists and kept their names in mind. After the war he could supplement his memory with a document compiled and distributed by the Office of Naval Research, listing 2000 young scientists in the war effort. Late in 1945 he sent Research Laboratory executive engineer Dudley Chambers and physicists Albert Hull and Kenneth Kingdon out to recruit—to Los Alamos, to the Metallurgical Laboratory at the University of Chicago, to the Radiation Laboratory at MIT, and to the Underwater Sound Laboratory at Harvard. By March 1946 he could announce that 60 scientists had joined the Research Lab's staff since the end of the war, including 20 from the Manhattan District atom bomb project.

GE's research was on the verge of a major change in both scale and nature. Ground had been broken for the construction of a new laboratory—not a squat, brown-brick office building in the midst of an industrial plant like the current home, but a structure specially designed for laboratory use, and placed on the landscaped grounds of what had formerly been the estate of a patent-medicine king.

Already, at the existing lab, the nature of the work was changing. Small-scale work by small teams was giving way to large-scale projects. And many of these projects ranged much further beyond GE's immediate needs than would have seemed possible or desirable before the war.

Combustion and gas dynamics laboratory in the late 1950s. In this facility research was conducted on combustion processes and phenomena occurring at flight speeds of Mach 5.

For example, in the years before the war, Irving Langmuir and his colleagues Vincent Schaefer (whom Langmuir had brought from the machine shop to the laboratory and who had developed into a scientist) and Katharine Blodgett (the first woman PhD to work in an industrial research laboratory) did research using simple tools in the classic tradition of string and sealing wax. Some of their major discoveries grew out of industrial needs. One of their effects led to the discovery of the monomolecular liquid–on–liquid layers, now known as Langmuir—Blodgett films. This discovery had been partly inspired by a request for a better lubricant for meter bearings.[6]

In the fall of 1946, Schaefer followed up on some of his wartime work on precipitation static in airplanes and discovered how to turn simulated supercooled clouds (actually water vapor in a food freezer) into a simulated snowstorm. A few weeks later, he repeated the same experiment in the actual atmosphere, dropping dry ice from an airplane to create the first manmade snowstorm. Within two more years, Langmuir and Schaefer became the leaders of a government-sponsored program on weather-modification experiments, the Project Cirrus. Cirrus involved dozens of people, years of field tests, nearly 300 aircraft flights over cloudy sites from the Adirondacks to Florida, and from New Mexico to Puerto Rico.[7] The little science of Langmuir's prewar laboratory had become big science, indeed, after the war.

Similar changes in scale occurred in other fields. In solid-state physics, GE had made only the smallest and most tentative explorations before the war. Wartime research brought GE researchers into the field permanently. Postwar advances, especially the inven-

tion of the transistor at Bell Laboratories, led GE to launch major efforts in semiconductor science and technology. Before the war, the Research Laboratory's efforts in high-energy physics had been concentrated almost entirely on x-ray apparatus, an important GE business. But by 1945 the lab possessed one of the world's highest-energy electron accelerators, a 100-meV betatron, and was making it available to scientists studying elementary particles. Then, using a 70-meV synchrotron completed at the lab in 1947, GE researchers Herbert Pollock, Robert Langmuir and Frank Elder made the first observation of synchrotron radiation.[8] But no field of GE's research showed more clearly the change from prewar to postwar attitudes than nuclear physics.

The peaceful atom

In 1939, Irving Langmuir had come back excited from a meeting of The American Physical Society. Speakers there had told of recent experiments in Germany that had breathed life into the 40-year-old dream of extracting power from the atom. Uranium atoms, it appeared, could split into fragments, releasing enormous amounts of energy. Langmuir's enthusiasm convinced lab director Coolidge to assign two scientists to the problem. Early in 1940, Kingdon and Pollock extracted from a sample of uranium chloride a few micrograms each of uranium's two isotopes, U^{235} and U^{238}. Scientists at Columbia University waited eagerly to test whether the rarer of the isotopes, U^{235}, would split when bombarded by neutrons, giving off both energy and more neutrons, which, in turn, would split other U^{235} nuclei.

But the urgency of the question was not enough to override totally the spirit of economy that still gripped the lab in

the aftermath of the great depression. "Get the excursion fare to New York," Kingdon instructed. So Pollock rode in the low-cost early-morning coach, carrying with him a substantial fraction of the world's supply of separated uranium isotopes.[9]

In Kingdon and Pollock's prewar laboratory, the sensitive equipment for the separation of uranium isotopes had to be shut down by day for fear that transient voltages from nearby trolley-car lines would upset it. By 1946, the nation's atomic energy effort commanded areas of specially designed equipment, running day and night behind a wall of secrecy and high priority, with the energy of the Tennessee Valley and the Columbia River and the skills of some of America's strongest technology-based corporations at its command.

By the end of the war, General Electric had established a Nuclear Investigations Group, consisting of a dozen scientists and engineers at its Research Laboratory. By April 1946, the company had agreed to take over responsibility for the nation's plutonium-producing plant at Hanford, Washington. Within a few more months, the government had agreed to fund a $36-million laboratory located near Schenectady and run by GE, dedicated to the peaceful use of nuclear power. Its staff, which reached 650 in 1947 and 1000 in 1950, quickly filled the available space in the Research Laboratory's downtown headquarters, and expanded into an abandoned tank plant in Schenectady and a demobilized radar test station in a rural area north of the city.[10]

It included veterans of the GE Research Laboratory and new faces, too. Theorist George Placzek arrived to head the theory group and was known

Betatron. Ernest Charlton (left) and Willem Westendorp in 1950 with the 100-MeV betatron they helped design in the early 1940s.

GENERAL ELECTRIC

to work out equations with lipstick on the mirror of his new Schenectady home. Henry Hurwitz, one of the most promising young Los Alamos theorists, joined him, as did Harvey Brooks, already recognized[11] by his colleagues as "a brilliant mathematician and a very dynamic fellow." Later, Brooks became one of America's leading science policymakers. The new staff underwent a crash course in nuclear power taught by some of the world's leading experts. Hans Bethe of Cornell, Ernest Lawrence of Berkeley and Eugene Wigner of Princeton—all of whom were already, or were soon to become, Nobel laureates—came to Schenectady as consultants.

Not all the hopes for the laboratory were realized. It had been created to develop a civilian nuclear reactor for power production. In the cold war climate of the late 1940s, and at the urging of Hyman Rickover of the Navy, it was redirected into the national effort to develop nuclear propulsion systems for submarines. Its researchers made notable innovations in nuclear power, proposing the use of uranium oxide fuel and zircalloy cladding in nuclear fuel elements. They designed the sodium-cooled nuclear reactor for the second US nuclear submarine, the Seawolf, and in this the laboratory served as a graduate school in nuclear engineering for a Navy lieutenant named Jimmy Carter. Soon the laboratory near Schenectady grew into one of the nation's two laboratories dedicated to the development of nuclear reactors for the propulsion of naval vessels.

It may not have become what General Electric had initially intended it to be: an arm of its Research Laboratory aimed at fundamental research that would give the company a grip on an emerging commerical use of atomic power. But the experience had not caused GE's management to become disillusioned with the shift its Research Laboratory was making into larger-scale, more exploratory research efforts. GE would proceed to undertake basic research programs in a remarkably wide range of areas—experimental high-energy physics, thermonuclear fusion, superconductivity, polymers, solid-state physics, information theory, fundamentals of metallurgy and ceramics, and more. In the postwar period, it would stand alongside Bell Laboratories as one of the world's preeminent industrial performers of fundamental research.

Funding research

GE's redirected laboratory would pay for its work in a new way. Before the war, the lab had raised about three-fourths of its annual budget from the company's operations, with the rest coming from corporate headquarters. To raise the money, the laboratory's principal goodwill ambassador, executive engineer Laurence Hawkins, would visit the company's different operations. Trained as an engineer and a patent attorney, Hawkins hid behind the cowlick and guileless face of an aging Huck Finn, a salty vocabulary, a mastery of bargaining and an unmatched familiarity with General Electric.[12]

Hawkins would annually ask each GE department to allocate a certain amount of money to the Research Lab. Some departments, such as Lighting or X-ray, had gained so much benefit from the lab's past work that they usually agreed to the request with no strings attached. Others agreed only to fund specific projects. Still others refused to pay anything. However, if the lab had not obtained research money up front, and if it then proceeded to invent something a department could use, it had the right to sell that invention to the department. The selling price would be set to recover the costs that had gone into the research and development. But checking up on those costs was impossible for the outsider. So the lab often succumbed to temptation and padded the cost with what it called "dead horses"—the carcasses of past projects that had not yet been written off.

This practice created an atmosphere of suspicion between the lab and some of the other operations, and could seriously delay the innovation process. The case of the mercury switch illustrates the problem.[13]

In the 1930s, an ingenious engineer at the Reseach Lab, John Payne, had conceived of a way to make a silent electric wall switch for homes by using as the electric contact a tilting vial of mercury. The help he got from his scientist colleagues—for example, from Suits he learned the characteristics of the electric arc; and from ceramist Louis Navias he obtained a ceramic container—enabled him to reduce it to practice. Payne even designed a machine to make the switches. But when Hawkins presented the package and its accompanying bill to the manager of GE's Small Apparatus Department, that manager thought the price was too high and turned it down. Then he called in one of his own engineers and asked him to reinvent the product and process. That engineer failed. A couple of years later, Small Apparatus had to admit defeat, call Hawkins back in, and sign on the dotted line. In other words, GE had lost valuable lead time in the marketplace and years of engineering effort due to a family squabble about internal bookkeeping. Lab direc-

Some discoveries and inventions at GE

1882   General Electric Company is formed by combination of Edison General Electric and Thomson-Houston Company.

1900   GE Research Laboratory is established under Willis R. Whitney.

1908   William D. Coolidge renders tungsten ductile, thus paving the way for modern electric illumination.

       Coolidge demonstrates first practical and safe x-ray tube.

1915   Saul Dushman develops first high-voltage vacuum rectifier.

1917   Albert W. Hull invents the magnetron (later used for radar transmission in World War II).

       Development of portable x-ray tube made possible by self-rectifying Coolidge tube.

1928   First public demonstration of radio transmission of photographs.

1941   Research Laboratory in cooperation with University of Illinois builds a 20-MeV betatron electron accelerator and begins work on a 100-MeV betatron.

1946   Vincent Schaefer performs the first successful cloud-seeding experiment, resulting in the first manmade precipitation.

1947   Herbert C. Pollock, Frank R. Elder and Robert V. Langmuir discover synchrotron radiation.

1950   Robert N. Hall invents the "p–i–n" structure, a method of alloying the metal indium with the semiconductor germanium to make a p–n junction, the basic element in power rectifiers and some transistors.

1952   Robert N. Hall proposes the theory of electron–hole recombination in semiconductors, a theory later elaborated by Shockley and Read of Bell Laboratories and since known as the Hall–Shockley–Read theory.

1954   Francis P. Bundy, H. Tracy Hall, Herbert M. Strong and Robert H. Wentorf Jr., invent the first reproducible process for making diamond.

1957   Robert H. Wentorf Jr invents borazon, cubic boron nitride, a manmade material second only to diamond in hardness.

1957–62   William C. Dash and Arthur W. Tweet invent a method for growing dislocation-free crystals of silicon.

1960   Ivar Giaever discovers superconductive tunneling, both a scientific phenomenon of major importance and the basis for the invention of new types of high-speed electronic switches. For this discovery he shares the 1973 Nobel prize for physics.

1962   Robert N. Hall and colleagues invent the semiconductor laser, which produces coherent light by the combination of electrons and holes at the junction of p-type and n-type semiconductor material.

tor Suits believed that this incident—on top of many similar but less pronounced ones—should signal the end of the old funding system.

In 1946, Suits and GE president Charles E. Wilson devised a new funding method. Business operations would no longer have the choice of whether or not to pay for the Research Laboratory's work. Instead, the corporate office would levy an assessment on each GE business, a sort of technology tax consisting of a fraction of a percent of sales. The exact fraction was initially set at a uniform level. But within a few years, the businesses less dependent on technology complained and subsequently convinced the corporate office to assign a different percentage to each business, based on a formula that took technology intensiveness into account. These "assessed funds" came to cover about two-thirds of the lab's expenditures. The rest came from additional contracts with the businesses or with external agencies, mainly in the Federal government.

The businesses now paid for their research in advance, so they had added incentive to see how much they could get out of the lab. And the laboratory had increased its autonomy in the choice of how to put this large amount of no-strings-attached corporate funds to work.

Trying industrial research

New fields of research and new sources of money would have meant little if GE could not have attracted

Israel Jacobs and Charles P. Bean
studying the magnetization of small iron particles at low temperatures in studies of superferromagnetism in the 1950s.

GENERAL ELECTRIC

good scientists to do the work. Why would a postwar job candidate have given industrial research a try? A look at some people who did make that choice suggests some possible answers. Consider, for example, three physicists who had been graduate students together at Ohio State back in 1935. For Francis Bundy, Volney Wilson and Herbert Strong, studying physics had offered a temporary haven from the Depression. Each of them had come to science through boyhood hobbies and curiosity rather than through a clear sense of career. But as they acquired degrees and raised families, careers became necessary.

"I'll teach for four or five years," Francis Bundy remembers[14] thinking when he got his doctorate. He was off to Athens, Ohio, for an instructor's job in the physics department of Ohio University on a salary of $1800 a year. Teaching shunted research into his spare time, and meager equipment kept his personal studies simple. He analyzed the way broken-off chimneys fall and spent summers studying cylinder wear in diesel engines.

Then came the war and an "out-of-the-blue" invitation from Frederick V. Hunt of Harvard for Bundy to come to Cambridge and join the Underwater Sound Laboratory. This implied an education in electronics and acoustics, contact with industrial firms such as General Electric and, above all, participation in the "biggest and best scientific melting pot that ever occurred." The war ended and, in 1946, the recruiters

descended on Harvard, Dudley Chambers of GE among them. The trip to Schenectady that followed included a drive out to the new laboratory site on the Mohawk and lunch with Langmuir. ("He had been my guiding star since high school," Bundy recalls. "I had to pinch myself to see if it was real.") He went home to consider offers from Harvard, Texas and International Nickel. But he looked back on his four years of mission-oriented interdisciplinary work with outstanding people and decided that the GE Research Lab might offer more of the same. "The experience of working with these guys had been such an education," he decided, "that I really ought to go over to the GE lab for a couple of years."

Volney Wilson had left Ohio State University in 1937 for the University of Chicago to study cosmic rays with Nobel laureate Arthur Compton. The war swept him into Chicago's Metallurgical Laboratory, a part of the atomic bomb effort. When the world's first controlled chain reaction went critical on the squash court under the stands of Chicago's Stagg Field, he was there as head of the instrument group. He did more work on instruments at Los Alamos.[15]

After the war, Arthur Compton gave Wilson some advice: Why not look over the GE lab, where Compton was a consultant? Wilson arranged an interview and was impressed, especially with the old guard. "I think it was Coolidge and Hull," he says. He got an offer, and accepted.

For Herb Strong, the path was longer. By the time he completed his dissertation in 1936, he already had a full-time job developing new surgical dressings for the Kendall Company. He moved on to a textile plant in Rhode Island. "It bothered me a lot that I was getting away from physics," he recalls.[16] "The work didn't have enough challenge—there wasn't room for a physicist." Visiting Bundy, he admitted to some envy about his friend's GE opportunity. Bundy put in a word, and Strong soon got a message to contact the GE recruiter at the next American Physical Society meeting. By the end of 1946, the three contemporaries at Ohio State were back together again at General Electric.

Other members of the new guard were equally attracted by the opportunities to do some real science and remain in the type of climate they had found attractive during the war. The bait Suits dangled before Los Alamos nuclear physicist Henry Hurwitz was the chance to do pure research. "Shockley at Bell labs made no bones about it that their job was to support the communications industry," he remembers.[17] "I was hired [by GE] to do high-energy physics." Jim Lawson of MIT's Radiation Laboratory got to know Suits during the war when both were working on radar problems. Many conversations and a couple of visits gradually warmed him toward GE, until he decided that "I might try it for a couple of years." Metallurgist Joe Burke went first from Los Alamos to

Manmade synchrotron radiation was observed for the first time in the 70-Mev synchrotron at the GE Research Laboratory in 1947. The radiation is visible as a bright spot on the left side of the "doughnut."

the University of Chicago, but he found the pay too low and the city constricting, and found a new job at GE.

A look at a sample of 33 scientists with PhDs who joined the research staff of the GE Research Lab in 1946 shows typical members of the new guard as about 30 years old, married, and just starting families.[18] All of the new researchers hired in 1946 were men (though in 1945, the lab had doubled its number of female researchers by hiring Edith Boldebuck, who held a PhD in chemistry from the University of Chicago). Almost all of the newly appointed researchers had gone to the best American graduate schools—Harvard, Princeton, Berkeley, Chicago, Caltech, Michigan, Wisconsin, Johns Hopkins, and Rochester, for example. About half were physicists; the rest were chemists or metallurgists. Most of them had worked in a government-sponsored research and development program during the war, with Los Alamos and the MIT Radiation Lab most strongly represented.

The move into industry did not always go smoothly. Volney Wilson went to GE in the hope of turning atomic energy into a benefit to humanity.[11] "I sort of felt it to be one way to justify my working on the bomb, if I could also make a peaceful application," he remembers thinking. But within two years, he was being shifted back into military work: "I was quite a pacifist still, and so this irritated me, and so I quit." Wilson continued:

I wrote a big, long nasty letter to Suits and then Suits called me to come and see him. I was going to resign from the company and, as I

walked into the office, he was talking to someone on the phone and just said 'yes, yes.' And I'm quite sure it was Albert Hull, and I think it was Hull who persuaded Suits to treat me gently. And so, I did stay, and Suits was extremely generous; he told me to spend all the time I wanted to read and decide what I'd like to do.

Not all of Wilson's colleagues made the adjustment. Of the 1946 sample of 33 recruits, 12 had left GE within five years. Not all should be assumed to have left disillusioned. It was a seller's market for scientists, and some simply got a better offer.

But the majority stayed. Francis Bundy and Herb Strong, for example, moved directly into the lab's Mechanical Investigations Section, an interdisciplinary group tackling varied problems in the development and testing of machinery and propulsion systems. They began with some fairly straightforward tasks: analyzing the vibrations of a misbehaving turbine generator; using spectroscopy to measure the speed of rocket exhausts; or designing a new refrigerator insulation. But within five years they were embarked on what would be one of the most interesting exploratory efforts in industrial research: the production of diamond in the laboratory.

The new guard that they exemplified, and that the recruit of 1946 was being asked to join, was part of a new generation of American scientists. It was the first to be educated by US research universities on a par with those of Europe. This group had experienced the Depression of the 1930s, and

appreciated the security of a good job. Many members had no burning desire to teach, and no unbreakable attachment to the amenities of a university. Most had experienced during the war the challenges and rewards of mission-oriented work. They were used to the idea that a scientific project might have a useful output. And, above all, they saw GE as a way station, not necessarily a final destination.

The nation had helped to bring them there. A war had sold science to the public. And that same war had gathered thousands of America's brightest young scientists and engineers into a small number of temporary laboratories. The demobilization of these laboratories gave industry an unprecedented opportunity to harvest bumper crops of talented researchers. At GE this new guard dominated its laboratory until the 1960s, when the faith in research was tested and found wanting. The outcome was a re-emphasis on results—though by no means simply a return to the old strategy. Instead, a new generation of leaders interpreted the relationship of science and technology differently, and took different practical actions. But that is another story.

GE's historical role

General Electric, a pioneer in prewar research, changed the scope and style of its Research Laboratory in 1945 as part of the national trend toward big and visible science. Between 1945 and 1950, it put a new emphasis on projects carried out by large teams in important areas of research such as atomic power and rainmaking. To support that thrust, its total staff grew from 630

GENERAL ELECTRIC

Nobel laureates Percy W. Bridgman, Harold Urey, and Irving Langmuir (from left) at the opening of the General Electric Research Laboratory at the Knolls site in 1950.

people (160 of them scientists or engineers) in 1945 to 950 (225 of them scientists or engineers) by 1951. It spun off a new atomic power laboratory with a staff of more than 1000 people. It adopted a new method of funding research that made it more independent of the whims of GE business operations than ever before. Under its old name, and without publicly renouncing the policies that had led to past success, it became essentially a new laboratory.

How might understanding GE's postwar experience help reshape interpretations of the history of modern American science? First, it cautions us against speaking of "the" industrial research laboratory—the embodiment of industry's perception of the relationship between research and results. The laboratory run by Whitney and the one run by Suits were two very different organizations, in both purpose and practice. Research by Margaret Graham indicates that similar changes occurred at RCA.[19] Did they happen at AT&T, Eastman Kodak, Du Pont or the other technology-based giants?

Second, the change at GE highlights an important but rarely commented-on effort by government to established industries resulting from World War II research. When the war ended, demobilization freed many scientists who had participated in an intense and often inspiring experience in the type of interdisciplinary, goal-oriented work in which industry specialized. The new research orientation of industrial laboratories offered an inviting blend of the emphasis on science that the recruit had experienced in a university, and

the teamwork and sense of purpose characteristic of the wartime laboratories.

And, third, General Electric's new outlook in research helped amplify the new national view of research. That view proved no more permanent than the ones that preceded it. But while it held sway, it supported work in industry that led to major technological advances ranging from new applications of semiconductors in electronics to the synthesis of diamond and new ways of making plastics. And on the national scale, that new view led to the creation of institutions (such as the National Science Foundation) and practices (like the large-scale sponsorship of research by the military) that continue to shape the support of research in America.

References

1. C. G. Suits, Interview with GE News Bureau, 14 March 1946, on file at GE R&D Center, Schenectady, New York.

2. E. W. Layton, Conditions for Technological Development, in *Science, Technology and Society, a Cross Disciplinary Perspective*, I. S. Rosing, D. de Solla Price, eds., Sage, Beverly Hills, Cal. (1971).

3. K. Birr, *Pioneering in Industrial Research*, Public Affairs Press, Washington (1957). For other perspectives on prewar industrial research, see G. Wise, PHYSICS TODAY, July, 1976, p. 9; and L. Hoddeson, PHYSICS TODAY, March, 1977, p. 23.

4. C. G. Suits, Biographical File, GE R&D Center.

5. C. G. Suits, G. K. Harrison, L. Jordan, eds., *Applied Physics: Electronics, Optics, Metallurgy.* Science in World War II Series, Little Brown, Boston (1948).

6. G. Gaines, Thin Solid Films, **99**, ix (1983).

7. B. S. Havens, J. E. Jiusto, B. Vonnegut, *Early History of Cloud Seeding*, New Mexico Inst. of Technology, Socorro, New Mexico (1978).

8. H. C. Pollock, Am. J. Phys. **51**, 278 (1983).

9. H. C. Pollock and K. H. Kingdon, Pioneering in the Atomic Age, GE Res. Lab. Bull. (Summer, 1965) p. 9.

10. J. R. Stehn, The Story of the Knolls Atomic Power Laboratory, Memo JRS-5, Knolls Atomic Power Laboratory, 24 April 1952; R. G. Hewlett, F. Duncan, *Atomic Shield*, US AEC Washington (1972), especially p. 120.

11. Interview with V. Wilson, 28 October 1977.

12. L. Hawkins, Biographical File, GE R&D Center.

13. Interview with C. G. Suits, 28 April 1981, GE R&D Center.

14. Interview with F. Bundy, 19 September 1975, GE R&D Center.

15. Interview with V. Wilson, 28 October 1977, GE R&D Center; V. Wilson, "Recollections of the Early Days of the Atomic Age," Address to the Northeastern New York Section, Am. Nuc. Soc. 7 December 1967.

16. Interview with H. Strong, 10 June 1980, GE R&D Center.

17. Interviews with H. Hurwitz, Jr, 7 April 1978; J. Lawson 10 July 1980; J. Burke, 20 March 1981, GE R&D Center.

18. Information on new employees at the GE Research Lab taken from the "Who's New" column of the GE Research Lab News for 1946.

19. M. Graham, The Corporate Laboratory as Entrepreneur: the RCA Experience, paper delivered at the Annual Meeting of the Society for the History of Technology, 21 October 1983. □

History of medical physics

Physicists applied many of their most important discoveries to medicine immediately, laying the foundation for today's radiation therapy, nuclear medicine and diagnostic radiology.

John S. Laughlin

PHYSICS TODAY/JULY 1983

First medical cyclotron in a US hospital. Photo shows Michel Ter-Pogossion with his cyclotron at the Mallinckrodt Institute of Radiology, Washington University, St. Louis, Missouri. The machine was installed specifically for medical research and applications.   Figure 1

On the occasion of the 25th anniversary of the American Association of Physicists in Medicine, it is appropriate to look at the history of the field of

John S. Laughlin is chairman of the department of medical physics at Memorial Sloan-Kettering Cancer Center in New York.

medical physics. This is not to imply that the history of the AAPM and the history of the field of medical physics are synonymous, but the latter does help us understand the former: By examining the history of physicists' scientific contributions to medicine, we will see how it was natural for the growing numbers of physicists involved

to form a suitably oriented physical society.

Many tools and methods first developed in the physics laboratory are now used on a routine basis in medical laboratories and offices. For instance, one now finds atomic absorption units and spectrophotometers in most biomedical laboratories. The physician uses instruments and procedures derived from an understanding of mechanics, heat, light and sound to examine the eyes, to measure hearing, to perceive respiration, and to make many other basic measurements of the body.

While the term "medical physics" is broad and refers to the application of principles of physics to any aspect of medicine, for historical reasons the AAPM and similar organizations in other countries have been largely concerned with the uses of radiation—in diagnosis, treatment, research, and protection in medical institutions. In this article we will trace the development of the basic radiation physics that underlies much of today's medical physics, and then we will look separately at the historical development of two major subfields of medical physics—radiation therapy and nuclear medicine. The article beginning on page 36 of PHYSICS TODAY, July 1983, covers another important subfield of medical physics, diagnostic radiology.

Medical physicists use the term radiation to cover not only ionizing radiation but some nonionizing radiations, the entire electromagnetic spectrum, and sound as well. Most of the applications of radiation physics have been in the clinical disciplines of radiation therapy, diagnostic radiology and nuclear medicine, but there have been many applications in radiobiology and radiation protection as well. Scientists sometimes refer to much of the work in diagnostic radiology and nuclear medicine as medical imaging. This terminology can have the disadvantage of focusing attention on anatomical structure, but not on studies of physiological and biochemical function with which these disciplines are also concerned.

Before we consider some of the applications of physics in radiation medicine, let us follow the historical development of the basic physics that came to be applied in hospital laboratories and clinics. Most of this work took place in university laboratories, but some of it was done in industry. We will take a more or less chronological approach, which may lead to some discontinuities, but will illustrate inadvertent interdependences of seemingly unrelated developments.

Limitations of space prevent discussion of the role of physicists in radiation biology and radiation protection, the study of which commenced on an anecdotal basis in 1896 and systematically in 1901 with the description of physiological effects of radium by Henri Becquerel and Pierre Curie.[1] The attention of radiological physicists to these matters over the decades has

contributed to the excellent record of successful radiation control in medical practice. The biological effects of radiation are probably better understood, and certainly more fully characterized, than are the effects of other common agents in our environment. Our emphasis on radiation should not be taken to minimize the significance in medicine of fundamental research carried out in areas of biophysics not involving radiation, for there is much important work underway on topics such as nerve conduction, the biophysics of large molecules, and biomedical engineering.

Radiation physics emerges

Wilhelm C. Roentgen's discovery in late 1895 of the highly penetrating "new rays" was certainly a major scientific finding that has had a continuing impact in medicine. Working with a Hittorf–Crookes tube at the University of Würzburg, he observed fluorescence in crystals of barium platinocyanide located at too great a distance from the tube to fluoresce due to the known properties of cathode rays; these had been described earlier by Philip Lenard. Roentgen systematically investigated the penetration of these new rays in different materials, recorded their absorption shadows on photographic plates, and determined that their intensity decreased inversely as the square of the distance from the tube. His discovery attracted immediate attention, and within two weeks he made a personal demonstration at the request of Kaiser Wilhelm II. The vitality of communication was such that the details of Roentgen's discovery were described in Paris in late January 1896, at a meeting of the French Academy of Sciences.

This finding led in turn to the discovery of natural radioactivity by a physicist in the Paris audience, Henri Becquerel. He assumed that the fluorescence of the glass tube wall produced the x rays. Becquerel postulated that the intense phosphorescence of crystals of potassium uranyl sulfate exposed to sunlight might also be a source of x rays. But when he developed photographic plates upon which he had placed the uranium sulfate, without the benefit of sunlight during a rainy period in Paris, he observed the same darkening as with sunlight. He recognized and reported that there was a spontaneous and continuous emission of radiation from the uranium sulfate crystals.

Marie Curie extended this discovery in her study of the nature of "Becquerel rays" for her doctoral thesis. She assayed many materials for evidence of

emission of radiation by measuring the conductivity of air with a piezo-electrometer previously constructed by her husband, Pierre. She independently discovered and reported the "radioactivity" of thorium in 1898. Through extensive chemical separations correlated with her emission measurements, she identified (and named) the new elements polonium and radium. It is interesting, and a tribute to the perception of the investigators involved, that although there is no direct relationship between the production of x rays and of natural radiation, the discovery of one set in motion events leading to the discovery of the other only 114 days later.

In 1897, Joseph John Thomson, who had postulated that the canal rays in his gaseous discharge tubes consisted of discrete particles with a negative charge, was able to report to the Royal Society that the masses of the negatively charged particles in a cathode-ray beam are about $1/1837$ those of hydrogen ions. Although he was more uncertain as to the magnitude of this ratio than the digits imply, he had identified the electron. It is of interest that Thomson, and his student Ernest Rutherford, studied the ionization current in a gas as a function of x-ray exposure and collecting voltage. They published in 1896 the first "saturation curve," which is one of the important characterizations of any ionization chamber employed in dosimetry.

Subsequently, at the time of his discovery of the atomic nucleus in 1911, Rutherford asked a young investigator in his laboratory at Manchester to carry out an assignment that led to the founding of nuclear medicine. He asked Georg Hevesy to separate radium D (the isotope Pb^{210}) from "all that nuisance of lead." Although Hevesy was unsuccessful after a year of effort, he reached the significant conclusion that one could use radium D as a "radioactive indicator" for the presence of lead.

About a decade after Rutherford had postulated the existence of a neutral nuclear particle, scientists at Giessen observed a penetrating radiation produced when alpha particles from polonium hit boron or beryllium. This induced radiation was even more penetrating than the gamma rays from radium. Irene Curie and Jean Frederic Joliot verified the discovery and found further that the penetrating radiation expelled energetic protons from hydrogenous material. They considered this essentially a "Compton effect" of gamma rays on protons. James Chadwick at Cambridge extended the experiments and—in line with Rutherford's earlier prediction—postulated in 1932 that the highly penetrating radiation consisted of neutral nucleons, which he named neutrons.

Shortly thereafter, on 10 February 1934, Joliot and Irene Curie announced in a note of fewer than 600 words in *Nature* the discovery of artificial radioactivity. They had bombarded boron with alpha particles from polonium to obtain nitrogen-13, and they had bombarded aluminum to obtain phosphorous-30; they observed the exponential decay of the new isotopes and measured their half-lives. They also produced ammonia labeled with nitrogen-13, a compound my colleagues and I produced on a medical cyclotron 37 years later in clinically useful quantities for imaging.[2]

By World War II, the artificial production of radionuclides was well understood, and they were being employed extensively in biological and clinical studies. The absorption of x rays via Compton scattering, coherent scattering, the photoelectric effect, pair production, and so forth, had been studied and characterized as functions of energy, atomic number and density. The diagnostic use of x rays was far advanced and equipment for this purpose was highly developed. Curative therapy with x rays was largely limited to superficial lesions because x-ray generators were limited to a few hundred kilovolts. However, doctors were employing radium effectively by placing it inside the body.

Let us look now at the connection between all the physics that we have discussed and the development of the fields of radiation therapy and nuclear medicine.

Radiation therapy

Radiation therapy includes both the administration of radiation from sources external to the body and the placement of encapsulated radioactive sources in the body. In the latter method, known as brachytherapy, sources in molds or placques are located on superficial lesions, inserted interstitially in tissue of or near the lesion, or placed into natural body cavities.

Within months after Roentgen's discovery of x rays, they were applied externally with therapeutic intent. Probably the first such application was by Emil H. Grubbé, a Chicago manufacturer of incandescent lamps and Geissler and Crookes tubes. Grubbé was also a second-year medical student, and had two cases, one neoplastic and one inflammatory, referred to him for treatment in his factory commencing in late January 1896.

Even with the further development of the Coolidge-type high-vacuum, heated-cathode x-ray tubes and reliable

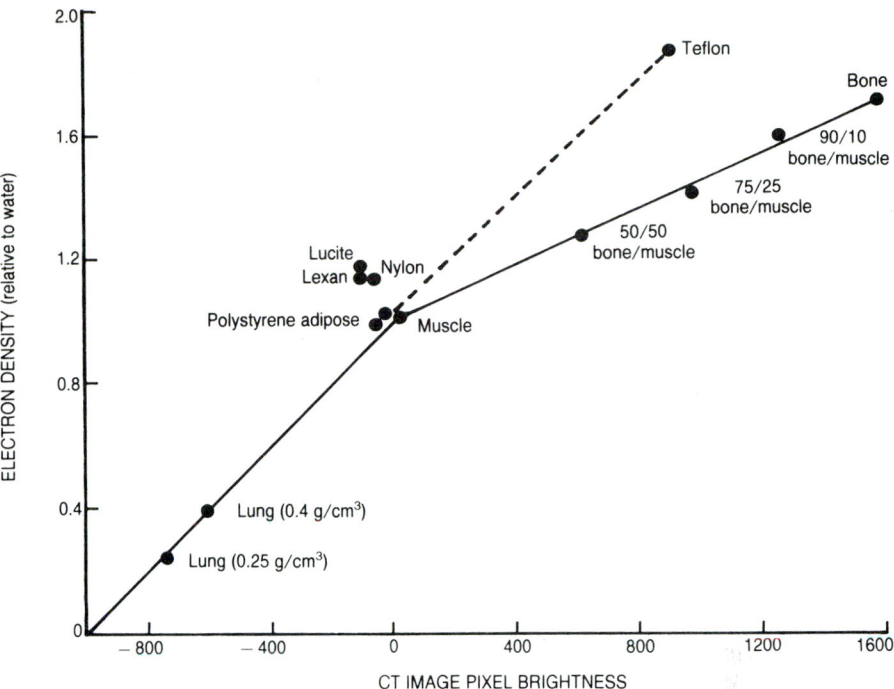

Calibration curves showing the response of an x-ray tomograph to various test objects. One must convert computed tomography images to electron density images to predict dose distributions for x rays and electrons. The artificial substances indicated are used for calibration. (Courtesy of Mary E. Masterson, see reference 22.)   Figure 2

power supplies capable of a few hundred kilovolts, curative external radiation treatment remained limited to the more superficial lesions. It was not until physicists provided the means for accelerating electrons to energies of several million electron volts that physicians had the opportunity to use ionizing radiation for curative treatment at any site in the body.

Brachytherapy physics. Doctors first applied radiation internally through the use of radium encapsulated in needles and tubes, radon gas encapsulated in gold capillary tubing, and radioactive glass "seeds" enclosed in gold tubing. The science of the use of radium needles was developed early in many centers, particularly at the Curie Institute in Paris, the Christie Clinic in Manchester, England, the Memorial Hospital in New York, and the M. D. Anderson Hospital in Houston.

In the period 1913–17, Harvard physicist William Duane developed[3] "seeds" containing radon gas. Duane also developed a radon "plant" and a variety of techniques for using radon needles in applicators, packs and implants. In New York, Gioacchino Failla developed a radon plant of improved design at Memorial Hospital and provided radon in glass seeds or in gold capillary tubing for interstitial implants. He collaborated closely with surgeons and designed a variety of radium applicators.[4]

Radiologists in the 1930s, including Ralston Paterson and Herbert Parker

at Manchester, developed[5,6] a system governing the location of radium needles. This system achieved uniform distributions of doses by using specific nonuniform distributions of sources for lesions of different sizes and configurations. At Memorial Hospital, Edith H. Quimby developed[7] an alternative system in which a uniform source distribution produced a nonuniform distribution of dose.

Although radium and radon are still employed for brachytherapy, they have been replaced largely by cobalt-60 and cesium-137 for needles and by iridium-192 and iodine-125 for seeds. The pioneering work in the use of these isotopes took place at several institutions, including Ohio State University, Memorial Hospital in New York City and the Henri Mondor Hospital near Paris.[8] Automatic computation methods developed in the 1950s at Memorial Hospital, M. D. Anderson Hospital and other institutions provide the dose distribution throughout the volume of the implant rather than at a few points.

Megavolt external sources. From the early days of radiation therapy, radiologists knew that adequate treatment with external x rays required acceleration of electrons to energies of many millions of electron volts. Originally, all differences of potential were achieved by electrostatic methods or by high-voltage step-up transformers. The insulation of the system had to sustain the full potential difference corresponding to the desired energy.

There was a clear need for an alternative approach to acceleration that would circumvent the block to adequate energies imposed by requirements of insulation and voltage enhancement. In 1924, the Swedish physicist Gustaf Ising proposed[9] a resonance method of acceleration, in which a limited potential is applied repeatedly to a given charged particle, which thereby accumulates much energy. His proposed apparatus required a linear increase in velocity with energy and was therefore not suitable to electrons of adequate energy. In 1928, Rolf Wideröe, a Norwegian physicist employed by the Brown–Boveri Company of Switzerland, reported[10] experimental achievement of resonance acceleration of sodium and potassium ions. He employed a linear array of three cylindrical electrodes with two 15-cm gaps, and an oscillator operating at a frequency of a little over 1 MHz with potential differences of 20–50 kV. In the same paper in which he reported this experiment, Wideröe proposed the "beam transformer," essentially the magnetic induction portion of the betatron, but without adequate provision for retaining the electrons in an orbit.

In 1930, Ernest Lawrence described the principle of circular magnetic resonance at a meeting of the National Academy of Sciences. He indicated that he had conceived this technique for circular resonant acceleration after reading Wideröe's paper on linear resonance acceleration. Two years later, Lawrence and Milton Stanley Livingston reported a working model that produced protons with energies of 80 000 eV using an accelerating potential of no more than 1000 volts. Stability of the hydrogen-ion trajectory in the median plane is essential and was achieved by the focusing action of a radial component of the magnetic field. As the ions attain relativistic velocities, their masses increase and they lose their phase relationship with the radiofrequency accelerating field. This limits the conventional cyclotron to approximately 25 MeV in the case of protons. One can exceed this limit by various stratagems as in the synchrocyclotron or in the cyclotron with an azimuthally varying magnetic field. The term "isochronous" is applied to the latter cyclotrons because the accelerated particles circulate at a constant frequency of revolution, even at relativistic energies.

Hospitals have used cyclotrons, such as the one shown in figure 1, for the production of radionuclides, for clinical studies of neutron therapy, for proton irradiation of the pituitary and for treatment of certain superficial lesions. Some clinical studies have used heavy-ion accelerators as well.

In 1940, Donald W. Kerst of the University of Illinois developed[11] the betatron. In addition to providing for an increasing magnetic flux to induce electrons to accelerate, Kerst shaped the pole faces to establish a stable equilibrium orbit. He also injected the electrons tangentially near this orbit. His first betatron operated at 2.3 MeV, the second at 20 MeV and the next at 300 MeV. The accelerated electrons had a relatively monoenergetic spectrum, and their energy was easy to vary. The target for x-ray production was mounted on the injector assembly and had such minute dimensions that the radiation field was sharply defined with no penumbra.

In 1948, a graduate student at the University of Illinois developed a glioblastoma, a serious malignant lesion in the brain, and it was decided to attempt localized irradiation following surgery. Radiologist Henry Quastler and Kerst organized this first treatment with high-energy x rays. They rapidly developed methods of dosimetry, monitoring and collimation of the fixed horizontal beam, and carried out[12] a treatment of 30 beams, or "fields." Although they delivered a tumoricidal dose out to the margins of the lesion, the patient eventually succumbed. Postmortem examination revealed no viable neoplastic cells (cells with uncontrolled growth) in the irradiated region.

The Allis–Chalmers Manufacturing Company developed a commercial version of the betatron with many improvements for reliable medical use, and in 1949 it installed the first unit at the University of Illinois College of Medicine. Its use in high-energy x-ray treatment commenced in 1950, and its use in electron-beam treatment in the 6–22-MeV range began the following year.[13] Workers at this facility and at a similar one at the Saskatoon Cancer Clinic in Canada developed many aspects of the basic radiological physics for the therapeutic use of high-energy electrons and x rays.[14,15]

During World War II, there was considerable development of high-frequency rf power oscillators at Stanford University and in England, and by 1948 microwave medical linear accelerators had been designed. An 8-MeV unit was installed in 1952 at the Hammersmith Hospital near London, followed by a 4-MeV unit at Newcastle-upon-Tyne in 1953 and a 6-MeV clinical unit at Stanford in 1956. By 1982, in the United States alone, hospitals were using approximately 700 linear accelerators and 35 betatrons for cancer treatment.

As a consequence of the development of electron accelerators, particularly the betatron and the linear accelerator, the physical potential for optimum radiation treatment with high-energy x rays and electrons came into being in the decade following World War II. This provided a challenge to the conventional radiation therapy procedures in the orthovoltage range (200–400 kilovolts, peak): No longer could local reddening of the skin be a treatment guide because the maximum deposition of energy now occurred far below the surface for deep-seated lesions. With

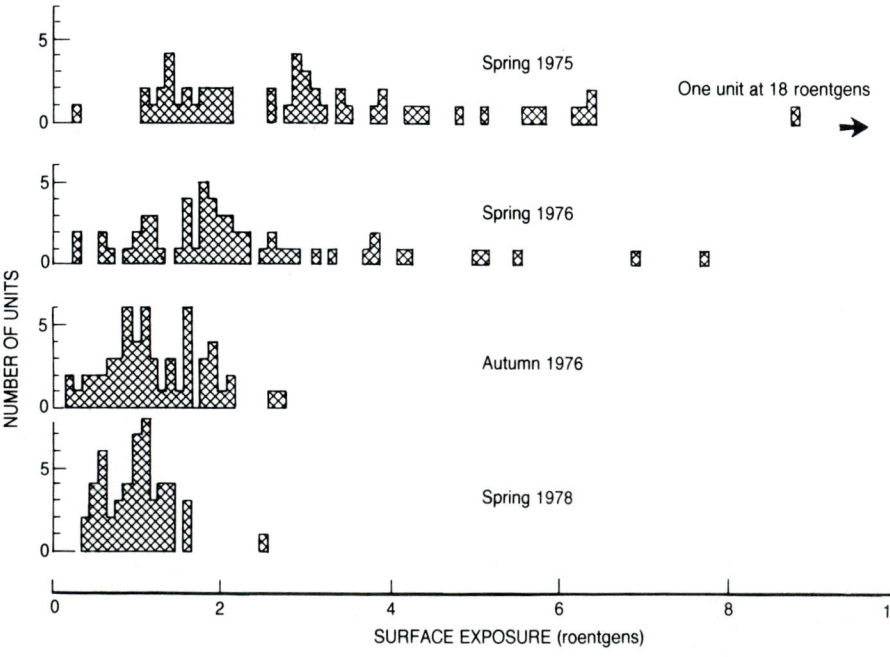

Mammographic surface exposure at 27 radiology departments involved in a project sponsored by the National Cancer Institute. The Regional Centers for Radiological Physics made these measurements over a three-year period. Substantial decrease in exposure was obtained on the basis of the measurement program.   Figure 3

the ability to concentrate the radiation dose came the responsibility and necessity to locate the target region as accurately as possible, to plan the treatment in three dimensions, and to deliver the treatment precisely. Dosimetry problems far more difficult than those of orthovoltage x rays had to be solved, particularly for the use of electrons.

The use of cobalt-60 was advocated for several years for both externally and internally administered radiation treatment.[16] In 1946, William V. Mayneord, chairman of the physics department of the Royal Cancer Hospital in London, brought three discs of cobalt-59 to Canada for irradiation in the neutron reactor at Chalk River. The three resulting cobalt-60 sources

were sent for testing to Harold E. Johns in Saskatoon, Ivan Smith in London, Ontario, and Gilbert Fletcher at the M. D. Anderson Hospital in Houston. Clinical deployment of teletherapy units containing cobalt-60 sources began[17] in 1951.

Dosimetry. The calculation or measurement of the energy deposited per unit mass of tissue for x rays, electrons and heavier particles has occupied the attention of many radiological physicists during much of the past half century, and the ensuing published literature can undoubtedly be measured in tons.

Most radiation users since Roentgen have used the ionization method of dosimetry, which remains the most widely used method to date. The fun-

damental work of Louis H. Gray of Cambridge relates ionization in a gaseous cavity to the energy deposited in the immediately surrounding wall or medium. The calculations require knowledge of the atomic constituents of the gas and wall, the energy spectrum of the secondary electrons, the average energy deposited per ion collected and the stopping power of the gas and wall as a function of energy. The "Bragg–Gray" law, with refinements in different applications, remains fundamental to dosimetry based on ionization. Ionization chambers have been designed with collection parameters that suit them for use in x-ray therapy; some have been designed for diagnostic energies (figures 3 and 4), some for electrons, and so on. The unit of absorbed dose is the Gray. One Gray is 10^4 ergs/gram, or 100 rads.

The Fricke ferrous sulfate dosimeter gives accurate results also. It uses ultraviolet light to measure the amount of ferric ion produced by the oxidation of ferrous ions during irradiation. Although ferrous sulfate can be used in different configurations and has essentially the same density and absorption characteristics as water, it has a relatively low sensitivity, requiring doses of a few thousand rads. Upon the recommendation of the American Association of Physicists in Medicine, the National Bureau of Standards commenced in 1967 a very helpful intercomparison service to electron-beam users based on mailed vials of ferrous sulfate.

The absorbed-dose calorimeter provides a sensitive and accurate basic calibration system. The technique, which my colleagues and I deveoped at Memorial Sloan-Kettering Cancer Center in the mid-1950s, involves a thermally isolated wafer surrounded by a homogeneous absorbing medium, both of whose temperatures are monitored with thermistors that are part of a bridge. The unit is calibrated electrically by passage of a known current through the wafer. Units are fabricated of carbon and of polystyrene for calibration of x-ray and electron beams. A unit constructed of a tissue equivalent containing adequate hydrogen has operated in the field for several years calibrating proton and neutron beams.

Other technologies employing diodes, radiographic film or the thermoluminescence of lithium fluoride or calcium fluoride are used widely and successfully as secondary methods.

Radiation treatment planning is vital to the exploitation of the high dose concentrations that are possible with megavolt x rays and electrons and with advanced internal sources. To achieve a tumoricidal dose and yet avoid irreparable damage to unavoidably irra-

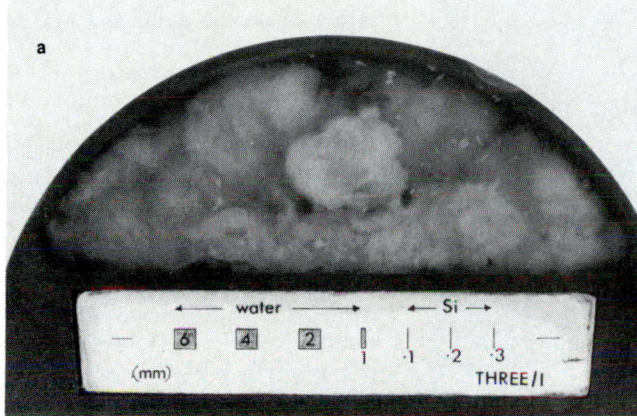

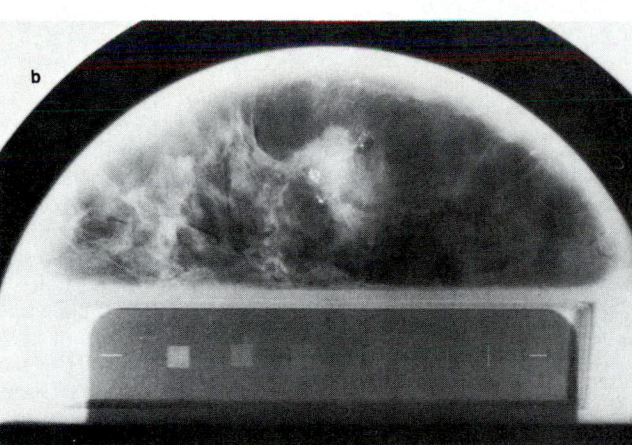

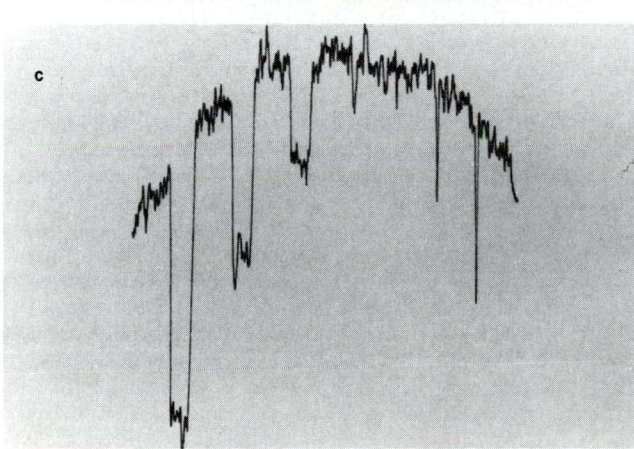

"Phantom" for representing a human breast with carcinoma. Tissue-equivalent material is employed together with an actual excised lesion (**a**). A physics "test strip" is embedded in the material. This unit permits physicists and radiologists to compare radiographs (such as **b**) made with different techniques. The microdensitometer trace (**c**) of a radiographic image of the test objects shows contrast and resolution.   Figure 4

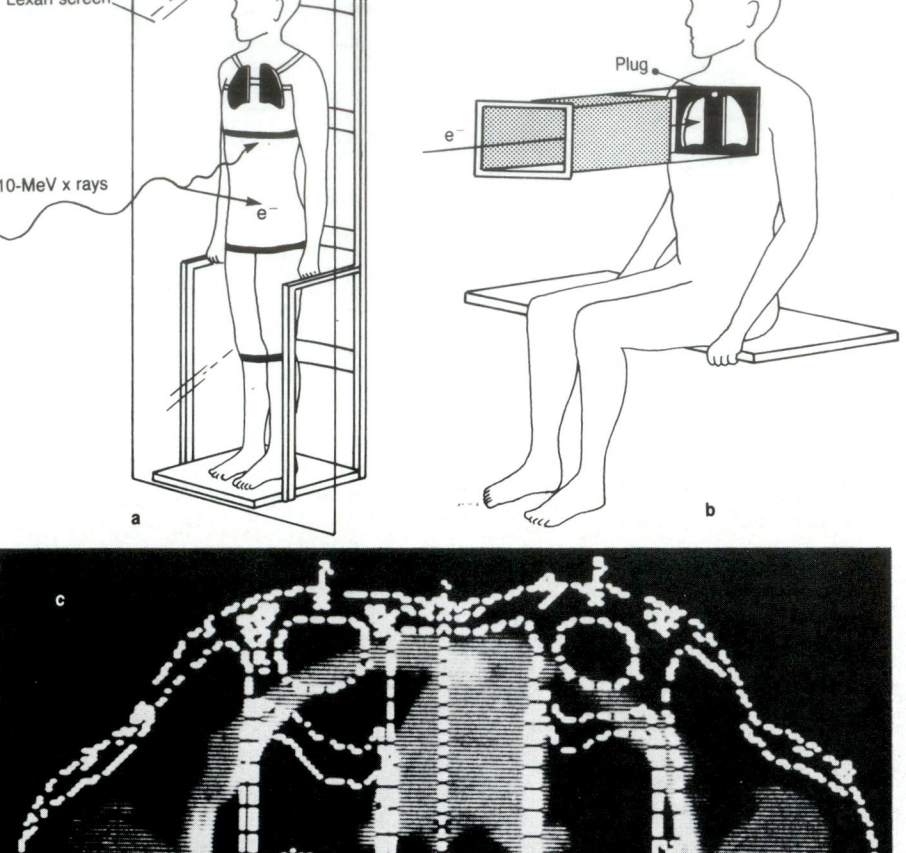

Irradiation strategy intended to minimize exposure of the lungs of patients undergoing total-body irradiation prior to bone marrow transfusion. **(a)** Patient, with lungs protected by anterior and posterior lead blocks, is exposed to megavolt x rays. **(b)** Electron radiation exposes the anterior and posterior regions surrounding the lungs. **(c)** Isodose contours on a cross-sectional map show reduced lung dose after irradiation with both megavolt x rays and electrons. (Courtesy of G. J. Kutcher, Memorial Sloan-Kettering.)   Figure 5

diated healthy tissue, the radiation therapist must achieve a specified tumor dose within narrow limits of uncertainty for a given treatment time. Therapists who design controlled clinical studies usually require that the specified tumor dose be achieved within 5%. Many steps in the treatment process can affect achievement of a given tumor dose, including
▶ the accuracy of calibration of the therapy department's basic dosimeter
▶ the accuracy of its use in calibrating the output of the radiation source at all of its energies, field sizes, angulations and distances
▶ the completeness and accuracy of the measurement of the three-dimensional distribution of dose in a relevant medium
▶ the completeness and accuracy of treatment planning
▶ the accuracy with which the treatment is delivered.

Failure to accomodate inhomogeneities adequately in treatment planning may affect the actual tumor dose substantially; an error in delivering any beam of radiation on any day of therapy may negate care in the rest of the sequence of treatments. The therapy plan may require modification during treatment if check-up examinations reveal changes in the configuration of the tumor or its surrounding tissue.

Radiation therapists need accurate three-dimensional information on the anatomy of the patient in the region to be irradiated. They obtained this from radiography and tomography originally, with an increasing contribution from ultrasonography and diagnostic nuclear-medicine scans. During the last decade, computed tomography has increasingly contributed such information for many patients. Those who plan treatment look forward to the availability of nuclear magnetic reso-

nance imaging, in which the data are obtained initially on a three-dimensional basis. We should note that the density distributions in a computed tomography scan are obtained with x rays of much lower energy than those employed in therapy. To plan therapy, it is necessary to convert the CT numbers to those that are proportional to electron densities (figure 2).

With knowledge of the contours of the patient in the plane of the beam, one can design compensators to accomodate surface irregularities. One can tilt the plane of the customary flat isodose contours by designing attenuation wedges for the beam or bolusing material to place near the surface of the patient. Bolus is flexible material with x-ray attenuation properties similar to tissue. The design of compensators and wedges, together with the number of beams employed and their size, shape, weighting, energy and angle, determine the pattern of local energy deposition in the complex structure of the various tissues of the patient. Controlling such patterns requires not only accurate data, but automatic computation technology with highly specialized software.

Other techniques. In addition to the regular techniques used in radiation treatment, many specialized and experimental procedures involving radiation physics are under study. For instance, in some cases of leukemia, mostly in children, in which bone marrow transplantation is prescribed, total body irradiation is necessary to help avoid a graft-versus-host reaction. Many of the patients treated with total body doses of the order of 1000 rads develop interstitial pneumonitis, which is usually fatal. This has been countered in two ways: reduction of dose to the lungs while maintaining a high total body dose, and fractionating the total dose over time. Reduction of the lung dose is accomplished by designing lead shields for each individual patient to attenuate the x rays from the linear accelerator to the extent that the lung dose is about half of the prescribed midline pelvic dose. The tissue regions anterior and posterior to the lung, which contain some of the target hematopoietic blood cells that might contribute to the graft-versus-host reaction, are then irradiated by electrons both anteriorally and posteriorally. The complementary use of electrons and photons makes it possible to provide adequate total body irradiation with reduced dose to lung tissue (figure 5).

In addition, there is evidence that lung cells have a single-dose survival curve with a broad shoulder indicative of substantial repair following sublethal exposure, whereas both the normal and leukemic components of the hematopoietic system have a narrow

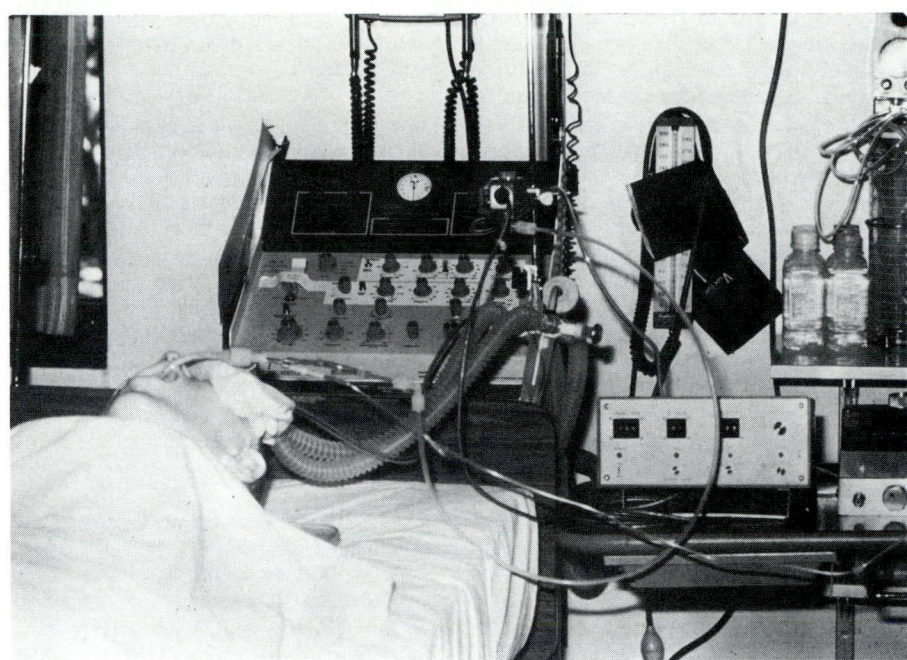

High-frequency jet ventilation system is an example of medical physics engineering. This system provides low-volume breaths to the patient at rates of 60–999 breaths/minute. The small white cabinet to the right of the patient is the electronic controller, which activates a high-speed solenoid valve that gates a blended air–oxygen mixture into the patient's upper airway via a specially designed humidifying injector. The high-velocity gases exiting the injector entrain additional gases from the conventional mechanical ventilator at the head of the patient's bed. There are a variety of alarm and monitoring functions. The system avoids excessively high and damaging peak airway pressures. Certain pathological pulmonary conditions that are unmanageable using conventional mechanical ventilation are manageable with this system. (Courtesy Saul Miodownik, Memorial Sloan-Kettering.)   Figure 6

shoulder, if any, on their single-dose survival curves. Thus, fractionation of the dose accentuates the therapeutic ratio between the target cells of the hematopoietic system and those of the lungs. Survival of children undergoing total body irradiation increased remarkably with introduction of this system.

Other specialized techniques include intraoperative treatment, in which the lesion is irradiated while exposed during surgery, and hyperthermia, the use of elevated temperatures in addition to radiation.

A final example is the use of fast neutrons and other heavy particles for direct irradiation of cancer. The first neutron therapy was at the University of California in 1936, where researchers produced a beam of neutrons by accelerating deuterons in the cyclotron to 16 MeV and directing them at a beryllium target. John Lawrence, Ernest's brother and a clinical scientist at the University of California, used the neutrons in biological studies, primarily on the survival of rodents, while physicist Paul Aebersold intensively investigated the collimation of the neutrons. Robert Stone, a physician at the University of California, undertook clinical trials with the neutron beam following the dosimetry base established by Lawrence's biological studies. During the period from 1937 to 1943, 226 patients were treated with a frac-

tionated schedule. Stone decided, in view of the severe late effect, that neutrons were not suitable for radiation therapy.

Subsequently, Gray advocated a reinvestigation of the role of neutron therapy. He had discovered the "oxygen effect," in which the level of a cell's oxygenation affected its sensitivity to x rays. Accordingly, the effect of neutrons should be more uniform. Further, studies carried out by radiobiologist Jack Fowler demonstrated that with neutrons there was substantially less repair between fractions than with x rays, making the biological effects of neutron doses more additive. Gray and Fowler thus suggested that Stone's total doses may have been too high. Several medical centers in various countries are now conducting clinical trials with cyclotron neutron beams.

The macrodose distribution (the spatial distribution of radiation energy absorbed per unit mass) of neutrons is not comparable to that achievable with high-energy x rays or high-energy electrons. The attractiveness of neutrons stems from their lack of dependence on the level of cell oxygenation and the hope that their use would result in decreased recurrence. Protons, negative pi mesons, and heavier nuclei are also undergoing clinical trials. Doctors at the University of California and at

the Massachusetts General Hospital in Boston are using protons, primarily for pituitary irradiation and for superficial head and neck lesions. This work[18] makes use of the superior macrodose distribution properties of protons and pions and involves highly advanced treatment planning. A group at the Swiss Institute of Nuclear Research in Villigen, Switzerland, is doing a clinical study of negative pion irradiation. To focus pions, Stanford University's physics department designed a system that directs 60 pion beams at a single center. Electrons or protons strike a target, producing negative pions, which cryogenic magnets focus in 60 channels into the patient. The Swiss laboratory built a similar system, the "piotron," in 1979 for use with the synchrocyclotron at Villigen, which accelerates protons to energies up to 600 MeV. The first patient was treated in November 1980, and several have been treated since.

The use of cyclotron neutrons and other heavy particles is still in a research stage. For the immediately forseeable future, megavolt x rays and electrons will continue as the preferred radiation in view of the very good macrodose distributions and clinical results that they give.

Physics in nuclear medicine

Following the discovery of artificial radioactivity by Joliot and Irene Curie in Paris in 1934, Hevesy, who had already suggested using natural radionuclides as indicators of the distribution of elements when he was with Rutherford, studied the distribution of injected phosphorus-32 in various organs of the rat, and deduced its continuous turnover in the skeleton. In 1936 John Lawrence carried out the first injection of a radioactive isotope in a patient for therapeutic purpose with administration of phosphorus-32 to a patient with chronic lymphatic leukemia. Shortly after that, he administered phosphorous-32 to a patient to treat polycythemia vera, a blood disease in which red cells proliferate excessively. At the same time, his colleague Joseph Hamilton was carrying out pioneer experiments with radioactive sodium in patients and in normal human subjects. In the early 1940s, doctors began treating patients with thyroid cancer by taking advantage of the concentration of radioactive iodine-130 and iodine-131 in functioning metastatic lesions.

In the mid 1950s, Rosalyn Yalow and Solomon Berson developed the radioimmunoassay procedure for insulin, based on the principle of competitive binding by antibodies of natural and labeled hormones. This method is the basis of an increasing number of assays in diagnostic and physiological research and is used in hospitals throughout the world. (See her article in PHYSICS TODAY, October 1979, page 25.) Yalow, a nuclear physicist, received the Nobel Prize for Medicine in 1977 in recognition of her contribution.

From 1951 to the present there have been major instrumental developments with respect to rectilinear scanning, gamma cameras, single-photon emission tomography and positron emission tomography. The next article in this issue describes these developments in medical imaging.

Although cyclotron-produced radionuclides were used for biomedical research almost from the time of the operation of the first cyclotrons, and although Ernest Lawrence dedicated his 60-inch unit as a medical cyclotron, such use was not general. The British Medical Research Council's authorization for the establishment of a cyclotron at the Hammersmith Hospital, in accordance with a program proposed by Gray, was an early recognition of the need for devoting a cyclotron exclusively to biomedical use. This machine has been used extensively for pioneer research in the production of a large number of radionuclides for medical use, including radioactive gases for pulmonary studies.

The next cyclotron exclusively for medical use was installed in 1965 in a hospital at the Washington University School of Medicine by Michel Ter-Pogossian, who is shown in figure 1. This machine is also of conventional design and was built by Allis-Chalmers. It is capable of accelerating deuterons to 8 MeV, and it has been used extensively for research on short-lived radionuclides, with particular emphasis on oxygen-15 and carbon-11, employed as labels for biologically significant compounds. Another Allis-Chalmers cyclotron was installed in 1967 at the Massachusetts General Hospital, where physicist Gordon Brownell uses it to produce positron emitters for his metabolic studies. Also in 1967, I installed a cyclotron at the Memorial Sloan-Kettering Cancer Center. This prototype unit, designed and built by The Cyclotron Corporation, is isochronous because of an azimuthally varying field, and it accelerates helium-3 ions as well as protons, deuterons and helium-4 ions. We had sought an isochronous unit to have the versatility for radionuclide production afforded by different nuclear reactions as well as the higher cross-sections available with helium-3, which it accelerates to over 23 MeV.

The use of a variety of biologically significant compounds labelled with cyclotron-produced radionuclides of oxygen, carbon and nitrogen—the elements commonly involved in human metabolic processes—is becoming important in the non-invasive study of

organ and tumor function. L-glutamate labeled with nitrogen-13 is useful in visualizing a number of human tumors. In patients with bone tumors, changes in nitrogen-13 L-glutamate scans during chemotherapy are useful in the evaluation of the response of solid tumors to chemotherapy. Scans of patients and volunteers using other amino acids labeled with nitrogen-13, such as valine or leucine, indicate the utility of these compounds in studies of metabolic processes in the liver, myocardium and pancreas. Red blood cells, labeled with carbon-11 monoxide via inhalation of the radioactive gas have been used to assess changes in tumor vascularity following radiation therapy. Alpha-aminoisobutyric acid, a non-metabolized amino acid, has been successfully labeled with carbon-11, and its distribution in patients indicates it may be useful for metabolic studies. Fluorodeoxyglucose is being used[19] in the study of cerebral and myocardial function. Celebral studies are also being carried out with molecular oxygen-15 and with water or carbon monoxide labelled with oxygen-15. Cardiac studies are using[20] carbon-11 palmitate, oxgyen-15 water and oxygen-15 carbon monoxide.

Imaging with nmr. Medical physicists have been active in the development of imaging techniques in diagnostic radiology, as well as in the development of quality-assurance procedures. Some of these imaging techniques include digital subtraction angiography, x-ray tomography, ultrasound imaging, xeroradiography, computed axial tomography and nuclear magnetic resonance. In the case of ultrasound, it is interesting that the transducer is based on the piezoelectric effect, discovered[21] by Jacques and Pierre Curie in 1880.

Nuclear magnetic resonance imaging currently is generating much excitement in the medical imaging community. Because the rf photon energies used are much lower than those of x rays—about 10^{-7} eV versus about 10^5 eV—nmr imaging promises more information at less risk to the patient. Unlike ionizing radiation, the rf electromagnetic field detected in nmr contains information about molecular bonds but does not have enough energy to break them. As far as is known, the radiofrequency intensities and the magnetic field strengths used at the present time are not hazardous. Currently, imaging is done almost entirely with hydrogen, this being both the most abundant element in the body and the most sensitive to nmr.

We have seen that radiation physics has made important contributions to solving biomedical problems in medical institutions. As further research and applications extend the benefits of ionizing as well as non-ionizing radiation,

the need to continue the highly productive alliance of physics and medicine will increase.

* * *

Criticism of this manuscript by Professor W. G. Myers is much appreciated. The assistance of K. Pentlow has been important, and contributions by T. Ho and G. J. Kutcher are gratefully acknowledged.

References

1. H. Becquerel, P. Curie, C. R. Acad. Sci. (Paris) **132**, 1289 (1901).

2. W. G. Monahan, R. S. Tilbury, J. S. Laughlin, J. Nucl. Med. **13**, 274 (1972).

3. W. Duane, Boston Med. Surg. J. **177**, 789 (1917).

4. G. Failla, Am. J. Roentgenol. **20**, 128 (1928).

5. R. Paterson, H. M. Parker, Brit. J. Radiol. **7**, 592 (1934).

6. M. C. Tod, W. J. Meredith, Brit. J. Radiol. **11**, 809 (1938).

7. E. H. Quimby, Am. J. Roentgenol. **31**, 74 (1934).

8. N. Simon, ed., *Afterloading in Radiotherapy: Proceedings of a Conference held in New York City, May 6–8, 1971*, DHEW publication number (FDA) 72-8024 (1972).

9. G. Ising, Ark. Mat. Astron. Fys. **18**, 1 (1924).

10. R. Wideroe, Arch. Elecktrotech. **21**, 387 (1928).

11. D. W. Kerst, Phys. Rev. **60**, 47 (1941).

12. H. Quastler, G. D. Adams, G. M. Almy, S. M. Dancoff, A. O. Hanson, D. W. Kerst, H. W. Koch, L. H. Lanzl, J. S. Laughlin, D. E. Riesen, C. S. Robinson, V. T. Austin, T. G. Kerley, E. F. Lanzl, G. Y. McClure, E. A. Thompson, L. S. Skaggs, Am. J. Roentgenol. **61**, 591 (1949).

13. R. A. Harvey, L. L. Haas, J. S. Laughlin in *Proc. 2nd National Cancer Conf.*, volume 1, American Cancer Society, New York (1952), page 440.

14. J. S. Laughlin, *Physical Aspects of Betatron Therapy*, Thomas, Springfield, Ill. (1958).

15. H. E. John, E. K. Darby, R. N. H. Haslam, L. Katz, E. L. Harrington, Am. J. Roentgenol. **23**, 290 (1949).

16. W. G. Myers, Am. J. Roentgenol. **60**, 816 (1948).

17. H. E. Johns, Int. J. Radiation Oncology Biol. Phys. **7**, 801 (1981).

18. A. E. Wright, A. L. Boyer, *Advances in Radiation Therapy Treatment Planning*, AAPM Medical Physics Monograph No. 9, American Institute of Physics, New York (1983).

19. M. E. Phelps, S. C. Huang, E. J. Hoffman, C. Selin, L. Sokoloff, D. E. Kuhl, Ann. Neurology **6**, 371 (1979).

20. M. E. Ter-Pogossian, J. O. Eichling, D. O. Davis, M. J. Welch, J. M. Metzger, Radiology **93**, 31 (1969).

21. J. Curie, P. Curie, C. R. Acad. Sci. (Paris) **91**, 294 (1880).

22. M. E. Masterson, C. L. Thomason, R. McGary, M. A. Hunt, L. D. Simpson, D. W. Miller, J. S. Laughlin, J. Soc. Photo-Opt. Instrum. Eng. **273**, 308 (1981). □

THE SITE CONTEST FOR FERMILAB

More than 20 years ago 46 states fought to become the home of the largest particle accelerator of the era, the 200-BeV machine. The AEC decision has been haunted by political questions ever since.

Catherine L. Westfall

PHYSICS TODAY/JANUARY 1989

On 10 November the Department of Energy announced that the Superconducting Super Collider will be located on the Blackland Prairie corn and cotton fields around Waxahachie, Texas. The decision ended a competition entered by 25 states more than a year earlier. The context for the SSC was reminiscent of the way a site for Fermilab was chosen in 1966. Inded, the procedure invented for selecting Fermilab was copied for the SSC, right down to the preliminary run-off refereed by a panel appointed by the National Academy of Sciences and the visits by a government team to the final few sites in the competition. Even the complaints from local citizens and the rumors of political influence that were rife during the Fermilab competition had their counterparts in the SSC contest.

For Fermilab, the suspense ended on 16 December 1966. On that day Glenn T. Seaborg, a distinguished chemist at the University of California at Berkeley who was then chairman of the Atomic Energy Commission, announced that the next high-energy accelerator would be built in Weston, Illinois, just west of Chicago. In the argot of high-energy physicists at the time, it was called the 200-BeV machine—though the abbreviation for billion electron volts soon became GeV, for gigavolts, which is the current usage. The AEC's choice of Weston, made over five other sites in the final run-off, was greeted with cheers in the Midwest. The runner-up, according to Seaborg, was a location near the University of Wisconsin at Madison, and, Seaborg remembers, the next best site was near Denver, Colorado.

Still, objections were raised to the winning location.

Catherine L. Westfall is an assistant professor of the history of science and technology at the Lyman Briggs School, Michigan State University. This article is based on her PhD dissertation.

Some of the nation's most experienced accelerator builders at the Lawrence Radiation Laboratory at Berkeley, having spent more than two years designing the 200-BeV machine, were angry that they had been deprived of the traditional prerogative of creating the facility at the site of their choice. Much more outspoken, however, was a nonscientific group, the National Association for the Advancement of Colored People. For months before the announcement the NAACP had criticized the Illinois record on civil rights, pointing out that of all the states in the final heat, Illinois alone lacked fair housing legislation.

Conventional wisdom in Washington at the time had placed the decision in the White House Oval Office. According to this view, President Lyndon B. Johnson directed the AEC members to select Weston as part of a political deal he had made with Everett M. Dirksen, the influential Illinois Senator who was the Republican minority leader. "Proof" for this speculation came when Dirksen suddenly switched from opposition to support of a fair housing bill a few months after the site selection was announced. If additional support for this notion were needed, the pundits asserted, it was that the President wanted to give Paul Douglas, a fervent supporter of the accelerator for his home state of Illinois, a parting gift on retirement from the Senate. And, anyway, so the story went, everybody knew that Johnson made such decisions himself, because he saw science and technology in terms of what they could do to solve problems—military, political, economic, any kind. After all, that was the way he viewed the space program.

The idea that Johnson was decisive in the choice of Illinois was accepted by many physicists who were surprised that neither Berkeley nor the Brookhaven National Laboratory had won the 200-BeV machine. Both centers were celebrated for their competence in designing

Illinois site covered 6800 acres of cornfields and dairy farms, along with an incorporated village of some 90 single-family prefabs. The village had neither shops nor schools for its population of 380. In fact its name, Weston, was also the name of a considerably larger and older town about 100 miles downstate. Soon after the AEC's decision was made to build the 200-BeV accelerator at the site, a real estate dealer put up the sign in the photograph and hoped to develop the area with more houses. Instead, the state bought the land and houses and conveyed these to the US government. About 12 farmhouses on the site were moved into the housing area. Most of the houses are still in use for visiting scientists and storage buildings in a cluster known as Fermilab Village. (Photo courtesy of Fermilab.)

and building particle accelerators. Even so, if a poll had been taken of the community's opinions at the time, it is virtually certain the results would show that most physicists were pleased that the public spectacle of site selection, with its petition drives and its protest demonstrations, had ended so that the new laboratory could proceed.[1]

Indeed, the time had come for a large, high-energy accelerator. Ideas were popping up everywhere. One called for a cascade of synchrotrons to boost the particle energy—a concept suggested in the 1950s by Robert R. Wilson and others and first developed by Matthew Sands and his coworkers at Caltech. In 1963 a panel appointed by the AEC and the President's science adviser recommended that the US develop higher energy machines with storage rings. The panel, headed by Norman F. Ramsey of Harvard, and including many of the leading high-energy physicists as well as the president of the National Academy, Frederick Seitz, urged the AEC to authorize construction of the 200-BeV machine at Berkeley "at the earliest possible date." Although designing the accelerator posed no major technical difficulty, planning and organizing the expensive, one-of-a-kind laboratory was more problematic. Physicists debated various management schemes. In particular, young experimentalists, such as Leon Lederman, who is now director of Fermilab, pushed for assurance that the new facility would be a "truly national" laboratory with open access to all high-energy physicists.

The pressure for a truly national, high-energy physics center led to demands from various political and academic quarters for a nationwide competition to locate the new

machine. What was the site-selection procedure? Why was Illinois chosen? Who really made the decision? Such questions have vexed physicists and science policymakers for decades. Similar questions have been asked for the SSC project.

Winning the big machine

By the early 1960s momentum for building more powerful high-energy accelerators in the US was driven by an exciting series of achievements in the field of particle physics, especially the discovery of dozens of new particles. Proposals for such machines were stimulated by such technological advances as the discovery of phase stability, the development of alternating-gradient or "strong" focusing magnets and the use of beam stacking. For a decade, support for particle physics had been provided by the Office of Naval Research, the AEC and the newly formed National Science Foundation. The rising budgets were justified in political circles by the accumulation of Nobel prizes by US high-energy physicists as well as by advances in science, education and technology.

In 1960, when planning began for a machine in the 200-GeV range, about 20 US universities housed machines capable of more than 50 MeV. Several universities had developed different approaches for designing accelerators. Largely because of the ingenuity and influence of such dominant figures as Ernest Orlando Lawrence and Isidor I. Rabi, an elite group of accelerator designers held sway at Berkeley and at Brookhaven. Beginning with the Berkeley synchrocyclotron and synchrotron, then continuing with Brookhaven's Cosmotron, the Berkeley Bevatron and the Brookhaven Alternating Gradient Synchrotron, major

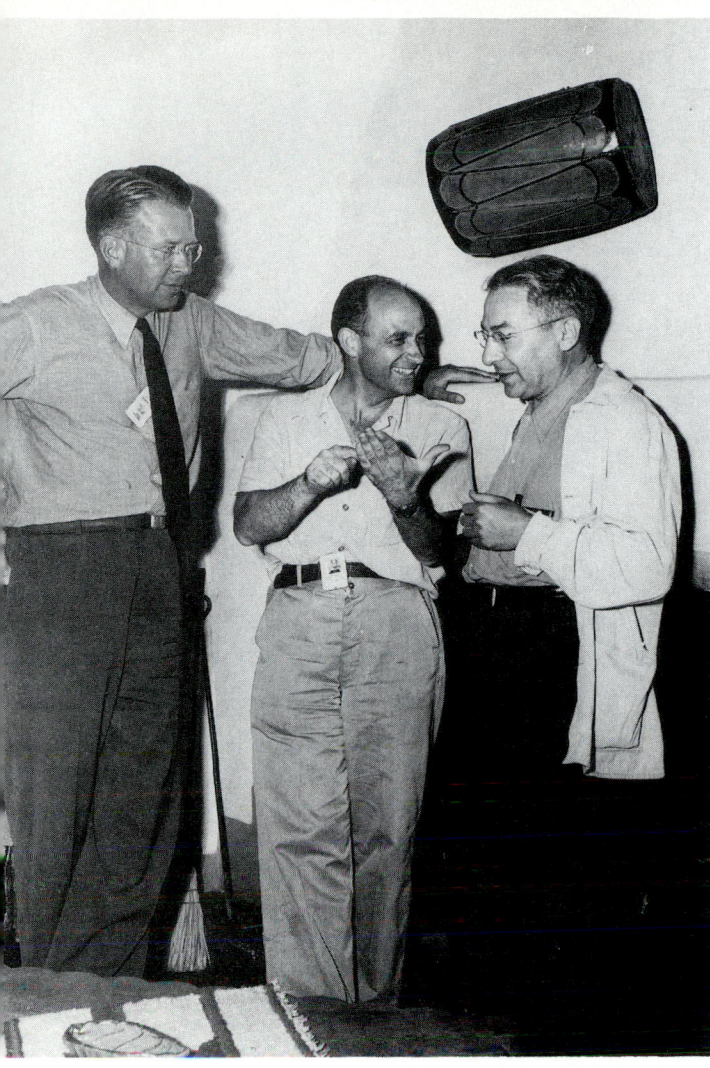

Famed physicists Ernest O. Lawrence (left) and I. I. Rabi (right) were longstanding rivals when it came to locating major particle accelerators in the 1950s. In April 1969 the AEC indicated that the new accelerator laboratory in Illinois would be named for Enrico Fermi (center), who had taught at the University of Chicago, 30 miles away. (Photo courtesy of AIP Niels Bohr Library.)

synchrotron projects had alternated between the two labs. When the Cambridge electron accelerator and the Stanford linear accelerator were built, doubts increased that large machines would go up anywhere but on the East and West coasts.

Lawrence's successor, Edwin M. McMillan, was confident that Berkeley would build and manage the next high-energy accelerator. His claim to the facility was based on Brookhaven's acquisition of the AGS, which, in 1960, had accelerated protons to 30 GeV. McMillan's expectation was further bolstered in 1963, when the AEC gave Berkeley funds for the preliminary design of the 200-BeV accelerator. In a 1962 report to the AEC, McMillan's group estimated that such a machine would cost $263 million. Just then, the AEC was handed a $148-million proposal from the Midwestern Universities Research Association, a group formed in 1954 to champion the interests of high-energy physicists in the region. Determined to develop a unique concept and thereby gain entry into AEC's accelerator budget, MURA physicists and engineers designed a particularly ambitious accelerator, the Fixed Field Alternating Gradient Synchrotron, a 12.5-GeV, high-intensity accelerator capable of producing

proton beams at least 100 times more intense than those in the Bevatron or Cosmotron. The 1963 Ramsey panel recommended going ahead with the MURA machine only if it did not delay the authorization to build the 200-BeV accelerator.

By 1964, however, events overtook the plans of MURA as well as those of the labs at Berkeley and Brookhaven. In consequence of a more stringent Federal budget, brought about largely by the escalating Vietnam War, as well as by increasing cost estimates for accelerators and diminishing enthusiasm for high-intensity machines in the wider physics community, MURA was denied construction funding. Midwestern physicists protested the decision to members of Congress and to politicians in the region. Their best chance to operate a world-class high-energy accelerator had been scuppered, they argued. Midwestern politicians, in turn, complained that their area had not received its fair share of research funding and held up MURA's case as a *cause célèbre*. The combined force of both groups produced considerable political pressure for a Midwestern accelerator, undermining McMillan's hopes to build the 200-BeV accelerator in California.

Many physicists believed they could obtain funding for only one accelerator in the $250-million price range. Moreover, many young experimentalists complained that most large laboratories, and Berkeley in particular, favored inside users. In June 1963, for instance, Lederman wrote an informal report with the title "The Truly National Laboratory," using the initials TNL as a pun on Brookhaven National Laboratory (BNL), which he claimed was not functioning as a truly national facility. Lederman insisted that large labs, including the proposed 200-BeV accelerator, should be accessible *as a right* to all high-energy physicists with competitive proposals.

In response to such demands, the Ramsey panel had proposed that the facility should be managed by a corporation with nationwide representation. The panel's recommendation received strong support from prospective users outside Berkeley and Brookhaven orbits. McMillan, however, insisted that a national management group would prove difficult to assemble and would waste time in reaching consensus. Furthermore, he said, such a group was not needed to ensure wide access.

Managing the new machine

Ironically, the impetus for a new management organization was advanced by the formidable funding difficulties anticipated for the proposed machine. On Capitol Hill and in the physics community there were advocates of more immediately practical projects, such as nuclear power reactors. Some decried the rise in appropriations for basic research and blamed the high cost of building and operating accelerators. Seitz worried that such complaints would fragment the physics community, which was already taking sides over the MURA debacle. Without the unified support of the physics community, Seitz and others reasoned, the chances would be small for getting the necessary funds to design and construct the accelerator.

Therefore, when the Berkeley lab's own advisory committee was unable to agree on a management scheme in November 1964, Seitz intervened. On the strength of his considerable influence as the academy president, Seitz called a meeting of 25 university presidents on 17 January 1965 to discuss various options for managing and administering a new accelerator laboratory. The meeting started a train of events that culminated with the formation of the Universities Research Association, which would build, manage and operate the 200-BeV accelerator under contract with the AEC.

AEC commissioners who made the decision to place the 200-BeV proton machine in Weston, Illinois, pose for this official AEC photograph. Left to right: Samuel M. Nabrit, Wilfred E. Johnson, Glenn T. Seaborg, General Manager Robert E. Hollingsworth, James T. Ramey and Gerald F. Tape.

The participants in the January meeting discussed the possibility that URA would not only manage the new laboratory but might also choose its site. In March 1965 Seitz and Seaborg conceived an alternate plan for selecting the location. To shelter the AEC from the full impact of the inevitable sore feelings of losers, an Academy Site Evaluation Committee was formed, under the chairmanship of Emanuel R. Piore, vice president and chief scientist of IBM. Members of the committee were Robert Bacher, Harvey Brooks, Val Fitch, William Fretter, William Fry, Edwin Goldwasser, G. Kenneth Green, Herbert E. Longenecker and Kenneth Reed, as well as such outstanding national figures as Crawford H. Greenewalt, president of E. I. du Pont de Nemours Inc, and John W. Gardner, president of the Carnegie Corporation of New York and director of several large companies.

The plan called for the AEC to solicit proposals and submit to the committee only those applications that passed the minimal requirements. Seaborg's letter to Seitz on 2 March 1965 included a set of criteria the AEC had collected over the years from studies of possible locations for a large accelerator. Among the criteria: suitable geology, sufficient power and water at reasonable cost, enough land for possible expansion, accessibility to a major airport and transportation hub, proximity to a cultural center that included both a major university and a research and development base, and availability of technical and office staff in the region.

Choosing the best site

Before the AEC signed a contract with the academy, the Subcommittee on Research, Development and Radiation of Congress's Joint Committee on Atomic Energy held four days of hearings on the high-energy research program. The subcommittee chairman, Melvin Price, a Democrat from Illinois, and other members seemed particularly concerned that the AEC had made no commitments about the future location of the new machine. When the Congressional subcommittee insisted that the AEC take final responsibility for the politically sensitive competition for the site of the accelerator, Seaborg agreed that the commissioners would make the final decision. The

academy committee, it was made clear at the hearings, would only provide a short list of finalists from the proposals submitted to it by the AEC.

Despite the careful arrangements made by Seaborg and Seitz, the selection process caused headaches from the beginning. An immediate problem was the unexpected number of proposals. The AEC received 126 proposals consisting of more than 200 sites in 46 states. A group of Midwestern physicists spearheaded a proposal for Madison, Wisconsin, the site of the former MURA headquarters. Brookhaven submitted a proposal and Berkeley submitted proposals for two sites, one near Sacramento in the Sierra Nevada foothills and another at Camp Parks in the Livermore-Amador Valley. Proposals were also sent from several AEC complexes, including those at Hanford, Washington, and at Oak Ridge, Tennessee.[2] Because of the widespread belief that the accelerator lab would bring political prestige and economic benefit to any nearby towns, a large number of proposals were submitted by citizens groups with no connection at all to the high-energy physics community or to the AEC. All the proposals were championed by local politicians, who immediately envisaged the virtues of winning the prize for their constituents.

Seaborg was soon faced with another problem: interference from the White House. When the AEC screening group skimmed 35 sites from the bulk of site candidates, the President intervened, insisting that Austin, Texas, be included on the preliminary list. To satisfy Johnson, extricate the AEC from the uncomfortable position of eliminating sites with minimal requirements, and increase the base of support in Congress, the AEC subsequently forwarded proposals for 85 sites to the Piore committee. While this move strengthened the position of the AEC and increased funding prospects for the accelerator, it placed a considerable burden on the academy panel. Moreover, the AEC attracted criticism because of the enlarged list, including an angry public statement from a leading member of the Joint Committee, Representative Craig Hosmer, a California Republican. Hosmer charged that the AEC had "bungled" siting arrangements, which were now "mired in a mammoth pork barrel."

Such criticism simply increased the efforts of the academy's Piore committee to make sure that it narrowed the field fairly and credibly. Believing that it was important to visit all 85 sites and worried that the Piore committee was already overburdened, Seaborg dispatched teams of AEC technical experts to the sites. Each team included at least one high-energy physicist. After each visit a report about the site was sent to the Piore committee, and, in most cases, committee members interviewed the AEC team leader. The committee's final report said the members made about 25 site visits "to get first-hand impressions" and "to sharpen their own insight about the factors important in site selection." To obtain additional advice and insight, the committee formed two panels of specialists: one examined the geological suitability of sites and another the physical attributes affecting construction and operation, such as power, water and climate.[3] While all this was going on, the Piore committee continually amended and refined the site criteria. At one point, Piore polled members of the physics community to get advice on site criteria, and the AEC presented a new set of criteria for consideration.

As the Piore group began evaluating all the accumulated data in late 1965, the members soon realized that the final choice would hinge on some intangibles. Many of the sites could be characterized by minimal risk of geological instability, but the least expensive construction sites were not necessarily the best. Indeed, in its final report the committee argued that "the ultimate cost of the project depends less on the physical features of the site than on the competence and ingenuity of the physicists and engineers who will design the facility and be responsible for its operation." The January 1966 report of the academy panel studying the physical attributes of the sites reinforced this finding. Because "application of ingenuity and money can overcome most physical disadvantages of a particular site," the panel claimed, "a prime consideration is the attractiveness of the site and its environment for the permanent staff," and "this consideration should not be subordinated to the physical ones."[4]

Deciding on a few choice sites

Once this was agreed to, the Piore committee found it relatively easy to narrow the field to a few choice sites. These included, among others, the two locations in California as well as the Brookhaven lab, two sites in the Chicago area and places near Denver, around Madison, and close to Ann Arbor, Michigan. According to notes taken by an AEC representative at a full committee meeting in November 1965, Piore had made the point that because of the expense of in-depth site evaluation, which could run as high as $1 million, his committee was concentrating on the best sites. When questioned about this by Seaborg, Piore admitted the committee was not "using the point systems at all," and only applying ratings "to check their judgments and conclusions reached by other means."[5]

Some AEC members expressed disappointment that so few AEC facilities were included among the choice candidates. They also seemed concerned that, in the desire to name a site that could be defended against technical and political attacks, so many proposals were being dismissed so quickly.

One physicist on the committee, Goldwasser of the University of Illinois, stated in a letter to Piore on 13 October 1965 that political considerations were clearly "involved in this decision." Unless the list of finalists was "quite universally accepted" by the AEC as well as by physicists, the public and the President, the exercise of site selection was "fruitless."

Such concerns led the AEC and the Piore committee to prepare charts ranking the strengths and weaknesses of various sites. In addition, although committee members seriously considered choosing their preferred site from the list of choice sites, they ultimately decided to submit a list of several sites with top qualities. They were pleased that these were widely dispersed across the country. Goldwasser remembers that the committee believed it would allow the AEC to consider the "political reality" when determining the preferred site.

In March 1966 the Piore committee submitted its list of finalists. In explaining the basis for its recommendations, the committee report noted that the members first determined whether "a given site had suitable physical properties," then "assigned paramount importance to the considerations that affect the recruiting of personnel for the national accelerator laboratory and the participation of the nation's high-energy physicists." The report said that although none of the sites was found to be "ideal," the committee found several sites that, "in general," satisfied all important requirements. Brookhaven and one of the California sites, the Sierra Nevada location, were on the list; the other California site, Camp Parks, was ruled out because the geology experts had found that the area could be affected by earth tremors and quakes. The list of finalists also had Ann Arbor, Michigan; Denver, Colorado; Madison, Wisconsin; and two sites in Illinois—South Barrington and Weston.

From the outset of the selection process, Seaborg made a point of keeping the White House and Congress informed—an action considered politically prudent because opposition by either would most likely jeopardize the project's ultimate funding. By mid-June 1966, Seaborg had ironed out a plan for participation by the White House and the Bureau of the Budget in site selection. Seaborg's notes from a 13 June meeting with the President's special assistant, Joseph A. Califano Jr, indicate that the AEC chairman would discuss prospective sites with Charles L. Schultze, director of the Bureau of the Budget, "at a stage just before the commissioners have made their final decision, but when the possible identity of the site is becoming known." Seaborg would then "get in touch with . . . Califano to get a feeling for the timing of the announcement and any possible White House reaction to the choice."[6]

In this way, Seaborg could keep the Budget Bureau and White House informed and give each a chance to argue against the AEC choice before a public announcement was made, thus avoiding the carnage that would certainly result from naming a site that the White House found unacceptable. At the same time the AEC was free from direct intervention while the decision was being made, and the commissioners would be able to deny, in good conscience, that politicians were dictating the choice of the final site.

While these steps were being taken, the AEC also quickly made arrangements to judge the site list provided by the Piore committee. Plans were made for site visits by AEC staff, with at least one commissioner taking part. The visits enabled the proponents to plead their cause directly. At these visits, the local advocates provided supplementary information and the AEC members were able to obtain more data. Thus, local utility companies were asked to provide data on projected power costs and the governors of the six states were asked to clarify the commitments made by local groups and politicians. With this information, the AEC staff prepared detailed reports to help the commissioners make their decision.[7]

In the midst of the AEC's deliberations, promotion efforts for individual sites reached a near frenzy. For

On an inspection trip to the Weston site in April 1966, AEC Chairman Seaborg spoke to the residents. To his left were Illinois Governor Otto Kerner Jr, village president Arthur Theriault and Illinois Senator Paul H. Douglas. (Photo courtesy of Fermilab.)

instance, the Long Island Association of Commerce and Industry announced on 15 April that its research "verified that all the competing sites have, as of this date . . . launched a major political program." The Long Island group itself organized meetings with local politicians and with education, labor and business leaders to lobby support for the project. Similar efforts were made for other sites. In Illinois, after a 13-day campaign, a citizens group presented Governor Otto Kerner with 6727 signatures of residents and landowners supporting the Weston site. Kerner had already endorsed efforts to make Illinois the new home for the accelerator, calling it the "greatest scientific prize in this century."

Indeed, all six final sites had their ardent proponents. Every member of the California delegation to the House of Representatives, with the exception of Hosmer, who reserved judgment, signed a letter to Seaborg insisting that "on the merits" the California site was "the obvious and only possible choice." The Colorado promoters were led, not surprisingly, by Governor John Love. The Madison site was warmly endorsed by Governor Warren Knowles, by the Wisconsin State Chamber of Commerce and by the Madison Federation of Labor. Michigan Governor George Romney declared in a letter to Seaborg that his state could "match or excel others in any requirement that may be considered."

An AEC tally of Congressional inquiries, compiled when the agency began its final evaluation, indicated that the Illinois site had especially strong support, with 27 House members backing it. The New York and Michigan sites came in with 20 and 19 members, respectively. Congressmen in Illinois, Wisconsin and Michigan, unlike those favoring other sites, lobbied for their regions, as well as for individual sites, further amplifying the strength of Midwestern support.

Despite the dramatic campaign efforts, some citizens protested against building the accelerator at their localities. The strongest complaints came, paradoxically, from Illinois. On 5 April, just as the AEC was beginning its final evaluation, the State of Illinois withdrew the South Barrington site from competition, noting strong community opposition to building the accelerator in the affluent Chicago suburb. *Science* reported that South Barrington residents feared that the influx of physicists would endanger the community's "moral fiber." In contrast, residents of the village of Weston actively campaigned for the project, proclaiming that it would bring prosperity to their community. However, farmers close to the Weston site complained vehemently, and a petition with 114 signatures opposing the use of the site was sent to the AEC. As one irate farmer explained in a letter to President Johnson, local farmers considered it "dastardly" to place such a facility on "some of the richest farming soil in the world."[8]

Many physicists also had strong views about where the machine should go. Commissioner Gerald F. Tape, who acted as the liaison between the AEC and the physics community, remembers that many Berkeley and Brookhaven physicists acted as though the accelerator could not be built anywhere except at their labs. The commissioners saw the matter differently. In their view, as Tape remembers it, the decision to have a site competition could not be defended if the location of the design team was the determining factor. How could the AEC mount an

Cost estimates of six finalist sites for 200-BeV accelerator

Site	Projected basis (in thousands of dollars)		Projected basis plus excluded costs (in thousands of dollars)		Judgment of engineering staff about projected cost range
	Min.	Max.	Min.	Max.	
Ann Arbor	251 200	297 100	253 200	299 100	Closer to max.
Brookhaven	263 900	285 300	264 100	285 800	Average
Chicago	246 500	293 400	247 400	294 400	Closer to min.
Denver	237 700	243 800	239 800	247 900	Average
Madison	243 100	248 300	244 900	250 100	Average
Sierra	239 810	246 330	240 410	247 330	Closer to max.

Table adapted from J. A. Erlewine, memo to G. T. Seaborg, W. E. Johnson, S. M. Nabrit, J. T. Ramey, G. F. Tape, 21 November 1966, box 139, Seaborg Collection, DOE Archives.

expensive, time-consuming site selection process, said Tape, "and then turn around and say the machine goes to Berkeley or Brookhaven because of proven competence," a fact well known at the onset?

Thus, it would have been difficult to defend the choice of either Brookhaven or the California site unless an extremely convincing case could have been made without reference to the expertise of local design groups, which considerably weakened the chances of both laboratories. Physicists at both laboratories sensed this, but in light of their growing list of disappointments, McMillan and his staff were particularly upset. Tape concluded: "Old habits had to change, and it was a painful time for all of us."

The AEC soon faced another dilemma. The agency had hoped to make a final site decision as early as July 1966, but in late June Seaborg received a letter from Clarence Mitchell, director of the Washington Bureau of the NAACP. Mitchell argued that Illinois had failed to pass legislation to enforce open occupancy laws and had a history of housing discrimination. Any hope for an imminent decision vanished as the AEC mounted an extensive campaign to investigate civil rights compliance at the six final sites. By the end of July, proposers were asked to provide assurances from local governments, labor unions, real estate associations, and citizens groups that minorities would not face discrimination in the communities surrounding the proposed sites. At the same time, the AEC asked for judgments on the six communities from the Equal Employment Opportunity Commission, the Community Relations Service of the Department of Justice, the President's Committee on Equal Opportunity in Housing, the Commission on Civil Rights, the Civil Service Commission, and the Office of Federal Contracts Compliance of the Department of Labor.[9]

In November 1966, after eight months of deliberation, the commissioners were ready to make the final decision. By this time, the AEC staff had summarized site information, including staff studies of operation and construction costs, into a working paper, "200 BeV Summary." It was placed into the hands of the commissioners just before the final decision was announced. Although this document did not give detailed information about the status of civil rights, such information had been summarized for the commissioners in August. While the commissioners may have discounted this evidence and made the decision on some other basis, or might have let others influence the decision, as is often claimed, the summary provides a solid foundation for judging the sites on the AEC's own criteria.[10]

The November summary revealed many similarities among the six sites. When land suitability was judged, all sites had sufficient acreage, could be turned over to the AEC at no cost, and had adequate subsurface and surface soils to support the proposed accelerator while the land was in use for grazing. All were within an hour's ride to communities of reasonable size. All had adequate power and water supplies for the accelerator. And, most important, the annual operating costs for all sites ranged from $63 719 million for Ann Arbor to $65 045 million for Brookhaven, a difference of only 2%.

Weighing some critical factors

Despite the similarities, there were critical differences. The Weston and Brookhaven sites showed less promise for future expansion than the other locations, and the Madison site had elevation differentials that exceeded the 100-foot limit set by the criteria. Construction cost comparisons (see the table above) showed that if all costs were considered, the accelerator would be most expensive to build at Ann Arbor, with Brookhaven a close second. The least expensive place was Denver, with the California site a close second. When the accessibility factor was compared for users, Weston and Ann Arbor topped the list, with an estimated average travel time of 3.5 hours for all groups and accessibility in one day for 77% of users in major cities. The California site had by far the poorest rating, with an average travel time of almost 7 hours and only 23% accessibility. Denver also ranked poorly, with an average travel time of almost 5 hours and only 23% accessibility.

The August memorandum giving the status of civil rights at the six sites showed that all sites were similar in many ways. As the report explained: "All six states have fair employment practices laws with enforcement provisions." In addition, the Equal Employment Opportunity Commission considered "the employment attitude at all sites generally progressive." The AEC staff concluded that discrimination was unlikely at schools, hospitals and stores at all sites. Nonetheless, the report stated that Illinois had "no fair housing laws" and generally "adverse comments" had been received about two site areas: "the Ann Arbor NAACP expressed the opinion that Negroes suffer discrimination in that area; civil rights leaders in DuPage County express doubt that assurance of nondiscrimination would be honored." Evaluating such information posed a challenge for the commissioners. Although some site information, for example construction costs and estimated travel times, could be easily ranked, the commissioners had to find a way to weigh the political and social factors that could easily kill the project. They were

Breaking ground for the linac at the site of Fermilab on 1 December 1968 were Robert R. Wilson, who was named laboratory director in March 1967, and Seaborg (at rear). To about 1000 physicists and politicians who attended the ceremony, Seaborg said: "Symbolically we could say that the spade that breaks ground on this site today begins our deepest penetration yet into the mysteries of the physical forces that comprise our universe." (Photo courtesy of Argonne National Laboratory.)

also under considerable pressure to produce a site that could be defended to both physicists and politicians as undoubtedly best.

Geographical considerations aside, many politicians and the Budget Bureau would surely see Denver as the most attractive site because the accelerator could be built there at the least cost. However, the Piore committee had judged, and the AEC had concurred, that construction costs were not as important as factors that would influence staff recruitment and use of the facility.

The most straightforward measure of the use of the facility was accessibility to outside users. When this factor was given top priority, the Ann Arbor and Weston sites topped the list. This was most likely to please active experimentalists in much of the country and eliminate the California and Brookhaven sites, which clearly scored highest in the potential to recruit staff because of the large staffs already working at those labs. But the commissioners would have had a hard time defending the choice of either Brookhaven or the California site if staff availability were given the prime consideration. Moreover, both sites got poor ratings in other important criteria: Brookhaven was the second most expensive site and the California site the least accessible.

According to Seaborg's records, on 29 November the commissioners narrowed the list of sites to Weston and Madison. On 7 December, after making sure that the Budget Bureau and the White House knew a decision was imminent, the Commission selected the Weston site. As recorded in Seaborg's diary, Chet Holifield, a California Democrat who was chairman of the Joint Committee on Atomic Energy, quizzed him on how the site was chosen. Seaborg noted his answer: the commissioners had "added up the positive factors of each site and it seemed that Chicago came out ahead." In a report entitled "Basis for the Selection of the Chicago (Weston) Site for Location of the 200-BeV Laboratory," dated 17 January 1967, the AEC defended its choice by emphasizing its accessibility, which would "assure that all available talents can be readily and speedily brought to bear on its design and use, even though many contributing scientists may never be actual members of the laboratory's staff."

In recent interviews, Seaborg, Tape, James T. Ramey and Samuel M. Nabrit said they were troubled by Illinois's civil rights problems at the time of their decision but they ultimately decided that Weston's advantages more than offset this matter. Although cost comparisons were taken into account, in the end accessibility was considered most important. Weston's easy accessibility clinched their decision.

While they knew that the choice of a Midwest site was politically expedient, if only to make up for MURA's unfortunate demise, the commissioners were pleased that Weston won on its merits, not as a consolation prize. Long before the winning site was announced, some doubted that the AEC would make the final decision. In November 1965, more that a year before the decision was made, UCLA Vice Chancellor Carl York predicted that Johnson

would determine the site. Well aware of this skeptical opinion of the agency's role in determining the site, Seaborg began denying such accusations the day the decision was announced. He still continues to insist that the commission received no pressure or interference from anyone in making the final choice. Seaborg's statement was recently confirmed in interviews with Tape, Ramey and Nabrit, the other surviving commissioners who served during this period.

Even so, tales of President Johnson's role still persist. A few months after the decision was announced, *Newsweek* repeated the more specific rumor that "Senate GOP Leader Everett Dirksen...might be brought around" to support a fair housing bill pushed by Johnson "by the prospect of a $375-million nuclear accelerator to be built back home in Illinois." When the bill came to a vote the next year, Dirksen did indeed switch and both houses of Congress passed it in April 1968. In *Poliscide*, a 1976 treatise decrying the acquisition of land for the accelerator, Theodore J. Lowi and Benjamin Ginsberg quote Senator Douglas saying he "got a solid promise with something of an escape clause" that the accelerator would be located in Illinois. The idea that Johnson was willing to promote a Midwestern site was particularly believable in light of his well-publicized January 1964 letter to Hubert Humphrey after the defeat of MURA, stating his "strong desire to support the development of centers of scientific strength in the Midwest...."[11]

That Illinois was Johnson's preference also has been repeated by other sources. In 1980, David Z. Robinson, who worked at the White House Office of Science and

Technology in the mid-1960s, asserted: "The AEC sent proposals for six finalists to the President and Johnson personally picked a site in Batavia, Illinois." This remark was used by W. Henry Lambright, a Syracuse University political scientist, in his examination of *Presidential Management of Science and Technology: The Johnson Presidency.*"[12] In a recent interview of Robinson, however, he said it was common knowledge in Washington that Johnson preferred Illinois as the accelerator site to woo Dirksen and reward Douglas. Robinson says he was not directly involved in the decision making and could have been misled. For example, as Tape has suggested, it would have been easy to interpret AEC efforts to keep the White House informed as evidence that the President made the final decision, especially since Johnson did intervene in the agency's selection of sites to be presented to the Piore committee.

Knocking down the rumors

It is easy to see how the Douglas-Johnson and Dirksen-Johnson stories gained currency in the Washington rumor mill. Because the AEC's decision met the demands that Midwestern physicists and politicians had expressed for more than two decades, the criticism was not surprising. Moreover, among the White House watchers the suggestion that Johnson contributed to the final choice seemed consistent with the President's past performance.

Yet the written record and the participants' memories speak otherwise. Official AEC files and Johnson's Presidential papers, as well Seaborg's own records, substantiate the oral assertions made to the author of this paper that the commissioners made the final decision without political pressure and on the evidence found in the site data and reports of various panels. The commissioners considered the site for the 200-BeV machine at 49 of their meetings between July and December 1965. They also had many discussions with members of Congress, state governments and local officials and conferred twice with the Piore committee in that period. From January through June the following year, the commissioners discussed the matter at another 39 meetings and made several site visits. When questioned about the events leading to their decision, each of the four commissioners provided details consistent with the written record and with the testimony of the others.

The record reveals that Johnson was informed of the status of decision-making from the outset and was given the opportunity to intervene—just before the site was announced on 16 December 1966. Seaborg says he sent word to the President in early December, informing him that the announcement was imminent. Word came from the White House that Johnson wanted to be informed of the choice the night before the announcement and then Seaborg was advised that the President did not want to know the AEC's decision before others were told. "Contrary to rumors that have circulated," Seaborg has written, "the President didn't exert any pressure on the AEC and left the choice of the site among the six finalists entirely to our discretion."

This does not mean, of course, that national political considerations had nothing to do with site selection. Once the site contest was announced, it acquired a political dimension. After all, a \$264-million project of whatever sort is a significant public investment. Such a project, in particular a scientific instrument that promises to bring social prestige as well as economic benefit, is likely to attract the interest of the public and, as a consequence, the attention of politicians. In the case of Fermilab, the situation was intensified by the ardent lobbying of powerful Midwestern politicians, who saw the contest as a prime opportunity for obtaining increased Federal funding for their area and for developing a high-technology center in Illinois. From the moment of President Johnson's intervention in the site selection process in 1965, through the careful deliberations of the academy's Piore committee and the politically savvy maneuvering of the AEC commissioners, both the participants and observers recognized the importance of political considerations.

The selection also was influenced by pressures in the physics community, and such pressures, not outside factors, gave rise to the site competition. Both the URA management organization and the site selection contest were devised to help unify the physics community to support the project. By choosing Illinois, the AEC commissioners appeased Midwestern political forces as well as Midwestern particle physicists who had been campaigning for a world-class accelerator in their region for 22 years.

⋆ ⋆ ⋆

The author wishes to thank Lillian Hoddeson of the University of Illinois for her cheerful scholarly guidance, as well as May West, Adrienne Kolb and other librarians at Fermilab for their help in archival research and the many scientists who granted interviews, reviewed drafts and provided historical documents—in particular Edward Lofgren, Emanuel Piore, Glenn T. Seaborg and Frederick Seitz.

References

1. For a more detailed account, see C. Westfall, PhD dissertation, Michigan State U. (1988). This paper had its roots in an earlier paper by L. Hoddeson, Soc. Stud. Sci. **13**, 1 (1983).
2. Atomic Energy Commission, "Wide Distribution Shown in AEC List of Proposals for 200 BeV Accelerator," press release, 9 July 1965. J. T. Ramey to G. T. Seaborg and others, 23 July 1965, files of G. T. Seaborg, Lawrence Berkeley Laboratory.
3. National Academy of Sciences, report of the Site Evaluation Committee, March 1966. G. T. Seaborg to F. Seitz, 13 September 1965, files of G. T. Seaborg, Lawrence Berkeley Laboratory. E. Piore to E. Lofgren, 1 November 1965, files of E. Lofgren, Lawrence Berkeley Laboratory.
4. National Academy of Sciences, report of the Panel of Accelerator Scientists, 25 January 1966.
5. P. McDaniel, notes on NAS Site Evaluation Committee, 22 November 1965, files of G. T. Seaborg, Lawrence Berkeley Laboratory.
6. G. T. Seaborg, record of meeting on 13 June 1966, files of G. T. Seaborg, Lawrence Berkeley Laboratory.
7. Atomic Energy Commission, summary notes of briefing on progress report on 200-BeV site analysis, 10 May 1966, files of G. T. Seaborg, Lawrence Berkeley Laboratory.
8. T. B. Husband to L. B. Johnson, 5 May 1966, Seaborg Collection, Department of Energy Archives, Germantown, Md.
9. G. T. Seaborg, record of conversation, 13 July 1966, files of G. T. Seaborg, Lawrence Berkeley Laboratory. G. T. Seaborg, diary, 15 September 1966, files of G. T. Seaborg, Lawrence Berkeley Laboratory.
10. J. Erlewine to G. T. Seaborg and others, with enclosure, 200-BeV summary, 21 November 1966, Seaborg Collection, Department of Energy Archives. H. Traynor to G. T. Seaborg and others, 31 August 1966, Seaborg Collection, Department of Energy Archives.
11. Newsweek, 27 February 1967, p. 28, quoted in T. J. Lowi, B. Ginsberg, *Poliscide*, Macmillan, New York (1976), p. 101. L. B. Johnson to H. H. Humphrey, 16 January 1964, Secretariat, Department of Energy Archives.
12. D. Z. Robinson, in W. T. Golden, *Science Advice to the President*, Pergamon, New York (1980), p. 158. T. J. Lowi, B. Ginsberg, *Poliscide*, Macmillan, New York (1976), p. 79. W. H. Lambright, *Presidential Management of Science and Technology: The Johnson Presidency*, U. of Texas P., Austin (1985), p. 62. ∎

Physics Today and the spirit of the Forties

When the magazine emerged in 1948, it reflected the hopes for science in the postwar period and focused attention on the new opportunities and environment for research.

Charles Weiner

PHYSICS TODAY/MAY 1973

Great expectations were in the air for the future of physics when the first issue of PHYSICS TODAY appeared in May 1948. The applications of physics in World War II had created a new social environment for science in terms of public attitudes and government financial support. Among physicists themselves excitement was brewing over new experimental discoveries and theoretical interpretations in solid-state and particle phenomena. At the same time, new instruments of unprecedented power and size were offering high hopes for probing both the smallest particles of matter and the largest dimensions of the universe. The birth of PHYSICS TODAY as a communication link among physicists, and between them and the larger community, reflected these events of the postwar years. The emergence of the new magazine in May—like the dedication in June of the 200-inch Palomar telescope and the public announcement in July of the discovery of the transistor—was the culmination of a process that had been set in motion years earlier.

In the midst of the intense involvement of science in World War II, US physicists had started to explore what their postwar role might be. Despite their busy wartime activities, many leaders of the profession speculated about the new social environment that was sure to develop from their own and the government's "discovery" of the utility of the concepts and techniques of physics and the skills of physicists. They began to plan for the future of physics and its financial support, within specific academic settings as well as on a national scale.[1] The time was also ripe for evaluating the effectiveness of the organizational structure of the physics community itself, especially because of the increasing specialization that had evolved during the prewar decade since the American Institute of Physics was founded in 1931.

Planning for postwar physics

These issues had been discussed at a small meeting on "Problems of Physics in the Postwar Period," held in Philadelphia in 1944 under the sponsorship of the National Research Council. One result of the meeting was a recommendation that the Institute should take steps to bring increased organizational unity to those whose primary scientific interest was in physics. By 3 January 1945 the Institute's Policy Committee presented a plan to accomplish this aim, noting that:

"A very significant change is that interest in the science of physics is now much more widespread than formerly. It is no longer so much concentrated in academic circles and extends into a host of industries and into the border ground of other sciences. The number of academic, institutional, and industrial workers who identify themselves with physics has approximately doubled in the past decade, and the postwar era promises a much greater expansion."[2]

The committee proposed a reorganization of the Institute and the publication by AIP of a journal that would bring to the attention of all physicists "subjects of general interest or far-reaching importance. It should aid the specialist to keep in contact with the broad progress of the science. It should bring to individual physicists news about his colleagues and about events, announcements, and legislation of general interest to physicists. It should, in short, serve as a broad unifying influence."[2]

On the occasion of the 25th anniversary of that journal it seems appropriate to look back on some of the events covered in its first issues. Today's observers and participants will easily recognize that many of these events in the postwar transition period for science have had a profound impact on scientific work throughout the past quarter century.

The postwar trends were examined in the first issue by Vannevar Bush, who had been one of the major architects of the wartime scientific effort and of the subsequent steps to define new government–science relationships. Bush expressed pleasure and surprise that the new public interest in physical science—which stemmed from the dramatic practical applications in the form of radar, proximity fuses and the atom bomb—was not limited to its applied aspects but extended to basic science as well. He observed: "There are indications that this broadening of the approach also is having the effect of ensuring public interest, and hence eventual support, not only in the physical sciences from which most of the war implements were evolved, but also

---

Charles Weiner is professor of history of science and technology at Massachusetts Institute of Technology and was the director of the Center for the History of Physics at AIP at the time of the 25th Anniversary of PHYSICS TODAY when this article was written.

PHYSICS *today*

TRENDS IN AMERICAN SCIENCE by Vannevar Bush . . . . *See Page 5*

VOL 1 NO 1 ● MAY 1948

First cover of PHYSICS TODAY in May 1948. This photograph of Robert Oppenheimer's porkpie hat, tossed on the Berkeley synchrocyclotron, symbolizes the early postwar period, when theorists eagerly awaited experimental results from the new particle accelerators.

in science generally." Despite the relative lack of publicity for the medical advances applied during the war, he noted, "the interest of the public today seems to be fully as much in the biological sciences, which are the basic source of medical advances, as in physics or chemistry." [3]

Evidence of public interest and support included the Congressional discussions of legislation to establish a National Science Foundation. Bush's July 1945 report *Science, the Endless Frontier* called for extensive federal support of science and recommended the establishment of a government foundation for support of basic research. After legislative disputes about its administrative structure were resolved, the NSF was created in 1950. In 1948, Bush told the readers of the

first issue of PHYSICS TODAY:

"... If science legislation of this general type had been enacted in the middle thirties, I think ... it would have been heavily weighted in the direction of aid to inventors, pilot plants for new processes considered likely to provide an immediate public benefit in goods or materials, and applied affairs generally. The discussion in the last Congress, however, revolved almost entirely about the support of basic science, and there seem to be no dissenters from the thesis that this is the stage most suitable for federal support, leaving applied science to government laboratories and industry. It is devoutly to be hoped that this point of view will continue, and that a highly representative Foundation will soon be

established and actively on its way." [4]

New sources of funds

While the NSF legislation was tortuously evolving, the burden of federal support for basic research was taken up by the Office of Naval Research, as part of a program initiated in the fall of 1945. By November 1948 ONR was spending twenty million dollars annually to support about 500 unclassified basic research projects in the physical sciences at universities throughout the US. That month, the ONR program was described in the PHYSICS TODAY article "Investment in Basic Research" by Emanuel Piore, who directed the ONR Physical Sciences Division. Piore noted that "support of a broad, long range, basic research program by the federal government in laboratories outside the government, is in a sense a new venture." He explained that "ONR tries to stimulate creative scientific activity to insure the existence of a broad base for applied research and development. Thus, support is given to the traditional source of creative thinking—the universities—stimulating their normal function of research and graduate instruction." [5]

Nuclear physics received the largest amount of ONR support; it accounted for thirty percent of the Physical Sciences budget during the fiscal year 1947–1948. This support helped make possible the proliferation of a new generation of large accelerators that were under construction or coming into operation at a number of US universities. Electrostatic and linear accelerators, cyclotrons and synchrotrons were being dreamed up, tuned up or turned on in 1948, most of them paid for—and the bills were large—by ONR and the new Atomic Energy Commission.

These aspects of the rise of big physics stimulated the song "Take Away Your Billion Dollars," by Arthur Roberts, printed in that same November issue. According to Roberts, he wrote the song in 1946 "when it seemed as though every physicist was inventing, building. or projecting a new and larger machine, and while plans for the Brookhaven Laboratory were being formulated. The AEC was not yet in existence, and all financing for new machines was being thought of as from the Armed Forces. This appeared to many people a dangerous situation ..." [6] His musical comment was:

"... *Now in my lab we had our plans, but these we'll now expand.*
Research right now is useless, we have come to understand.
We now propose constructing at an ancient Army base,
The best electronuclear machine in any place,—Oh,

*It will cost a billion dollars, ten bil-
   lion volts 'twill give.*
*It will take five thousand scholars
   seven years to make it live.*
*All the generals approve it, all the
   money's now in hand.*
*And to help advance our program,
   teaching students now we've
   banned . . .*
*This machine is just a model for a
   bigger one, of course,*
*That's the future road for physics, as
   I'm sure you'll all endorse.*

*And as the halls with cheers resound
   and praises fill the air,*
*One single man remains aloof and si-
   lent in his chair.*
*And when the room is quiet and the
   crowd has ceased to cheer,*
*He rises up and thunders forth an
   answer loud and clear:*
*'It seems that I'm a failure, just a
   piddling dilettante,*
*Within six months a mere ten thou-
   sand bucks is all I've spent.*
*With love and string and sealing wax
   was physics kept alive,*
*Let not the wealth of Midas hide the
   goal for which we strive.—Oh*

*Take away your billion dollars, take
   away your tainted gold,*
*You can keep your damn ten billion
   volts, my soul will not be sold . . .' "*

Era of the big machines

What the newly completed machines "could and should do" was the focus of a conference of 150 accelerator specialists held at MIT in 1948. Most were experimentalists concerned with high-energy particle accelerators, but a number of theorists were also present. E. Alfred Burrill, reporting on the conference in the September 1948 issue of PHYSICS TODAY, observed a division between those physicists concerned with high energy and those concerned with higher energy: The greater precision obtainable in the relatively lower energy range was needed for studies of nuclear structure; the study of nuclear forces required ever increasing energies.[7]

A major stimulus to interest in high-energy machines had occurred in February 1948 when the Berkeley 184-inch synchrocyclotron produced the first artificially made mesons, from 380-MeV alpha particles. Meson physics had begun to take off in 1947, when C. F. Powell and G. P. S. Occhialini discovered pions in cosmic rays. The production of mesons in the laboratory promised new research opportunities in a field that was providing fundamental challenges to the persistent attempts at formulating a theory of nuclear forces. Increased particle energies could also provide data relevant to the new theories of quantum electrodynamics. The accelerator specialists were aware of these theories, because, as Burrill pointed out in his report, "the theoretical physicists present at this conference were fresh from their sessions at the Pocono Conference in the early spring and essentially echoed its conclusions. There was hope of discovery and anticipation of what would happen when the particle energy is increased." [8]

The Pocono Conference, held in April 1948, involved 27 participants who spent four days in a Pennsylvania mountain environment talking about the theoretical implications of the discovery of new particles and of the precise measurement of atomic energy levels. With J. Robert Oppenheimer as chairman, the conference was spon-

The Shelter Island conferences brought together leading theorists —and an occasional experimentalist. Here we see (left to right) Willis Lamb, Abraham Pais, John Wheeler, Richard Feynman, Her- man Feshbach and Julian Schwinger considering a problem at the first conference, which took place in 1947. The photo seen here, now at the AIP Niels Bohr Library, is from Abraham Pais.

Dedication of the 200-inch Hale telescope took place in June 1948, one month after the beginning of PHYSICS TODAY; planning had begun in 1928. Photo from reference 14.

sored by the National Academy of Sciences. It was a follow-through of a similar conference held the previous year at Shelter Island, New York, which Oppenheimer later described as "the first serious and intimate conference after the war." [9] These conferences brought together leading theorists and, according to the participants, were of immense importance because they focused attention on the outstanding problems facing theoretical physics, and helped to solve some of them. One participant, Richard Feynman, described the Pocono conference in the second issue of PHYSICS TODAY: After discussing the discovery of pions (which were then called "heavy mesotrons") and their controlled production and use in experiments at Berkeley, he reported:

"Faced with all this wonderful confusion of new particles decaying into one another, the theoretical physicists admitted that they were unable to bring appreciable order into the picture, and certainly not to predict what kind of particle would be discovered next, or any new properties for particles already discovered. The future of these problems lies almost completely in the hands of the experimenters." [10]

Feynman then gave an account of the past year's progress in understanding some observed discrepancies from theory. Willis Lamb had reported his work on precise energy levels of hydrogen at the Shelter Island Conference, and I. I. Rabi had reported on the magnetic moment of the electron at the same meeting. These results pointed up the difficulties in the theory of quantum electrodynamics. It was at the Pocono meeting that Julian Schwinger and Feynman each presented and defended his own ideas on the subject. Just after the meeting Oppenheimer received a letter from Sin-itiro Tomonaga in Japan who described

his progress on these problems. [11] (In 1965, Schwinger, Feynman and Tomonaga shared the Nobel Prize for this work.) Feynman's summary of the 1948 conference symbolized the atmosphere of the late 1940s:

"The conference showed that just as we were apparently closing one door, that of the physics of electrons and photons, another was being opened wide by the experimenters, that of high-energy physics. The remarkable richness of new particles and phenomena presents a challenge and a promise that the problems of physics will not be all solved for a very long time to come." [12]

The cover of the first issue of PHYSICS TODAY was also an expression of the spirit of the time: a photograph snapped after Oppenheimer had tossed his familiar pork pie hat on a portion of the 184-inch synchrocyclotron at Berkeley. Construction of the machine had started in 1940 when The Rockefeller Foundation agreed to provide a grant of $1 150 000. Private foundations had been the major source of financial support for physics research in the US before World War II. Cyclotrons, in particular, attracted foundation support because of their social potential through production of radioisotopes for use as tracers and for experiments in radiation therapy. But the Rockefeller Foundation support of the 184-inch machine in 1940 was a harbinger of the support accelerators were later to receive from government agencies. As Warren Weaver of the Foundation noted at the time, it was viewed as "the definitive instrument for the investigation of the nucleus—the infinitesimally small—just as the 200-inch telescope is viewed as the definitive instrument for the investigation of the universe—the infinitely great." [13]

The 200-inch telescope began its metamorphosis from dream to reality in 1928 when George Ellery Hale convinced the Rockefeller-supported International Education Board to provide $6 000 000 for the project. After a long period of planning and construction of this optical instrument of unprecedented size, which was interrupted by the war, the 200-inch Hale telescope was dedicated atop Palomar Mountain on 3 June 1948. [14] Hopes for the new instrument were expressed by Lyman Spitzer in his September 1948 PHYSICS TODAY article, "The Formation of Stars":

"Perhaps when the 200-inch telescope probes further into the secrets of space, and when further progress in experimental and theoretical physics increases our understanding of the processes at work between the stars, we may then outline with more assurance the detailed steps by which supergiant stars may be forming almost before our very eyes." [15]

Growth of solid state

A month after the Palomar dedication, an event in New York opened yet another era in physics. Although the 1 July press conference announcing the point-contact transistor rated only eight sentences in *The New York Times,* buried among programming changes in "The News of Radio" column, PHYSICS TODAY thought it was more newsworthy:

"A semi-conductor has been used for electronic amplification in the Bell Telephone Laboratories. For years the only flexible amplifier available, the vacuum tube, has been an important tool, not only in radio, telephony, and industrial control but in physical research as well. Now a fundamental study of certain problems of solid state physics has provided a new amplifier which seems suited for a variety of practical applications. Developed by John Bardeen and Walter H. Brattain under a general research program initiated and directed by William Shockley, the transistor, as it is called, is a semi-conductor triode which can be used as an amplifier, an oscillator, and in other ways in which vacuum tubes perform." [16]

Like the 200-inch telescope, the transistor's origins went back to the late 1920's, when the new concepts of the quantum theory of matter were applied to the theory of metals. Wartime semiconductor studies, undertaken in connection with the development of radar, contributed importantly to greater understanding of the subject and to the availability of pure germanium and silicon. All of these developments helped set the stage for the Bell Labs project, which was launched late in 1945 and by December 1947 had produced the first transistor.[17]

At the time of the public announcement of the transistor, solid-state physics was flourishing on many fronts. A spring conference at Shelter Island on low temperatures "tried to interpret low temperature theory, and make some sense out of superconductivity and liquid helium," according to John Slater who reported on the meeting in the August PHYSICS TODAY.[18] Subsequent articles on the subject in 1948 included ones by Laszlo Tisza and by David Shoenberg. At the same time the newly formed Solid State Division of The American Physical Society was providing communication channels for the increasing number of physicists in the field.

Vannevar Bush, in his inaugural-issue article, foresaw that the growth in research might create some problems for the scientific community:

"What a pickle this whole movement is getting us into in regard to publication of scientific results . . . " [19]

Readers of PHYSICS TODAY in 1948 also learned about "The Challenge of Industrial Physics" and about the needs and techniques for "popularizing science." In each case they were reminded that the traditional academic approaches to research and teaching needed to be more flexible: "Industrial work requires a breadth of viewpoint and a necessity for cooperation which is sometimes considered unnecessary in a university;" [20] and, "If science wants a mass radio audience, it must compete for it using radio, not classroom techniques." [21] The latter statement introduced John Pfeiffer's article "Science on the Air" in July 1948 in which he expressed concern that the public's interest in science was due to the role of science in the war and that they "will continue to associate it with uniforms unless they learn to appreciate its peacetime meanings—and the power of the methods behind the discoveries." [22]

Growth and its consequences

These illustrations highlight the circumstances surrounding the emergence of PHYSICS TODAY. Now, years later, we can appreciate well the spirit of Bush's concluding remarks in the inaugural issue. His expectations for the growth of research were coupled with concern for the nature of this growth and its effects on the scientific community and on the society in which it funcitons:

"It would be still more interesting, if anyone could accomplish it, to examine what sort of a world all this is leading us into, quite apart from the nature of possible future war, with digital calculators expanding our mathematical power, and enzymes capable of all the chemical reactions that evolution ever found useful."

Ending on an optimistic note, Bush observed:

"But it is sufficient for the moment to note that we are on our way, that public support of science apparently rests on a deep-rooted conviction that the public has apparently a sound conception of what science really is and what parts of it can best be furthered by public support, that the present expansion carries with it an emphasis also on advanced training, and that thus far at least the movement has proceeded reasonably soundly and free from regimentation of fundamental science. The results in a decade or two, if the trend continues, should be exciting." [19]

The seeds that were taking root in 1948 have now borne their fruit. We do not need to be reminded of the wealth of knowledge and insight into the physical universe that has unfolded during the past years, nor of the vast social impact of the applications of physics research during the same period. The changes in the social environment are also apparent. Now the expectations of increasing government support for basic research have given way to concern for establishing priorities, both within science and in relation to overall national needs. Amidst current expectations and concerns, it would be interesting and useful

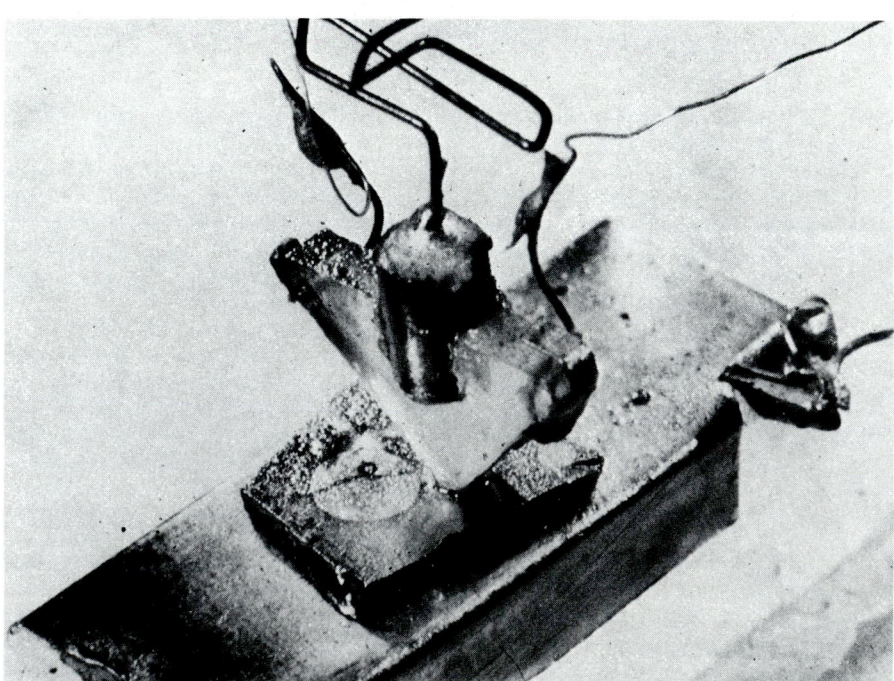

Point-contact transistor. Bell Telephone Laboratories announced the new device at a press conference in July 1948, during a time of rapid growth for solid-state physics.

to examine and attempt to understand what has happened in the past years, while we ponder, as Bush did, about the future of science in the next decades.

References

1. Postwar planning for science is discussed and documented in *The Politics of American Science: 1939 to the Present* (J. L. Penick, C. W. Pursell, and others, eds.) MIT Press, Cambridge (1972).

2. "Preliminary Report of the Policy Committee on the Reorganization of Physics," AIP Archives at the AIP Center for History of Physics; H. A. Barton, "The Early Years," in PHYSICS TODAY 21, no. 5 (1968), page 66.

3. V. Bush, "Trends in American Science," in PHYSICS TODAY 1, no. 1, (1948), page 5.

4. V. Bush, "Trends in American Science," in PHYSICS TODAY 1, no. 1 (1948), page 6.

5. E. Piore, "Investment in Basic Research," in PHYSICS TODAY 1, no. 7, (1948), page 6.

6. A. Roberts, PHYSICS TODAY 1, no. 7, (1948), page 17.

7. These postwar developments in nuclear physics are discussed by many of the major participants in *Exploring the History of Nuclear Physics* (C. Weiner, ed.), AIP Conference Proceedings no. 7 (1972).

8. E. A. Burrill, "The Accelerator Conference," in PHYSICS TODAY 1, no. 5, (1948), page 15.

9. J. R. Oppenheimer, "Thirty Years of Meson Physics," in PHYSICS TODAY 19, no. 10, (1966), page 57.

10. R. Feynman, "Pocono Conference," in PHYSICS TODAY 1, no. 2, (1948), page 9.

11. S. Tomonaga to J. R. Oppenheimer, May 1948, Oppenheimer Papers, Manuscripts Division, US Library of Congress, Wash., D. C.

12. R. Feynman, "Pocono Conference," in PHYSICS TODAY 1, no. 2, (1948), page 11.

13. Quoted in H. Childs, *An American Genius: The Life of Ernest Orlando Lawrence*, Dutton, New York (1969), page 299.

14. The events leading up to the 200-inch telescope dedication are documented in *The Legacy of George Ellery Hale* (H. Wright, J. Warnow, C. Weiner, eds.), MIT Press, Cambridge (1972).

15. L. Spitzer, "The Formation of Stars," in PHYSICS TODAY 1, no. 5 (1948), page 11.

16. PHYSICS TODAY 1, no. 4 (1948), page 22.

17. C. Weiner, "How the Transistor Emerged," in IEEE Spectrum 10, no. 1, (1973), page 24.

18. J. C. Slater, PHYSICS TODAY 1, no. 4 (1948), page 22.

19. V. Bush, "Trends in American Science," in PHYSICS TODAY 1, no. 1 (1948), page 39.

20. H. A. Robinson, "The Challenge of Industrial Physics," in PHYSICS TODAY 1, no. 2 (1948), page 5.

21. J. Pfeiffer, "Science on the Air," in PHYSICS TODAY 1, no. 3 (1948), page 20.

22. J. Pfeiffer, "Science on the Air," in PHYSICS TODAY 1, no. 3 (1948), page 24. □

Chapter 2
Statesmanship in Physics

As the physics profession has grown, its relations to the rest of society, to the other sciences and to those between its own various branches have multiplied, along with the problems involved in these relations. The people who have addressed themselves to these problems, and to their solutions, may be called statesmen of physics. Not all statesmen of physics have been physicists, and certainly not all physicists have been statesmen. Some have, but we find few feature articles on such activity in *Physics Today*. One of the most outstanding statesmen of the "intramural" relations of physics in America, F. K. Richtmyer, died in 1939, before *Physics Today* existed. Richtmyer's career was truly unique in American physics for versatility and service, according to the National Academy Biographical Memoir, which goes on to say "In addition to a full record as a teacher and investigator Richtmyer had served as president of three of the member societies of the American Institute of Physics, was the editor of two of its journals, had been a vice-president of the American Association for the Advancement of Science, representing Section B, and had taken a leading part in most of the important committees in his field. To replace him it has been necessary, as one faced with part of the responsibility remarked, to 'enlist an army'." The three societies he had headed were the Optical Society of America, the American Physical Society, and the American Association of Physics Teachers. He was also very active in AIP governance, was the editor of a series of well known physics books, the author of a long-time standard *Introduction to Modern Physics*, and prominent in his research fields, optics and x rays.

Of those whose statesmanship we are able to represent in this volume some names are more familiar today than others. We begin with Arthur Gordon Webster, founder of the American Physical Society, who is not now widely known. Neither is John Trowbridge, who was responsible for the Jefferson Physical Laboratory and the development of research physics at Harvard. On the other hand Harlow Shapley, the third man considered here, is a great name in American astronomy, and his influence through the Harvard summer school is only a sample of his statesmanship; his obituary, also included in this section, indicates the scope of his activities. Niels Bohr is a famous name in physics indeed, but Finn Aaserud shows us he was also an effective manager of the Bohr Institute during a period of its transition from atomic to nuclear physics during the 30s. Next, Spencer Weart spells out the pivotal role of Jean Perrin in the reorganization of French science during the same period.

Back on the American scene there are only obituaries for two of the men who coped with the problems faced by the physics community during the Great Depression: John T. Tate and K. T. Compton. We are fortunate to have two pieces on K. K. Darrow. Well before his 36-year stint as secretary of the American Physical Society he was a familiar figure at meetings, usually sitting in the front row, and often asking penetrating or conceptually difficult questions. Later, his work in arranging meetings seemed effortless, but his own story shows this appearance was a matter of style, a style well described just after his retirement by J. H. Van Vleck.

Few statesmen of physics have been featured as such in *Physics Today* articles, but for some there are obituaries that convey some idea of the undeniable mark they have left on American science. Three are E. U. Condon, Harlow Shapley, and I. I. Rabi. And, finally, since the separate societies have tended to be parochial, activities of a general nature were often carried out under the auspices of the American Institute of Physics. It is therefore appropriate to conclude with a picture summary of AIP's first 40 years. The summary is introduced by H. A. Barton who, on the recommendation of Richtmyer, was the first Director of the Institute.

Chapter 2. Statesmanship in Physics

Arthur Gordon Webster, founder of the APS

He was a man of many talents and considerable renown in his day who helped establish physics education in America, but his life ended in a tragic suicide.

Melba Phillips

PHYSICS TODAY/JUNE 1987

Fewer than one in a hundred present members of The American Physical Society could name its founder. Carl Barus, head of physics at Brown University for many years and a member of the committee Arthur Gordon Webster organized in 1899 to discuss the possibility of forming a physical society, wrote many years after that meeting, "The foresight and chief credit . . . must be assigned to the tireless activity of Prof. Webster and it is to be hoped that the Physical Society may some day commemorate the event in his honor." According to Ernest Merritt, who was the first secretary of the society and later became its president, All of us who remember those days are agreed that Webster thoroughly deserved the title that was often given him—'father of The American Physical Society.'"

Webster, professor of physics at Clark University for many years, was a colorful figure, quite well known in his time, who dedicated his life to research and teaching in physics while he also pursued his interests in languages and writing. His report to the secretary of the Harvard class of 1885, on the occasion of the class's 25th reunion in 1910, describes in very simple but sure words the joys and satisfactions of a life of single-minded dedication to one's calling: "My life has been entirely devoted to scientific work, which I have thoroughly enjoyed. I come in contact only with advanced students, and have ample time for my own researches. My life has been totally uneventful, unmarred by accident or sadness. I have hardly been ill since leaving college, a result of the use

of the gymnasium then and since, and the avoidance of athletic contests. My scientific work has been rewarded by election to the National Academy of Sciences, the American Philosophical Society, and the American Academy of Arts and Sciences in Boston."

This seems like an apt description of the life and career of the founder of a learned society, but the circumstances of Webster's death leave us baffled and incredulous: Webster killed himself on 13 May 1923, using a gun he had obtained a day before, ostensibly for use in his research laboratory. Why should so fulfilled and fruitful a life have ended in suicide?

I will discuss in this article Webster's physics career and his contribution to the founding of the APS, and will examine in detail the circumstances— the professional uncertainties and financial insecurity—that led him to the tragic decision to end his life.

Early life and career

Webster, the only son of William Edward Webster and Mary Shannon Webster, was born on 28 November 1863, in Brookline, Massachusetts. On his father's side he was descended from an Englishman, John Webster, who settled in Ipswich in the 1630s, but there was also some Scottish blood in the ancestry, hence the middle name Gordon; from his mother he inherited a strain of Irish blood.

Webster prepared for college at Newton High School and entered Harvard in 1881. After graduating at the top of his class in 1885, he stayed at Harvard for a year as an instructor in mathematics and physics and spent the four following years abroad, mostly at the University of Berlin, as Parker Fellow. In Berlin he studied with Hermann von Helmholtz and took his PhD in 1890

with an experimental dissertation directed by August Kundt. On his return to the United States Webster accepted a position at the promising new graduate school Clark University, as docent under Albert A. Michelson. When Michelson left for the University of Chicago in 1892 Webster became assistant professor and head of the physics department at Clark. He was promoted to full professor in 1900, a rank he held until his death. In 1889 he married Elizabeth Monroe Townsend, daughter of Captain Robert Townsend of the United States Navy. They had two daughters and a son.

Webster's most notable scientific contributions were to electromagnetism, acoustics and, toward the end, ballistics. He is credited with introducing the concept of acoustic impedance; an early work, completed in 1893, "An experimental determination of the period of electrical oscillations," won for him in Paris the Elihu Thomson prize of 5000 francs, in competition with widely known physicists such as Oliver Lodge and R. T. Glazebrook. He also published several papers on pure mathematics. His approach toward physics was primarily mathematical, but he had a marked talent for doing experiments. According to Joseph Sweetman Ames of Johns Hopkins, who had known him since their Berlin days in 1886, "He was as much interested in what one may properly call the engineering side of his subject as in the purely physical one, and his ability was so great that there was no practical field in which he could not venture with great profit to all concerned." Although he kept well informed on the developments that revolutionized physics—x rays, radioactivity, the electron—and lectured on these to his students, Webster did no research on

Melba Phillips is professor emeritus at the University of Chicago and now lives in New York.

WEBSTER

these subjects. During World War I, in the course of his tenure on the Naval Consulting Board, he became a leading authority on ballistics and for several years contributed papers to the National Academy of Sciences on the theory and practice of gunnery.

At Clark University, according to the *Dictionary of American Biography*, Webster developed "a systematic and comprehensive course of lectures on mathematical physics which was unsurpassed in scope and thoroughness by any corresponding course offered elsewhere." Three excellent texts arose from these lectures: *The Theory of Electricity and Magnetism* (1897), *The Dynamics of Particles and of Rigid, Elastic and Fluid Bodies* (1904) and *The Partial Differential Equations of Mathematical Physics*, not quite complete at his death in 1923 and published in 1927 after final editing by S. J. Plimpton of Worcester Polytechnic Institute. These books played an extremely important role in advancing physics education in America for they—especially the books on electromagnetism and dynamics—were the first comprehensive treatises on these subjects by an American.

Webster's formal teaching was limited to graduate students, both at Clark and when he was a visiting lecturer at other universities. He successfully trained 27 doctoral students at Clark, and his influence on young physicists extended well beyond that of instructor and dissertation adviser. According to Walter G. Cady of Wesleyan University, who took his PhD in Berlin in 1900 and first met Webster at a meeting in New York of the American Association for the Advancement of Science soon after he returned to the US, "Webster was noted for the kindly interest he took in the younger men who were just coming up in the profession. Many times when a beginner on the verge of stage fright had nervously read a paper of no great importance, which no one else cared to discuss, Webster would think of something complimentary or encouraging about it." Undoubtedly, as Edwin H. Hall of Harvard also remarked, few Americans have done more to promote the higher study of physics in this country.

Webster was devoted to physics above all, but he also had a great talent for languages. There are stories of his giving an address in modern Greek, after taking pains to learn how the language differed from the classical version he had learned in school. He was fluent in several European languages and on numerous occasions was the official American spokesman and representative at conferences in Europe. At the International Congress of Arts and Sciences held in connection with the Universal Exposition at St. Louis in 1904, Barus recalled, " [I] had the honor of being the speaker for physics at the Congress, charged with the duty of presenting a succinct account of the progress of the whole of contemporary physical thought. It was the first time I had ever addressed a large audience and I was a bit anxious. The ordeal, however, was far less severe than I had expected, and less exacting than A. G. Webster's accomplishment in translating and interpreting, *pari passu*, the papers of French physicists like Langevin, into English."

Besides his contribution to higher education, Webster also worked intensely to promote science among nonscientists. He wrote articles for many magazines including the then popular *Review of Reviews* and *The Nation*. His contributions consisted not only of

Honorary degree recipients at the 20th anniversary of Clark University (1909) included R. W. Wood and A. A. Michelson, at the extremes of the front row, and Ernest Rutherford, the big man next to Webster toward the left in the second row. The impressive figure in the center front is Vito Volterra, the Italian mathematical physicist. The bearded man in the upper right is Carl Barus. Robert H. Goddard, then a graduate student at Clark, is the balding young man behind Webster in the upper left.

pieces on science and reports of scientific meetings but also letters to the editor, some exhibiting a robust sense of humor. For example, on 4 August 1911, a sweltering day in Worcester by his own account, he wrote three separate letters to *Science* magazine, each in response to a letter published there. In one he agrees with an earlier writer that the atmosphere for science in Washington is rotten, but adds that it is also "infested with a most dangerous parasite, the *red-tape-worm*!" In another letter: "I hope this letter may provoke discussion, but I do not wish to take part in it. Like all brave anar-

chists, I wish merely to explode the bomb, and then run like...!" And in the third letter he raises the question, "Which is worse, the English of scientists or of politicians?" In a quite different vein, his article "Education and learning in America" (*Science Monthly* **11**, 419, 1920) is very serious, almost solemn, in the way it deplores, among other things, giving our universities over to athletics.

Founding the APS

According to Merritt, "In the years between 1890 and 1900 the need of a society where physicists could get to-

gether for discussion and the presentation of papers was frequently mentioned. All physicists realized the need. But the one who was most active in the movement which ultimately brought about the organization of the physical society was Professor Arthur Gordon Webster."

Webster's contribution to the APS began with his forming the committee—comprising, besides himself, Ames, Barus, William F. Magie (Princeton University), Edward L. Nichols (Cornell University), Benjamin Osgood Peirce (Harvard University) and Michael Pupin (Columbia University)—that sent out a call for a meeting to discuss and, if possible, to organize a physical society. The meeting was held at Columbia University on 20 May 1899, and Webster, as secretary *pro tem*, sent a notice of the new society to *Science*. He had already obtained permission from Henry A. Rowland (Johns Hopkins University) and Michelson, neither of whom attended the initial meeting, to nominate them for president and vice president, respectively. The presidential terms were two years at the beginning, and Webster was third in this succession, after Michelson.

The physical society that Webster founded fulfilled well its function as a forum for presenting research papers. Gradually, the APS Council also began to raise and vote on policy issues. Webster probably brought up the first of these on 24 February 1900, for on that day he was made chairman of a committee to draw up "a memorial to Congress in the name of the Society, favoring the establishment of a Bureau of Weights and Measures, in connection with the United States Coast and Geodetic Survey." The government, acting on recommendations from several scientific societies including the APS, established the National Bureau of Standards in 1901.

In spite of the prominent role he played in founding the APS, Webster did not gain support on many issues. In a letter to Elizabeth Laird of Mount Holyoke College dated 20 November 1905, in answer to one of hers, he wrote, "I have often tried to get the Physical Society to take up pedagogical questions, but without success." Although the council formulated in 1907 an explicit policy that "all pedagogical matters lie outside of the Physical Society," Webster apparently must have continued to raise such matters, for the council appointed in 1915 a committee to consider "the extension of the influence of the Society among teachers of physics." This committee, consisting of George W. Stewart (University of Iowa), Webster and W. S. Franklin, moved for adoption three recommendations on 22 April 1916:

▶ The establishment of student membership

▶ A special *Physical Review* subscription rate for members of societies that are interested in physics teaching

▶ The appointment of an APS representative "who shall prepare for each issue of *School Science and Mathematics* a record of some of the most interesting achievements in physics."

Only the third recommendation was promptly put into effect and Homer L. Dodge was appointed "for the purpose of presenting various items of research in physics" to the editorial board of *School Science and Mathematics*, the most influential journal of the time for physics teachers. The choice of Dodge, who had been an assistant for Webster's lectures on mathematical physics at Columbia in the summer of 1913, suggests the role Webster must have played in drawing up these recommendations and having them adopted by the APS Council. (Dodge later became the first president of the American Association of Physics Teachers.)

Webster participated actively in the actual meetings of the society, which were usually held at Columbia University. H. W. Farwell, a beginning graduate student at Columbia in 1906, remembered: "Some of the older men were always alert to point out flaws or give praise, as the occasion demanded. If any one of those meetings passed without numerous comments from Arthur Gordon Webster we young folks felt something was wrong." And Lyman J. Briggs of the National Bureau of Standards wrote to Karl Darrow in 1949: "The two most colorful physicists in the early days of the Society were Prof. A. G. Webster and Prof. W. S. Franklin. They seldom missed a meeting and they almost invariably had something to say about each paper. Webster had a brilliant mind and his

keen analysis of a paper in his booming voice was something to remember."

It was inevitable that some people should find such an outgoing individual abrasive. His frankness may well have been hard to take on occasion, but in the words of A. Wilmer Duff of the Worcester Polytechnic Institute, "there was a quality of naive sincerity about his occasional impulsive speech that, while it did not always prevent temporary resentment, did usually avert anything like permanent hostility." We know relatively little of Webster's personal and social life, but there is considerable evidence of his charm. Pupin, who was Webster's contemporary in Berlin, recalled: "During a short visit in Paris, in 1887, Webster and I made the acquaintance of many Serbian students who were studying there.... I never visited Belgrade without taking away with me many cordial greetings for Webster from these acquaintances of many years ago. I often heard them say to me: 'If Americans are like Webster then it is no wonder that you prefer to live in America.'... When he stood up for right and justice and truth he was fearless and full of fight, and he reminded you of the Massachusetts men who fought at Bunker Hill. When you addressed yourself to his sympathy he was as mellow and as gentle as the gentlest saint in heaven."

Webster at Clark University

Clark University was founded by Jonas G. Clark, a Massachusetts farm boy who became a multimillionaire as a successful merchant in California after the gold rush. He began shipping goods to San Francisco in 1851 and moved there in 1853. When he retired to Worcester, his home city, he set out to realize his dream of founding a university. He chose a group of distinguished trustees, who selected G. Stanley Hall, a brilliant psychologist at Johns Hopkins, as president. But Hall and Clark differed in their ideas about the university. It was Hall's dream to create a great graduate institution, while Clark really wanted a "college where boys of limited means . . . could obtain an education at low cost." Although Clark agreed to have the graduate school started first, he was not impressed with Hall's plans in spite of the fact that

when the university opened in 1889 the faculty was unequaled by any other university in the country—Michelson headed the physics department, the anthropology department included Franz Boas, and Charles O. Whitman, head of biology, was also director of the Woods Hole Marine Laboratory. Because of his differences with Hall, Clark restricted the funds he gave for the university and, after 1892, furnished no more money during his lifetime; Hall, on the other hand, could not fulfill his promises. Many faculty members were unhappy at this situation, and news of their discontent soon got around. In 1892—the year of "Harper's raid"—William Rainey Harper, president of the newly founded University of Chicago, offered positions at better salaries to a number of professors at Clark, so that, in the words of the 1937 history of Clark University, "at the end of the academic year 1892, but two men of full professorial rank remained."

Jonas Clark died in 1900, willing his fortune to the university, but on condition that a college be established under a different administration from that of the graduate school. This was done in 1902. Facilities, including a good library, were shared between the college and the graduate school, and so were some members of the faculty. But the low tuition did nothing to make the college prosperous, and despite the infusion of its founder's money, financial difficulties were inevitable. In 1920 Hall resigned as president of the university, and the trustees decided to combine the two schools. They chose Wallace W. Atwood to lead the joint institution. Atwood had been the author of a series of very successful school textbooks in geography. He started a new graduate department—in geography— set up a summer school, with an emphasis on geography, for schoolteachers and turned Clark into a very different institution.

Because of Webster's courses and his reputation as a physicist, the physics department at Clark was very highly regarded. Good students came to Clark, and other universities turned to it for suggestions and recommendations to fill positions in physics. For example, in 1896 D. W. Hering of New York University got permission to en-

Webster in academic dress.

gage an assistant in physics, and we learn from the NYU archives that "Clark University had at that time the reputation of giving, under Professor A. G. Webster, perhaps the best training in physics in the United States; and to Clark Dr. Hering turned for his assistant." Hering chose Thomas W. Edmondson, who had come from England with bachelor's degrees from both London and Cambridge universities, had been a fellow at Clark from 1894 to 1896 and got his PhD there in 1896. And the Clark physics department retained until Webster's death a good measure of its prestige. Webster apparently loved Clark: He turned down the offers he received from other institutions—he was wooed by the University of Illinois in 1909 and gave the address at the opening of the new physics laboratories there that year—and continued to head the physics department despite his relative isolation in research and Clark's diminishing commitment to research and graduate education. Robert Hutchings Goddard, generally acknowledged as the father of modern rocketry in this country, was an alumnus of Clark and took his physics PhD there in 1911.

Jonas Clark's will had declared that "the said university in its practical management, as well as in theory, may be wholly free from every kind of denominational or sectarian control, bias or limitation, and that its doors may be open to all classes and persons, whatsoever may be their religious faith or political sympathies, or to whatever creed, sect, or party they may belong." For many years, the atmosphere at Clark could be characterized as liberal, but the "red scare" of the early 1920s, with its Palmer Raids, although it was directed primarily at aliens and labor,

also seriously affected educational institutions, and few more sensationally than Clark University. Early in 1922 the Liberal Club, a student organization, invited as a speaker Scott Nearing, an economist and sociologist who was a socialist. The students had obtained permission from Atwood, but the president arrived at the hall halfway through the evening and stopped the lecture. The incident received nationwide publicity. There is no record of any connection between this event and Webster, but Atwood's arrogant patriotism must have disturbed Webster: There exists a newspaper report that "Dr. Webster was in controversy with President Wallace W. Atwood over the barring of the magazine *The Nation* from the library of the university." Webster had been a frequent contributor to *The Nation*, especially in reporting international scientific conferences, but he would have spoken out in any case, for according to G. Stanley Hall, "he spoke out his mind in the press, in faculty, political, and other meetings, and even in the American Academy of Sciences. He, better than anyone I ever knew intimately, illustrated academic freedom to the fullest."

Atwood, acting as if to make Clark the kind of college that Jonas Clark may have originally envisioned, abolished the graduate department of mathematics, forcing two full professors into retirement. There were rumors that the entire graduate school was to be abolished and that physics would go next. In 1922 Princeton University conferred the honorary degree of Doctor of Science on Webster, but at Clark he was in danger of losing his job. He had signed a one-year contract with Clark not too long before his suicide, his son related shortly after the tragedy, but that contract came so late in the academic year that one can have little doubt that Webster's future was uncertain and that there was little support for his research in acoustics and ballistics.

Physics, too, had changed in the previous two decades. Several new and exciting areas of research had opened up and no one person, no matter how learned and gifted, could be expert in all aspects of the subject. President Hall's account of his last conversation with Webster—"he portrayed his experiences with this drama of struggle and readjustment in his own department, which led him as we all know to focus and become our leading authority on sound and ballistics, and renounce the leadership in the fascinating new fields opened up by x rays, metatomic physics, and relativity"—suggests that the fear of losing his position must have

been traumatic for Webster. What seems to have been his conscious decision to become a leader in the classical mathematical physics rather than follow the new developments—most of which were happening in Europe—probably now made it difficult for him to secure a position elsewhere. That Webster could not face up to this adversity is entirely understandable, but it is a pity that he found himself in such a situation, for even though he did not make many very original discoveries, he was certainly highly esteemed by some of the most renowned physicists of his day. After Webster's death, for example, J. J. Thomson wrote that Webster had taken a very active part in the lectures Thomson gave at the sesquicentenary celebration at Princeton and "showed that he possessed an intimate knowledge of the latest developments in both Pure Mathematics and Physics," and Owen W. Richardson remarked: "None of those who, like myself, had the privilege of being associated with him will ever forget his great geniality, his quick mind and his forceful methods of expression. Any scientific gathering which secured his presence was assured of success."

The scientific world found Webster's suicide one of the most shocking and astonishing things that could have happened. The event was noted in scientific circles both here and abroad, and was the subject of a special feature of the Sunday *New York Times*, with statements from Pupin and George Pegram; Harvard's Edwin Hall wrote a five-column obituary in *Science*. But there was only a brief—though moving—obituary in the *Physical Review*, written by a former student, Gordon S. Fulcher, then editor of the *Physical Review*. In retrospect it seems that The American Physical Society should have taken official note of the death of its founder, but he was not memorialized by its Council, and I have found no mention of him in the programs of membership meetings. It seems time for us to remember our institutional roots after so many years!

* * *

Almost all the source material for this article can now be found in the Niels Bohr Library and the archives of the AIP Center for History of Physics. It has been a pleasure to add to this documentation through the kind generosity of Stuart Campbell, university archivist at Clark University, to whom I owe special thanks. It is also a pleasure to thank the entire personnel at the Center for History of Physics for their ongoing assistance. Most quotations in the article are taken from Arthur Gordon Webster, 28 November, 1863 -- 15 May 1923: In Memoriam, Louis N. Wilson, ed., Publications of the Clark University Library, vol. 7, no. 4 (1924).

□

How the Jefferson Physical Laboratory came to be

The first building in America dedicated to physics opened its doors 100 years ago: "furnished in the plainest possible manner, but provided with everything which intelligent forethought could plan."

Gerald Holton

PHYSICS TODAY/DECEMBER 1984

A hundred years ago, the first building in the western hemisphere designed for research and teaching in physics opened its doors. The consequences of such an event are of very different interest to different groups. The physicist will ask about the advances made in Jefferson Lab, and in nearby structures added later, by its faculty, students and collaborators: Lyman lines and broken symmetries; dimensional analysis and nuclear magnetic resonance; the muon and the 21-cm line; tests of the equivalence principle and of quantum electrodynamics; the acoustics of buildings and of violins; precise mass spectra, and the phase diagrams of hundreds of substances; the theory of magnetism and quadrupole moments; medical uses of particle beams and determinations of the structure of the ionosphere; the Duane–Hunt law of x-ray emission, and Russell–Saunders coupling; the research and teaching of Edwin C. Kemble, America's first quantum theorist; and the latest in math-ematical physics, condensed-matter theory or elementary-particle interactions.

An industrialist might be more interested in other advances, achieved in Jefferson Lab or spun off from work done there: communication-engineering devices and computer design; industrial diamonds and atomic clocks; the large-sheet polarizer; the light-weight phonograph pickup; piezoelectric and magnetostriction devices, and so on.

An educator or administrator may be more interested in the large number of students, including some 900 living alumni, who received their advanced degrees in physics from its faculty, and in the further outreach through many widely used textbooks at all levels, from high-school projects to advanced monographs.

The historian, however, is likely to ask about this building—this somewhat time-worn *grande dame* in need of refreshments: How did it come to be? Why at that time, and why in that place? We shall see that the construction of the building at Harvard was really a rather improbable event, and also that the story is not without some contemporary parallels.

In terms of the physics of their time, most scientists were certainly not clamoring for such a structure in the late 1870s and early 1880s. That period was a sort of halfway house between great accomplishments: on the one hand the grand syntheses of the past decades—the law of conservation of energy, and Maxwell's treatise joining electricity, magnetism, and light—and on the other hand, the excitement to come a decade or two later, with x-rays, radioactivity and the electron. While the Jefferson Lab was being planned and built, physics was really waiting for something to happen to put it to good use. This impression is borne out by glancing through the volume for 1884 of the main journals of the day: With rare exceptions, the articles do not rise above the level of useful but pedestrian investigation.

Gerald Holton is Mallinckrodt Professor of Physics, and professor of history of science at Harvard University.

The Jefferson Physical Laboratory at Harvard University soon after it opened in 1884. It allowed physics instruction to take place not by rote and by books, but in the laboratory, "with objects and instruments in hand."

An imposing temple of science

When this temple of science was opened it looked quite imposing—210 feet long and four stories high. As President Charles W. Eliot of Harvard described it,[1] the building was "furnished throughout in the plainest possible manner, but provided with everything to facilitate physical research which intelligent forethought could plan."

But it also must have looked rather overdesigned in terms of the manpower available in the US. There was only a small handful of top quality men: Henry A. Rowland, J. Willard Gibbs, and young Albert A. Michelson, none of them at Harvard. In the whole country there were, during the period of 1870 to 1893, fewer than 50 physicists in America who had an average rate of publication of even one paper per year or higher. Of these, less than half had PhDs, some from Yale, Johns Hopkins, or Harvard, but mostly earned at foreign universities. Moreover, most university presidents were not interested in furthering research. As Eliot put it,[2] that activity would require "fanatical zeal" if one's main obligation, namely "regular and assiduous class teaching," were not to be neglected. Francis Walker of MIT agreed, saying[3] as late as 1889: "Our aim should be the mind of the student, not scientific discovery, not professional accomplishment."

There was also not yet a professional physical society nor a national physics journal. There were hardly any jobs for trained physicists except in teaching, and that would have to be either elementary physics in the classical college curriculum (a mandatory course of long standing, with hapless Harvard freshmen getting their Aristotelian physics at least as early as 1642), or else physics taught with an eye to applied science in schools set up separately from the College, such as Harvard's Lawrence Scientific School (established in 1847), Yale's Sheffield Scientific School (1847), and MIT (1865).

But in the search for historical causation, one should not underestimate the role that a concatenation of benign accidents can play. President Daniel C. Gilman's leadership in starting The Johns Hopkins University in 1876, with the first systematic program in "advanced study and research," might by itself not have been an urgent example to anyone else. President Eliot's mission to bring to Harvard a modern-day education, an enthusiasm he had first discovered as an instructor at MIT, also did not point inevitably to the realization of a research laboratory at Harvard. John Trowbridge, if he had been a professor at some other university, might not have perservered against the odds he found at Harvard. On arriving as an assistant professor in 1870, he certainly did not find Joseph Lovering, the senior professor in the department, an ally in his plans to make Harvard an institution for research as well as teaching. As Edwin H. Hall, who joined in 1881, later put it:[4] "I doubt whether Professor Lovering ever made an original experiment, or any experiment not required for his lectures. . . . As a young man he had been a student of Divinity, and as a college professor he seems to have felt no more called upon to extend the domain of physics than as a preacher he would have felt obliged to add a chapter to the Bible. . . . I once proposed to him, probably about 1884, that we should drop a certain textbook from

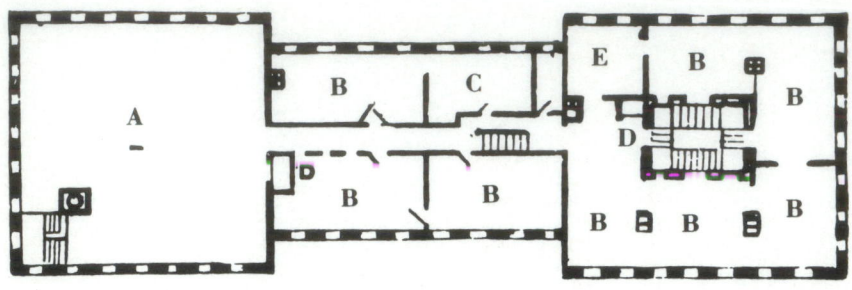

THIRD FLOOR

A Elementary laboratory; **B** Special investigations; **C** Library; **D** Elevators; **E** Photographic chamber.

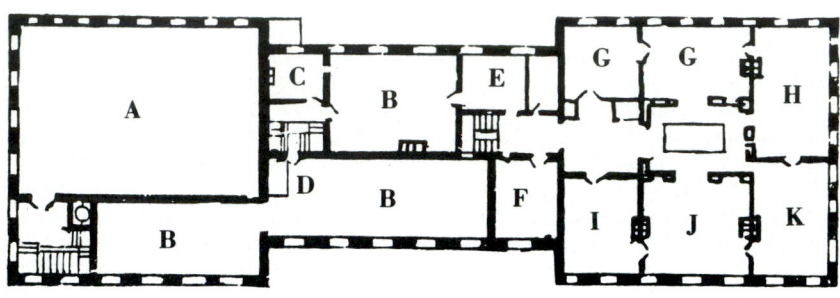

SECOND FLOOR

A Lecture room; **B** Cabinets; **C** Professor's room; **D** Elevator; **E** Professor's room; **F** Library; **G** Optical room; **H** Rumford lecture room; **I** Sound laboratory; **J** Special investigations; **K** Chemical laboratory; **R** Recitation room.

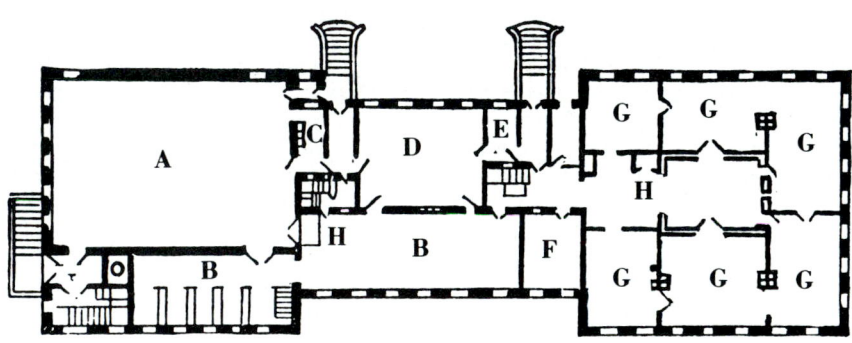

FIRST FLOOR

A Space under lecture room; **B** Cabinets; **C** Preparation room; **D** Recitation room; **E** Professor's room; **F** Balance room; **G** Special investigations; **H** Elevators.

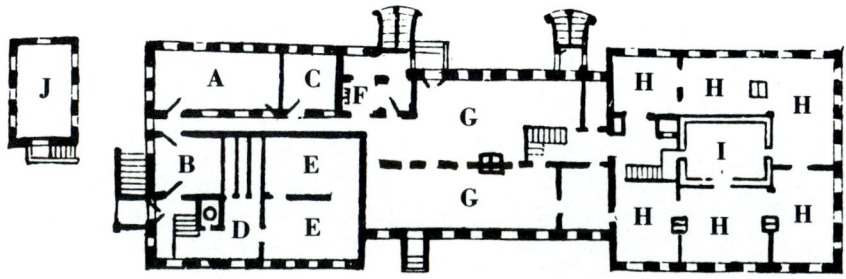

BASEMENT

A Workshop; **B** Forge; **C** Battery room; **D** Fire room; **E** Coal; **F** Mercury room; **G** Receiving room; **H** Special investigations; **I** Constant-temperature room; **J** Engine room.

our list of admission requirements. 'Why?,' he asked. 'Because,' I replied, 'it is behind the times.' 'That is just why I like it,' he said, 'bringing it up to the times means putting in a lot of improper matter.' "

Trowbridge changed all that. From 1877 on, we find him planning and scheming brilliantly to have a laboratory built in which to teach physics with laboratory exercises, "with objects and instruments in hand." Moreover he insisted that physics instruction required immersion in research as well. Harvard had already an herbarium, a Museum of Comparative Zoology and an astronomical observatory. Trowbridge asked that the analogy be extended to the case of physics.

A campaign document

The important break came when Johns Hopkins acquired its first large shipments of European scientific equipment, mostly bought by Trowbridge's friend, Henry A. Rowland, for investigations and not, as Rowland put it, "for amusing children." Adopting a strategy not new then and not forgotten since, Trowbridge wrote to Gilman: "I shall use your list as a campaign document" in raising an endowment for a research laboratory at Harvard. When Trowbridge published a collaborative paper surveying laboratory apparatus in the United States, he could show that the newly established Johns Hopkins University owned almost seven times as much physics apparatus as did Harvard.

By 1880 Trowbridge had been given his hunting license to seek the funds for

The Jefferson lab in 1984, a hundred years after it opened. "A somewhat time-worn *grande dame*, in need of refreshments," its exterior remains almost unchanged—the red brick had originally been a problem because of its slightly magnetic iron oxide content.

construction and endowment of operations. An anonymous "friend of the University" came forward to give $115 000 on the condition that another $75 000 would be raised to cover running costs. The anonymous donor turned out to be the Boston businessman Thomas Jefferson Coolidge (class of 1850). The name of the Jefferson Laboratory honors Coolidge's ancestor, the President who was not only a vocal supporter of science in America but also a publishing contributor to it. The endowment for running costs was raised through the generosity of Alexander Agassiz. Thus, as so often in the US, the project became a reality when a person with a vision met both an administrator who provided encouragement and philanthropists who recognized a moment of leadership.

When the building opened in the fall of 1884, its very design and "plainest possible" furnishings—down to the unpainted brick walls inside—were statements opposing the ornate European style of laboratory construction. Standing 300 feet from the nearest street at the northern end of the campus, outside the built-up area of Harvard Yard, the big red building must have seemed like some curious fortification. Inside, separated by a one-foot gap from all main walls, a 75-foot-high free-standing tower served as the mounting wall for the galvanometers and other electrical measuring equipment of the time. To make them swing true, all magnetic building materials had been kept to a minimum—although the brick itself turned out to be slightly magnetic, and thus vitiated

much of the work that went into making the rest nonmagnetic.

'What is electricity?'

A building is the tangible metaphor of an idea in its architect's mind. In this case, it embodied Trowbridge's passionate vision of what the physics of the future would bring and how to reach it. That conception came out most clearly in his talk, "What is electricity?," which Trowbridge gave within days of the opening of the building.[5] He began:

I must express my conviction that we shall never know what electricity is, anymore than we shall know what energy is. What we shall be able probably to discover is the relationship between electricity, magnetism, light, heat, gravitation, and the attracting force which manifests itself in chemical changes. We have one great guiding principle which, like the pillar of cloud by day and the pillar of fire by night, will conduct us, as Moses and the Israelites were once conducted, to an eminence from which we can survey the promised scientific future. That principle is the conservation of energy. . . . The ancients had a god for every great manifestation of nature—a god of peace, a god of war, a god of the land, a god of the sea. Fifty years ago scientific men were like the ancients. There was a force attached to every phenomenon of Nature. . . . But what we are to have in the future is a treatise which will show the mechanical

relations of gravitation, of so-called chemical attracting force and electrical attracting force, and the manifestations of what we call radiant energy.

Just how were these relations to be found? This was where the warrant for the new laboratory entered. "Let us strive with the most powerful instruments we have, to survey the promised land which is undoubtedly to be the possession of those who come after us." In this way, we shall "see the relations of electrical and magnetic attraction to the attraction of gravitation and to what we call chemical attraction." Indeed, all the forces of nature would be as one. The old monistic dream of science would triumph at last.

For example, consider thermoelectricity. "I have often thought that the jostling, so to speak, of these ultimate molecules of two metals at definite temperatures might form a scientific unit of electro-motive force in the future science of physical chemistry. Look at the great field for investigation there is in the measurement of what we call electro-motive force." Some future "great generalizer, like Sir Isaac Newton," would surely come along and use our store of patiently gathered data, to lay bare and explain the "ultimate motions of the molecular world."

Moving on to still more daunting problems, Trowbridge said, "We seldom reflect that gravitation is as great a mystery as electrical attaction." What is the relation between them? "I have often asked myself . . . can not the refined instruments and methods of the electrical science of the present aid us

The Harvard physics department around 1900. Among the department members shown here, sitting with instrument makers and staff on the steps of Jefferson Lab, are: (top row, second from left) Wallace C. Sabine, John Trowbridge (third), Edwin H. Hall (fourth), Benjamin O. Peirce (fifth); (second row, third from left) Theodore Lyman; (bottom row, second from left) George W. Pierce.

in more prominent lines of research? Should we not expect that, when two balls of copper, for instance, are suddenly removed from each other, a difference of electrical potential should manifest itself?"

What is called for here is "a delicate electrometer." To be sure, so far experiments along this line have given negative results. But that is only a challenge to our ingenuity: "We are like blind men in a great field of energy striving to ascertain the configuration about us with only three senses—the galvanometer sense, the electrometer sense, and the voltameter sense." If one could add to these senses, one might "become sensible of every change among atoms and molecules." And venturing to the very edge of the precipice, Trowbridge added: "Suppose that the quick passing of what we call life from the body into another shape or state of existence should be sensible as a reaction in electrical and magnetic effects. . . ."

All these dreams had been thwarted so far by the same obstacle. "We have arrived at that stage in our study of electricity where our instruments are too coarse to enable us to extend our investigations. . . . Is not the physicist of the future to have instruments delicate enough . . . to discover beats of light, as we now discover those of sound—[or] an apparatus which will measure the difference of electrical potential produced by the breaking up of composite grouping of molecules?"

At the end of his talk, Trowbridge confessed he had failed to answer his question, "What is electricity?" To be sure. But he had made a confession that now sounds curious indeed in the setting of his physics, yet is still perfectly recognizable in the context of ours. The chief task was to find a way by which to glimpse experimental evi-

dence of the unification of forces. And for that purpose, he thought, sensitive detectors are the key. The research wing of the Jefferson Lab was designed in every detail with that aim in mind, the free-standing tower serving to push down the noise level, what he called[6] "prejudicial vibrations" set up by the wind or the movement of faculty and students. Indeed, the whole building was to be a giant detector, to test his premature grand-unification ideas.

A vision for the future

Trowbridge's dream was not an unreasonable one if one considers it an extrapolation of the trajectory of 19th-century physics—from Alessandra Volta, Hans Christian Oersted, Michael Faraday and André-Marie Ampère to James Clerk Maxwell, and from Count Rumford, Julius Robert Mayer and James Prescott Joule to Hermann von Helmholtz. Nor was Trowbridge alone in his passion. His friend, Rowland, at Johns Hopkins, addressing the Electrical Conference in September 1884, spoke[7] of the need for a theory of matter that would combine electricity, magnetism, light, gravitation, heat and chemical action: "It forms the great problem of the universe . . . which *looms* up before us, and before which we stand in awe." To attain results, one condition had to be fulfilled: "Let physical laboratories arise."

If the vision of the future of science was one source of energy behind the project, a second source reinforced it powerfully. To put it simply, it was national pride, hurt by the conditions that had kept research down. The astronomer Simon Newcomb had said in 1874, good scientists in America are thought to be "necessarily of transatlantic origin." In a famous lecture[8] of 1883, "A plea for pure science," he had spoken for a growing number who were

warning Americans not to glory only in the applications of science. The Age of Electricity had brought the first telephone in 1876, the first good arc light in 1878; in 1879, Thomas Edison's incandescent bulb had lasted a full 40 hours; and all along there were improvements in telegraphy. The excitement caused by these applications was high, and the average man's life was made more pleasant by them. But Rowland thundered that scientists missed in all this the vigor of high ideas:

We are tired of mediocrity, the curse of our country. . . . We are tired of seeing our countrymen take their science from abroad. . . . Shall our country be contented to stand by, while other countries lead in the race? Shall we always grovel in the dust, and pick up the crumbs which fall from the rich man's table, considering ourselves richer than he because we have more crumbs, while we forget that he has the cake which is the source of all crumbs? Shall we be swine, to whom the corn and husks are of more value than the pearls? If I read aright the signs of the times: I think we shall not always be contented with our inferior position. From looking down we have almost become blind, but may recover.

He found it outrageous that this small and thinly populated country should at that time have 400 institutions that called themselves colleges or universities. "The whole earth could hardly support such a number of first-class institutions. The curse of mediocrity must be upon them, to swarm in such numbers. There must be a cloud of mosquitoes, instead of eagles as they profess." Barely five percent of them had 20 or more faculty members in all

Charles W. Eliot, president of Harvard University at the time the Jefferson lab was built. He espoused the relatively novel view that education should include science and modern languages as well as classics and theology.

John Trowbridge joined the Harvard physics department in 1870. By 1880, he had persuaded Eliot that physics should be taught in a laboratory; funds were raised and the laboratory was built in four more years.

fields, or an endowment for all purposes of over $500 000. Fewer still had more than 300 students in all fields.

Worst of all, research-minded physicists were poorly dealt with. "Life is short: old age comes quickly, and the amount one pair of hands can do is very limited. . . . I know of no institution in this country where assistants are supplied to aid directly in research; yet . . . there are many physical problems, especially those requiring exact measurements, which cannot be carried out by one man." Large, well-equipped, well-staffed physical laboratories must be created, on the model of the endowed observatories that have advanced astronomy so well. "The time has even now arrived when such a grand laboratory should be founded." Then indeed a sound science of physics will "arise among us, and make us respected by the nations of the world. . . . [I] have the feeling, common to all true Americans, that our country is going forward to a glorious future, when we shall lead the world in the strife for intellectual prizes as we now do in this strife for wealth."

Great expectations

When the Jefferson Lab opened its doors, both scientific and patriotic hopes must have been focused on it with great expectations. Here at last was a facility with several faculty members (Lovering, Trowbridge, Benjamin O. Peirce, Edwin H. Hall, Harold Whiting and chemist Wolcott Gibbs). There were assistants and students, well-appointed machine shops, an engine room, a battery room, an optical room, a constant-temperature room, a chemical laboratory, a mercury room, a photographic chamber, not to speak of lecture halls, recitation rooms, a library, instrument cabinets, and professors' rooms.

In truth, for the first few years the fruits reaped at the lab were not outstanding. But there was splendid space there, equipment and hands and funds; some very talented students; and an unflagging sense of mission that kept Trowbridge active for decades.

By the time he retired as director in 1910, the faculty included Wallace C. Sabine, George W. Pierce, Theodore Lyman and Percy W. Bridgman, all of whom had done their graduate work in the Jefferson Lab (with Lyman and Bridgman getting their bachelor's degrees at Harvard as well).

The building—as all good buildings—proved adaptable to the constantly changing and unforeseeable needs of the research carried on in it. Sabine was able to use the constant-temperature room and the lecture hall for his acoustical studies. The battery room could, early in this century, deliver the potential differences needed for Lyman's spectroscopic discharge tubes. The isolated central tower—less than ideal for the sensitive galvano-magnetic experiments for which it was intended—served well to reduce the vibrational noise when Robert Pound and Glen Rebka came to measure the gravitational redshift of γ rays. In the machine shop, staffed with fine craftsmen such as Charles Chase, presses could be built with which Bridgman achieved record pressures.

The physicist as both teacher and superb researcher—a novel idea when Trowbridge had proposed it for the endorsement of Eliot and Coolidge—had become a reality by the early part of our century in a growing number of centers. The groundwork had been laid for this laboratory, together with similar institutions elsewhere in this nation, to play its role as physics in America was entering on its maturer phase—still propelled by the enthusiasms to which Trowbridge and Rowland had confessed.

References

1. C. W. Eliot, Annual Report of the President of Harvard University, 1883–84, p. 43.
2. C. W. Eliot, in *The Development of Harvard University*, S. E. Morison, ed., Harvard U.P., Cambridge, Mass. (1930), p. 378.
3. C. W. Eliot, Inaugural Address as President of Harvard College; reprinted in *A Turning Point in Higher Education*, N. M. Pusey, ed., Harvard U.P., Cambridge, Mass. (1969), p. 21.
4. E. H. Hall, in *The Development of Harvard University*, S. E. Morison, ed., Harvard U.P., Cambridge, Mass. (1930), p. 277. In addition to Hall's essay, there are several other useful sources on the Department's history: L. I. Aronovitch, "Towards a New Knowledge of Nature: Physics at Harvard University, 1870–1910" AB Honors Thesis in History and Science (1983); S. Goldberg, "History of Physics at Harvard University, 1907–1912," Cambridge, Mass., 1962 (unpublished manuscript); [T. Lyman] *The Physical Laboratories of Harvard University*, printed at Harvard U.P., Cambridge, Mass. (1932); K. R. Sopka, "Physics at Harvard during the Past Half Century: A Brief Departmental History" (unpublished manuscript). The literature on American science in the 1880s is large and the scholarship very strong. A good bibliography for a first look can be found in D. J. Kevles, J. L. Sturchio, P. Carroll, Science **209**, 27 (1980).
5. J. Trowbridge, *The Popular Science Monthly* November 1884, p. 76 ; address before Section B, AAAS, at the "Electrical Conference," Philadelphia, 4 September 1884.
6. J. Trowbridge, Science **5**, 230 (1885).
7. H. A. Rowland, Address of 8 September 1884, in *The Physical Papers of Henry Augustus Rowland*, Johns Hopkins U.P., Baltimore, Md. (1902), p. 619.
8. H. A. Rowland, Science **2**, 242 (1883); delivered at the AAAS meeting, 15 August 1883.

□

The Harvard summer school in astronomy

The annual conferences that Harlow Shapley organized from 1935 through 1942 fostered the growth of astrophysics in the US, and were the model that inspired similar sessions on other campuses.

David H. DeVorkin

PHYSICS TODAY/JULY 1984

Professional astronomy in the 1920s and 1930s was a science in transition. Modern relativistic cosmologies were born, and three major areas of modern physics became central to progress in astronomy: quantum mechanics, nuclear physics and relativity.[1] The old empirical and qualititative methods of spectroscopic astronomy, methods that generated vast amounts of systematic knowledge of the spectra of the Sun and stars and the dynamics of stellar systems, were being supplemented by rational and quantitative methods. These new techniques promised to reveal not only the compositions of the Sun and stars, their sources of energy and their ages, but also the origin and ultimate fate of the universe.

Relatively few classically trained astronomers were at first able to take an active and creative part in the new astrophysics. In the late 1920s, a general awareness grew among astronomers that future progress in their field lay in the application of modern physical theory to the many problems that intrigued spectroscopic astronomers. But who was capable of doing

David DeVorkin is associate curator for the history of astronomy, in the space science and exploration department at the Smithsonian Institution's National Air and Space Museum, in Washington, D.C.

this? In the early 1930s, the few institutions in the United States offering a doctorate in astronomy were producing an average of 10 PhDs per year.[2] Most of the graduate students wrote dissertations on classical subjects tied closely to the observational programs of their parent institutions and very far from astrophysics, the forefront of astronomy.

Harlow Shapley, the young director of the Harvard College Observatory, firmly believed at the time that in the future, "astronomy wasn't going to be only run by astronomers, but in the sort of new astronomy we are going to need physicists."[3] Until Otto Struve, the new director of Yerkes Observatory, launched a search in the 1930s for staff to strengthen training in theoretical astrophysics at the University of Chicago, he felt that the US had only one center for such training, and that was under Henry Norris Russell at Princeton. Russell was Shapley's teacher and a pivotal transition figure in the maturing of modern astrophysics. He lectured widely on the central role of physics in the new astronomy, stating in 1927 that astronomy had become a branch of physics, and that the boundary between the fields was practically nonexistent.

When Shapley succeeded Edward C. Pickering as director of the Harvard

College Observatory in 1921, he quickly exploited the rich legacy of spectroscopic data that had been amassed there over the past four decades. While Pickering's staff had collected vast quantities of stellar spectra and brightnesses, Shapley collected staff and increased institutional focus and prominence in areas now identified with modern astrophysics. At Harvard and Radcliffe, he established an astronomy graduate program, which in 1923 attracted the brilliant Cambridge University student Cecilia Payne, and later brought in Donald Menzel as an instructor who also became a center of attraction for students desiring training in modern astrophysics.

Shapley's staff-building in the 1930s yielded impressive results. But these were the depression years, and it was far from clear that he was going to achieve the economic stability necessary to maintain the constant flow of professional visitors from abroad, the staff of young superstars, and funding for the best students. Shapley therefore looked for innovative schemes that would, on the one hand, increase Harvard expertise in modern astrophysics, but would also increase the prominence and prestige of Harvard astronomy, and thereby further ensure continuing support.

In this article I focus on one such

Informal discussion during an afternoon tea at the Harvard summer school in astronomy, 1939. From left to right in the foreground are Bart Bok, Justice Felix Frankfurter, E. Lindsay, Jan Oort and Harlow Shapley. (Photography by Dorothy Davis Locanthi, AIP Niels Bohr Library, Dorothy Davis Collection.)

innovative scheme—the Harvard summer school in astronomy, which Shapley established in 1935 after several unsuccessful attempts. What Shapley's plans and motives were for the school, how he gained support for it and how it was structured help to illustrate the state of modern astronomy and astrophysics at Harvard and in the United States generally during the mid-1930s.

Also, the origins, development and influence of the Harvard summer school in astronomy, which was the first of its kind in the United States, reveal one way an emerging hybrid discipline sought to achieve the status of a fully integrated science. And finally, recalling this episode in the history of astronomical education and professional development aids in understanding the phenomenon of the growth of such summer schools and conferences, which are quite common today in hybrid disciplines.

No published accounts of the lectures, addresses or general proceedings of the Harvard summer school have been discovered. This article is based, therefore, upon surviving records of the summer school in Shapley's papers at Harvard, and upon extensive interviews with the many summer school participants, conducted in recent years under the auspices of the American Institute of Physics Center for History of Physics.

Origins of the summer school

Shapley used the University of Michigan summer school in theoretical physics as a model for the Harvard summer school. Established in 1927 by Harrison M. Randall, chairman of Michigan's physics department, the Michigan summer school became famous as a place to learn about the latest advances in modern physics through lectures by distinguished and active practitioners, and through intense leisure activities and informal living arrangements, all designed to bring people together to talk physics.[4] This model was tailor-made for the highly social Shapley, who, with Bart Bok and Harry H. Plaskett, attended one of the Michigan sessions and came away with the conviction to implement the idea at Harvard.[5]

Possessing both superabundant energy and legendary charisma, Shapley had a long and successful record of getting support for visiting scientists to lecture on their specialties. In 1929, Edward Arthur Milne from England, Svein Rosseland from Norway and Albrecht Unsöld from Germany gave lectures at Harvard on their stellar atmosphere research. Milne was in the United States to attend sessions of the Michigan summer school, and Shapley persuaded him to stop at Harvard. Unsöld was in the US to exploit Mount Wilson's spectroscopic plate collection and solar instrumentation.

Shapley was not a theorist, although he naively dabbled in occasional theoretical speculation. He liked to associate with the illustrious theorists who visited Harvard and he worked hard to get them to stay, as did Struve at Chicago. While the visits arranged by Shapley proved to be memorable, they were sporadic and piecemeal. His own attendance at the Michigan summer school in the late 1920s convinced him that institutionalizing these visits by distinguished theorists was the best way to develop a strong theoretical component at Harvard.

Plaskett, a junior Harvard faculty member from the Dominion Astrophysical Observatory in British Columbia, and a spectroscopist in need of a firmer understanding of physical theory, encouraged Shapley to reproduce the Michigan model at Harvard. Harvard did have a general summer program of courses for nonspecialists and graduate-level research courses. During these depression years, however, even the popular undergraduate courses were running a deficit—a poor situation because the summer school had to operate on a profit-making basis. Still,

Shapley initiated a plan in 1932 to invite several specialists in the summer, "as an entering wedge" for the summer graduate school.[6] But the plan failed.

After a poor showing that summer, the director of the summer school decided that the astronomy offerings should be dropped in favor of more profitable courses. Shapley sarcastically pointed out that by the same argument the summer school itself was a luxury, and so was the observatory and the entire astronomy program, so they should all be dropped. Shapley objected to the profit motive, and felt that the summer school was weak because of it. He pointed[7] to the prestige of the Michigan summer school and several other schools around the country, and added that Harvard could easily build up "the leading summer school of graduate astrophysics on the planet" at an expense of some $3000.

Shapley wished to provide continued advanced training in the summer months so that the observatory would not become dormant. Harvard College Observatory possessed the largest photographic and spectroscopic plate collection in the world—a vast and unique resource for stellar and galactic astronomy. Astronomers worldwide sought summer research appointments at the observatory, and for years Harvard's financial resources were sufficient to bring visitors for extended periods of time. With the depression, these resources disappeared, and Shapley knew that a program such as a Harvard graduate summer school in astronomy would offer replacement funds to bring in these visitors, who in turn would greatly enrich Harvard projects and prestige by their presence and collaborative efforts.

In the fall of 1934, Kirtley Mather, a geologist and strong campus ally of Shapley, became the new director of the Harvard summer school of arts and sciences. Through Mather, Shapley was finally granted the necessary funding, and plans went ahead for a summer school in astronomy in 1935. Shapley felt that with the requested $3000, primarily for stipends for outside lecturers, and "some vigorous volunteer help on the part of our own staff . . . it should not be difficult to make the Harvard department and Observatory a regular summer mecca for astronomers of the eastern colleges and universities. . . ."[8]

Shapley saw "glory, publicity, service, and scientific advance" in the graduate summer school, and believed it to be a pioneering effort in astronomy, as well as an example for other Harvard departments. Aside from examination of this rhetoric, one can get some indication of Shapley's goals by

looking at whom he invited to lecture in the seven years of the summer school.

The first session

In 1935, Shapley invited Struve, Jan Oort, Antonie Pannekoek and Ira S. Bowen to be his primary outside lecturers. Struve, director of the Yerkes Observatory, practiced spectroscopic astrophysics and advocated the blending of astronomy and theoretical physics. Oort, of Leiden, made the first observational demonstration of the differential rotation of our galaxy. Pannekoek, the oldest summer-school participant, was legendary in statistical astronomy and spectroscopy. Only Oort was unable to attend the 1935 sessions, although he came in later years.

These three invitees were astronomers. The fourth, Bowen, was approached because, as Shapley noted to Mather, he was "more physical than astronomical." Bowen specialized in optics and laboratory spectroscopy. A student and associate of Robert A. Millikan, he had in 1927 discovered that nebulium, the enigmatic strong spectral feature of nebulae first detected in the 1860s, was caused by forbidden transitions of ionized oxygen and nitrogen. MIT also invited Bowen to participate in its summer conferences in spectroscopy that year, so Shapley hoped to share Bowen with MIT, and saw the liaison as a fruitful step that would be good for astronomy, physics, Harvard and MIT. He noted to Mather: "The MIT Group will in turn feed into our longer and more serious astrophysical efforts."

Struve's presence would be beneficial to many of the Harvard staff, including Menzel, Fred Whipple, Theodore Sterne and Cecilia Payne (Payne-Gaposchkin after 1934), all of whom were interested in stellar atmospheres. Both Bowen and Menzel were concerned with the nature of nebulae. Pannekoek could aid in Bok's research on galactic structure and Whipple's interest in the spectrophotometry of Cepheid variables. Shapley told Bowen that Payne-Gaposchkin and Menzel would both follow Bowen's lectures closely, but that because most astronomers were relatively untutored in physics, some elementary groundwork had to be laid.

Shapley also needed assistant lecturers and demonstrators. He turned to past students and also brought in promising junior faculty from other universities and observatories. His former students were delighted at the prospect of a summer at Harvard. Shapley offered a chance "to gather here in a pleasant group a lot of sources of astronomical inspiration and information. . . ." To this, Peter Millman replied[9] that he was very anxious to

keep studying and learning, "and in a place like Toronto where there is no group interested in astronomical research I think that something like [the] advanced summer school is very important." Toronto had just installed a 74-inch reflector, so it was not due to lack of instrumentation that Millman longed for Harvard.

To increase enrollment, Shapley advertised in *Science* and in such astronomical publications as *Popular Astronomy* and *The Telescope*. Notices even appeared in local newspapers. Clearly, he hoped to attract many teachers of astronomy who could benefit from renewed contacts with practicing astronomers and physicists.

The summer school's first year included courses on the structure of the Milky Way, organized by Bok, "Cosmic Physics" with Bowen and Menzel, "Astrophysical Problems" with Pannekoek, and "Stellar and Interstellar Problems" with Struve. There were also general courses for undergraduates, refresher courses, as well as a series of topical seminars—Shapley's famous "Hollow Squares" colloquia—and popular evening lectures given in conjunction with the Amateur Telescope Makers of Boston. The "Hollow Squares" seminars, named for the arrangement of the tables, centered around short presentations of work in progress, or recent articles in the literature, and were free-for-alls that allowed for the fruitful exchange of ideas. They have been recalled by a good number of participants as memorable and valuable experiences.

Shapley put Bowen and Menzel together as co-lecturers on "Cosmic Physics"—the application of atomic theory to the study of gaseous nebulae and to the interpretation of stellar and nebular spectra. Menzel had already contributed considerably in this area, and throughout the remainder of the decade, he collaborated with a number of younger Harvard staff and students to produce a long and seminal series of studies on ionization phenomena in gaseous nebulae. Of course, Shapley made no secret of such manipulations. In writing to potential outside lecturers, he often noted their value to the research interests of his own staff.

Soon after the close of the 1935 sessions, Mather asked Shapley about his plans for the next year, noting the importance of 1936 as Harvard's tercentenary. Shapley responded[10] that it would be no difficulty to bring a distinguished body of astronomical visitors to Harvard again because

The 1935 Summer School of Astronomy apparently made a great impression among American and European astronomers. Five prominent European astronomers

Harlow Shapley. (From AIP Niels Bohr Library, Shapley Collection.)

have already slyly hinted as to their availability, and the comments that are coming in from American astronomers have been most gratifying. It was an astonishingly good job. It should help to indicate to other departments that the name and reputation of Harvard are both assets in any attempt to do first-class graduate work in the summer school.

Endorsements from 1935 session participants were indeed positive, although surviving letters all postdated Shapley's remarks to Mather. From Pannekoek, Shapley received a warm letter[11] in late October supporting the school,

... where young astronomers gather to hear and discuss some important selected parts of science in full, as a unit, whereas in ordinary conferences of Astronomical Societies the attention is divided in[to] a big number of small separate points. I am glad I was able to attend to this experiment.

Part of Shapley's success in getting astronomers to participate derived from his talent to make them feel their specialties were being respected. To Frank Schlesinger, director of Yale University Observatory and a major figure in positional astronomy, Shapley indicated that in 1936 the astrometric offering would expand to include lectures by Peter van de Kamp, then of the University of Virginia and an ardent disciple of Schlesinger's photographic techniques. Schlesinger, who participated in the 1935 sessions, in turn heartily endorsed[12] the summer school as a "feast of thoroughly cooked dishes," and expressed personal satisfaction at being one of the cooks.

While the primary accomplishment of the 1935 session was to increase contact with modern issues in observational astrophysics, it also succeeded in bringing attention to the existence of Harvard's astronomical programs. The following years did see an increase in the modern physics offerings, but the emphasis remained within observational areas, even though many unique interdisciplinary topics were developed.

Subsequent sessions

For the 1936 sessions, Shapley planned a multidisciplinary course to attack the problem of the cosmic time scale. Thus he invited lecturers in physics, geology and astronomy. He asked Alfred C. Lane from Tufts to speak on the age of the Earth, Georges Lemaître to lecture on the expanding universe and derived cosmological time scales, and Knut Lundmark, a Swedish specialist in observational cosmology and a general encyclopedist, to talk on various aspects of cosmology and the history of astronomy. Others invited included Schlesinger and Russell and, from the Harvard staff, Whipple to talk on the theory of orbits, Sterne on theoretical modeling of stellar interiors and stellar time scales, and Bok to speak on the general problem of time scales, especially the problem of the vast differences between time scales derived from different measurements—disruption times of star clus-

James Baker's summer school diary—annotated excerpts

Student's view. James Baker took this photograph during the 1935 summer session. It shows Bark Bok (center) engaged in conversation, with Otto Struve (left) and Peter Millman (right) looking on. (Photograph courtesy of Owen Gingerich.)

Twenty-year-old James Baker arrives at Harvard from Louisville, Kentucky, on the first of July, 1935. After a stop at the local YMCA for a map and a shower, he walks to the observatory and immediately runs into Bart Bok, who appears "very keen and speaks very quickly but with a peculiarly Hollandish accent. He is very interesting." Within hours Baker attends lectures by Donald Menzel on planetary nebulae, and by Bok on stellar statistics: "He began with some fierce-looking equations, although I believe it's mostly my ignorance of the notation and symbols." That evening there is a meeting at the observatory; next week there will be a joint spectroscopic conference at MIT and a picnic at Oak Ridge.

2 July: "Things are happening here so rapidly that I can hardly keep up...." Baker registers for Otto Struve's course, "Stellar and interstellar problems," noting "It's going to be very technical." "All four of the lectures have assumed a pre-knowledge of the subjects.... Dr. Bok is certainly very nice... very witty... he was about to give me $15 to register ... but Miss Mohr found out that I can have my money anytime." Baker also attends lectures by Ira Bowen and Antonie Pannekoek.

3 July: "... Dr. Bok hunted me up ... and provided a desk.... Everyone treats me with courtesy and consideration.... This afternoon I attended the first colloquium and had tea beforehand.... I walked around the campus.... The place is covered with antiquity and majesty."

5 July: "I wonder what I may do research upon? ... Among the students one hears of comparators, calculating machines, densitometers, photoelectric photometers, and other modern improvements. The atmosphere here at Harvard is one of the main charms.... We had tea and cakes again at the colloquium and talked a great deal...."

6 July: "I'm sure that I shall want a brief vacation after Summer School is over. I've never studied just one subject for such a long time before."

13 July: "Books are almost entirely disregarded and original papers are read. I attended a beautiful lecture on the local cluster by Dr. Bok ... and attended a ... colloquium after tea. The colloquium was excellent. Subject: 'The Color Temperatures of the Stars,' by Dr. Pannekoek. I certainly admire this old astronomer of about 65. He is probably the best lecturer here and has done an enormous amount of work. He and Dr. Bok spoke Dutch today."

15 July: "Dr. Bok gave me a book today which he said wasn't on my book list. I'm becoming very much interested in the lectures on the Milky Way."

17 July: "I've just been to a very interesting colloquium on the time-scale, from which we concluded that our atoms are something like 3×10^{10} years old.

... [At] the class conference... I was instructed to learn about nebulous stars and to begin some kind of research upon them."

18 July: "After reaching magnificent MIT, we listened to about six lectures on spectroscopy. (Even the blackboards go up and down by electric motors and the lighting is excellent.) Afterwards a few of us visited the spect. lab. downstairs. One room contained a 21-foot circular spectrograph and another a ten meter spectrograph completely free of the building, a building within a building. We saw interferometers ... photography, all elements and sources of excitement, 60 000 line gratings in the original, etc."

19 July: "Meeting men who have been myths in the past to me is common p[lace] now. Present today [at the picnic] were Dr. K. T. Compton, Struve, Bok, Menzel, Bowen, Boyce, Wolf, Mack, Whipple, and other dignitaries. Dr. Bowen, famous for identification of nebular lines, drove me 30 miles back to Cambridge, and discussed ordinary affairs with his wife and me."

24 July: "My original research went to pieces on nebulous stars because the looked-for abnormal reddening was due to. Dr. Struve's not reading the whole article and the author's misuse of the term 'color-excess.' Now I have a very good problem on determining the distribution of cosmic calcium clouds in the Galaxy, [to] stellar distances of 300 [parsecs] by the galactic rotation method." Struve advises Baker in this research, but Bok cautions that Plaskett obtained no results by this method.

28 July: Menzel advises Baker on courses for the fall. "I shall probably take Astrophysics, Stellar Statistics, Introduction to Astronomical Research, Theoretical Physics, and either Complex Variables or Theory of Functions...."

3 August: "... a most unusual and enjoyable day. All of us obtained rides for the 50-mile journey [from the observatory to the summer beach home of a friend of Shapley's for archery, baseball, deck tennis and swimming]. After watching Dr. Shapley sit down by himself on the lawn, take out his bottle of rubbing alcohol and put in a variety of ants (300 species in Massachusetts), we ate lunch."

19 August: "Dr. Shapley gave me a nice project in working on Beta Lyrae.... I afterwards spoke to Dr. Bok who told me that I had gotten an A in my course.... I also purchased my first paper for 10 days. I knew nothing of Will Rogers' or Post's deaths nor of the impending diplomatic crises."

*   *   *

Excerpted from letters and diaries compiled by Elizabeth Baker and James Baker, 9 August 1983, and kindly made available by Owen Gingerich.

FIGURE 1.

Harvard summer school, 1936 session. Seated, left to right: Loring B. Andrews, Donald H. Menzel, Fred L. Whipple, Knut Lundmark, Leon Campbell Sr, Harlow Shapley, Meg Nad Saha, Theodore Sterne, Rupert Wildt, Jenka Mohr, Paul Merrill, Sergei Gaposchkin. Standing, left to right: Henrietta Swope, Carroll Anger Rieke, Florence Campbell Bibber, Leah Allen, James Cuffey, Helen Dodson, James G. Baker, Carl Seyfert, Alice H. Farnsworth, Harold Lane, Helen L. Thomas, Wallace Eckert, Lois Slocum, Frances W. Wright, Jesse Greenstein, George Z. Dimitroff, Rita Paraboschi (Mrs. Cuffey), Herbert Grosch, unidentified, Elizabeth Baker, Barbara Cherry (Mrs. Martin Schwarzschild), John Evans, Charlotte Klein, Daniel Norman, William Calder, Frank K. Edmondson, Richard Emberson, Richard Leary, Leo Goldberg, Dorrit Hoffleit, Rebecca Jones, Bart J. Bok, Sidney W. McCuskey, Samuel L. Thorndike, (Bancroft Sitterly or Martin Schwarzschild?), Arthur Sayer. (Photograph courtesy of R. S. Choudhury. Identifications courtesy of James Baker, Owen Gingerich and Martha Liller.)

ters, rates of stellar evolution, the Hubble constant.

This second year was the most active and hectic year of the summer school's short life. Plans continually changed. Lemaître could not attend, and a difficult situation arose when Harvard imposed a loyalty oath upon all outside lecturers, demanding that they pledge allegiance not only to the United States, but to the Commonwealth of Massachusetts. Many initially declined; Schlesinger of Yale noted that if Connecticut were to go to war with Massachusetts, he would have to side with his own state. Shapley was humiliated by this outrageous requirement, and eventually had it waived, but not before Lane was dropped from the roll. Replacing Lane was William D. Urry of MIT, a specialist in meteorite ages and a frequent collaborator with those at Harvard interested in meteoritics, including the Estonian peripatetic genius Ernst Öpik. Another replacement was the recent German emigré, Rupert Wildt from Göttingen, who was one of the first of many Europeans to enjoy the support of the Harvard summer school in seeking a haven from the iniquities of the Nazis.

To observe Harvard's tercentenary in proper style, Shapley planned commemorative ceremonies and symposia. He invited the American Astronomical Society to have its 1936 summer meetings at Harvard, and asked many people to stay the entire summer. Shapley invited Meg Nad Saha, primarily as a representative of India to the tercentenary, secondarily as a participant in the summer school, and finally as a commentator at the AAS meetings. Shapley thought it was most important to bring Saha, "who is the father of the Saha Theory and therefore the grandfather of about forty Harvard Observatory papers and five hundred from elsewhere."[13] Saha, the first to apply the theory of ionization to solar and stellar atmospheres in 1920, was an attraction for many younger astronomers, as well as for Russell, who was to lecture on the composition of the stars, and for Pannekoek, who lectured on the stellar temperature scale. According to Shapley, Saha would be in good company at Harvard that summer, as Niels Bohr, Werner Heisenberg, Albert Einstein, Arthur Compton, Arthur Eddington and Millikan would be there. While we do not know how many of these people actually came, we do know that Shapley did not hesitate to advertise the possibility as an inducement for others to attend.

Saha saw the opportunity to spend the better part of the summer at Harvard as a happy thought. He was isolated in Allahabad, and was not able to pursue astrophysics there. This opportunity to travel rekindled many of his research interests. En route to Harvard, Saha attended Bohr's conference on nuclear physics at Copenhagen in June 1936, and brought with him to Harvard a transcript of the conference. At the last moment, Shapley asked Saha to give a colloquium on the proceedings of that conference. In those years, astronomers were becoming aware that the key to understanding the energy source of the Sun and stars was to come from nuclear physics. Several at Harvard, including Bok and Sterne, were interested in such problems, as were many of Harvard's visitors and short-term lecturers, such as Martin Schwarzschild and S. Chandrasekhar. After 1938 and Hans Bethe's announcement of a fusion mechanism that explains the power of the Sun, Shapley made sure that a representative from nuclear physics was in attendance at the summer school. He couldn't bring Bethe, who had his pick of choice summer research appointments, but in 1939 he did manage to bring Robert Marshak, one of Bethe's young students and collaborators.

As in 1935, the 1936 summer sessions dealt heavily with spectroscopic astrophysics. With people from Mount Wilson, Yerkes, Harvard and elsewhere in attendance, a fair fraction of the country's spectroscopic workers capable of applying physical theory were together discussing mutual interests.

Henry Norris Russell and Harlow Shapley traveling to the 1938 International Astronomical Union meeting in Stockholm. (Photograph by Dorothy Davis Locanthi, American Institute of Physics Niels Bohr Library, W. F. Meggers Collection.)

ogy, looked forward to returning to Harvard in 1937, chiefly because it would give him a chance to become more familiar with observational cosmology, specifically observational aspects of general relativity. He and Shapley corresponded frequently on various matters of mutual interest, which widened during the summer of 1937. Bok recalls that Robertson's lectures and contributions to the Hollow Squares seminars were stimulating highlights. Robertson later noted to Shapley that his summer contacts rekindled an interest to examine the two-body problem in general relativity, the luminosities of receding galaxies and the dynamics of stellar systems.

The 1937 sessions, now called "Summer Conferences," came through on a balanced budget. An elated Mather was thus shocked to learn that no astronomy would be offered in 1938, because all astronomers would be traveling to Stockholm for the International Astronomical Union general assembly. However, the conferences resumed in 1939, and continued through 1942, when the war disrupted all normal activities of the Harvard College Observatory.

In the last three years of the summer school's existence, Robertson, Slater and Marshak attended. Lyman Spitzer, then of Yale, discussed his recent revelations that ruled out the encounter theory as an explanation of the origin of the solar system and forced astronomers to rethink how the system could be formed at all. Russell visited again and talked on many different topics, and many prominent displaced scientists from Europe lectured. Starting with the 1937 session, Menzel became the general manager of the program. His participation ensured a strong, consistent focus on the application of techniques of modern physics to problems in astrophysics.

Influence of the summer school

From what we have seen, it is clear that the summer school was not founded for the presentation of formal papers, or for the formal review of current research. Its purpose was advanced training, and retraining, of professional astronomers, students and teachers. However, the summer school was far more complex in motive and structure than this. It was a product of three conditions: the problem of ensuring the survival and continued health of a scientific institution during hard economic times, the remarkable blend of Shapley's social and professional styles, and the need for classically trained astronomers to learn the techniques of modern physics.

Harvard graduate student James Baker's letters, partially abstracted in the box, Figure 1, and the correspondence

Another theme of the AAS meetings and tercentenary sessions was a "Joint Session of Mathematicians and Astronomers," again organized by Shapley. Here, Eddington discussed "cosmological constants," and Tullio Levi-Civita spoke on "relativistic problems of several bodies." The heavy dose of mathematical cosmology proved quite stimulating to Bok and to Princeton cosmologist Howard P. Robertson, and encouraged Shapley to include a course by Robertson on the theory of relativity in two following summer sessions.

Shapley noted to himself early in October 1936 that he wanted Robertson to return to teach an extensive course on mathematical relativity. Shapley asked John Slater of MIT to introduce quantum mechanics to astronomers, and Brian O'Brien of Rochester to review stratospheric exploration techniques and ultraviolet solar research. Saha also visited in 1937, and both he and O'Brien discussed in detail their hopes for balloon-borne stratospheric solar observatories. Shapley also invited Charles Edward Kenneth Mees of Kodak, a longtime friend of astronomy, to give lectures on photographic theory.

Robertson, who was already a recognized figure in mathematical cosmol-

and lecture notes of other participants, reveal that the summer school sessions were intense and happy times—a dream world for young students suddenly immersed in the intense Harvard atmosphere of study and play. Expertise on modern physics was readily accessible, as were the best spectroscopic data for projects assigned or self-created. And the Harvard atmosphere indeed survived, year-round, during these otherwise hard years.

In 1937, Shapley prepared for Mather a summary of the value of the summer school. He observed that it was unique in astronomy, bringing together leading astronomers, physicists, college instructors and students for mutual enrichment and refreshment. All participants, felt Shapley, gained insight into both old and new problems through the many technical discussions and informal contacts. Shapley was supported by many testimonies from those who attended. Pannekoek came away with an increased understanding of the physics behind the stellar temperature scale and a better appreciation for the physical meaning of stellar spectra. Saha's excitement for astrophysical problems was rekindled. The summer school helped Marshak gain access to computational resources at MIT and in New York that could further his studies of stellar structure, and in 1940 Bok reveled in William W. Morgan's successful defense of his new two-dimensional spectroscopic classification for stars in discussions with Menzel, Payne-Gaposchkin and Russell.

Shapley pointed out that new research programs emerged from the weekly "Hollow Squares" discussions, as well as from the numerous contacts visitors had with each other and with the data resources available at Harvard. Thus Robertson looked anew at various astronomical phenomena in the light of relativity theory, while others used the Harvard plate vault during their visits, and acknowledged its value in subsequent research.

Beyond gaining training in, and increased awareness of, new research programs, some participants, mainly the younger ones, enjoyed and profited from professional exposure at the summer school. One astronomer recalls that after he gave a lecture with Russell in the audience, Russell recommended him for the vacant directorship at Sproul Observatory: Thus Peter van de Kamp got the resources that allowed him to embark upon his life-long astrometric study of binary systems.

The most elusive legacy of the Harvard summer school, but possibly the most lasting, was the great emotional feeling of collegiality the experience offered. Harvard astronomy under

Shapley was unique. Upon returning to India after the 1936 session, Saha wrote[14] to Shapley: "After two months at your observatory, one gets a bit spoiled and would very much like to have the characteristic Harvard, or—should I say the Shapley atmosphere!" The remarks of at least a dozen physicists and astronomers interviewed support this statement.

Bok recalled that on the average 50 to 60 people were drawn each year to all or part of the summer session, and that most were astronomers because few physicists had begun to appreciate the amenability of astronomical problems to physical explanation. It was usually impossible to tell how many people were actually in attendance at the summer sessions, because so many were constantly dropping in for shorter informal visits. Among those officially enrolled, according to surviving records in the Harvard University archives, astronomers did comprise the majority. One can identify some 65 astronomers who participated in some way during the first three years. Of these, 27 were from Harvard, 9 were from overseas, 13 were from large astronomical institutions and 16 were from smaller teaching institutions. Fourteen states were represented, as was Canada. The most prominent names were Europeans, and in addition to the 65 astronomers, one can identify 13 physicists.

The Harvard summer school in astronomy did not survive World War II. The ever-widening war changed many activities and priorities. In 1942, Harvard University went on a 12-month academic schedule, and special summer sessions designed for wartime training replaced all regular summer-school activities.

After the war, Shapley's political and social activities drew him away from the observatory, and his administration became more distant and indirect. Many activities ceased to function with their pre-war vigor, and one of the casualties was the summer school. It reverted to its original pre-1935 character, offering a small collection of introductory courses and a few directed research programs.

The legacy of the Harvard summer school in astronomy is not only the research it fostered as a forum for the introduction of modern physics into classical spectroscopic astronomy, but also the post-war summer schools founded elsewhere by Harvard participants. Organizers of conferences at Michigan, Berkeley and elsewhere recall the Harvard summer school as a memorable and intensely human episode in the maturing of modern astrophysics in America.

* * *

I would like to thank Karl Hufbauer for suggesting this study. Both he and Owen

Gingerich provided much useful information and commentary. The archives at Harvard University, the California Institute of Technology and the American Institute of Physics Center for History of Physics kindly made available letters and material from oral-history interviews.

References

1. Recent studies by historians of this transition period include: R. W. Smith, "The Origins of the Velocity–Distance Relation," Journal for the History of Astronomy (JHA) **10**, 133 (1979); R. F. Hirsh, "The Riddle of the Gaseous Nebulae," Isis **70**, 197 (1979); J. M. Crelinsten, *The Reception of Einstein's General Theory of Relativity by American Astronomers, 1910–1930*, unpublished PhD dissertation, University of Montreal (1981); K. Hufbauer, "Astronomers Take up the Stellar Energy Problem," Historical Studies in the Physical Sciences **11**, 277 (1981); D. H. DeVorkin, R. Kenat, "Quantum Physics and the Stars (I): The Establishment of a Stellar Temperature Scale," JHA **14**, 102 (1983), "Quantum Physics and the Stars (II): Henry Norris Russell and the Abundances of the Elements in the Atmospheres of the Sun and Stars," JHA **14**, 180 (1983). Astronomers and historians have written numerous review papers on this period. See David H. DeVorkin, *The History of Modern Astronomy and Astrophysics: A Selected, Annotated Bibliography*, Garland Press, New York (1982).

2. R. Berendzen, M. T. Moslen, in R. Berendzen, ed., *Education in and History of Modern Astronomy*, Ann. New York Acad. Sci. **198**, 46 (1972).

3. Bart Bok Oral History Interview, AIP Center for History of Physics, page 328.

4. C. Weiner in D. Fleming, B. Bailyn, eds., *The Intellectual Migration*, Harvard U. P., Cambridge (1969), page 190.

5. Reference 3, page 332.

6. Letter from Harlow Shapley to Antonie Pannekoek, 8 November 1934, Harlow Shapley Director's Correspondence, Harvard University Archives (HSDC-HUA).

7. Letter from Harlow Shapley to N. Henry Black, 22 September 1932, HSDC-HUA.

8. Letter from Harlow Shapley to Kirtley Mather, 24 September 1934, HSDC-HUA.

9. Letter from Harlow Shapley to Peter Millman, 24 October 1934; letter from Millman to Shapley, 4 November 1934, HSDC-HUA.

10. Letter from Harlow Shapley to Kirtley Mather 7 October 1935, HSDC-HUA.

11. Letter from Antonie Pannekoek to Harlow Shapley, 22 October 1935, HSDC-HUA.

12. Letter from Frank Schlesinger to Harlow Shapley, 10 December 1935, HSDC-HUA.

13. Letter from Harlow Shapley to Frank Schlesinger, 26 December 1935, HSDC-HUA.

14. Letter from Megnad Saha to Harlow Shapley, 5 November 1936, HSDC-HUA. □

Niels Bohr as fund raiser

Bohr's dealings with the Rockefeller Foundation were decisive in bringing about the successful reorientation of his institute from atomic to nuclear physics in the 1930s.

Finn Aaserud

PHYSICS TODAY/OCTOBER 1985

Nuclear physics came of age as a discipline in the 1930s. In 1932, the "miraculous year" of nuclear physics, physicists discovered two new particles, the neutron and the positron, developed revolutionary particle-accelerator equipment and split nuclei for the first time by manmade machines.[1] The study of the nucleus soon developed into an independent field, becoming the central area of research in theoretical physics. During the mid-1930s, Niels Bohr's Institute for Theoretical Physics at the University of Copenhagen experienced a successful transition to the new field of investigation, consolidating its position as an international Mecca for theoretical physics research. In early 1936 Bohr proposed his revolutionary compound-nucleus model, and in late 1938 the Bohr institute's cyclotron, a major device for provoking nuclear reactions, was the first such apparatus to go into operation in Europe. Since then, Bohr's institute has been an international leader in theoretical nuclear physics research.

What was the background of Bohr's successful redirection of his institute from atomic quantum theory to nuclear physics? In this article I argue that Bohr's activities as an institute director in dealing with changes in funding conditions for international basic science, occasioned in particular by new policies of the Rockefeller Foundation, played a crucial role in inducing the change. The discussion sheds light on the little-investigated relationship between scientific and administrative matters at Bohr's institute. It brings forth an important aspect of Bohr's activities that has commonly been ignored in favor of his unique abilities as a scientist and as a teacher for several generations of the most eminent theoretical physicists.

The 'Copenhagen spirit,' 1921–33

In their reminiscences, physicists who worked at Bohr's institute between the world wars emphasize the unique "Copenhagen spirit" of scientific investigation.[2] They remember this spirit as, first, an unlimited freedom, in a completely informal atmosphere, to pursue whatever problems in theoretical physics that they considered most urgent. The second aspect of the Copenhagen spirit, as they recall it, was that this pursuit took the form of intense discussions between Bohr, the acknowledged master, and the most promising, yet young and unestablished, physics students visiting his institute from several countries. Dependent upon a human sounding board on whom to try out his ideas, Bohr encouraged some visitors at his institute to become his "helpers"—that is, to take part in his own thinking process. The Copenhagen spirit thus comprised complete freedom of research pursued within a division of scientific labor between Bohr and the cream of international theoretical physics students.

In the physicists' accounts, the freedom reflected in the Copenhagen spirit is implicitly contrasted with the constrictions of administrative work. Yet, although rarely noted by the reminisc-

ing physicists, Bohr was involved from the outset in matters of administration; in particular he worked hard for several years to secure the funds necessary to establish his institute in 1921. In fact, Bohr's justification for creating an institute for theoretical physics—to establish a union between physical theory and experiment—made it particularly necessary for him to involve himself in questions of funding, and he made a substantial effort to formulate elaborate—and successful—funding proposals toward this end. Bohr thus established what amounted to a research program for his institute: theoretical and experimental work on the manifestations of electrons orbiting the atomic nucleus.[3]

Bohr's substantial prestige as a scientist and institute director allowed him to avoid difficulties in obtaining support, first from such Danish agencies as the Carlsberg and Rask–Ørsted Foundations, and later from an American agency, the Rockefeller-funded International Education Board.

The IEB, conceived of and led by Wickliffe Rose, a long-time Rockefeller philanthropy administrator, was explicitly devoted to "making the peaks higher" in contemporary international, particularly European, basic science research. Pursuing an elitist "best science" policy of exposing the most promising postdoctoral students to the very best institutions of research, it considered general quality and prestige, rather than specific projects or research goals, as criteria for funding.[4] Bohr's application for an expansion of his institute was accepted even before Rose set out in 1923 on an extensive tour to survey the needs of European basic science. This grant constituted the IEB's first support to any physics

Finn Aaserud is director of the Niels Bohr Archive in Copenhagen. He was formerly associate historian at the American Institute of Physics, Center for the History of Physics. He is the author of *Redirecting Science: Niels Bohr, Philanthropy, and the Rise of Nuclear Physics*.

Niels Bohr with the Cockcroft–Walton high-voltage equipment at his institute, in the late 1930s. This was the first large-scale nuclear physics apparatus to be installed at the institute. (Niels Bohr Institute photograph, courtesy AIP Niels Bohr Library.)

institution. In obtaining this support, Bohr was able to confirm the extent and importance of his prestige, as well as to consolidate the union between theory and experiment he had advocated from the institute's establishment. Furthermore, the IEB's emphasis on quality and prestige allowed the freedom—contained in the Copenhagen spirit—to take up any question without regard for immediate utility.

Although support for building and equipment made up the largest part of the IEB's budget, its substantial fellowship program was the backbone and major justification for all of its basic science funding. This program supplied Bohr with eminent younger collaborators such as Werner Heisenberg and Pascual Jordan from Germany, Samuel Goudsmit from Holland and George Gamow from the Soviet Union. Together with the Rask–Ørsted Foundation, which offered a similar fellowship program, the IEB provided the economic means for the division of scientific labor between Bohr, the international leader of theoretical physics, and his young and eminent, but unestablished, colleagues. Arriving in Copenhagen to put the finishing touches on their scientific education, the visiting physicists were strongly motivated to discuss their research problems with the acknowledged master, as well as to work as Bohr's helpers. The division of labor between Bohr and his younger collaborators, then, was reinforced by Bohr's active exploitation of the existing funding opportunities. Bohr's involvement in physics and administration were not in conflict, but reinforced one another.

The spirit in action. The lack of emphasis in physicists' recollections on Bohr's administrative work is now easier to

Wickliffe Rose. Rose led the Rockefeller-funded International Education Board from 1923 to 1927. (Courtesy Rockefeller Archive Center, Pocantico Hills, New York.)

understand. Destined to stay in Copenhagen for a limited period of one to two years, without any prospect for advancement in Danish science, the young visitors at Bohr's institute saw no reason—and were not encouraged—to involve themselves in matters of administration. The physicists' concentration, in their recollections, on disinterested scientific discussion hence reflects their own situation and concerns at the time, rather than a complete portrait of Bohr's actual activities.

Nevertheless, during the late 1920s and early 1930s, the day-to-day theoretical discussion represented by the Copenhagen spirit was crucial in determining the development of science at Bohr's institute. In his address of appreciation to the Danish Academy of Sciences and Letters in late 1931 upon being chosen as the resident of the honorary Carlsberg Mansion, Bohr went as far as to express regret that living on the institute's premises had involved him in theoretical discussions at the expense of planning and following experimental research. He looked forward to being able to make plans for his institute from a distance. Yet, during the following couple of years Bohr continued to concern himself mainly with fundamental questions in quantum physics. Of these, the extension of quantum theory to the relativistic domain was considered particularly urgent. Many physicists expressed frustration about the meager progress in this direction. Bohr, however, pursued the problem with characteristic zeal and optimism. Thus, while he followed closely the events of the miraculous year, 1932, of nuclear physics, he scrutinized their general implications for his study of quantum theory, without seeking to change the long-established union between theory and experiment at his institute.

Bohr's rising interest in biological questions, first expressed publicly in 1929, was also oriented toward theory and discussion. His interest in biology came out of his concern with the epistemological implications of quantum mechanics; he pursued the subject in discussions with fellow theoretical

Warren Weaver. Weaver was the Rockefeller Foundation's natural sciences director from 1932 to 1955. (Courtesy Rockefeller Archive Center.)

physicists, with no intention of bearing upon the experimental work at his institute. Hence, in spite of Bohr's original emphasis on a union between theory and experiment, from the second half of the 1920s through 1933, the crucial element in forming the concerns and activities at his institute, even outside theoretical physics, was the Copenhagen spirit, which emphasized open theoretical discussion rather than the initiation of experimental projects. By early 1934 the concerted scientific reorientation toward nuclear physics had yet to take place.

Help to the refugees, 1933–1934

In 1933 the Rockefeller Foundation instituted an emergency program for the academic refugees from Nazi Germany. Contrary to traditional funding policy for international basic science personnel, this program centered on established scientists, and thus did not serve to sustain the division of scientific labor between Bohr and his younger helpers. Nevertheless, Bohr did not hesitate to appeal to the new program to further research at his institute. The resulting departure from the Copenhagen spirit would eventually hasten the transition to nuclear physics.

Bohr had been exposed immediately to the problem of the physicist refugees, most of them Jewish, after Adolf Hitler's accession to power in Germany in January 1933. He was continually informed about developments by his Danish assistant Ebbe Rasmussen, who just before Hitler's takeover had gone to Berlin on a one-year fellowship from the Rockefeller Foundation. Early on, Rasmussen urged Bohr to use his influence to help the young and less established refugee physicists, who were hit particularly hard. During the following months Bohr also solicited reports on developments from German physicist colleagues.

In October he joined the executive board of the newly established Danish Committee for the Support of Refugee Intellectual Workers, which like other European national refugee committees concentrated on helping the less well-known refugees. Through this committee, as well as through the traditional

fellowship programs, Bohr was able to create a temporary haven for young physicist refugees at his institute. Consequently, Bohr's personal priorities in handling the refugee problem, as expressed before his exposure to the Rockefeller Foundation's emergency program, did not affect the division of scientific labor at his institute, where the Copenhagen spirit continued to serve as the main inspiration for physics research.

In early May 1933, at the beginning of an extended visit to the United States, Bohr urged Max Mason, president of the Rockefeller Foundation, which in the meantime had taken over the IEB's fellowship program for young natural scientists, to abolish the requirement that fellowship recipients should have a position to return to. Without such action, Bohr reasoned, the fellowship program would not provide sufficiently effective help for the hardest-hit refugees. It was in the course of Bohr's American visit, however, that the Rockefeller Foundation decided upon its emergency program toward the academic refugees.

The new program sought to establish positions for the most established and renowned of the academic refugees, with emphasis "on the preservation of scholarship rather than on personal relief for scholars."[5] The American Emergency Committee in Aid of Displaced German Scholars, established at about the same time and collaborating closely with the Rockefeller Foundation, explicitly excluded aid "to younger German scholars of outstanding promise."[6]

In spite of Bohr's and the Rockefeller Foundation's different approaches to the refugee problem, Bohr did not hesitate to make the best of the foundation's new program for his institute and for two of his closest physicist colleagues. Before his departure from the United States, he had already met with the Rockefeller Foundation's natural sciences director, Warren Weaver, for this purpose. Bohr's subsequent formal application for salaries for James Franck and George Hevesy to work at his institute for three years was approved in January 1934. The promi-

nence of these two scientists in the international physics community, as well as their different personalities, would prove crucial for the subsequent developments at Bohr's institute.

Franck and Hevesy. Franck was a prominent experimental physicist who had quit his prestigious professorship at Göttingen University soon after Hitler's accession to power. Although Jewish, Franck was not affected by the racial laws, owing to his engagement in front-line combat during the First World War. Nevertheless, he was unable to accept the regime's actions against his Jewish colleagues.[7]

After their first meeting in 1920, Bohr and Franck had developed a close personal and scientific relationship. Franck, who was three years older than Bohr, was a strong admirer of Bohr's theoretical work and devoted a substantial part of the following years to the experimental verification of Bohr's theories. In the process he studied increasingly larger and more complex physical systems. Having started with individual atoms, Franck was investigating processes on the molecular level by the time he moved to Copenhagen.

Hevesy, although an experimenter like Franck, represented a significantly different approach to scientific research. Having a background in physical chemistry, he tended to apply the most advanced techniques of modern physics research not to answering urgent questions of theoretical physics, but rather to obtaining more precise information in a variety of fields. Thus he employed his knowledge of radioactive isotopes—acquired during an extended apprenticeship at Ernest Rutherford's laboratory in Manchester from 1911 to 1913—to developing the radioactive indicator technique, which enabled him to identify and measure minute quantities of elements embedded in other material.

Bohr and Hevesy had developed a close friendship based on mutual admiration during their shared period in Manchester in 1912, shortly after Bohr had completed his doctoral work in Copenhagen. Hevesy, less than three months older than Bohr, was more extroverted, standing up publicly for

Groundbreaking. Niels Bohr takes a symbolic step toward the expansion of his institute, *circa* 1935. (Courtesy Niels Bohr Institute.)

Bohr's views at a time when they were not generally accepted. Whereas Bohr admired Hevesy's more practical bent, Hevesy for his part considered Bohr's theoretical intuition unique.

Nuclear work begins, 1934

In the difficult period that Hevesy's home country, Hungary, underwent following the First World War, Bohr invited his friend to work at his newly established institute. Although Hevesy provided an important verification of Bohr's atomic theory of the periodic system by codiscovering the element hafnium in 1922, this event was only his first step toward developing x-ray techniques for the study of minerals. In the 1920s Hevesy devoted most of his time to designing and applying such x-ray techniques, having momentarily exhausted the radioactive indicator technique, owing to the small number of naturally occurring radioactive isotopes.

As a more marginal effort in Copenhagen, Hevesy also sought to apply his radioactive indicator technique to investigating the exchange of some elements in living plants and animals. There is no indication that Bohr at the time expressed interest in this work, which Hevesy pursued in collaboration with other prominent Danish scientists and institutions. The work shows Hevesy's independence and his different approach to scientific research.

After six years in Copenhagen Hevesy accepted a call to become professor and director of the Institute for Physical Chemistry at Freiburg University in Germany, where he intensified the research efforts he had begun in Copenhagen. In the summer of 1933 he decided to retire from this position

owing to the situation provoked by the Nazi takeover in Germany. He kept postponing his departure, however, and did not move permanently to Copenhagen until September 1934.[8]

Before the arrivals of Bohr's two prominent friends, neither Bohr, Franck nor Hevesy had specific plans for what kind of research the newcomers would pursue in Copenhagen. Upon his arrival in early April 1934, after a semester in the United States, Franck set out to study the fluorescence of green leaves—a natural extension of his previous researches, which required a minimum of equipment. However, Bohr saw quickly that the need of the two established experimenters—former professors and institute directors—for their own equipment, assistants and research projects could be used as an opportunity to follow up the experimental investigations of the atomic nucleus begun in March by Enrico Fermi and his collaborators in Rome.

After the discovery by Irène and Frédéric Joliot-Curie in Paris that some elements could be made "artificially" radioactive through bombardment with protons, the Rome physicists had set out to bombard systematically all chemical elements with neutrons. Because neutrons, unlike protons, would not be repelled electrically when approaching the positively charged nucleus, nuclear reactions could be obtained at smaller energies, and hence at lower cost.

The new experimental developments consolidated the successes in nuclear physics during the miraculous year of 1932. They distinguished themselves from the latter events, however, in having an immediate effect on experi-

mental work at Bohr's institute. Although such experiments had never before been reported from Copenhagen, and despite Franck's lack of experience in nuclear physics, Bohr wrote Heisenberg in the second half of April 1934 that Franck was directing experimental investigations at the institute along the same lines as the work on radioactive phenomena in Paris and Rome. Bohr subsequently arranged for Otto Robert Frisch, a young refugee physicist and nuclear experimenter, to come to the institute as Franck's assistant. While absent on a trip to the Soviet Union in early summer, Bohr induced Franck and Hevesy to correspond about the acquisition of Cockcroft–Walton high-voltage equipment for nuclear investigations.

By the end of the year only one publication had come out of the Copenhagen neutron bombardment program. During this period Franck complained in his correspondence about his lack of expertise, and about the appearance in *Nature* of his own findings—arrived at independently by other physicists—before he had had a chance to write them down for publication himself. Nevertheless, in December Bohr reported in a letter to his colleague Ralph Fowler in Cambridge that "the whole laboratory is busily occupied with research on nuclear physics under the effective guidance of Franck and Hevesy."[9]

In short, during 1934 Bohr's concern for physics in relation to his institute changed from a theoretical interest in the fundamental problems of quantum physics to an effort to introduce new experimental research on the atomic nucleus. This change cannot be understood in terms of developments in

Bohr's institute after its expansion in the 1930s. (Courtesy Niels Bohr Institute.)

physics alone. Bohr had followed closely the experimental study of the nucleus from the outset. However, as long as the funding policy for international basic science did not change, research pursued in the Copenhagen spirit did not provoke a change in the institute's research effort. Although Bohr showed immediate enthusiasm for the developments in nuclear physics in 1932, a concerted change in the emphasis of the research at his institute was triggered only by his willingness and ability to take advantage of the Rockefeller Foundation's new approach to the international funding of elite scientists represented by the emergency program for the academic refugees.

Physics tools and biology, 1934–35

While the Rockefeller Foundation's emergency program toward the academic refugees worked against the Copenhagen spirit by bringing two renowned physicists to Bohr's institute, the same foundation's substantially more ambitious "experimental biology" funding program challenged the precept of undirected "best science" itself. This program sought specifically to place biological research on what the foundation viewed as the more solid basis of mathematics, physics and chemistry. The new program was an even more profound counterpoint to the previous funding policy, still practiced by other agencies, that had sustained the Copenhagen spirit. Nevertheless, Bohr quickly took advantage of the new program to continue the transition to nuclear physics at his institute.

At the end of the 1920s the Rockefeller philanthropists had begun a far-

reaching reconsideration of their basic science policy. The educationally motivated best-science policy of Rose's International Education Board came to be seen as too broad, and hence too scattered and expensive. When it was decided, in 1927, that the Rockefeller Foundation would take over the IEB's responsibilities for science funding, Rose chose to leave Rockefeller philanthropy for good. It took time, however, for the foundation's new natural sciences division to decide upon a well-defined alternative policy. Only when the physicist Weaver was appointed as its director in late 1931 did a reorientation of the foundation's "major program" of science funding begin. The new program was soon to emerge as "experimental biology."[10]

In the course of intense internal meeting activity and committee work, especially during 1933, the new funding policy crystallized. Instead of supporting individuals and institutions on the basis of their general prestige and quality, the Rockefeller Foundation decided to concentrate on "the application of experimental procedures to the study of the organization and reactions of living matter."[11] By encouraging work in biology using modern theoretical concepts and experimental techniques from mathematics, physics and chemistry, the foundation hoped to advance the putatively backward life sciences. Economic hardship induced the foundation during 1934 to increase the number of "project grants" at the expense of the previously more common general-purpose grants. Under Weaver's leadership the Rockefeller Foundation set out on an unprecedented attempt to guide international basic science in a preferred direction.

Even before the Rockefeller Foundation's new major program was defined with any rigor, Weaver was eager to add prestige to his division by attracting the renowned Bohr to his funding program. Thus, although no biologically oriented work had ever been done at Bohr's institute, Weaver made the institute his first stop in Copenhagen during a grand tour in the spring of 1932 of "essentially all of the university centers of western and south-western Europe."[12]

Meeting with Weaver in New York a year later, Bohr suggested the possibility of bringing together at his institute one, two or three "able and thoroughly trained young men in mathematics, physics or chemistry, who under B[ohr]'s direction turn their attention to some quantitative phase of important biological problems."[13] In April 1934, when Wilbur Earle Tisdale and David Patrick O'Brien from the Rockefeller Foundation's Paris office visited Copenhagen to discuss specific research proposals, Bohr was equally vague about the kind of biological activity he would like to see funded. His statements thus reflected his own philosophical and general interest in biology, which was in conformity with the Copenhagen spirit of open and disinterested scientific discussion. Notwithstanding their eagerness to find a way to support Bohr, the Rockefeller Foundation's officers judged that Bohr was too far from suggesting, as required by the major program, a specific experimental project. Hence, Bohr was not at the moment considered a candidate for support.

By the time of Tisdale's next visit to Copenhagen in October, Bohr's concern for general theoretical questions in

James Franck (center) and George Hevesy (right) with Niels Bohr, on the porch of Bohr's institute in the 1930s. (Courtesy AIP Niels Bohr Library, Margrethe Bohr collection.)

physics was being replaced by his enthusiasm for Franck and Hevesy's work in experimental nuclear physics. Tisdale found that Bohr's attitude toward experimental biology had also changed dramatically. While previously "Bohr's interests [in biology] were mainly on the philosophical side," Tisdale noted[14] in his diary, "in recent months he has come to feel that it may be possible to do more effective work at the present time in connection with definite problems." In effect, Bohr suggested that the possibility of producing radioactive isotopes artificially made Hevesy's radioactive indicator technique suitable for biological problems.

Bohr proposed that high-voltage equipment, of the kind Franck and Hevesy had previously corresponded about for other reasons, be installed at his institute for the production of biologically useful isotopes. He proposed further that August Krogh, a prominent Danish physiologist and Nobel laureate in medicine, join the project as a third person, to provide the required biological expertise. Obviously, Bohr had come to perceive the Rockefeller Foundation's experimental biology program as a possible means to acquire equipment for experimental nuclear physics research. At the same time he was able to define a research program for Hevesy in direct continuation of the latter's previous concern and work.

Applications approved. During the next few months Bohr conducted detailed negotiations with the Rockefeller Foundation, seeking to define a research project acceptable for funding. By December he had become more confident about the foundation's particular emphasis on his own scientific renown. Although he had previously stressed the importance of Krogh's biological expertise, Bohr now simply presented Hevesy's study of the elements' "fate in the body" as the central theme, noting for the first time an especially promising artificially induced radioactive isotope—phosphorus.

When Bohr, in late January 1935, applied for the first time in almost ten years to the Carlsberg Foundation for substantial new and expensive equipment, he argued that acceptance of the current application was a condition for

continued support from the Rockefeller Foundation. He did not, however, mention that the support from the Rockefeller Foundation was for biology, not physics. Instead, he motivated his request for a fully equipped high-voltage laboratory by noting the general shift in emphasis in theoretical physics from an interest in the atom's orbiting electrons to an interest in its nucleus. Unlike the Rockefeller Foundation, the Danish agency had not changed its funding policy, and Bohr may have judged that the new work in biology was irrelevant from the point of view of the Carlsberg Foundation. In any case, Bohr's application was approved in early February on the condition that running expenses for the high-voltage laboratory be obtained from other sources.

Bohr referred to this condition toward the end of February, when he finally formulated an elaborate letter of application to the Rockefeller Foundation. In this application Bohr did not even mention Krogh's name. Instead, Bohr projected[15] in general terms the "cooperation between [his institute] and Danish biological institutions to utilize the new possibilities for the investigation of fundamental problems in biology opened by the recent advances in atomic physics." Bohr was clearly becoming increasingly aware of the Rockefeller Foundation's emphasis, within the framework of constructing a physics-based biology, on his own prestige rather than on the specific proposals of biologists. Yet, as the application did not formulate a well-defined biological project, Hevesy, when delivering it personally in Paris, found it pertinent to assure[16] Tisdale "that this project, involving as it does so much of physics, is completely oriented towards bio-physical problems and is in no wise an attempt on Bohr's part to obtain equipment to permit him to in any wise compete with the Rutherfords, the Lawrences, and others who are working in the field of pure physics."

Although the changes in the Copenhagen experimental-biology proposals from October 1934 to February 1935 did arouse suspicion among some Rockefeller Foundation officers that the project was not sufficiently planned, or even that it might contain too little biology, the foundation granted support in April. As requested, the foundation provided funds for a cyclotron, as well as for assistants, equipment and running expenses for experimental biology for five years. In changing his attitude toward the Rockefeller Foundation's experimental-biology program, Bohr had not only introduced at his institute new research outside the domain of physics; he had also been able to obtain advanced equipment for experimental nuclear physics research.

Transition consolidated, 1935–39

The year 1935 brought a revolution in the economic situation of Bohr's institute. In addition to the support from the Carlsberg and Rockefeller

Foundations, the Thrige Foundation, established from the profits of the largest electrical firm in Denmark, provided the electromagnet for the institute's cyclotron. For Bohr's 50th birthday on 7 October, Hevesy masterminded what became known as the "radium gift"—funds for the acquisition of 600 milligrams of radium, contributed by some 16 Danish firms and foundations.

In this and following years Bohr also obtained support from agencies emphasizing medical and biological, rather than physical, research. Thus he applied to a Danish agency for funds to pursue "atomic-physical investigations as preparations for physiological experiments."[17] Moreover, he applied successfully to both the Carlsberg and Rockefeller Foundations for additional support for the cyclotron. Finally, in addition to obtaining increased running expenses from the Carlsberg Foundation, Bohr instigated collaboration in 1938 with the Danish Cancer Committee to build equipment for the treatment of cancer.

Bohr's appeals for support for biological research were not merely a tactic to secure the transition at his institute to nuclear physics. Although continuing to do work mostly on induced radioactivity during his first two years in Copenhagen, the prolific Hevesy collaborated with Krogh and published papers with two other prominent Danish scientists and institutions on the exchange of phosphorus in animals and plants. By 1937 Hevesy had turned all his energies toward applying his radioactive indicator technique to biological problems. In addition to expanding the collaboration with other scientists and institutions, Hevesy developed his own small group of younger collaborators at Bohr's institute. Although Hevesy and his collaborators employed the same experimental apparatus as the physicists, in doing biological experiments they worked on entirely different problems, and there was little communication between the two groups. While Hevesy quickly developed a reputation as the world's expert in his line of work, the visiting physicists do not remember him as participating in the kind of open discussion represented by the Copenhagen spirit. Not only the subject matter, but also the more project-oriented style of Hevesy's research differed substan-

tially from the physicists' approach.

Thus, in the latter half of the 1930s the physics research at Bohr's institute continued to be pursued mainly through open and intense theoretical discussion. Bohr made this continuity possible by leaving it to Hevesy to take care of the Rockefeller Foundation's demand for a specific project. The work of the physicists dealt increasingly with problems relating to the atomic nucleus. In early 1936 Bohr presented his revolutionary "compound nucleus" model, and in the same year the majority of published papers from Bohr's institute was devoted for the first time to nuclear problems. This emphasis remains today.

Although Franck left Bohr's institute in mid-1935 to take up a position in the United States, Frisch, who had originally arrived to become Franck's assistant, stayed for almost five years and was crucial in carrying out the transition to nuclear physics.

The installation of the Cockcroft–Walton equipment supported by the Carlsberg Foundation was conceived and directed during the early stages by the German experimental physicist Arthur von Hippel, Franck's son-in-law. First, in the spring of 1938, it supplied x rays for cancer therapy, and then, around New Year 1939, it served as a particle accelerator for nuclear studies. The cyclotron was built by Lawrence Jackson Laslett, who was lent for the purpose from Ernest Orlando Lawrence's laboratory in Berkeley. It went into service in November 1938 as the first working cyclotron in Europe.

At about this time Frisch and his aunt Lise Meitner, who had just escaped Nazi Germany, were able to explain the process of nuclear fission on the basis of Bohr's conception of the nucleus. This explanation—occasioned by the experimental findings of Meitner's previous collaborators in Berlin, Otto Hahn and Fritz Strassmann—led to a

Otto Robert Frisch. Bohr arranged for this young refugee physicist and nuclear experimenter to come to his institute as Franck's assistant. (Courtesy Niels Bohr Institute.)

The Copenhagen spirit in action. Bohr faces Wolfgang Pauli (back to camera), Lothar Wolfgang Nordheim, an unidentified man and Léon Rosenfeld. (Courtesy AIP Niels Bohr Library, Landé collection.)

concrete research program at Bohr's institute employing the new experimental equipment. Physicists used the high-voltage equipment, and subsequently the cyclotron, to study fission, with Bohr taking an active part in both the theoretical and experimental investigations. In 1939, on a visit to Princeton, he coauthored a seminal article on "The Mechanism of Nuclear Fission" with his younger collaborator John Archibald Wheeler. By this time the union between theory and experiment in physics at Bohr's institute had definitely become based upon nuclear physics, and the institute maintained its place as an international Mecca for theoretical physics research.

References

1. D. J. Kevles, Phys. Teach. **10**, 175 (1972); C. Weiner, PHYSICS TODAY, May 1972, p. 40; R. H. Stuewer, *Nuclear Physics in Retrospect: Proceedings of a Symposium on the 1930s*, Univ. Minnesota P., Minneapolis (1979).

2. The original meaning of the term referred to the conceptual approach to physics at Bohr's institute; see W. Heisenberg, *The Physical Principles of the Quantum Theory*, Dover, New York (1930), preface. The more recent meaning of the term can be obtained from numerous physicists' reminiscences, but see in particular L. Rosenfeld, *Niels Bohr: An Essay Dedicated to Him on His Sixtieth Birthday 1945*, North Holland, Amsterdam (1949); V. Weisskopf in *Physics in the Twentieth Century: Selected Essays*, MIT, Cambridge, (1972), p. 52. The concept is discussed at some length in P. Robertson, *The Early Years: The Niels Bohr Institute 1921–1930*, Akademisk Forlag, Copenhagen (1979), p. 152.

3. As a notable exception, P. Robertson, *The Early Years: The Niels Bohr Institute 1921–1930*, Akademisk Forlag, Copenhagen (1979) stresses the administrative aspect of Bohr's work at the institute.

4. G. W. Gray, *Education on an International Scale: A History of the International Education Board, 1923–38*, Harcourt Brace, New York, (1941); R. E. Kohler, Minerva **16**, 480 (1978).

5. "The problem of the refugee scholars," p. 3, typewritten, undated report, circa 1940, in the Rockefeller Foundation Archive (RFA), Pocantico Hills, New York, record group 1.1, series 717, box 1, folder 6 (1.1,717,1,6).

6. *Report of the Emergency Committee in Aid of Displaced German Scholars*, 1 January 1934, p. 11, pamphlet contained in RFA (1.1,717,1,1).

7. A. D. Beyerchen, *Scientists Under Hitler: Politics and the Physics Community in the Third Reich*, Yale U. P., New Haven (1977), p. 15.

8. For an apt description of Hevesy's personality, see H. Levi, *George de Hevesy*, Adam Hilger, Bristol (1985).

9. Letter from Bohr to Fowler, 12 December 1934, Bohr scientific correspondence, microfilm 19, section 2, deposited in the Niels Bohr Archive, Copenhagen, the AIP Niels Bohr Library, New York, and elsewhere.

10. R. E. Kohler, in *The Sciences in the American Context: New Perspectives*, N. Reingold, ed., Smithsonian Institution, Washington, D.C. (1979), p. 249; P. Abir-Am, Social Studies of Science **12**, 341 (1982).

11. "Report of the Committee of Appraisal and Plan," p. 58, in RFA (3,900,22,170).

12. W. Weaver, *Scene of Change: A Lifetime in American Science*, Scribners, New York (1970), p. 65.

13. Warren Weaver diary, 10 July 1933, RFA, record group 12.

14. Wilbur Tisdale diary, 29 October 1934, RFA, record group 12.

15. Letter from Bohr to Tisdale, 22 February 1935, RFA (1.1,713,4,47) (transcribed version), and Bohr general correspondence, special file, microfilm 6, section 1 (BGC-S 6,1), Niels Bohr Archive, Copenhagen (Bohr's carbon copy).

16. Letter from Tisdale to Weaver, 27 February 1935, RFA (1.1,713,4,47).

17. Letter from the Nordic Insulin Foundation to Bohr, 22 March 1934, BGC-S 3,4. □

JOHN TORRENCE TATE

1889-1950

PHYSICS TODAY/AUGUST 1950

John Torrence Tate, Professor of Physics in the University of Minnesota, died of a cerebral hemorrhage on May 27. An earlier attack in December, 1949, had kept him away from work for some time, but he had recovered to the point of working several hours daily, and he attended the Washington meeting of the American Physical Society, to the great pleasure of his fellow members.

Professor Tate, or Dean Tate, as he had come to be called, or Jack Tate to his closer friends, was one of the physicists who took a large part in the initiation and organization of the American Institute of Physics. No one saw more clearly than he the need of, and the possibilities of, a central organization through which physicists and the several societies representing physics could cooperate for the advancement of physics. He quite naturally was one of the initial members of the Governing Board of the Institute and through successive renominations by the American Physical Society remained a member until his death. At the beginning it was seen that the Institute would need experienced counsel on matters relating to scientific publications. Tate was so obviously qualified that he was asked to become the Adviser on

Publications of the Institute and served in that office until his death. When Dr. Karl T. Compton in 1936 declined re-election to the chairmanship of the Institute, Dr. Tate was the obvious choice of the Board for the chairmanship. As the second Chairman of the Institute he served with distinction from 1936 to 1939.

Born in Lenox, Adams County, Iowa, on July 28, 1889, John Tate spent many of his early school years in New York living with relatives. Taking an interest in chemistry, he was graduated from the Dewitt Clinton High School and then went to the University of Nebraska where, majoring in physics, he attained the BS degree in 1910, MA in 1912. For the PhD degree he studied at the University of Berlin, receiving the degree in 1914. He returned to the University of Nebraska as an Instructor in Physics, 1914–15, and was made an Assistant Professor the next year. In the following year he went to the University of Minnesota as an Instructor, becoming an Assistant Professor, 1917–18; Associate Professor, 1919–21; Professor, 1921–37; then Dean of the College of Science, Literature and Arts, 1937–43 (though he was on war leave 1941–43). Since it was his desire to return after the war to his scientific work rather than to administration, he resigned the deanship in 1943 and became research Professor of Physics.

Dean Tate was a fellow of the American Physical Society, the Optical Society of America, and the Acoustical Society of America, and was a member of the American Association of Physics Teachers and the Society of Rheology, thus holding membership in all five of the member societies of the Institute of Physics. He was a member of Sigma Xi, Phi Beta Kappa, and of other scientific and scholarly organizations and of the Cosmos Club in Washington and the Century in New York.

His own scientific work was mainly on electron impact phenomena in gases and electron interactions with matter, though he was a scholar of wide knowledge and interests in physics, theoretical and experimental. In the guidance of students in research, he was stimulating and tireless. His influence in the advancement of physics, through his own researches and through the students trained under his tutelage, has been notable. While he was successful in educational administration, he loved physics better and was gratified to be able, after his distinguished war service, to go back to research. His personal devotion contributed much to the development of the Department of Physics at Minnesota.

One of Dr. Tate's greatest services to physics was his accomplishment as Managing Editor of the American Physical Society. He edited The Physical Review from 1926. It was then a journal of 2250 comparatively small pages a year. By 1949, with its larger pages and more compact type-setting it had grown to four times that volume with three times the number of subscribers thanks to the great increase in the number and activity of American physicists, and is now the leading research journal of physics in the world. As Editor, Tate has had to be concerned with the rising cost of publication and how to meet it, as well as with responsibility for the more scientific aspects of editing, in which he has been open-minded,

impartial, and helpful to contributors. In 1929 Tate launched The Reviews of Modern Physics for the Physical Society, a quarterly journal, the conception and editing of which have been such that it has an almost unique record for a scientific periodical in that the subscriptions rather more than meet the whole cost of the publication. In 1931 Professor Tate sensed the need of an additional periodical for the publication of research papers relating to applied physics, and gained the support of the American Physical Society for the publication of a new journal, Physics, which later, in 1936, was taken over by the Institute of Physics to become the Journal of Applied Physics.

Professor Tate's academic life was interrupted by his national service through two wars. In the first world war he was engaged in war research as a 1st Lieutenant in the Signal Corps, U. S. Army. In the second world war, he served in the very responsible post of Chief, Division 6 of the National Defense Research Committee, in charge of research and development of anti-submarine and sub-surface warfare devices, equipment, and methods. His headquarters were in the Empire State Building in New York and his duties involved responsibility for scientific supervision of the activities of numerous contractors with the Office of Scientific Research and Development for work on anti-submarine and sub-surface warfare. Among the larger contractors employing thousands of scientists and engineers were Harvard, Columbia, and California universities, Bell Telephone Laboratories, and Submarine Signalling Corporation.

After the war, from 1946 to 1949, he served as Chairman of the Board of Governors of the Argonne National Laboratory, one of the regional laboratories of the Atomic Energy Commission.

Professor Tate married Lois Beatrice Fossler of Lincoln, Nebraska in 1917 who died in 1939. Their son, John Torrence, Junior, has just been appointed to the Fine Instructorship at Princeton. On June 30, 1945, Professor Tate married Madeline M. Mitchell, who as Manager of Publications of the American Institute of Physics from its start until 1945, had a large part in organizing the work of the Institute and in the successful efforts to acquire and equip the present home of the Institute.

Professor Tate did not fail to receive high honors in his lifetime. To cite some, he was made president of the American Physical Society for 1939, elected to membership in the National Academy of Sciences and the American Philosophical Society, and received honorary degrees of Doctor of Science from the University of Nebraska in 1938 and from Case Institute of Technology in 1945. He was awarded the Medal of Merit by President Truman, the citation reviewing the war services already referred to, and also King George's Medal for Service in the Cause of Freedom in recognition "of the valuable services rendered . . . to the Allied Cause".

John Tate had the respect and affection of physicists everywhere, and of many other colleagues and friends, as a man of learning and wisdom, and of unfailing good will. The loss of his counsel, which was widely sought and always modestly and helpfully given, will long be felt.   GEORGE B. PEGRAM

KARL TAYLOR COMPTON,

1887–1954

PHYSICS TODAY/SEPTEMBER 1954

IN the death of Karl Taylor Compton on June 22, 1954, American physicists lost one of their most stalwart colleagues, friends, and benefactors; the world of science lost one of its proudest examples of what a scientist should be and do; the nation lost a scientist-statesman of unusual insight and wisdom; and the world lost a human being of rare perception, integrity and charm.

A mere recapitulation of Karl Compton's activities would give no indication of his real stature and achievements, and of the affection in which he was held by all who came in contact with him. In the words of J. R. Killian, Jr., his successor as President of the Massachusetts Institute of Technology, "The response of his friends, his associates, and even those casually acquainted with him to Karl Compton's personality was invariably one of spontaneous pleasure in a personality completely free of guile, a personality sensitive in perception, emanating goodness and wisdom and always generous and benevolent in human relations." This freedom from guile, which was immediately apparent to all with whom he came in contact, is probably the principal key to an understanding of Compton's character. His complete integrity was so apparent to all who worked with him that his mere presence was sufficient to increase cohesive activity in any group in which he was found. It was impossible for him to believe ill of any human being, and virtue radiated tangibly from him.

Karl Taylor Compton was born in Wooster, Ohio, on September 14, 1887, to Dr. Elias Compton and Otelia Augspurger Compton. His father was a professor in and Dean of the College of Wooster, and the entire family became famous in educational fields. His brothers were Arthur H. Compton, who became equally well known as a physicist, and served as Chancellor of Washington University in St. Louis, and Wilson M. Compton, who became President of Washington State College, while his sister Mary became the wife of Dr. C. Herbert Rice, President of Forman Christian College in Lahore, India. Young Karl attended the College of Wooster, receiving the degree of Bachelor of Philosophy in 1908, and that of Master of Science in 1909. He became an instructor in chemistry there for a year before entering the Graduate School at Princeton, where he obtained the degree of Doctor of Philosophy *summa cum laude* in 1912. He then served for three years as an instructor in physics at Reed College in Portland, Oregon, and in 1915 was called to an assistant professorship at Princeton.

During World War I Dr. Compton was an engineer in the U. S. Army Signal Corps, engaged in the development of submarine detection devices. Later he became an officer of the Research Information Service and an associate scientific attaché to the U. S. Embassy in Paris. His further career divides itself into three major portions: first, that in which he was teacher and research worker in physics, specializing in electronics and spectroscopy, from 1912 to 1930; second, that in which he served as the ninth president of MIT, from 1930 to 1949; and the third, from the early 1940's until his death, when his major effort was centered in a statesmanlike attack on national problems of science and defense.

Compton returned to Princeton after the first World War and by 1919 had risen to a full-professorship. He served in this post for eleven years, and was chairman of the department of physics for the last year of his residence. When called to the presidency of MIT he had risen to the front rank of scientists, with an enviable reputation as an educator of outstanding physicists and as a contributor of important research results. His laboratory at Princeton was a mecca for young men interested in the then newly developing field of electronics, and his guidance of the thesis work of a large number of young doctors of philosophy made Princeton one of the outstanding centers of research in fundamental electronics in the 1920's.

Before being asked to take the presidency of MIT Dr. Compton was asked to make a report on the Institute's physics department, and his reaction to his visit there was typical. While effective in a restricted sphere, the department had been built up mainly as a service organization for teaching young engineers, and the research carried on by its faculty was largely of the applied variety to be expected in an engineering school of that era. The numbers of students who specialized in physics were relatively small. Dr. Compton, as a professional physicist, felt that a much greater contribution to the education of engineers and to the cause of science could be made by a department that had, in addition to its service function, a primary professional function of training physicists who could compete on the national scene with those produced by outstanding universities. When asked to accept the presidency of MIT he replied that he would not be interested in being associated with an institution whose physics department had so small a research budget as that then provided. He was told that, if he accepted the presidency, this would be a matter for his own decision. Later he appreciatively remarked that it had not been emphasized that it would be up to him to raise the needed money.

The task of guiding the Massachusetts Institute of Technology and of building it in a new direction was a challenge which Compton could not resist, and during the 19 years of his presidency he greatly broadened, deepened, and enriched the MIT curricula, adding new courses and departments as these seemed needed for the better education of scientists, engineers, architects, and industrial administrators. Under his guidance MIT grew from an outstanding engineering school to a university which, though of limited objectives, specialized in some twenty disciplines arranged in five Schools.

Before Compton's call to the duties of administration made him reluctantly give up his teaching and activities, he had published more than 100 technical papers, mainly in the fields of photoelectric phenomena, thermionics, fluorescence, the dissociation of gases, and spectroscopy in the extreme ultraviolet. His transition from physicist to administrator, which was made willingly but not without some internal struggle, was a great sacrifice. When he first came to MIT he took the very firm stand that he must allot himself at least a day a week in which to carry on his own research. He and a research associate brought with them from Princeton a large vacuum spectrograph, which they had together designed. On arrival in Cambridge, they discussed the course which this research was to follow, and then Compton became immersed in his new duties as president, after assuring his colleague that he would be available every Thursday to help in tracking down the wave lengths of elusive spectral lines. Unfortunately his administrative duties were so overwhelming that several weeks went by with no visits to the laboratory. Finally his young assistant telephoned, and Dr. Compton promised faithfully to put in an appearance on the following Thursday. He did so appear, and spent the afternoon happily wiping oil off the insides of a vacuum pump, which he disassembled for cleaning. Unfortunately the presidential duties soon closed in so heavily that this was his last appearance as an active participant in the Laboratory.

The third phase of Compton's career was that of a combination scientist-statesman and statesman-scientist. He contributed his time generously to public activities, and although possessed of an athletic and vigorous constitution, for a dozen years prior to his death kept himself constantly at the edge of exhaustion by his inability to refuse any request to serve in a worthy cause.

During World War II Dr. Compton was a member of the National Defense Research Committee, set up in June of 1940 and later expanded as a part of the Office of Scientific Research and Development, directed by Dr. Vannevar Bush, who before going to Washington as president of the Carnegie Institution had been vice-president and Dean of Engineering at MIT under Compton. Compton was in charge of that division of the NDRC which dealt with radar, infrared signalling, and optical devices of all sorts. He was the prime mover in the development of the great Radiation Laboratory at MIT, one of the most successful of wartime laboratories. His intimate knowledge of science and scientists, his deep awareness of the problems of national defense, and the universal confidence in his judgment, statesmanship, perception, and impartiality, made men willing to work under his aegis who might have hesitated when summoned by lesser men.

Dr. Compton was a member of the War Resources Board in 1939 and 1940 and the Baruch Rubber Survey Committee during 1942. He was chairman of the Radar ad hoc Committee of the Joint Committee on New Weapons of the Joint Chiefs of Staff during 1942–45, and served on the Advisory Staff of the Chief of Ordnance Military Training 1942–46. In 1943 he went to England as chairman of the U. S. Radar Mission to the United Kingdom. Also during 1943 he was appointed chief of the Office of Field Service of the Office of Scientific Research and Development, and travelled to Australia and New Guinea as a Special Representative of the Secretary of War to investigate weapons problems in the Southwest Pacific area at General MacArthur's request. In 1945 he was a member of both the Advisory Board of the Chemical Warfare Service and of the Secretary of War's Special Advisory Committee on the Atomic Bomb, and served on a special Scientific Intelligence Mission to Japan, landing in Tokyo among the first civilians after the surrender terms were signed. In 1945–46 he served as chairman of the Executive Committee of the Research Board for National Security, and in 1946 was named chairman of the Joint Chiefs of Staff's Evaluation Board on the Atomic Bomb Tests. He served as chairman of the President's Advisory Commission on Universal Military Training in 1946–47. He was chairman of the Research and Development Board of the National Military Establishment from 1948 to 1949, when his health required him to lessen his activity in affairs of the nation.

Compton's work brought him many honors, including the Rumford Medal of the American Academy of Arts and Sciences; the Hartley Public Welfare Medal of the National Academy of Sciences; the Washington Award of the Western Society of Engineers; the Lammé Medal; the Thacher E. Nelson Award; the Hoover Medal, presented in 1950 to a "great leader in engineering education for distinguished public service"; and the Priestly Memorial Award "for distinguished contribution to the welfare of mankind through physics". In recognition of his services during World War II, Compton was awarded the Presidential Medal for Merit. The citation states that the importance of the work he did might be said to have made him personally responsible for hastening the termination of hostilities. He was named Honorary Commander of the Civil Division of the Most Excellent Order of the British Empire and a Knight Commander of the Order of St. Olav. In 1951 the president of France advanced him in rank from Chevalier to Officer in the French Legion of Honor. Throughout the years he received more than thirty honorary degrees, including doctorates from his two alma maters and from Harvard and Cambridge Universities.

Characteristic of Dr. Compton's influence in science is the part he played in the inaguration of the American Institute of Physics. By 1931 the numbers of physicists in the United States had increased to such an extent that incipient fission was beginning to be apparent. Groups of physicists interested in optics, in acoustics, in thermodynamics, and especially in applied physics, had somewhat different interests from those interested in the newer fields of electronics, spectroscopy, and quantum physics in general, and were feeling inner stirrings toward the founding of new professional societies. Compton recognized the desirability and inevitability of such fission, but felt also the need for coordination and unity, with the result that he became a leading spirit in the founding of a new amalgamation of societies. Thus he encouraged unity in diversity, with the result that American physics has been able to make an orderly growth with due attention to the interests of physicists of all sorts. The program of publication of the research results of American physicists is as a result probably in better shape than in any other country.

As the first chairman and one of the principal founders of the American Institute of Physics in 1931, Dr. Compton did more than any other single person to ensure the success of the Institute, and he remained, after serving five years as its chairman, one of its most influential and respected elder statesmen. His sound judgment and reliable decisions set the course of its future development, and throughout the remainder of his life he gave unstintingly of his time when this was needed.

Dr. Compton was a member of numerous societies, including the National Academy of Sciences, the American Philosophical Society, the American Academy of Arts and Sciences, the American Association for the Advancement of Science (of which he was president in 1935), the American Chemical Society, the American Institute of Electrical Engineers, the American Society for Engineering Education (president in 1938), the Franklin Institute, the American Physical Society (president 1927–30), and the Optical Society of America. His fraternities were Alpha Tau Omega, Phi Beta Kappa, Sigma Xi, and Tau Beta Pi.

Dr. Compton is survived by his widow, Margaret Hutchinson Compton; by two daughters, Mrs. Bissell Alderman and Mrs. Carroll W. Boyce, and a son, Charles Arthur Compton. He affected profoundly the lives and careers of many fellow humans. Even those who saw him only in passing salute his memory, and join in gratitude at having known the influence of so great and good a man.

George R. Harrison

Jean Perrin and the reorganization of science

A group of French scientists, dissatisfied with the bureaucratic organization of higher education, made use of the shifting politics of the 1930's to set up the CNRS.

Spencer Weart

PHYSICS TODAY/JUNE 1979

Around 1930 a new generation came to the leadership of science in France, determined to overhaul it from top to bottom. Already in their fifties or sixties by the time they achieved leading positions, they had long suffered under the direction of superannuated professors, and they had seen French physics, once first in the world, sink toward obscurity. It was their deeply feared rivals, the Germans, who had made and welcomed many of the exciting new discoveries in quantum theory and relativity. The new generation in France, led by the physicist Jean Perrin, now launched a campaign for science. Within a few years they managed to create an organization of a new kind—one that not only revived French science, but that also served as a model for science organizations in many other countries. The story of this effort is interesting not simply because it is an important part of the history of modern science, but also because it can help us to see the social and ideological forces that may help or hinder any reorganization of science.

Blocked careers

Since the nineteenth century French science had been embedded in a flawed system of higher education. Nearly all French scientists, like most scientists in other countries at that time, were employed in the first place as teachers. And in France, nearly all teachers were employees of the government. Salaries and promotions, strictly determined by the Ministry of National Education, were tangled in a web of inspectors' evalua-

Spencer Weart is director of the Center for History of Physics at AIP.

tions, letters of recommendation from friends and functionaries, committee reports, and gossip. Aside from minor funds controlled by the Academy of Sciences and similar bodies dominated by aged conservatives, the operating expenses of most laboratories were dictated along traditional lines by government bureaucrats. These officials worked closely with a small circle of established professors, some of whom had done little productive research for many decades or were not even scientists at all. Research was supported by diversions of funds, virtually illegal, from the teaching budget. Favoritism and nepotism were common and old school loyalties taken for granted when people were chosen for a professorship.

There was little room for new blood. During the 1920's the number of science students rose to about double the level that held before World War I, but the number of university science professors actually declined between 1914 and 1925, and by 1930 their number was still barely above the prewar level. The freeze on jobs for scientists was only one of many cases where France was locked in place, its economy, politics and society all stalemated.

An example of the students whose careers were blocked was Frédéric Joliot. When Joliot had first considered making his career in science, his physics teacher, Paul Langevin, had told him that such ambition was risky: Unless Joliot made a notable discovery, he could never go far. Joliot nevertheless wanted to try his luck in science, so Langevin got him a part-time job at Marie Curie's Radium Institute. Here Joliot met Madame Curie's

daughter, Irène, and the pair were soon married. There were some in the Paris academic establishment who suspected that Joliot was seeking to use the well-trodden path of family favoritism to advance his career. He was called the "prince consort." Wounded by the gossip, Joliot set out to prove his value as a physicist. He won his doctor's degree in 1930 and began to look for a full-time position in the university system. But no jobs were available. The Radium Institute could not keep him on as a student assistant indefinitely, and he and Irène now had a young daughter. Joliot began to consider leaving science for a career in industry.

The group who had recently begun to take over the leadership of French science

Bastille Day, 14 July 1935, marked a strong leftward shift in French politics and led to the formation of a Popular Front that came into power in 1936 and put into effect many of the reforms in science administration advocated by Jean Perrin (above) and his colleagues. The photo on the left shows the assembly of supporters of the Popular front in the Place de La Bastille before the parade.

were determined to endure such weaknesses no longer. The movement for change found its chiefs among the directors of the Institute for Physicochemical Biology, a private research establishment founded in Paris in the 1920's through the generosity of Baron Edmond de Rothschild. In their weekly meetings the directors began to discuss the larger problems of French science. The institute operated outside the educational bureaucracy; could this principle somehow be extended? As Jean Perrin, one of the directors, explained, they decided to press for a new system under which "the various categories of researchers should correspond to the hierarchy of the university system . . . each with the corresponding salary, so that a scientist could make a

career entirely in Research with absolutely no other obligations." The sweeping plan sought to create an entirely novel division of the government, a research corps, operating directly under the minister of national education and in no way subject to the university bureaucracy. The principle of setting up research entirely apart from education could be traced back through the Institute for Physicochemical Biology to the Radium Institute and the old Pasteur Institute, but this was the first time people had seriously proposed to create such a system wholesale.

A scheme for reorganization

One spur to this change was the scientists' experience in World War I. During

the battles they had worked successfully as military research teams under a central administration of scientists. They were proud of the important part they had played in the victory, and felt they could do still more for the nation in time of peace, if their efforts could be mobilized. The French also saw that similar wartime experiences had led to reorganization of scientific work in other countries, such as Germany and Belgium, and especially in the Soviet Union, where research had been granted high prestige and a centralized organization.

One of the main bearers of the good news from Moscow was Langevin. The son of an artisan, he had been converted to socialism when he was a student, along with his friend Perrin and others. He was moving more and more to the left, visiting the Soviet Union several times around 1930, entertaining visiting Soviet scientists in his Paris home, and presiding over a Committee for Scientific Relations with the USSR. He also welcomed leftist British scientists such as J. D. Bernal who, with an eye on the priorities Soviet researchers enjoyed, were beginning a campaign to honor, finance and organize scientific work in Britain.

Perrin took charge of the French campaign. He began in June 1930 by going to the Academy of Sciences, winning the support of members of all political views. Since no plan could be enacted unless key people in the bureaucracy were also won over, Perrin now set off on a round of office visits. A Nobel Prize for work on Brownian motion and a popular book on atoms had brought Perrin fame, and he retained the exuberance and vigor of his youth. He had always been able to charm people, men and women alike, with his good looks and mischievous wit. But now that his abundant curls and beard had turned white, he had begun to resemble less a cherub than an Old Testament prophet. He would need the force of this appearance, for he was arguing that science had a moral right to a better position. If scientists were properly supported, Perrin believed, then science, through the beauty of its discoveries, the power of its truth, and the practical application of its knowledge, could make humanity happy and free. It was this, he said, that was his religion. He was speaking less as a practical man than as a prophet.

First in line for persuasion was Perrin's administrative head in the university, the dean of the Faculty of Sciences, Charles Maurain. Maurain had known Perrin since the early 1890's when they had studied science together at the Ecole Normale. The Ecole Normale was a school whose small classes provided many of France's professors and other intellectual leaders, after they had learned to know one another well in years of grueling studies and schoolboy horseplay. Perrin had introduced Maurain to Langevin, Pierre and Marie Curie, and other scientists who formed a close-knit circle of friends. Maurain readily allied himself with Perrin's scheme for reorganizing science.

Next in line above Maurain was the powerful head of the university bureaucracy, the director of higher education, Jacques Cavalier. Intelligent but unoriginal, laconic, punctual, dressed always in black, he was one of those reliable officials who governed behind the scenes while ministers came and went. Perrin and Maurain spent four hours in his office presenting their case. Before he turned administrator, Cavalier too had been a scientist, an expert on the chemistry of alloys, and he had become friends with Perrin and Maurain when they were all

Marie Curie at the Radium Institute in 1934 (left).
Frederic Joliot and Irene Joliot-Curie in 1928, while they were students at the Radium Institute (next page).

students together at the Ecole Normale; so he was soon won over.

Perrin now had to go into governmental circles beyond the university community, and his first approaches had no success. Then the politician Edouard Herriot, another old friend and schoolfellow, invited him to give a speech to a teachers' group. Perrin repeated his usual arguments with unusual force:

As soon as you understand that Scientific Research is our only chance to create truly *new* conditions of existence, where human life will be ever more happy and more free, you will find it senseless, in the full meaning of the word, that until recently the various nations have not made any serious effort to prevent scientifically gifted men from being thwarted by material needs ... Our existence *at this moment* is certainly poor and miserable compared with the existence we would be leading, this very day, if these men

of genius had been allowed to develop. That loss is irreparable, but we can at least try to organize the immediate future.

After this address, Perrin recalled, Herriot embraced him and praised him to the skies, "but I told him that my soul was base and that I wanted some money." Cornering the politician for a long conversation, Perrin beseeched him to support the plan for an independent research organization.

Herriot was a chief of the Radicals, a party that held the center of balance in French politics. With the face of a peasant and the intellect of a professor of literature, he embodied the contradictions of his party. Although more interested in stability and prosperity for the middle classes than in sweeping change, the Radicals had inherited not only the name but the ideology of a group that at the turn of the century had hoped to see society transformed. These earlier Radi-

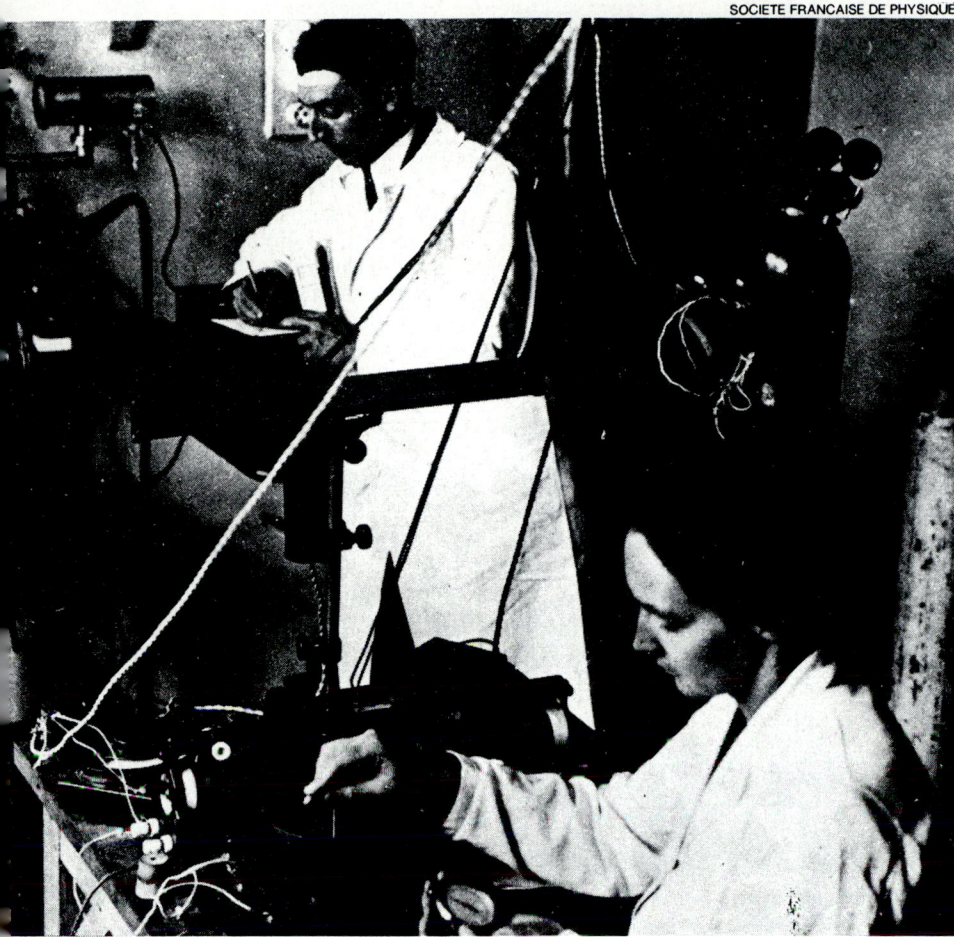

Paul Langevin, one of France's most respected physicists and a strong supporter of Perrin.

cals, who had included prominent scientists, had hoped that science would be able to unite all people, rising above political and religious "superstitions." They also expected that technology, based on new discoveries, would create prosperity for all, thereby eliminating struggle between the classes. Herriot had held these beliefs and spoken about them for years; now Perrin was holding him to his word. The politician gave in, and began a campaign in the press and parliament, pressing for a reorganization of French science.

Herriot, however, was not in power at that time, and the mood of the French parliament was conservative; the plan's success was still in doubt. Perrin continued to go about knocking on doors, sometimes making use of his old friend Marie Curie, who obediently followed him from office to office. She was a frail, pinched old woman draped in black, a legend of self-sacrifice. Despite her severe shyness, she could grow fierce when the future of her Radium Institute was at stake, and on her visits with Perrin to officials, both her arguments and her mere presence added to his message. One observer said she had the appearance and moral force of a Buddhist monk from Tibet.

Moral force can be a mighty lever in the material world, but it must have a material fulcrum to press upon. The scientists needed a new source of money. In 1930

France was prosperous, enjoying a budget surplus, but it was one thing to see money available and another to divert some of it into a new organization. A solution was suggested by one of Perrin's aphorisms: "A corps of men used to considering things from the standpoint of scientific research can render as much service in time of war as a number of army corps." This was a sentiment people of every political party could agree to—and a timely sentiment, for France was awake to the resurgence of Germany and the rise of fascism. Construction had just begun on the multi-billion-franc Maginot fortifications. Herriot managed to secure five million francs of that allocation, a tiny fraction of the total defense budget, for the "research corps."

To create a new institution would have required a protracted legislative struggle, so Herriot used an artifice: the money was turned over to the Caisse Nationale des Sciences, a minor state fund, independent of the bureaucracy, which till then had doled out honorific aid to aged and needy scientists. The Caisse Nationale was quietly reorganized along the lines of Perrin's original plan, with details drawn up by his friend André Mayer, a physiologist. Following established practice, the money was allocated by committees. Here Perrin's circle predominated. Of the ten members of the committee responsible for physical science, only three were outsiders; the

others were Marie Curie, Langevin, Maurain, Perrin, and three more Ecole Normale alumni who had been close friends of the group since the turn of the century. This was the sort of control over the financing of research that the group had been fighting for.

In 1931 the Caisse Nationale opened its first salaried positions for research. These came just in time for Joliot as he considered whether to give up trying to find a job in pure science. Perrin settled the matter by arranging a half-time research position for Joliot at the Radium Institute. He soon found a job for the other half of his time, and later Irène Curie also got a Caisse Nationale post. Now they had their chance to do research.

Attacking the Establishment

The rebirth of French science was still far from secure. The Caisse Nationale could easily lapse into the normal bureaucratic inertia of state organs. Worse, it lacked the authority to defend itself against the established powers—professors who ran their laboratories as personal fiefs, jealously independent in their research or lack of it. Such professors might seize the Caisse's funds and direct them back into traditional channels, so that the Caisse could end up supporting the system it was meant to supersede. Therefore, on Perrin's advice Mayer drafted another proposal, asking for an advisory council for the Caisse. Seemingly an innocuous administrative adjustment, this proposal was in fact the boldest of the moves by which Perrin's group edged toward their objective.

The old ways of managing French science were symbolized in two institutions, the Academy of Sciences and the Superior Council of Public Instruction. All the energy of Perrin's circle could not budge the Academy, a little club of eminent and aged gentlemen, encrusted with centuries of tradition. Fortunately, the Academy had no direct control over any funds but its own small endowments, so it could be circumvented. The Superior Council of Public Instruction was another matter. A formal assembly of senior figures of French learning, representing all the sciences and all the humanities too, the Superior Council had long experience as lawful adviser to the education ministry. While ministers could overrule the Council, they almost never risked the uproar that this would provoke. One of the ministers said that there was nothing that awed him more than his ritual semi-annual appearances before this body; he felt a "respectful fear" as he stood behind a little table with an ancient inkwell and green blotter, faced with row upon row of distinguished professors:

> It wasn't at all the anxiety that your decrees and ordinances would be rejected, it was the obscure feeling of being judged yourself, while you were talking, by those augurs covered with diplomas. Once back among themselves, in their institutes and laboratories, they would make commentaries, emit lapidary evaluations, which would weigh more heavily on your reputation than all your citations by parliament.

Representing every traditional pressure within the little world of Parisian professors, the Superior Council of Public Instruction could impel a minister to weaken the Caisse Nationale, and this threat could not be circumvented. It would have to be challenged head-on.

Perrin and Mayer planned to establish by law a new democratic assembly, a Superior Council of Research. Its members would not be drawn entirely from the old professorial establishments, but would in some cases come directly from the scientific disciplines themselves. Some representatives would be named by scientific societies and organizations, presumably on the basis of merit as scientists. Others would be elected only by researchers under forty years of age. To dilute the influence of the elderly even more, when any member reached seventy a new member would be chosen to join him on the Council. The Council, which would control the Caisse Nationale's money, would not only be a match for the conservative Academy of Sciences but would also be the juridical equal of the Superior Council of Public Instruction. In short, the scheme was aimed point-blank at the traditional domination of French science by cliques of aged professors.

As before, it was politics that gave the reformers an opening. In the elections of 1932 Herriot and his party won control of the government. Herriot's minister of national education was persuaded to endorse the plans for the research council, and in 1933 it was established, though not

UPI

without a lengthy battle. Among other obstacles was a condition imposed by the minister that Perrin's group obtain the approval of their old friend Cavalier, who had to arbitrate between numerous conflicting interests.

We can only guess at the arguments raised against Perrin, but we do have documentation for one of the difficulties, thanks to the chance survival of a few notes in a folder. Here we find a letter to Cavalier complaining that the Council, in which only one section out of seven would be given the social sciences (*sciences humaines*), would "make the sciences of man a little appendage to the others." Perrin wrote that he would agree to make changes but his colleagues would not, for "they want to take advantage of the current Minister" before the government changed again. There follows a memorandum by Cavalier on a visit from the mathematician Emile Borel, who insisted, "don't put too many literary people" on the Council. Finally everyone compromised, agreeing to two sections for the social sciences and related humanities. They got up a petition signed by over a hundred savants, and Cavalier gave his consent.

The reformers' next step was to make the research council the effective ruler of

all French scientific research, not only outside but even within the universities. Perrin and his friends persuaded the government to promulgate this change in 1935, when the Caisse Nationale des Sciences and another government science fund, the Caisse de Recherche Scientifique, were combined into a Caisse Nationale de la Recherche Scientifique (CNRS). This took over administration not only of the original Caisses' budgets but also of the universities' research budget and miscellaneous other funds formerly distributed by the educational bureaucracy. As a powerful and systematic organization for pure research, the CNRS had no precedent outside the Soviet Union, if there.

But the reorganization of French science could mean little unless the CNRS had money to spend, and when asked for money the government left the scientists worse off than before. The Great Depression had come, and the French government, like an old-fashioned owner of a family business, sought a balanced budget, cutting the salaries and pensions of those who depended on it. The scientists, like other state functionaries, suffered. So many professorships were eliminated that by 1936 the system of higher education had lost over 250 positions, while the research budget was slashed across the board. The tortuous advance of French science had come up against a wall. And the threat to French science seemed to be only one part of an attack on all democratic society.

Alliance with the Socialists

Every few months a new government would be formed, each as scandal-ridden and ineffectual as the last. A new revolutionary right, inspired by the success of fascism in Italy and Germany, was on the rise; in 1934 it brought rioting mobs almost to the doors of parliament. Among the scientists Langevin now took the lead. He told Bernal that he was no longer concerned with his personal contribution to science, for the first task was to defeat the fascists in order to maintain conditions where science would be possible at all. Langevin, who had long been active in various groups that supported civil rights or opposed war, now helped to found a "Committee of Antifascist Intellectuals." Thousands of scientists and others rallied to the Committee. Meanwhile the French Communist party asked the Radicals to join in a common front with the Socialist party. The Committee of Antifascist Intellectuals supported this move. At a meeting in June 1935, with Langevin presiding, the parties of the center and left began to organize a march for Bastille Day. That 14 July saw one of

Edouard Herriot (left), leader of the radical party, and an early supporter of Perrin's reforms.
Leon Blum (above), Socialist and leader of the Popular Front that came into power in 1936.

the grandest demonstrations in the history of France as marchers poured down the boulevards by the hundreds of thousands. Langevin was at the front, linking arms with the leaders of the socialist, Radical and Communist parties. The popular emotion remained a political force and pressed the parties to unite in a Popular Front, and the elections of 1936 swept the coalition into power.

The new premier was the head of the Socialists, Léon Blum. A cultivated man of letters with a walrus moustache and round glasses, Blum was a graduate of the Ecole Normale; he had known Perrin, Langevin and other scientists since the turn of the century, when they had all met for long discussions of socialism and other liberal causes. Blum's attitude toward science was characteristic of his mixture of Marxist principles and humanist thoughtfulness. Science, he said, was the main driving force behind the advance of society and could revolutionize the human condition. But without political intervention the wealth that science created would always be hoarded by the few. "It is under socialism," he said, "that science will really become a benefactor ... Science develops humanity's riches; socialism assures their rational exploitation and equitable distribution." Blum

promised that the combination of science and socialism, enriched by the arts and humanities, would lead humanity into a "paradisiacal state." This ideology even figured in his political speeches. In a radio address delivered in the heat of the 1936 election campaign, he said that under capitalism "science, the honor of our race, engenders nothing but waste, degradation and misery," but under socialism science will become a servant rather than a scourge, so that "well-being and leisure" will "increase without end."

All this was in complete harmony with the ideas of Perrin and his friends, all of whom supported the Popular Front. Perrin himself had been asked to address the vast crowd during the Bastille Day demonstration. Joining the socialists, he had declared that only a small privileged class prevented scientists from creating a millenial future. Perrin and Blum believed completely in the approaching day when every laborer could be like themselves, comfortable and secure, appreciating and supporting all the arts and sciences.

When the new government took office Blum rushed through bills legalizing collective bargaining, raising wages and so forth. Two particularly cherished laws

reflected the belief that the benefits of technology should be used to give everyone time to cultivate themselves: paid two-week vacations and a forty-hour work week. Perrin publicly defended all these measures.

While the liberal scientists supported the Popular Front's social programs, Blum willingly supported research. Perrin asked him to make official mention of the importance of science, and Blum replied that he could do even better: he meant to raise science to new status by creating an undersecretary of state for scientific research. The job was offered to Irène Curie, for in the last few years she and Joliot had made a series of discoveries in nuclear physics, culminating in the discovery of artificial radioactivity, winning the public's attention and esteem. Irène Curie sympathized with the Popular Front's politics, but she was artless, devastatingly frank, and with no taste for political maneuvering. She accepted Blum's offer only on the condition that she could soon return to her scientific work, and after enduring her quota of ministerial meetings she was replaced, by prearrangement, with Jean Perrin.

The old dream of Perrin's circle was realized: one of their number was in power. It was not that they believed that government should be run entirely by scientists and technical experts, for they were too humanistic and politically sophisticated for such fantasies. Rather, they felt that some scientists should be in a strong position to guide and promote scientific research, and then to put the results of this research to work. Since they were convinced that science was the main agent behind all social change, such power seemed to be quite enough for their purposes.

Triumph of the new system

Jean Zay, the new minister of national education, recalled that while in the government, Perrin "seemed naive and absent-minded, almost in the clouds, but in reality he was always attentive, precise, concentrated, if necessary crafty." With ardor, obstinacy, and patience Perrin set out to increase the funding of research. Perrin and Zay saw to it that the total state allocation for research leaped from eleven million francs in 1935 to twenty-six million in 1936 and thirty-two million in 1937. By 1939 the CNRS was providing partial or total support for about six hundred researchers, over half the scientists in France.

Whatever strains the swift growth may have caused were hidden by the traditional privacy of the French bureaucracy. But there were behind-the-scenes confrontations between the university "teachers" and the CNRS "researchers." The researchers usually won. Not only entrenched professors but some younger scientists grew wary of the power that Perrin and his circle had won over appointments and money. As one scientist recalled, "there were people who were not so much opposed to structures but who feared that the clan would get its hands on everything."

Perrin took advantage of his position as undersecretary of state to strengthen the CNRS. His friends warned him that the Superior Council of Research, which met only from time to time, was not enough to protect the CNRS on the bureaucratic battlefield against other organs of state; without a permanent administrative corps of its own, the CNRS could disappear at the next shift of government. Perrin determined to create such an administrative corps. Although he began modestly by asking for only a dozen people, there was still a sharp struggle in the French Senate. Perrin came up against Joseph Caillaux, the fierce old dictator of the Senate Finance Committee, who knew that in modern government a seemingly insignificant squabble over a new budget line can fix state policy for decades. Worse, Caillaux believed that technological breakthroughs disrupted employment and lay at the root of the Great Depression, and he was notorious for his demands that science and invention be reined in. But there were also a number of politicians who were ready to fight for the scientists. Perrin's group pushed through their proposal in the closing days of 1936.

The CNRS administrative corps was only one of Perrin's innovations. He created a technical corps of glassblowers, machinists, statisticians and the like, who by 1939 numbered about a thousand. He founded a museum of science in Paris, the Palace of Discovery. And he created or took over a number of laboratories, observatories and research stations, thenceforth to be administered by the CNRS. The scientists had finally won a formidable and autonomous research organization.

The influence of the CNRS on France and the world at large was profound, as may be seen for example in the later career of Joliot. By 1939 he was in charge of magnificent new laboratories paid for by the CNRS, and worked with colleagues supported like himself by the CNRS—notably Hans Halban and Lew Kowarski, Frenchmen whose foreign birth might well have barred them from any career under the old Ministry of National Education. In 1939 and 1940 Joliot and his colleagues, aided by special grants from the CNRS, formed one of the leading groups working on uranium chain reactions. Had the Second World War not intervened, they might well have built the world's first critical nuclear reactor. Every heavy-water reactor built can be traced directly to their work, and they were a key stimulus to much other work on nuclear reactors and bombs in various countries. After the war, Joliot became head of the CNRS, which began a great expansion. He then helped to found the French Atomic Energy Commission. Today France has a powerful nuclear armament and a vigorous nuclear power program, both well in advance of the programs of most other countries of comparable size, and this is all directly traceable to the support that the CNRS gave Joliot and his team. In other fields, the impact of the CNRS on France has been equally powerful.

The effect of the CNRS has also been directly felt abroad. Already in the late 1930's British scientists were closely watching the growth of the CNRS, and when in the 1940's the United States government began to plan its National Science Foundation, French science leaders were questioned about the CNRS. As the first major government science organization in a democratic country, and in some respects the first anywhere, the CNRS has had many lessons to offer. It is likely that we are not yet through learning from the ideals of Jean Perrin.

* * *

This article has been largely based on a chapter from a book by Spencer Weart, Scientists in Power, *published in 1979 in Cambridge by Harvard University Press and in Paris by Fayard. Full documentation and bibliography will be found in the book.* □

KARL KELCHNER DARROW
—Writer, Councilor and Secretary

Twenty-six years as secretary of the American Physical Society is the latest portion of a career that has included much graceful writing and a reputation for style that is admired everywhere.

PHYSICS TODAY/APRIL 1967

by John H. Van Vleck

THERE COMES A TIME when it is appropriate for an opera association to recognize not just the prima donnas and conductors, who receive the customary applause, but also the stage manager, who is usually hidden behind the stage but who is responsible that everything goes smoothly. The American Physical Society finds itself in essentially this position. Our stage manager, of course, has been Karl Darrow, who has just completed 26 years of service as secretary, twice as long a term as anyone else. Incidentally he held the same seats at the old Metropolitan Opera House for 48 years; so the operatic analogy seems particularly à propos. When things go smoothly, one takes the stage manager for granted, but once things go awry, he is the target of complaint. It is a tribute to the way Darrow has performed his duties as secretary of APS that he has never had to be its whipping boy.

In one respect, however, my operatic analog fails. The old Met had a constant capacity whereas the Physical Society has grown from a membership of about 4000 when Karl took over to the present figure of 23 000.

Another analogy might be made to a small-town grocer who over a period of years finally finds himself manager of a supermarket. The supply of small cartons in the form of ten-minute papers has proliferated over the years, with never—well, hardly ever—any congestion in the aisles. Also Karl has furnished a constantly enlarged supply of bulk packages, which we call "invited papers." He has seen to it that the different food values have been kept in proper proportion so that we have a well balanced physical diet

The author held the oldest endowed science chair in North America, which was occupied before him by P. W. Bridgman. He was on the faculty of Harvard University for most of his very distinguished career, and was awarded the Nobel Prize in 1977. This article is an adaptation of a January 1967 speech before an APS-AAPT meeting.

"ONE FOR ROOSEVELT, one for Andy Gump, the rest for the machine slate." Darrow announces election returns in his early days as secretary.

PHOTO BY BACHRACH

with a considerable choice of menus because of multiple sessions. He has kept ahead of changing conditions. In modern times the problems of obtaining proper hotel and meeting-room facilities have become increasingly complex and demanding, requiring that arrangements be made what I consider a geologic amount of time ahead.

Youth in Chicago

Most of you doubtless think of Darrow as indigenous to New York, but actually he spent the first third of his life in Chicago—except for a year in Europe after graduation from its university. Child psychologists claim that one's lifetime pattern is conditioned by very early environmental influences, and in Karl's case this thesis is well illustrated. The day on which he was born, 26 Nov. 1891, was, very appropriately, the day on which ground was broken for Cobb Hall, the first building of the University of Chicago, and he grew up near its campus. Presumably he acquired his avid fondness for classic art and architecture from contemplation in infancy of the construction of its collegiate-Gothic buildings and of the celebrated neoclassic buildings of the 1893 Chicago world's fair, which were a renaissance in American architecture after the atrocious General Grant period. I assume that at the age of two he saw the renowned dancer Little Egypt of this fair and that this event gave him his lifelong fondness for ballet.

During his youth from time to time he saw his uncle Clarence Darrow, the distinguished lawyer who defended the liberal cause in the Scopes trial. If you saw the play or movie, "Inherit the Wind," based on this trial, you saw a character representing Clarence Darrow. He was given the name Henry Drummond. Perhaps Karl's visits to Clarence Darrow explain his desire for legalistic precision in administration of APS affairs.

DARROW has been particularly known as a clarifier and elucidator of science in many books and articles.

Karl tells me that his mother presented his father with a watch on the latter's 53rd birthday and had the date engraved—28 Oct. 1899. This was the very day on which was held the first meeting of the American Physical Society. No doubt this coincidence, occurring when Darrow was seven years old, accounts for his being a physicist. (I am happy to report that he has presented the watch to the American Institute of Physics as a memento.)

Darrow took his PhD under Robert A. Millikan at the University of Chicago with an experimental thesis concerned with measurement of the ratio of the specific heats of hydrogen. Of course he could not have realized that explanation of the heat capacity of hydrogen by D. M. Dennison some ten years later would prove a landmark in the history of quantum statistics because of the role of nuclear spin.

Working and teaching

After receiving his PhD, Darrow joined the staff of the Bell Telephone Laboratories, with which he was associated for 40 years. There were, however, summer interludes during which he was visiting professor at Stanford, Chicago and Columbia. Also he taught for two spring semesters at Smith, during one of which he held the William Allen Neilson professorship. He showed himself to be a strong-minded man; for he resisted holy matrimony right through these two springtime sojourns in Northampton. Nevertheless, being secretary of the American Physical Society was more than he could take single handed, and in 1943, two years after he became secretary of our organization, he married Elizabeth Marcy. We are happy that during practically the entire period of Darrow's tenure as secretary, our speakers' table has been adorned and graced by Elizabeth.

At the Bell Telephone Laboratories Darrow's work consisted largely of digesting, abstracting and appraising the literature of physics. As Samuel A. Goudsmit recently emphasized in PHYSICS TODAY,[1] unless we have proper critical review articles, publications in physics will simply proliferate into fragmented so-called "research" articles so innumerable that they soon pass into oblivion, and then science will be smothered in its own debris.

So do not underestimate the value of a writer who coördinates and explains things so as to be more intelligible to others than the original researcher and his little group of specialists.

Darrow was a master at this sort of thing. I will give two examples of the impact of his writing. A few months ago an APS officer told me how, when he was an undergraduate at a southern college a little over 30 years ago, its physics department was so undersupported that it didn't even have *The Physical Review* in its library, and he was advised to consult Darrow's articles on current developments in physics in the *Bell System Technical Journal*, which somehow was being sent as a courtesy to the local municipal library. This the young student did, and this move started his reading of research journals in a career that has won him a Nobel prize and made him our current APS president, Charles H. Townes.

At a somewhat lower level, a recent letter from a high-school student asked the chairman of the Harvard physics department to write him "all that is known about quantum mechanics." W. H. Furry replied that the shipment of this information would require several freight trains and referred him as a starter to an article by Darrow in the *Scientific American*.[2]

Besides his numerous articles in the *Bell System Technical Journal* and other journals, Darrow is an author of four books, *Electrical Phenomena in Gases, Introduction to Contemporary Physics, The Renaissance of Physics* and *Atomic Energy*,[3] the last partially the outgrowth of his work during the war when he abstracted classified literature for the so-called "Metallurgical Laboratory" at the University of Chicago.

Honors and style

In recognition of his contributions to science, Darrow was made a Chevalier de la Légion d'Honneur by France and he holds an honorary doctorate from the University of Lyon. He is a member of the American Philosophical Society and the American Academy of Arts and Sciences.

If I were to single out any one word to summarize Darrow's unique role in the life of American physics, it would be "style." This attribute stems from

MRS DARROW was Elizabeth Marcy before her marriage in 1943 made her a frequent head-table guest at APS dinners and banquets.

his wide literary and cultural background. Too often scientists are branded as narrow-minded specialists, but Karl usually combines his conscientious attending of meetings with a visit to the local art gallery. If all savants were like him, C. P. Snow would never have been able to coin the phrase "The Two Cultures."

Karl Darrow's sense of style is evident in both his spoken and written words. Anyone who has heard any of Karl's colloquium talks, which he gave more frequently in the past than at present, will recognize them as models of clarity, perfect timing and subtle humor. His scientific writings have an elegance and manner of their own. These qualities are also evidenced in his preambles in the *Bulletins* describing the meetings of the American Physical Society. *The New Yorker* magazine at one time even reproduced his lucid and lively instructions as to what the absent-minded or unmetropolitan professor should do in case he found himself carried away on a Lenox Avenue rather than a Broadway subway train in trying to go to a meeting of the society at Columbia.

Since I am a lover of the Great

PHOTO BY MITCHELL VALENTINE

FACE AND STANCE have become familiar to thousands of physicists and teachers who have attended APS meetings arranged by Darrow during his term as secretary. He served in the office twice as long as anyone else.

talk at one of the society's dinners five years ago honoring 20 years of Darrow's secretaryship, I said, "There is something classic in Karl's writings in marked contrast to the rather barbaric English that, alas, too often appears on the pages of *The Physical Review*. In this regard I might compare them to the Roman grandeur of New York's Penn Station, which houses the trains he loves so dearly amid the rather polyglot collection of architecture that surrounds it." Now, alas, the classic Pennsylvania terminal is no more, and Lucius Beebe, aptly in my opinion, calls its demolition "the most barbarous act of vandalism in New York's long record of civic atrocities." In Karl's case the situation is somewhat better; for plans have been effected whereby he will be on the APS council and help arrange meetings for another three years, thereby giving us the benefit of continuity and his long experience and background.

A stroke of the Harvard crew is reputed to have said that Harvard was not an undemocratic institution—that he knew personally five of the seven men who rowed on his boat; two were too far away. Karl does much better. He tells me that he has known personally 53 of the 55 presidents the society has had, but two were too far removed in time.

I do hope he can be persuaded to put his memoirs in print. Whether the course of the American Physical Society under its new constitution will turn out more like the fine new Lincoln Center, with its grand opera and symphonies, or like the Madison Square Garden being built over the Pennsylvania railroad tracks, with its circuses, wrestling matches, etc., time alone can tell. One thing is for sure: Things will never be the same.

Northwest, I persuaded him in preparing the *Bulletin* for a meeting at Seattle to borrow a reference to the Indians from one of the ads of the Milwaukee railroad in the good old days when it operated the Olympian Hiawatha. I quote verbatim from what appeared in the preamble:

"The thirty-eighth President of the Society desires to call this *The* Washington meeting and to point out that, while the Indians sold Manhattan for twenty-four dollars, they fought for the Pacific Northwest."[4]

Two comments should be made in this connection. One is that when I attended this meeting at Seattle, the physicists at the University of Washington had made an erroneous calculation of the presidential succession and were crediting the remarks to Hans Bethe rather than to me, much to my annoyance. The other is that Karl had inserted these remarks in proof in the preamble, which went out under Joseph Kaplan's signature as the West Coast secretary. Kaplan told me that when some of his friends read

these remarks in the program, they said, "You're getting to write in a very sophisticated style—just like Karl Darrow."

The tuxedo was pressed

You probably wonder why I was selected to give a talk about Darrow at the January APS-AAPT meeting. It is simply because I have known Karl Darrow much longer than the current APS officers. We first really met 41 years ago at a meeting of the British Association for the Advancement of Science at Oxford. The big event was a reception given by the association president, then Prince of Wales, later Edward VIII. I was unable to attend because it was early-closing day, and I couldn't get my tuxedo pressed. Needless to say, Karl hadn't exhibited such lack of foresight and was there.

I remember that at that meeting Karl and I attended a performance of Bernard Shaw's "Heartbreak House." What is heartbreaking as one gets older is that things have to change—often not for the best. When I gave a

References

1. S. A. Goudsmit, PHYSICS TODAY **19**, no. 9, 52 (1966).
2. K. K. Darrow, Scientific American **186**, no. 3, 47 (1952).
3. K. K. Darrow, *Introduction to Contemporary Physics*, D. Van Nostrand, New York (1926) 2nd edition (1939); *Electrical Phenomena in Gases*, Williams and Wilkins, Baltimore (1932); *The Renaissance of Physics*, MacMillan, New York (1936); *Atomic Energy*, John Wiley and Sons, New York (1948).
4. Bull. Am. Phys. Soc. **29**, 3 (1954). □

My sixty years with The American Physical Society

In reminiscent mood, the Secretary Emeritus recalls some of the people and events important in the development of the society since the early years of this century.

Karl K. Darrow

PHYSICS TODAY/JULY 1974

In the year 1913, the last of the Golden Years before the first of the great convulsions that have afflicted the twentieth century, I was a graduate student at the University of Chicago. I was a Chicagoan by birth, and my parents and I lived about a mile from the University. This geographical accident was what caused me to be enrolled in a University having one of the strongest, perhaps indeed *the* strongest, of departments of physics in the country. No such circumstance was responsible for the presence of another graduate student of physics, Yoshio Ishida, whose name discloses his native land. I wish I knew why he had chosen the University of Chicago: probably his

In 1974 The American Physical Society celebrated its seventy-fifth anniversary. A special session at the society's joint meeting with the Optical Society of America in Washington D.C. in April 1974 marked the occasion with retrospective and prospective views of the APS and its role in the development of physics in America. Among them was Karl K. Darrow's recollections of his sixty-year association with the society, upon which this article is based.

Darrow (1891–1982) was a member of the society from 1913 and its Secretary from 1941 to 1966.

choice is proof that it had already a high reputation overseas.

One day Yoshio was standing nearby when I asked Robert Millikan whether I might join The American Physical Society—*the* Millikan, of course, already well on the way to fame and the Nobel Prize as the measurer of the electron charge by the "oil-drop experiment." Like "Little Audrey" in the now antiquated joke, Yoshio "laughed and laughed and laughed." It was evident that he thought that my question was presumptuous, and his laughter made me think that I had indeed made a *gaffe.* But Millikan evidently did not consider it too presumptuous, for within a few months I became a member.

Although meetings were held every year at the University of Chicago, my first out-of-town meeting of the Society—remember that "out-of-town" meant at that time, for me, "out-of-Chicago"—was held at Columbus. It was a joint meeting of the APS with the American Association for the Advancement of Science. Since this was a meeting of all the societies belonging to the AAAS, the conspicuous "retiring presidential address" was not that of the APS President, but that of the AAAS President, who was Charles William Eliot, long the President of Harvard University. Of all the words of wisdom that he uttered I remember

not one, but I still remember his dignified appearance as he stood upon the lecture platform.

A year later, the meeting of AAAS and APS was held in New York. On New Year's Eve I stood in a vast crowd at the corner of Forty-third Street and Broadway, and waited for the numerals 1916 to change to 1917 on the north facade of the Times Building. When they changed there was wild applause from the crowd. I did not join in it, but somberly reflected that 1916 had been the worst year of the twentieth century and that 1917 would probably be still worse. How right I was! Then I walked to Penn Station and boarded my train, not divining for a moment that the next New Year would find me a New Yorker. But back to New York I came in September 1917 (and from then until 1956 my services were to Bell Telephone Laboratories).

I do not recall when I first met Willard Severinghaus, my immediate predecessor in the Secretaryship of the APS. Probably it was at Chicago or Columbus. The first vivid memory that I have of him refers to the New York meeting of 1916–17. I was one of those who joined a tour of Fayerweather Hall, then the physics laboratory of Columbia University, and Severinghaus conducted the tour. Wishing to make a favorable impression on him, I asked him a question that I immedi-

ately realized to be foolish, as did he. A convenient failure in my memory has erased the details.

The Secretary's duties

It was at this same meeting that I watched Severinghaus performing what was then one of the chief tasks of the Secretary. During the sessions he would come from time to time to the room where the session was going on, pull down one of the upper blackboards, write upon it with frantic haste the names of members who had just brought to him the news that they intended to speak, and add the titles of their papers. This experience caused Council to institute the draconian regulation that no (contributed) paper could be accepted unless its abstract was in the Secretary's hands by the deadline date for the meeting. This regulation greatly simplified my life when in my turn I became Secretary of the Society.

One day early in December 1941 George Pegram told me that Severinghaus felt he had served as APS Secretary long enough and wanted to reserve his time for Columbia University where he was professor of physics. Then Pegram asked me whether I would be willing to accept the Secretaryship. I said that I should indeed be, but that first I would have to seek the approbation of Mervin J. Kelly, then President of Bell Telephone Laboratories. Pegram gravely nodded assent. What he knew and had not told me was that he had already discussed the matter with Kelly, who had given his full assent. Further, Kelly had given Pegram more than I suspect the latter had ever dared to hope for. Not only would Bell Telephone Laboratories continue my salary undiminished, but it would continue to pay my travelling expenses to APS meetings. This bonanza was welcome to the society and a blessing to me, for I retained my moorings in Manhattan while the Laboratories drifted off into the far countryside of New Jersey where most of its staff were happier. However, the Manhattan building of Bell Telephone Laboratories had not been completely vacated when I reached retirement age, near the end of 1956, though it was fast becoming deserted and lonesome. It was not till mid-December that I walked out of my old office, not quite happy to be going, for the ties of habit are strong. Thereafter I had no office but that of the APS at Columbia University. The Treasury of the society, like many a boy graduating from college, found out what it was to have to pay its own way so far as the expenses of the Secretaryship were involved. Pegram in earlier days had spoken in conversation, and sometimes at Business Meetings, of the society's "scandalous surplus." It melted fast as inflation proceeded, and the dues had to be raised.

Sessions, papers and abstracts

The principal task, or in common parlance "the big job," of the Secretary was to arrange the papers into sessions, preferably of ten to fourteen papers each. These were the "contributed" or "ten-minute" papers, so-called to distinguish them from the "invited papers" to which no statutory limit of length existed. The ideal was to set up each session in such a way that everyone who attended it would be interested in every paper in it, and not in any paper included in any other session simultaneous (or, as we often said, "parallel") with it. Like other "counsels of perfection" this ideal was unattainable, but I strove to come as close as possible to it. In the halcyon days when papers were so few that there was no need of simultaneous sessions, only the first problem existed; but this by itself was no mean one, and I had to draft into service such people as Robert Serber (for theoretical physics, naturally), James Rainwater, and William W. Havens, my successor who got part of his training in this way. Fortunately, at Columbia University there were two large student laboratories, one on each side of my tiny office, each of them with numerous large tables. At certain times these were empty of students, and in the last days before the deadline date for either an Annual Meeting or a Washington meeting these tables would be covered with abstracts so that they might have been supposed to be snow-drifted but that there was so much ink on the "snow." I do not know how this job is now handled at The American Institute of Physics. I have not asked.

Finally would arrive The Great Day when I packed all of the abstracts, clipped into groups each corresponding to a single session, into some kind of a satchel and, clutching the satchel as though my life depended on it, hailed a taxi and directed the driver to the AIP. There I delivered the packet into the very able hands of Wallace Waterfall, Ruth Bryans, Jean Doty, or one of their aides. These remarkable people knew all of the cabalistic signs whereby printers are instructed what kinds of type to use and what spacings to adopt in converting typescript into print and most particularly in converting formulas into print. It was a treat to see an abstract decorated with these signs, and a still greater treat to see it as it emerged in our *Bulletin* from the Lancaster Press.

I didn't read very many of the abstracts otherwise than in the swift and superficial way that usually sufficed to locate it in a not-too-inappropriate session. I was usually very conscious of the fact that much labor and energy and time might have gone into the work to which a particular abstract was devoted, and that the paring of an abstract by its author or authors to get its length within the two-hundred-word limit might have been a painful business. It was obvious that some abstracts were more important than others; many still remember the meeting at which Felix Bloch and E. M. Purcell

---

THE

AMERICAN PHYSICAL SOCIETY.

Minutes of the First Meeting.

MORNING SESSION.

The first meeting of the Society was held in Room 304 Fayerweather Hall, Columbia University, New York City, on Saturday, May 20, 1899, at 10:30 A. M.

The meeting was welcomed on behalf of the University by Professor M. I. Pupin.

On motion of Professor Pupin, Professor LeRoy C. Cooley was made temporary chairman, and Professor A. G. Webster was elected secretary of the meeting.

On motion it was resolved that a Physical Society be formally organized.

On motion a committee was appointed, consisting of Messrs. A. G. Webster, E. L. Nichols, W. F. Magie, B. O. Peirce, Wm. Hallock, and M. I. Pupin, to draft a constitution.

After a general discussion of the aims and policy of the Society it was voted that, except in special cases, the meetings be held in New York City ; that the meeting express the willingness of the Society to establish local sub-sections when demand shall arise ; and that it further express the sentiment of the Society to cultivate the closest relations with Section B of the American Association for the Advancement of Science, and to contribute by every thing in its power to the latter's success.

The Bulletin of The American Physical Society, volume 1, number 1, reported in its opening pages the first meeting of the APS, on 20 May 1899 at Columbia University.

simultaneously brought to the world the discovery of magnetic resonance. It was obvious that others were trivial, but they were not excluded. Nonmembers would sometimes send in abstracts totally irrelevant to physics, but the fact that a nonmember was debarred from presenting a ten-minute paper was sufficient for rejecting these. The fact that "physical" is a word of more than one meaning subjected us to experiences that chemical and astronomical and geological societies do not have. My wife remembers a headline in a newspaper of a medium-sized city which made it clear that its editor thought that "physical" had to do with physique and body-building. I wish I had that headline now. It read: "Body-builders Assemble for Meeting!"

Finally came the preparation of the Preamble. The word "preamble" was, I am nearly sure, introduced by myself into the vocabulary of the society. Into the Preamble I strove to insert, or others strove to insert, all that the members had to know about such matters as the names of the "official" hotels and their rates. In olden days I would take care to include instructions how to get from the hotels to the meeting-place by streetcar or bus, but as the members got more flush and their expense accounts more liberal, most people just took taxis. This worked very well for those going from the hotels to the meeting-places, but not always in the other direction, and many were the times when I stood on Connecticut Avenue near the National Bureau of Standards waiting for a streetcar or a bus to come along.

In the one-session-at-a-time days the President or the Secretary could be conspicuous by his absence. In the multisession days this handicap vanished, and I could disappear into the local art museum after leaving word with a few important people that I could be paged there. I never was. In fact I acquired quite an extensive knowledge of the museums of the US and Canada during my years in the Secretaryship. Among these is the superb museum of Kansas City, to which I paid the "ultimate tribute" once by going at my own expense from St. Louis to Chicago via "K.C."

The Council

Timing of Council meetings could be a delicate business. There seemed always to be at least one Councillor, or more than one, who would complain that the Council meeting was scheduled just at the time when some paper was being given that he particularly wanted to hear, or worse yet, when he himself was scheduled to give a paper. This simply could not always be helped. We finally circumvented this

Albert A. Michelson (left), the second president of APS, is seen in Marguerite O'L. Crowe's 1908 photograph with Robert A. Millikan, Henry G. Gale and Carl Kinsley, seated on the steps of the Ryerson Physical Laboratory at the University of Chicago.

by holding the Washington and New York meetings on the days before the scientific sessions commenced—this was likely to ruin Sunday, until the meetings were transferred the latter days of the week—and the Midwestern meeting at some specially chosen day in October so that Thanksgiving would not be ruined for the Midwestern members. A Chicagoan myself, I never realized how great was the antipathy in the Midwest to Thanksgiving meetings until I had to take responsibility myself for recommending a date to the Council.

The Council also has changed, and in my not-so-humble opinion, not for the better. It has expanded from a membership of about fourteen to twenty-five or thereabouts. Formerly it could meet in Pegram's office at Columbia; now it meets at the AIP, around a table so long that it seems to extend almost from the United Nations to the Rockefeller University. I, who used to know all of the members of the Council, now see people sitting at the council table whom I do not recognize. This is my loss, not theirs; but I remember that when the change was proposed, someone—luckily I have forgotten who—said that the old system favored "the élite." I remembered the story of a member of the Chambre des Députés in Paris, who was present when some radical measure was proposed, perhaps the extension of the franchise to eighteen-year-olds. The proponents said the then system favored the élite. Upon this another member of the

Chambre leaped to his feet and shouted "*Vive l'élite!*". And so say I.

Michelson and Millikan

The first President of The American Physical Society was Henry Augustus Rowland. Him I never saw, and he is one of the two Presidents who enjoy, or enjoyed, that distinction. The second President was Albert Abraham Michelson, whom I was to see often in due time, for he was professor and head of the department of physics at the University of Chicago—the titular head, that is to say: Millikan apparently was the real head. Michelson sat in a big office with a matte-glass panel on the door: he was supposed to spend his time cogitating on the problem of making bigger and better diffraction-gratings. Twice, and twice only, I dared to tap on that door. One occasion is without significance, but the other seems worth telling. A fellow-student and I differed on a point in electromagnetic theory that had figured in Michelson's lectures. We decided to abide by Michelson's arbitration, and my fellow-student was so unwise as to insist that we bet three dollars on the outcome. One of us rapped on the glass panel, and Michelson came and opened the door and received us with the grave courtesy characteristic of him. My fellow-student was first to put his case, and Michelson agreed with him. Then came my turn: I put my case, and Michelson reversed himself and agreed with me! We excused ourselves, and like a gentleman my colleague at once paid

HAVENS

PEGRAM

over the three dollars. I tried my best to refuse it, for I felt that the three dollars meant much moe to him than to me who was being supported by my parents; but I failed.

Millikan was a great physicist but not a good linguist. I, being in somewhat the opposite situation, sat and winced when he addressed a Colloquium at the University of Berlin in 1912. Genders annoyed him, as they have annoyed many another; and sometimes he would jocularly prefix *der die das,* all three of them, to a noun. This, I felt, made the audience friendly to him. To our Mexico meeting of 1950 he came, I think, partly at my urging, for I knew that his prestige would build up what I thought would be a small meeting (it wasn't). I do not recall that he ventured into Spanish; but I do recall that in a short allocution, which I was asked to translate, he said that he was glad that he had "put me on the skids." I did not choose to translate this remark literally into Spanish; but of course there were a great many Mexicans there whose English was perfect, or nearly so. They grinned.

Lord Rutherford had something against Millikan, and I never discovered what. I went to call on him once every two years in the 1920's and 30's, and I think he never failed to make some disparaging allusion to Millikan—except once. In 1937 I went with Harvey Fletcher to make my biennial call. Fletcher had been a student of Millikan's and admired him greatly, and so I warned him of what I thought would happen. It was unnecessary. Rutherford talked about the impending war, which he foresaw accurately—this in a period when most of us were still deluding ourselves that Hitler didn't

mean what he was saying. I never saw him again. But I am going rather astray from my topic.

Joint meetings

The first of our meetings in Mexico occurred while the Mexican Physical Society was being organized, and may have given a stimulus to it. The other three were joint meetings with the Sociedad Mexicana de Fisica, of which Manuel Sandoval Vallarta was deservedly the first President. Vallarta was the principal architect of these four meetings in Mexico, indeed the man without whom they never would have been. Those who were at our first Mexico meeting will never forget the great party for us given around midnight with supper, drinks and a Mexican orchestra, in the mansion of the Margains who were the gracious parents of María Luisa the gracious wife of Manuel. Later we had some joint meetings of SFM and APS in the US, but naturally none could rival the splendor of that first meeting and its ambient. Keys to the city were handed out at a formal luncheon at the first of our Mexico meetings. I am looking again at mine as I write this. I am reminded that one of our meetings in Mexico was a joint meeting of all three societies—SFM, APS and the Canadian Association of Physics. I don't think that we have had so many in the US, but there have been some in states of the USA that are contiguous to western Canada.

Joint meetings with European physical societies were for me an unattainable dream. The financial problems involved in raising funds to pay travelling expenses across the ocean were insoluble. At least, they were insoluble for me; if Bill Havens solves them I will hail his success. However, one French physicist once assured me that if a French scientist were invited to address the APS his Government would pay his way. This occurred so late in my career that I never put it to the test. Perhaps Havens will.

One European physicist of great distinction, a Nobel laureate in fact, once told me that no European should be expected to come to America if his way were not paid. I told him that I had crossed the Atlantic more than 45 times at my own or my father's expense (the total is now 64). He made no comment.

Commuting physicists

During most of "my" sixty years of The American Physical Society, George Braxton Pegram was a towering figure on the Council and in the society. He was also a towering figure in the department of physics of Columbia University, and in the administration of the university itself, where for a long

period he was Dean of the Faculty of Arts and Sciences. His home was in New Jersey, but at busy times he used a room in Butler Hall across the street from the University so that he could work late in the evening or begin work early in the morning—or both.

I came to New York City in 1917 to do war work at Bell Telephone Laboratories, or to be more precise, the Western Electric Company from which the Laboratories were soon to be separated. The job was to last "for the duration" as people then said, meaning the duration of the war. It lasted for 39 years and I am still here. During all this period I have lived within a mile of Columbia University. This was my father's wise decision: he wanted to be in an academic community, and so did I. With my parents I lived a couple of blocks from the Hudson River and with my wife I have dwelt beside the river. From our windows we see the ships departing for Europe, and when I come back I can see my windows as the steamship turns into its "slip." When the weather is not unfriendly I walk between my home and the University where my office used to be. This walk for many years constituted most of my exercise. My successor Bill Havens also lives beside the Hudson River, about twenty miles upstream. In distance he is much farther from his office than I was, but in time the difference is unimportant, for of course he drives. On his way he passes the hilltop where sits "The Cloisters," a great building in a pure medieval style containing a great collection of medieval art. It has an awe-inspiring chapel adjoining the cloister from which it takes its name. Havens can pause there on his way to work and ask for divine guidance: I have not been told how often he does.

□

Edward Uhler Condon 1902–1974

PHYSICS TODAY/JUNE 1974

The extraordinary career of Edward Uhler Condon, president of the American Physical Society (1946) and of the American Association of Physics Teachers (1964), ended with his death in Boulder, Colorado, on 26 March 1974.

Born in Alamogordo, New Mexico, on 2 March 1902, Edward Condon was one of the young Americans who made the pilgrimage in 1926 to Göttingen and Munich and grasped immediately the significance and power of the new quantum theory. As an undergraduate, Condon had worked as a reporter for the Oakland *Inquirer,* thinking he might pursue a career in journalism. But the intellectual challenge of physics, after a brief flirtation with chemistry, caught his fancy. When he returned from Göttingen, he worked briefly as a public-relations man for Bell Labs, lectured at Columbia and then embarked on an academic career that took him to Princeton, Minnesota, and back to Princeton, where he taught until 1937.

Like most great scientists, Condon made important contributions while still a student. The basis for his papers on the separability of electronic and vibrational motions in molecules (the Franck–Condon Principle) was in his Berkeley thesis. With R. W. Gurney, he was an early explorer of quantum-mechanical tunneling, applied to the phenomenon of alpha-particle radioactivity. In 1937, with Gregory Breit and Richard Present, he interpreted proton–proton scattering data and established the importance of charge independence in the strong nuclear interaction. His early solid-state theory work was the explanation of optical rotatory power, and later he studied semiconductor-contact potentials.

With Philip M. Morse he wrote the first English-language text on quantum mechanics (1929). With G. H. Shortley he wrote the *Theory of Atomic Spectra* (1936), still the primary treatise in the field. In later years the *Handbook of Physics,* which he edited with Hugh Odishaw, and his editorship of *Reviews of Modern Physics* demonstrated once again his facility for dealing with the full range of topics in physics.

These brief notes are an inadequate tribute to the side of Ed Condon's career that will have the most lasting value—his great contributions to knowledge through discoveries in physics. This side of his career is much more fully documented in *Topics in Modern Physics: A Tribute to Edward U. Condon* by Wesley Brittin and Halis Odabasi (Colorado Associated UP, 1971). Younger physicists who may wish to emulate Condon's courageous public record as an outspoken defender of truth, civil liberties and peace may lose sight of the monumental research contributions that won him the admiration of his fellow scientists and the respect of the public, which permitted him to make a major impact on public affairs.

The second phase of Condon's career began with his move to Westinghouse as associate director of research, just two years before the beginning of World War II in Europe. He brought Westinghouse into the nuclear age and earned an accolade from *Time* as "king of the atomic world." He served on the National Defense Research Committee during World War II, but was not present at his birthplace in Alamogordo when the Trinity explosion gave that small New Mexico town its second claim to fame.

With the war over, Condon became director of the National Bureau of Standards and, concurrently, science advisor to Senator Brian McMahon, chairman of the special Senate committee on atomic energy. McMahon was leading the forces for civilian control of the nuclear-weapons program and with Condon's active help saw success in the McMahon–Douglas bill, passed in August 1946. Condon believed deeply that civilian control over nuclear-weapons development and production was essential to avoidance of nuclear war. In the year of Condon's death, this issue may be reopened as Congress considers the Energy Research and Development Administration proposal which restructures the AEC.

At NBS, as he had at Westinghouse, Condon concentrated his attention on good science, stripping away administrative encrustations of the past, hiring the next generation of scientific leadership, pulling together programs (like building technology) of great potential benefit to the public. He built the NBS Boulder Laboratories. But soon these accomplishments were dwarfed in the public eye by the relentless attacks of Congressman J. Parnell Thomas and the House Un-American Activities Committee, which Thomas headed. The press picked up the phrase in a HUAC report (published on Condon's birthday in 1948) that stated, "It appears that Dr Condon is one of the weakest links in our atomic security." Privately, Condon described the impossibility of refuting such a charge with characteristically colorful language: "If you say I've got a wart on my nose, I can deny it. But if you just say I'm one of the ugliest men in town, all I can do is argue that I'm really quite pretty." Time and again, his security clearance status was reviewed

HEKA

CONDON

and re-established, only to be challenged again. Finally, in 1951, with his record cleared and with Parnell Thomas in Danbury Prison, convicted of taking kickbacks from his office staff, Condon left government to become head of research and development for the Corning Glass Works.

In October 1954, Condon's Navy clearance was again re-established in connection with government contract research at Corning. When the clearance was dramatically suspended by intervention of the Secretary of the Navy, the press reported that Vice President Nixon, a former member of HUAC, implied in campaign speeches that he had requested the suspension.

Ten years later, after Condon had taught at Oberlin two years and at Washington University for seven, he moved to Boulder, Colorado, as professor of physics and fellow of the Joint Institute for Laboratory Astrophysics. His security clearance was quietly restored, clearing his record once again.

What kind of man was he? Grace Marmor Spruch's profile in *Saturday Review* (1 February 1969) says it well: "The composite Condon is a moral, impassioned man, with a depth of concern for mankind not common in scientists; a man fiercely principled and anti-diplomatic; a man who believes and feels in sharp contrasts, who will let the world know his position without ambiguity. Fuzzimindedness is an anathema to him and he insists on saying so at every opportunity. But this rasping trait is wedded to an extreme generosity and kindness. Throughout his life he has given freely of his time, his counsel, his finances, and his home."

Watergate came as no surprise to Edward Condon, nor did its aftermath. I imagine he would like to have lived to see the outcome of the impeachment inquiry. But Condon understood and paid his share of the price of liberty. Somehow his idealism, his sense of humor and his inexhaustible energy made his relentless quest for a better world look like optimism. He was elected president of the American Association for the Advancement of Science during the height of his troubles with HUAC. He was president of the Society for Social Responsibility in Science (1968–69) and co-chairman of the National Committee for a Sane Nuclear Policy (1970). He was appropriately honored on his retirement from JILA and the University of Colorado in the summer of 1970 by the volume edited by Brittin and Odabasi mentioned earlier. Brittin relates a comment about Condon by E. Bright Wilson: "Sometimes I think he looks for trouble," Wilson said. Condon's comment: "It's not hard to find."

Sadly, brilliant scientists—who serve their country and principles, their love of truth and their fellow citizens with relentless determination and delightful good humor—are hard to find indeed.

Lewis M. Branscomb
*J. F. Kennedy School of Government,
Harvard University*

Harlow Shapley
1885–1972

PHYSICS TODAY/JANUARY 1973

Harlow Shapley, director emeritus of the Harvard College Observatory, died in Boulder, Colorado, on 20 October, two weeks short of his 87th birthday. In science as in politics, he never hesitated to oppose the mainstream of scientific or political thought whenever he felt morally or intellectually obliged to do so. His achievement in deriving the size and shape of the Galaxy and in locating the solar system far from its center was perhaps the most revolutionary development in cosmology since the time of Copernicus. Yet astronomy was only one of the many ways in which he expressed his love for nature and its inhabitants; for example, his classical studies of the behavior of ants and his rescue operation in behalf of refugee scientists were two others. It was as easy to be awed by his professional accomplishments as it was to be charmed by his warm, almost boyishly friendly personality, and by his delightfully pointed wit and good humor.

Shapley was born and raised on a farm in Missouri. He attended a one-room country schoolhouse and worked as a newspaper reporter for several years before enrolling at the University of Missouri, where he was attracted to astronomy by F. H. Seares. In 1911 he arrived at Princeton to study and work for the PhD with H. N. Russell, in time to assist him in fashioning elegant and definitive mathematical methods for deriving the elements of eclipsing binary-star orbits from observed light curves and to publish a volume of results that established him as a leading authority on double stars.

In 1914 Shapley was appointed to the staff of the Mt Wilson Observatory and launched almost straightaway into the work on stars in globular clusters, which revealed in 1918 that the size of the Galaxy was ten times larger than the generally accepted value, and that the sun was not central but far out on the periphery of the Milky Way. Six years later, when E. P. Hubble began to establish the scale of the universe, he used precisely the same techniques for measuring the distances of galaxies that Shapley had invented to determine the distances to globular clusters.

PHOTO: CHRISTIANE GALLET

SHAPLEY

In 1921 Shapley became director of the Harvard College Observatory, at the beginning of the decade that was to see the discovery of the expanding universe of galaxies, the derivation of the ionization equation, the birth of quantum mechanics, and the emergence of astrophysics as a quantitative science. He soon transformed the Observatory from a relatively isolated bastion of 19th-century classical astronomy, lacking a PhD program, into a busy and thriving international center for research and education in almost every aspect of modern astronomy, in which his own researches of the distribution of galaxies and of variable stars in the Milky Way and in the Magellanic Clouds were central.

Graduate work at the Observatory in the 1930's was an exciting, intellectual experience, and great fun besides. A young and brilliant faculty, supplemented by a steady stream of famous visitors, and Shapley's administrative skill and charisma, created a remarkable feeling of unity and camaraderie among the staff and students. The Hollow Squares over which he presided were memorable occasions at which one might learn the latest news of astronomy, from the discovery of a dwarf galaxy in Fornax to the latest theory of the identification of the coronal lines. Shapley enjoyed working with his associates, and he also liked to play with them; there was volleyball on Saturday afternoons, softball at picnics, and an incredibly abundant and varied menu of hospitality offered by the Shapleys at the Observatory residence.

In the postwar period, Shapley was in the forefront of successful political activity that led to civilian control of atomic energy, and to the establishment of the National Science Foundation and UNESCO. He was ardently internationalist in science and in his passionate belief in the essential unity of the human race. He had the courage, after World War II, to speak out against the Cold War and its proponents, and it is only now, 25 years later, that we perceive that he was opposing policies that may have led to the Vietnam disaster.

After retiring from the directorship of the Observatory in 1952, Shapley remained at Harvard as a professor until 1956 and continued to lead a busy and active life as a writer, lecturer and world traveler until two or three years before his death.

LEO GOLDBERG
Kitt Peak National Observatory

I. I. Rabi
1898–1988

PHYSICS TODAY/OCTOBER 1988

On 11 January 1988, I. I. Rabi, a creative scientist, an innovative statesman and a philosopher, died at the age of 89. He had received numerous awards and honors, including the 1944 Nobel Prize in physics for inventing the molecular-beam magnetic resonance method and for using it to measure the magnetic, electrical and structural properties of atoms, molecules and nuclei.

Rabi was born on 29 July 1898 in Rymanow, Austria, to an Orthodox Jewish family who soon thereafter moved to New York City, where they lived initially on the Lower East Side, but later in the Brownsville section of Brooklyn. He attended New York public schools and, as an avid reader, gained much of his education and interest in science through books borrowed from the public library. In 1916, after graduating from the Manual Training High School in Brooklyn, Rabi entered Cornell University. He started in electrical engineering, but graduated in chemistry. After three years of uninteresting jobs he returned to Cornell to do graduate work in chemistry; a year later he moved to Columbia University, and to physics. At Columbia, Rabi did his doctoral research on magnetic susceptibility with Albert P. Wills, but, characteristically, it was on a subject of Rabi's own choosing and employed a novel technique that greatly simplified the experiments. The day after he sent in his doctoral thesis, he married Helen Newmark, who remained his lifelong companion.

Rabi soon went to Europe on a traveling fellowship, where he worked intermittently with Arnold Sommerfeld, Werner Heisenberg, Niels Bohr and Wolfgang Pauli. The Stern–Gerlach experiment demonstrating the reality of space quantization had earlier sparked Rabi's interest in quantum mechanics, so he became a frequent visitor to Otto Stern's molecular-beam laboratory in Hamburg while working there with Pauli. During one of these visits Rabi suggested a new form of deflecting magnetic field, and Stern invited Rabi to work on it in his laboratory. Rabi's acceptance of this invitation was decisive in turning his interest toward molecular-beam research.

Rabi returned from Europe to join the faculty at Columbia and to begin atomic-beam research in his own laboratory. In 1931 he and Gregory Breit developed the important Breit–Rabi formula, which showed how the magnetic energy of an atom and its effective magnetic moment vary with the strength of the external magnetic field. These changes occur because the atomic configuration varies from the electron's being coupled primarily to the nucleus at a low external field to its being coupled primarily to the external magnetic field at a high field.

Using the Breit–Rabi formula and an atomic-beam apparatus with inhomogeneous magnetic fields, Rabi, Victor Cohen and others were able to determine the strengths of electron–nucleus interactions and the magnitudes of nuclear spins and magnetic moments. Rabi further improved the precision of the measurements by noting from the Breit–Rabi formula that the effective magnetic moments are zero at certain magnetic fields, which give a marked rise in the intensity of the undeflected atoms passing through an inhomogeneous field. By measuring these zero-moment magnetic fields, Rabi's students and associates determined a number of hyperfine interactions. Although the zero-moment method did not work for atoms with nuclear spin $\frac{1}{2}$, Rabi devised an alternative refocusing technique that did.

Rabi also showed that one could adapt the molecular-beam deflection method to measure the signs of nuclear magnetic moments by determining which transitions occurred when atoms went through a region of space in which the directions of the magnetic fields were successively reversed.

Rabi developed the theory of such transitions in his important paper entitled "Space Quantization in a Gyrating Magnetic Field" (*Physical Review* **51**, 652, 1937). In this paper Rabi assumed for simplicity that the applied field changed its direction ("gyrated") at a fixed frequency. As a result this paper has provided the theoretical basis for all subsequent magnetic resonance experiments.

Rabi initially applied his theory to fields that changed only in space rather than in time. A few months after the publication of that paper, following a visit by C. J. Gorter, Rabi directed the major efforts of his laboratory toward the development of molecular-beam magnetic resonance, with the magnetic fields oscillating in time. A molecular beam was deflected by one inhomogeneous magnetic field and refocused by a similar field. In passing between the two fields the molecules were subjected to a weak oscillatory magnetic field at frequency ν. When ν was equal to the Bohr frequency $\nu_0 = (W_i - W_f)/h$ (where W_f and W_i are the energies of the final and initial states), transitions could take place, with a consequent refocusing failure and a reduction in beam intensity. By measuring the beam intensity as a function of frequency one could thereby determine the spacing of the molecular energy levels.

The first successful molecular-beam experiment was that of Rabi, Sidney Millman, Polykarp Kusch and Jerrold R. Zacharias in 1938, which determined the nuclear magnetic moment of Li^7. Soon thereafter Jerome M. B. Kellogg, Rabi, Zacharias and I applied the method to molecular hydrogen and discovered a multiplicity of resonance lines, whose separation arose from the magnetic interactions of the nuclear moments with each other and with the magnetic field caused by the rotation of the molecule. The separations of the resonances for D_2 were much greater than could be attributed to such magnetic interactions, but could be fitted by assuming the deuteron had a nuclear electric quadrupole moment, that is, by assuming it was ellipsoidal like an American football, rather than spherical; such a shape would result from the existence of a previously unsuspected tensor force between the neutron and proton.

In subsequent years Rabi and his associates successfully applied the beam-resonance method to single atoms as well as to polyatomic molecules, and in such experiments they measured numerous nuclear spins, nuclear and atomic magnetic moments, atomic hyperfine interactions and nuclear quadrupole moments.

World War II interrupted Rabi's molecular-beam research from 1940 to 1945, during which time he was actively involved with the development of microwave radar. He headed the magnetron group at the MIT Radiation Laboratory, where he later became deputy director. He was particularly active in developing shorter wavelengths, first from 10 cm to 3 cm at MIT; later he initiated the establishment of the Columbia Radiation Laboratory, which pioneered in the development of 1-cm-wavelength radar.

During Rabi's early European travels he had become a great friend of J. Robert Oppenheimer, so Oppenheimer tried to persuade him to become deputy director at Los Alamos. Although Rabi declined because of his commitment to important work at MIT, he was a frequent visitor to Los Alamos as consultant, friend and adviser to Oppenheimer. Rabi's persuasive advice led Oppenheimer to

I. I. Rabi

abandon his original plans for a military establishment with uniformed scientists at Los Alamos in favor of a civilian research and development laboratory.

In his 1945 Richtmeyer lecture, delivered shortly before the end of the war, Rabi discussed the possible use of an atomic-beam magnetic resonance apparatus as the control element of an accurate clock. The *New York Times* report on this lecture is the first published account of atomic clocks, which have now become so accurate that they are the basis of the international definition of the second.

Following World War II, Rabi returned to Columbia to reestablish his molecular-beam laboratory. With his students John Nafe and Edward Nelson, Rabi successfully applied the magnetic resonance method to atomic hydrogen and discovered that the hyperfine separation due to the interaction between the magnetic moments of the proton and electron was slightly different from the theoretical expectation. This was the first indication that the magnetic moment of the

electron was different from the expected Dirac value, an observation later confirmed by Kusch's direct measurements of the electron magnetic moment. This experimental anomalous magnetic moment was the principal stimulus for the development of relativistic quantum electrodynamics, the first successful quantum field theory.

Another important molecular-beam development was the adaptation by Rabi and Harold K. Hughes of the resonance method to electric deflecting and oscillating fields. Subsequently improved by Rabi, John Trischka, Vernon Hughes and others, the electric resonance method has been used for many precise measurements of the spin-dependent internal interactions within molecules in specific rotational states.

Although most of Rabi's experiments were with molecular beams, he also participated with William Havens and James Rainwater in a neutron–electron scattering experiment that provided the first experimental evidence for the neutron–electron interaction.

Although Rabi's classroom lectures were often chaotic, he was a stimulating teacher who made his students think. He was an inspiring supervisor of PhD students whose research experiments were innovative and fundamental. Rabi gave his students freedom and independence while maintaining high standards for both the quality and the interest of the research. Rabi and his wife Helen were personally very helpful to his students and associates, most of whom remained lifelong friends.

Rabi's influence extended far beyond his own research through his membership on important committees, his many public lectures and his innovative proposals for new means of cooperation among institutions and nations. Discussions late in 1945 between Rabi and Oppenheimer led to the Acheson–Lillienthal–Baruch plan proposed by the US for the international control of atomic energy. One of Rabi's greatest disappointments was that this forward-looking plan, after initial favorable consideration, was never adopted by the United Nations.

Rabi was a member of the Atomic Energy Commission's General Advi-

sory Committee and joined with Enrico Fermi in writing a strong memorandum supporting the committee's controversial recommendation against a crash program for developing a hydrogen bomb. Later, Rabi became chairman of the committee and an eloquent defender of Oppenheimer in the AEC hearings that culminated in the removal of Oppenheimer's security clearance.

Rabi and Ramsey initiated the first proposals for Brookhaven National Laboratory and were early strong proponents of the construction of the Cosmotron. Later, with the model of Brookhaven in mind, Rabi pioneered in advocating the European collaboration that led to CERN. Rabi was the 1950 president of The American Physical Society and for a number of years was a highly effective chairman of the Columbia physics department; his critical and stimulating presence was clearly responsible for much of the greatness of that department.

Rabi initiated the International Conferences on the Peaceful Uses of Atomic Energy and was a principal speaker at them. Through his friendship with President Eisenhower, Rabi was largely responsible for the establishment in 1957 of the President's Science Advisory Committee and the Office of Special Assistant to the President for Science and Technology. For many years, Rabi was the US Representative to the NATO Scientific Advisory Committee, where he effectively advocated the establishment of the highly successful NATO-supported summer school and fellowship programs.

Rabi's carefully prepared public lectures were stimulating and presented fresh points of view, as illustrated by his words at the Fourth International Conference on the Peaceful Uses of Atomic Energy:

Real peace is more than the absence of violent war. To fulfill human expectations peace must be a condition which permits the release of the latent creative energies of all the people to the end of enhancing and elevating the quality of human life on this globe.

NORMAN F. RAMSEY
*Harvard University
Cambridge, Massachusetts*

AIP 40 YEARS

Michael Pupin, professor of physics at Columbia, congratulates first AIP director Henry Barton (photograph) on the founding of the Institute.

We are grateful to the Niels Bohr Library staff, who assembled the materials for this visual history. All photographs not otherwise credited are by courtesy of the Library or from the files of *physics today*.
—*The Editors*

Four decades of AIP

The American Institute of Physics began in October 1931, a time of great difficulty for our science. This was the decade of the Great Depression, and a widespread "stop-science" movement blamed us for society's problems. It was also a time of divisiveness within physics: In 1899 one group, the American Physical Society, could encompass all physicists, but separatism had given rise to five societies. The leaders of that time, men such as Paul D. Foote, George B. Pegram, F. K. Richtmyer and Karl T. Compton, conceived of the rather close federation that is AIP to bring physicists together again, to improve the relations between physics and the rest of society and, not incidentally, to serve as the publisher of the increasingly important US physics literature.

The Institute had its share of crises in its continuing efforts to cement the unity of the federated societies; like the nation, we had our Adams, Jefferson, Franklin and Washington; our own federalists and states' righters, and even a threat of secession. But we overcame these crises largely by our usefulness as a publisher, unifying the production of the existing journals and developing new ones as the need arose.

The Institute provides a staffed headquarters whose existence has proved useful in many other ways as well. Some of the advantages, as we shall see in the next eight pages, have included, in addition to publishing, increased administrative services for the member societies and the creation of special divisions to deal with education and manpower, with communication among ourselves and with our public relations. The establishment of *physics today* and the Niels Bohr Library are among the tangible results of AIP's efforts.

Physics shared the financial and employment famine of the 1930's, grew again the 1940's and 1950's, rode the boom of the 1960's and now again must tighten its belt. With the benefit of the time perspective of these cycles, we can be confident of continuing change. But it is up to the physics community and AIP to give the change some constructive guidance by reawakening the nation to the essential role of science and technology in the solution of today's major problems.

Henry A. Barton
Director, AIP 1931–1957

1930's
American Institute of Physics

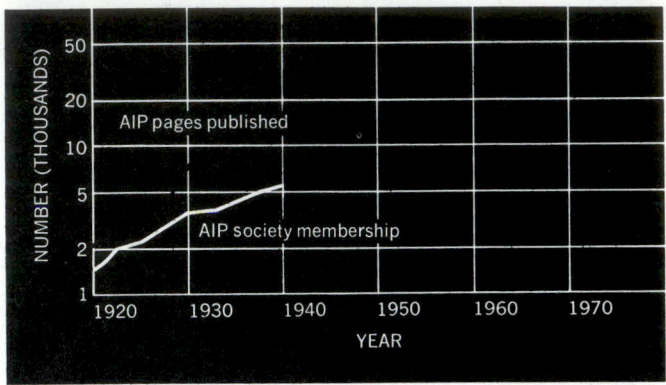

(graph)
NUMBER (THOUSANDS)
50
20
10
5
2
1
AIP pages published
AIP society membership
1920 1930 1940 1950 1960 1970
YEAR

Karl Compton, first chairman of the AIP Governing Board, tells of the importance of physics to industry (right) at a symposium "Science Makes More Jobs," organized in 1934 by AIP. The Institute, actively responding to those who blamed science for the Great Depression, broadcast the talks, along with letters from President Franklin D. Roosevelt (center) and Albert Einstein, nationwide.

KARL COMPTON

Van de Graaff generator. On 11 November 1931, the new AIP invited newsmen and scientists to a dinner at which Robert Van de Graaff first demonstrated his 1 500 000-volt "atom smasher." *The New York Times,* in a lengthy story, noted that the generator ". . .might be taken for two identical rather large floor lamps of modernistic design."

AIP guestbook. First to sign were Karl Compton, George Pegram (a founder of the Institute who wrote its constitution and was its first Secretary) and John Tate.

DATE	NAME
24 Oct 1931	Karl T. Compton
"	George B. Pegram
"	John T. Tate
20 Nov '31	Kurnc Keehan
	Frederick A. San
	F. W. Richtmyer
	Loyd A. Jo
	Eugene C. Bin
	Arthur Hfers
	W W Buff
	Wuder P. D.

February 12, 1934

My dear Doctor Compton:

The value to civilization of scientific thought and research cannot be questioned. To realize its true worth one has only to recall that human health, industry and culture have reached, in a century of scientific progress, a far higher state than ever before.

The idea that science is responsible for the economic ills which the world has recently experienced can be questioned. It would be more accurate to say that the fruits of current scientific thought and development, properly directed, can help revive industry and the markets for raw materials.

Very sincerely yours,

Franklin D. Roosevelt

THE SCIENTIFIC MONTHLY

APRIL, 1934

THE CONTRIBUTIONS OF SCIENCE TO INCREASED EMPLOYMENT[1]

SCIENCE MAKES JOBS

By Dr. KARL T. COMPTON

PRESIDENT, MASSACHUSETTS INSTITUTE OF TECHNOLOGY; CHAIRMAN SCIENCE ADVISORY BOARD; CHAIRMAN, AMERICAN INSTITUTE OF PHYSICS

THE idea that science takes away jobs, or in general is at the root of our economic and social ills, is contrary to fact, is based on ignorance or misconception, is vicious in its possible social consequences, and yet has taken an insidious hold on the minds of many people. Conscious of the fallacy of this idea, but

[1] A symposium on "Science Makes More Jobs" presented at a joint meeting of the American Institute of Physics and the New York Electrical Society at the Engineering Auditorium in New York on February 22. The address of Dr. Coolidge was broadcast from Schenectady by the National Broadcasting Company.

[2] A letter addressed to President Karl T. Compton, chairman of the American Institute of Physics, on the occasion of the symposium.

characteristically intent on their work and averse to publicity, the productive scientists of the country have thus far taken little or no part in discussions of the subject.

It has become evident, however, that the spread of this idea is threatening to reduce public support of scientific work, and in particular, through certain codes of the N.R.A., to stifle further technical improvements in our manufacturing processes. Either of these results would be nothing short of a national calamity —barring us from an advanced state of knowledge and standard of living and soon placing us at an economic disadvantage in respect to foreign countries

---

Picnicking on Barton's farm are George B. Pegram, Barton, John Tate and Madeline Mitchell. Tate, the University of Minnesota physicist who was editor of *The Physical Review*, unified the production and style of member-society journals (see samples); Mitchell, the editorial secretary of the Institute, headed Tate's new "editorial-mechanics" staff. By 1933, all the journals had uniform style, and the Institute could save money for the societies by ordering printing in large quantities. Publishing became one of AIP's most important services.

Member Societies of AIP

Founding members

American Physical Society
(founded 1899)

Optical Society of America
(founded 1916)

Society of Rheology
(founded 1929)

Acoustical Society of America
(founded 1929)

American Association of
Physics Teachers (founded 1930)

American Astronomical Society
(founded 1899, joined AIP 1966)

American Crystallographic
Association
(founded 1950, joined AIP 1966)

RESIDENCE

Cambridge, Mass.
Columbia University
Minneapolis, Minnesota
Hamden, Connecticut
Cambridge, Mass.
Pittsburgh, Pa.
Ithaca, N.Y.
Rochester, N.Y.
Easton, Pa.
Buffalo, N.Y.
Montclair, N.J.
Schenectady

1940's
American Institute of Physics

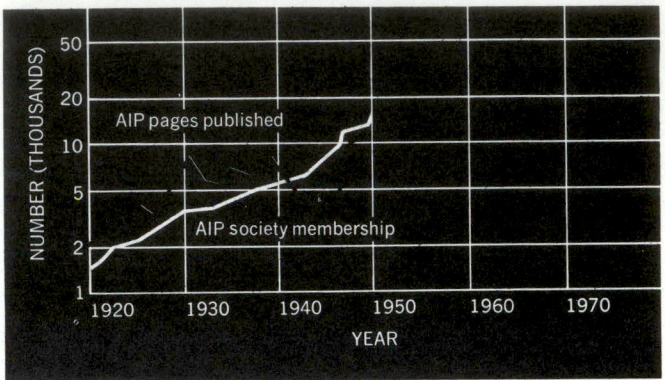

Shhh, Scientists at Work

▶ Atom-smashing research has now become secret at the University of California—including further developments of the half-built 4,900-ton cyclotron.

▶ "Voluntary censorship" of all U.S. scientific publications was announced by the National Academy of Sciences and the National Research Council. Reason: lest research data of indirect military significance fall under enemy eyes.

▶ There is so little new, non-secret research that *The Physical Review*, abstruse physicists' tradesheet, now appears only once a month instead of twice a month; and other once-plump learned journals are beginning to show their ribs.

TIME, March 16, 1942

PHYSICS INSTITUTE ACQUIRES BUILDING

Scientific Body Buys Former Dwelling in 55th Street for Headquarters

STUDIES POST-WAR WORK

Federation of Societies Has Used Rented Quarters in City Since 1931

The American Institute of Physics announced yesterday that it had acquired the five-story former residence at 57-59 East Fifty-fifth Street for national headquarters of the organization and its affiliated scientific societies.

The purchase was...

The Institute buys a home of its own—a brownstone on East 55th Street, New York.

Representatives of United States Government agencies, institutions of learning, the press, and various national associations are turning more and more frequently to the American Institute of Physics for information, advice, and cooperation from physicists as a national group. These appeals present opportunities for national service. It falls within the designated scope of the Institute to grasp such opportunities, but in order to do so it must first clearly understand the place of the science and the profession in the affairs of the nation and the world. Then it must accept the responsibility of speaking with authority about physics, of reflecting truly the wise consensus of physicists.

Acute Shortage Of Physicists Is Predicted

Warning that this country faces an acute shortage of physicists, which would be inimical to the war effort, the American Institute of Physics yesterday recommended a five-point remedial program to the Government.

The institute's war policy committee calls upon the Army, Navy and War Man-Power Commission to:

Expand facilities for turning out physics teachers, particularly in connection with Army and Navy training programs "in which it is essential to convey some knowledge of p...

At Los Alamos, New Mexico "top secret" research led to the atomic bomb. Ernest O. Lawrence, Enrico Fermi and I. I. Rabi (below) were among those at Los Alamos. As physics became militarily important, AIP recognized the need for a War Policy Committee (see statement and clipping above). After the war, in his 1945 Director's Report (right) Barton stressed the need for social responsibility, as dramatized by the bomb.

American Institute of Physics
—Report of the Director for 1945

February 23, 1946

I. General Situation

A. The Bomb

Unless there was born in 1945 a philosopher whose teaching will instill in mankind a capacity for responsibility sufficient for the atomic age, the actual opening of the era on July 15, 1945 was the most important event of the year. The unprecedented physics experiment at Alamogordo had perhaps more social and political importance than scientific value. Leaders of the atomic bomb project claim no considerable advance in fundamental physics from the immense effort and time expended. Rather do they fear that considerations of mass life and death may henceforth hamper the freedom of research and publication without which the development of science in the service of mankind will be impossible.

B. New Concerns and Old

As never before physicists must now concern themselves with public opinion, the civil and military affairs of state, and the machinery of international cooperation. This concern may be all too demanding of the best efforts of physicists. For granting that a surviving civilization is essential to the advancement of science, it may be equally true that the further advancement of science will prove to be essential to the survival of civilization. Certainly the sciences have, in the past hundred years, contributed more to public health, and progress and cult...

First issue of *physics today* **in May 1948.** The cover, showing J. Robert Oppenheimer's porkpie hat on a cyclotron, symbolizes the victory of civilian over military interests for control of atomic energy.

1950's
American Institute of Physics

Dedicating the new building. AIP outgrew its brownstone and bought its present home on East 45th Street (right). At the building-dedication ceremonies, Prince Philip of the UK presents (above) the first Compton award to Pegram; in the background is Elmer Hutchisson, who became AIP director when Barton retired in 1957.

Talent Scouts Go Hunting With Jobs for Scientists

By WILLIAM PERCIVAL,
Staff Writer.

When last seen today six technical recruiters in search of a Ph.D. were seen pursuing a likely young scientist down a side corridor of the Hotel New Yorker's "talent market" in a vain effort to give him a job.

The young man had made the near-fatal error of walking into the wrong recruiting suite and came within an ace of being signed up by an electronics firm when he really wanted to work on guided missiles.

The incident was another illustration of the fanastic demand for scientific brains that has grown with the ad...

...describes himself as "a sort of scientific talent scout," says the demand has turned the job hunter into the hunted.

"You don't interview the job hunter any more," he said. The job hunter interviews you and he wants to know all about wages, opportunities, living conditions and extra benefits before he'll go to work for you.

One young man the other day even inquired about the stability of my company. I was forced to tell him our capital investment only amounted to $500 million."

Happy hunting grounds for the recruiters is on the fifth floor of the...

...firms trying to give more than 1000 jobs to 225 graduate physicists who have indicated they might be in the market.

A Westinghouse Electric Corp. recruiter, one of 20 sent to New York for the meeting, says his company alone would like to hire 10 percent of the young physicists who will be given their doctorates this year.

A recruiter for the Glenn L. Martin Aircraft Corp. of Baltimore says he tries to interest young scientists in the challenge of working on the jet aircraft, guided missile or the earth satellite projects of the company.

An electro... ...

Corporate Associates of AIP meet at Arden House, N. Y.   The Associates became more active during the 1950's.

Translation journals. In 1955, before "Sputnik," the Institute began providing English-speaking physicists with translations of Soviet physics journals.

Visiting-scientists program in physics, administered jointly by AAPT and the AIP Education and Manpower Division, sponsors short visits of physicists to liberal-arts colleges without graduate programs in physics.

1960'S
American Institute of Physics

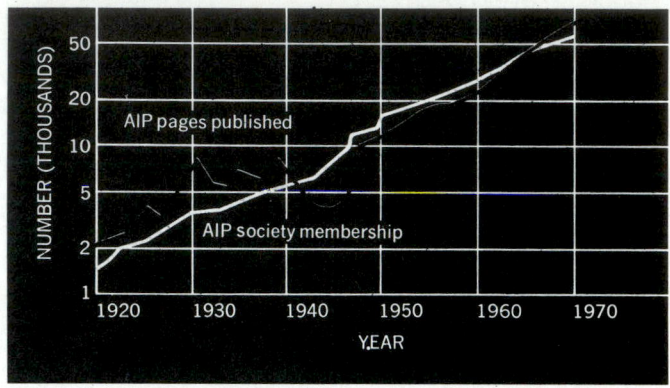

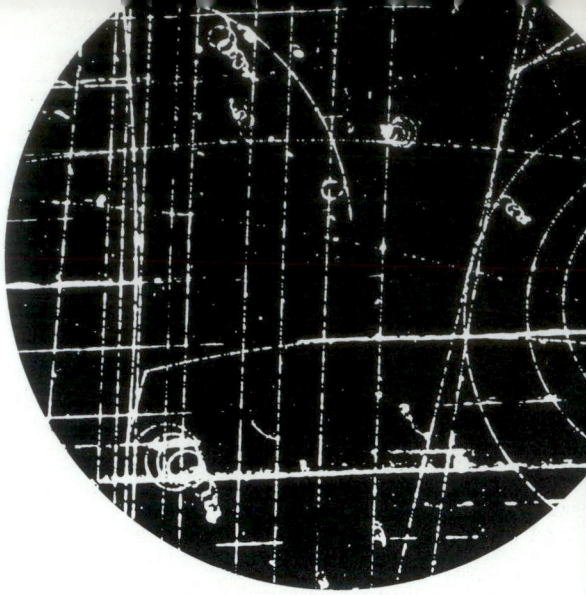

Omega Minus

FOR the better part of a wintry afternoon, we sat in the Karl Taylor Compton Boardroom of the American Institute of Physics, on East Forty-fifth Street, and listened to five physicists from the Brookhaven National Laboratory talk about the discovery, in January, of the subatomic particle called the omega minus. The talks were officially billed by the American Institute of Physics as a press conference, but, as it

turned out, the physicists lectured and the members of the press, including us, remained respectfully silent, except during a brief question-and-answer period at the end of the session. All this is somewhat by the way, though, for the occasion, whatever its name, was a fascinating one, and we found ourself increasingly interested not only in the physicists' discovery but in their reactions to it.

Four directors of AIP. Talking together are Ralph Sawyer, Van Zandt Williams (1965–66), Barton (1931–57) and Hutchisson (1957–65). Sawyer, chairman of the Governing Board until this year, was acting director twice; when Hutchisson retired and after the sudden death of Williams in 1966.

Information program. To help physicists keep up with the growing literature, AIP established an Information Division. One of its projects, the National Information System for Physics and Astronomy (NISPA) is exhibited at the 1971 Physics Show.

Funding cuts. As the decade ended, the Government reduced its support of physics research. These cuts, combined with the general economic recession, have caused unemployment and distress among physicists. (See Koch's accompanying article.)

Scientist Cautions Against Cut in Funds For Basic Research

A scientist told about 450 colleagues of the American Physical Society on Tuesday night it would be a "tragedy" should future governmental science appropriations be chopped to the detriment of basic research in favor of technological development.

Dr. Frederick Seitz, head of the physics department at the...

...when he said it took a minimum of $50,000 "to launch" each of the 6000 doctors of science American colleges and universities are turning out annually...

A recent "New York meeting." The annual joint meeting of APS and AAPT can be depended on to provide a challenge for the AIP Society Services Division.

Niels Bohr Library. J. Robert Oppenheimer speaks at the library dedication in 1962. The library is part of the Institute's Center for the History and Philosophy of Physics, which studies the recent history of physics.

COMPAS. The Committee on Physics and Society was formed in 1967. Commenting on the new committee, director Koch noted:

"... These contributions [of physics] have influenced our society and it is now important to evaluate [them] by a study of the role of physics in education, industry, and government, the direction of basic and applied research, the need for manpower, and the relationship of physics to the other sciences and the humanities..."

H. William Koch, director since 1967, notes increase in journal publication between 1956 and 1966 (far left). The Publications Division takes advantage of new composition methods; typewriter composition (left) has been used for some journals since 1969.

Chapter 3
____ Some Pathfinders of Physics_____

The history of physics is wonderfully rich in pathfinders, many of them commemorated in the subject itself by the names of laws, units, classic experiments, effects, equations, etc. The present selection is gleaned from *Physics Today* with the omission of articles reserved for other publications, as noted in the general introduction. This limitation does not lower the quality of what is offered here, and while the chapter may seem eclectic, it has the virtue of exhibiting numerous facets of physics.

We begin with J. Willard Gibbs, surely the greatest scientific pathfinder America has produced, one who can be claimed by chemistry and even mathematics as well as by physics. There is also Sabine, the founder of a whole new science, that of architectural acoustics. Next comes Hertz, a notable pathfinder in two major areas, most familiarly in electromagnetism, where his name is linked to those of Faraday and Maxwell, and again in the discovery of the photo-effect, which led to so much of modern physics. Michelson was a transition figure, the first American to win a Nobel prize; the two articles on him date from the centennial year of the definitive ether-drift experiment; John Stachel supplies an authorative view of the evidence as to the effect of those experiments on Einstein himself. And David Jackson spells out the course of Einstein's great path, special relativity, in its impact on theoretical physics.

The two articles on the beginnings of modern nuclear theory are closely related. Heisenberg's ground breaking paper (1932) on nuclear forces is seen as an attempt to *visualize* exchange energy in the nucleus, and A. I. Miller discusses more generally the role of visualization in the development of physical theory. Another pathfinder, Yukawa, proposed the fundamental the-ory of this exchange by means of a massive "quantum," the meson. Later he was able to unify the nuclear forces and the weak interaction involved in beta decay by invoking a second meson. The later pathfinders who discovered fission are also represented in this section, and Karen Johnson traces the path by which Maria Mayer found a satisfactory shell model of the nucleus.

C. V. Raman was a pathfinder not only in light scattering, where his name is best known, but also in such related fields as acousto-optics and its applications. He created numerous scientific research institutions in India, and brought them to world renown, and he trained generations of students in a variety of fields. Raman's strong and sometimes difficult personality was well known: "he was a supreme egotist" but he often demonstrated great cordiality and genuine humility. To read this article is to want to learn more about him (see Further Reading).

In the field of cosmology, Edwin P. Hubble was a pathfinder *par excellence*. His discovery of the red shift produced a truly revolutionary change in our world view, comparable to that of Copernicus, confirming theoretical ideas which had been hardly more than guesses. The cosmos has been called the poor man's accelerator, but high energy physics is usually done with synchrotrons. Edwin M. McMillan related the story of his finding the basic idea of phase stability which made that development possible, along with its independent (even somewhat prior) discovery by Vladimir Veksler. Finally in this section we have the colorful engineer-physicist Theodore von Kármán, who found new paths in the classical field of fluid mechanics even when modern physics had taken over center stage.

_____ Chapter 3: Some Pathfinders of Modern Physics _____

THE PHYSICS OF J. WILLARD GIBBS IN HIS TIME

A century and a half after his birth, Gibbs's work in thermodynamics and statistical mechanics stands out more clearly than ever. The historical origins of this work, however, remain hidden behind his austere and abstract presentation.

Martin J. Klein

PHYSICS TODAY/SEPTEMBER 1990

Who was the "Mr. Josiah Willard Gibbs of New Haven" who was appointed professor of mathematical physics by the Yale Corporation on 13 July 1871?[1] He was born on 11 February 1839, the only son and the fourth of the five children of Mary Anna Van Cleve Gibbs and Josiah Willard Gibbs the elder, a distinguished philologist and professor of sacred literature at Yale. Although the elder Gibbs worked in a field of learning very different from his son's, there were common features in their approaches. What was said of the father could well have been said of the son: "Mr. Gibbs loved system, and was never satisifed until he had cast his material into the proper form. His essays on special topics are marked by the nicest logical arrangement."

The younger Gibbs graduated from Yale College in 1858, having won a series of prizes and scholarships for excellence in Latin and mathematics. He continued his studies in Yale's new graduate school and received his PhD in 1863, one of the first scholars to be awarded this degree by an American university. In view of the work Gibbs is known for, it is a little surprising to learn that his doctorate was earned in engineering and his dissertation bore the title "On the Forms of the Teeth of Wheels in Spur Gearing." But as Gibbs pointed out in his first paragraph, "the subject reduces to one of plane geometry," and the thesis is really an exercise in geometry and kinematics. The dissertation went unpublished until 1947, when the centennial year of Yale's Sheffield Scientific School provided the occasion for a little book on his early work in engineering. The editor's description of Gibbs's dissertation could apply to almost any of his later, more famous works: "If he has a natural friendliness for

the niceties of geometrical reasoning, [the reader] will be rewarded with a sense of satisfaction akin to that felt upon completing, say, a book of Euclid; if he is not so endowed, he had perhaps better not trouble himself with the austerities of style and extreme economy (one might almost say parsimony) in the use of words that characterize the entire work."[2]

After receiving his doctorate Gibbs was appointed a tutor in Yale College, where he gave elementary instruction in Latin and natural philosophy (physics) for three years. Although he continued to work on engineering problems during this period, a paper he presented to the Connecticut Academy of Arts and Sciences in 1866 shows that Gibbs was already concerned with clarifying fundamental physical concepts. In August of that year Gibbs sailed for Europe, where he spent three years studying mathematics and physics. This would be his only extended absence from his native city. Gibbs spent a year each at the universities of Paris, Berlin and Heidelberg, attending a variety of lectures and reading widely in both mathematics and physics. He did not work as a research student with any of the scientists whose lectures he attended, a list that included Joseph Liouville, Leopold Kronecker, Heinrich Magnus, Hermann von Helmholtz and Gustave-Robert Kirchhoff. Nor is there any indication in the few notebooks that constitute the only record of his European studies that he had yet begun any research on his own or even decided what line he would try to follow in his later work. He was apparently satisfied to broaden and deepen the knowledge of mathematics and physics he had previously acquired, and to wait until after he returned home to choose the subjects of his own researches.

For two years after he came back to New Haven, Gibbs had no regular employment and his future was uncertain. He taught French to engineering students for a term, and he worked on an improved version of James Watt's governor for the steam engine. Gibbs was evidently able to manage adequately on the money he had

Martin Klein is the Eugene Higgins Professor of history of physics and professor of physics at Yale University. He is now the senior editor of the collected papers of Albert Einstein.

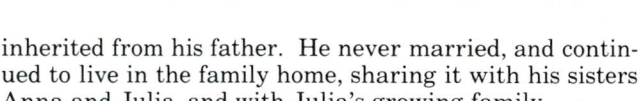

inherited from his father. He never married, and continued to live in the family home, sharing it with his sisters Anna and Julia, and with Julia's growing family.

This fortunate state of financial independence must have been known within the Yale community when Gibbs was appointed to the newly created professorship of mathematical physics in 1871. This knowledge may help to explain why the official record of Gibbs's appointment includes the phrase "without salary." The new chair was unendowed, but in any case Gibbs's teaching duties would be light, since the appointment was in the small graduate department. An offer from Bowdoin College in 1873 that would have given him $1800 a year and his choice between chairs of mathematics and physics did not tempt him to leave New Haven. Only in 1880, when he was on the verge of accepting a professorship at the new and appealing Johns Hopkins University, a university devoted primarily to graduate study and research, did his own institution offer him a salary. Even though Yale offered only two-thirds of what Hopkins would have paid, the advantages of remaining in his familiar surroundings, of being at home, convinced Gibbs to stay. He must also have been influenced by his colleagues' warm reaction to the threat of losing him. "Johns Hopkins can get on vastly better without you than we can," James Dwight Dana wrote him, and then added, "*We can not.*"

Gibbs's appointment to the chair of mathematical physics preceded his first published research by two years. This inversion of what we now take to be the normal order of events was not so extraordinary at the time. The 32-year-old Gibbs, with his brilliant college record, his doctoral degree, his demonstrated abilities as an engineer and his three years of postdoctoral study in Europe, had far more impressive qualifications for a professorship than most of his colleagues had had when they were appointed. Yale had every reason to express its confidence in Gibbs, who had, after all, been a member of this small academic community since his birth.

Fundamental thermodynamic equation

The new professor of mathematical physics published his first paper[3] in 1873. He had chosen thermodynamics as the subject of his research, and immediately demonstrated his mastery of that field. But why had Gibbs chosen thermodynamics? During his three years in Europe he had attended no lectures on the subject, nor had he listed in his notebooks any references on thermodynamics for future reading. Electromagnetism and optics drew more of his interest than anything else, judging by his notebook entries, and this is not surprising; they were certainly the most active areas for research in the 1860s. The first course he taught at Yale seems to confirm this predominant interest in physical optics. In the next academic year Gibbs lectured on potential theory, which hardly provided a natural step toward thermodynamics, even though he did use a text by Rudolf Clausius.

Gibbs's biographer and former student, Lynde Phelps Wheeler, conjectured that "the arousing of Gibbs's interest in thermodynamics is due at least as much to an interest in engineering applications of the subject as in its theoretical foundations." He points to Gibbs's work on the governor for the steam engine, which he dates to the early 1870s. Wheeler offers Gibbs's first paper in support of his interpretation, but Gibbs was never the sort of writer who tells his readers about the problems that prompted the work. He preferred a laconic, mathematician's style, making sure to say what was necessary for the logical structure of his argument but nothing more.

Gibbs began that first paper, "Graphical Methods in the Thermodynamics of Fluids," by remarking that although geometrical representations of thermodynamic propositions were in "general use" and had done "good service," they had not yet been developed with the "variety and generality of which they are capable." Such representations had been restricted to diagrams whose coordinate axes denoted volume and pressure, but he proposed to discuss a range of alternatives, "preferable...in many cases in respect of distinctness or of convenience." This modest beginning suggested that the paper to follow might break no new ground and that it would be of primarily didactic interest. But in the next few paragraphs, where he chose the basic variables and wrote down the equations that would underlie his argument, Gibbs quietly changed the direction of thermodynamics. In a few lines he arrived at what he named "the fundamental thermodynamic equation of the fluid,"

$$dU = T\,dS - P\,dV$$

The quantities involved are the temperature T, pressure P and volume V of the fluid, along with the fluid's internal energy U and entropy S. (I have not used Gibb's notation.) In 1873 this was by no means the standard way of developing thermodynamics. Even Clausius, who had formulated the second law in 1850, introduced the concept of entropy in 1854 and invented its characteristic name in 1865, never gave entropy a central place in his exposition of thermodynamics.[4] For Clausius and for his contemporaries thermodynamics was the study of the interplay between heat and work. For Gibbs, who eliminated heat and work from the foundations of the subject in favor of the state functions—energy and entropy—thermodynamics became the theory of the properties of matter at equilibrium.

Gibbs wanted to find "a general graphical method which can exhibit at once all the thermodynamic properties of a fluid concerned in reversible processes." To this end he considered the general properties of any diagram in which the states of the fluid were mapped continuously on the points of a plane. The fluid's thermodynamic properties would then be expressed by the geometrical properties of the several families of curves connecting states of equal volume, of equal temperature, of equal entropy and so forth. He discussed the temperature–entropy diagram, in many ways closely analogous to the familiar pressure–volume diagram. It had obvious uses for the engineer's analysis of heat engines, but its real advantage was that it "makes the second law of thermodynamics so prominent, and gives it so clear and elementary an expression."

Most interesting to Gibbs was the entropy–volume diagram, whose "substantial advantages over any other method" he analyzed in some detail. Although less suited to applications in engineering, this diagram was the natural accompaniment of the fundamental equation that expressed energy as a function of entropy and volume. Among other things, Gibbs pointed out how well this diagram expressed the region of simultaneous coexistence of the vapor, liquid and solid phases of a substance. This "triple point" corresponds to a unique set of values of pressure and temperature, but it occupies the interior of a triangle in the entropy–volume plane.

As a natural aspect of his geometrical investigation,

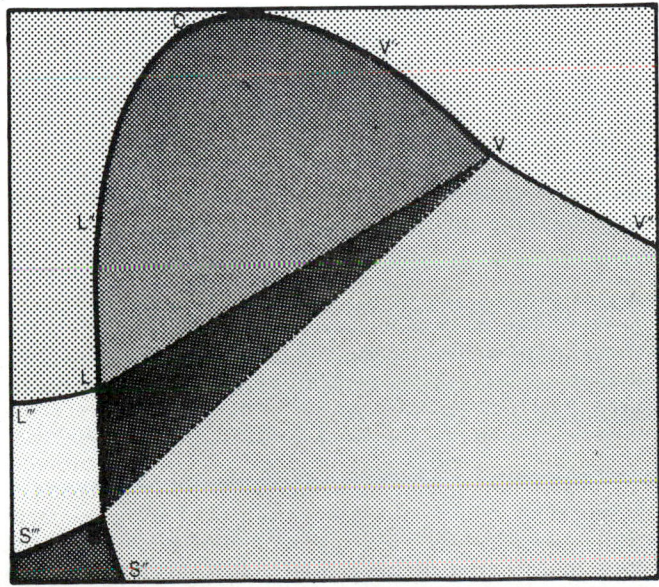

'Surface of absolute stability' projected onto the entropy-volume plane, adapted from Gibbs's 1873 paper.[3] The points S, L and V together represent the solid-liquid-vapor triple point. Inside the triangle SLV these states coexist in an equilibrium mixture. Between LL' and VV' is a liquid-vapor mixture, between SS'' and VV'' is a solid-vapor mixture, and between SS''' and LL''' is a solid-liquid mixture. Segments $L'''LL'$, $V'VV''$ and $S''SS'''$ form the boundaries representing the stable states of liquid, vapor, and solid, respectively. C is the critical point.

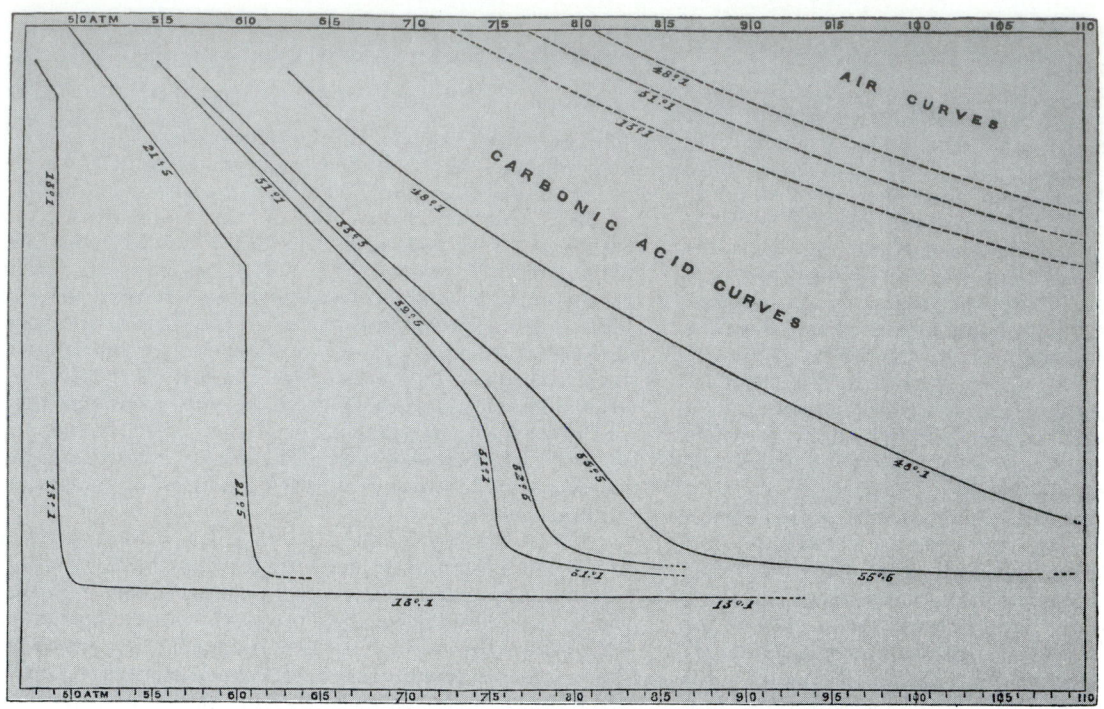

Pressure–volume curves for carbonic acid measured by Thomas Andrews and reported in his 1869 paper.[7] James Thomson argued that a minimum and maximum (rather than the straight horizontal lines that appear on the 31.1 °F, 21.5 °F and 13.1 °F isotherms) should exist for a given isotherm below the critical temperature.[9] The volume increases from bottom to top, and the pressure increases from left to right.

Gibbs examined those features of the families of curves representing thermodynamic properties that are independent of the choice of coordinates. Among these invariant features were the order of the curves of different kinds (isobars, isotherms, isentropics and so on) as they cross at any point, and the geometrical nature of these intersections, which could involve contacts of higher order.

But even after this sketch of the many fresh ideas in Gibbs's first paper, we are left with the question of what brought him to this study. Wheeler's suggestion of the importance of his engineering background can account at most for part for Gibbs's motivation. If we look at the literature of the time, we find a number of articles on thermodynamics appearing in the widely read *Philosophical Magazine* during 1872. They were largely devoted to the history, that is to say to the priorities, of the discovery of the second law. For all the lively, pointed and even angry words exchanged between Clausius and Peter Guthrie Tait (writing on behalf of William Thomson's claims), it seems likely that this controversy would have repelled rather than attracted Gibbs. It might have sent him to the library, however, to see what the shouting was all about.

One of the works discussed by Clausius in the article that touched off the controversy[5] was a little book by James Clerk Maxwell, his *Theory of Heat*,[6] published in 1871. It was Maxwell's neglect of his work that prompted Clausius to set the record straight by writing his own account of his earlier contributions. Gibbs might well have read Maxwell in any case, even without the priority dispute raging in the *Philosophical Magazine*. What Gibbs would have found in Maxwell's book, intended (according to the publisher) "for the use of artisans and of students in public and science schools," included Maxwell's discussion of recent developments that interested him in the area of heat. One such development was

Thomas Andrews's recent (1869) discovery of the continuity of the two fluid states of matter—liquid and gas—and of the critical point above which these states were indistinguishable. Whether Gibbs learned of Andrews's work from Maxwell or by reading Andrews's paper in the *Philosophical Transactions of the Royal Society*[7] we do not know, but he certainly knew about it when he wrote his first paper. There is a footnote referring to Andrews at the place in the text where Gibbs discusses the possibility of a second-order contact between the isobar and the isotherm corresponding to a particular state of the fluid. It reads, "An example of this is doubtless to be found at the critical point of a fluid."

Andrews did not theorize at all about the implications of his discovery. His paper offered neither a thermodynamic analysis nor a kinetic–molecular explanation. It seems to me that Andrews's discovery—a new, unexpected and general feature of the behavior of matter, as yet totally unanalyzed—would have been just the sort of thing to capture Gibbs's attention as that promising new professor of mathematical physics sought for a suitable subject on which to work. (Edwin B. Wilson, Gibbs's student, remembered his teacher telling him that "one good use to which anybody might put a superior training in pure mathematics was to the study of problems set us by nature."[8])

Entropy and equilibrium

Gibbs's physical interests are much more apparent in his second paper, which appeared only a few months after the first. Although its title, "A Method of Geometrical Representation of the Thermodynamic Properties of Substances by Means of Surfaces," might suggest that Gibbs was merely extending his geometrical methods from two dimensions to three, one does not have to read very far to see that he was doing something quite different. His emphasis was now on the phenomena to be explained, rather than on the methods as such. The problem Gibbs treated was the characterization of the equilibrium state of a material system, a body that can be solid, liquid, gas or some combination of these according to the circumstances. This time Gibbs used only one way of representing the equilibrium states, namely the surface described by those states in the space whose coordinate axes are energy, entropy and volume; this surface represents the fundamental thermodynamic equation of the body. He proceeded to establish the relationships between the geometry of the surface and the conditions for thermodynamic equilibrium and its stability.

Gibbs showed that for two phases of the same substance to be in equilibrium with each other, not only must they share the same temperature T and pressure P, but the energies, entropies and volumes of the two phases (per unit mass) must satisfy the equation

$$U_2 - U_1 = T(S_2 - S_1) - P(V_2 - V_1)$$

where the subscripts 1 and 2 refer to the two phases.

This equation provided the answer to a question that had puzzled Maxwell: What is the condition determining the pressure at which gas and liquid can coexist in equilibrium? Maxwell originally thought that the difference in internal energy between the two phases must be a maximum at the pressure where they coexist. He soon changed his mind on reading Gibbs's second paper.

When Gibbs analyzed the conditions for the stability of thermodynamic equilibrium states in his second paper, he arrived at a new understanding of the significance of the critical point. The critical state not only indicated where the two fluid phases became one, it also marked the limit of the regions of instability associated with the two-phase system—both the "absolute instability" of the states like that of a supercooled gas and the "essential stability" of states on the rising part of the Thomson–van der Waals loop.[9] The critical point is the common limit of both regions of instability, though it is itself stable. Gibbs's analysis also led him to a series of new explicit conditions that must be fulfilled at the critical point and that can serve to characterize it.

These two early papers were published in the *Transactions of the Connecticut Academy of Arts and Sciences*. Although the Connecticut Academy, of which Gibbs had been a member since 1858, was a local society centered in New Haven, it had been in existence since 1799 and had developed a regular program for the exchange of its *Transactions* with comparable journals published by some 170 other learned societies around the world. Although Gibbs could expect his work to be circulated far and wide by this route, he also sent copies of his papers directly to an impressive list of scientists at home and abroad. We do not know how many of those 75 or so scientists who received Gibbs's first two papers read them, but we do know of one crucial reader: This was Maxwell, who learned the proper definition of entropy through Gibbs's papers. Maxwell had misused the term, following his friend Tait, in the first edition of the *Theory of Heat*, but corrected his error in later editions.[10] Maxwell talked about Gibbs's work to Cambridge colleagues in several fields and wrote about it to others, recommending it highly. He especially appreciated Gibbs's geometrical approach, since he too preferred geometrical to algebraic or analytical methods. The thermodynamic surface introduced in Gibbs's second paper fascinated Maxwell so much that he actually constructed such a surface showing the thermodynamic properties of water and sent a plaster cast of it to Gibbs. He even discussed the surface at considerable length in the 1875 edition of his book; whether this discussion met the needs of the intended audience of "artisans and students" may be doubted.

In a lecture to the Chemical Society, Maxwell spoke about Gibbs's "remarkably simple and thoroughly satisfactory method of representing the relations of the different states of matter by means of a model," adding that by using this model—the thermodynamic surface—"problems which had long resisted the efforts of myself and others may be solved at once."[11]

Chemical potential and the phase rule

When the Connecticut Academy held its regular meeting in June 1874, the 20 or so members who were present heard a brief talk by Gibbs on the application of the principles of thermodynamics to the determination of chemical equilibrium.[12] This was not just another ordinary evening in New Haven. Gibbs then gradually worked out the consequences of the ideas he could only have sketched in his talk, and developed them into a long memoir, "On the Equilibrium of Heterogeneous Substances," published in due course by the academy. This book-length work, which appeared in two parts, in 1876 and 1878, surely ranks as one of the true masterworks in the history of physical science.

In this memoir Gibbs vastly enlarged the domain encompassed by thermodynamics, treating chemical, elastic, surface and electrochemical phenomena by a single, unified method. Gibbs described the fundamental idea underlying the whole work in a lengthy abstract, published in the *American Journal of Science*, which reached a much broader audience than the academy's *Transactions*: "It is an inference naturally suggested by the general increase of entropy which accompanies the changes occurring in any isolated material system that when the entropy of the system has reached a maximum, the system will be in a state of equilibrium. Although this principle has by no means escaped the attention of physicists, its importance does not appear to have been duly appreciated. Little has been done to develop the principle as a foundation for the general theory of thermodynamic equilibrium." It was just that development which Gibbs set forth in his memoir.

The general criterion for equilibrium could be stated simply and precisely: "For the equilibrium of any isolated system it is necessary and sufficient that in all possible variations of the state of the systems which do not alter its energy, the variation of its entropy shall either vanish or be negative." But to work out the consequences of this general criterion, and to explore their implications allowing for the variety and complexity that thermodynamic systems can have, was a major undertaking. From the outset Gibbs introduced the chemical potentials, those intensive variables that must have the same values throughout a heterogeneous system in equilibrium and that function much like the temperature and pressure. From these conditions on the intensive variables Gibbs derived the phase rule, which specifies the number of independent variables (degrees of freedom) in a system of a certain number of coexistent phases having a specified number of chemical components. This phase rule later proved to be an invaluable guide to understanding a mass of experimental material, but Gibbs did not single it out for special mention in his memoir. As Pierre Duhem once remarked, it took "a remarkable perspicacity" on the part of J. D. van der Waals, who first saw its power, to perceive the phase rule "among the algebraic formulas where Gibbs had to some extent hidden it."[13] And Duhem wondered how many more such seeds that might have grown into whole programs of research "had remained sterile because no physicist or chemist had noticed them

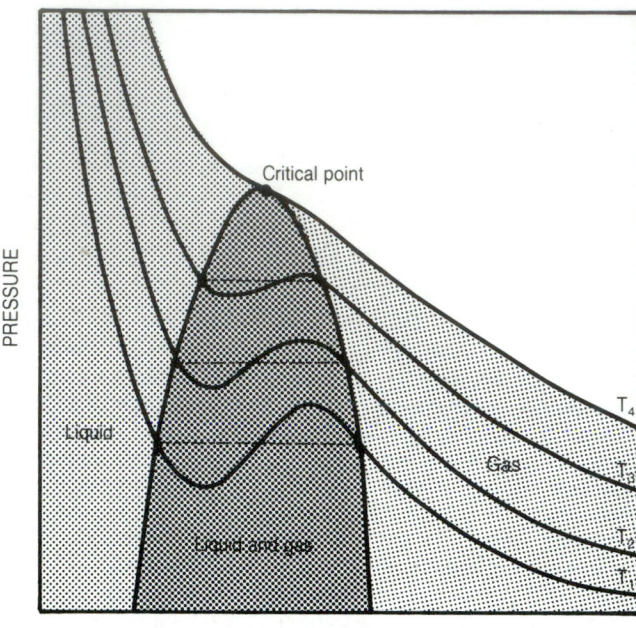

VOLUME

Pressure–volume isotherms
($T_1 < T_2 < T_3 < T_4$), as illustrated in a modern textbook, show minima and maxima below the critical temperature—the temperature below which the volume changes discontinuously at the equilibrium, or phase-transition, pressure. (The dashed lines cut across the "loops" at the phase-transition pressures). At the critical temperature T_4 the transition between gas and liquid is smooth and continuous. (Adapted with permission from Herbert B. Callen, *Thermodynamics*, Wiley, New York, 1960, p. 158.)

completing his thermodynamic studies. His vector analysis, his calculations of orbits and especially his investigations into the electromagnetic theory of light will have to go undiscussed here,[14] along with his new variational principle for mechanics and his famous note on the Fourier series. I do, however, want to say something about the development of Gibbs's ideas on statistical mechanics.

Gibbs's book, *Elementary Principles in Statistical Mechanics: Developed with Especial Reference to the Rational Foundation of Thermodynamics*, published in 1902 (Scribner's, New York) as one of the volumes in the Yale Bicentennial Series, stands alone among his works. It was his first and only treatment of the subject, with one minor exception, to be discussed shortly. Almost totally devoid of references to the literature, presenting the subject in a form quite different from that used by any other author, the book has some of the same quality—at once abstract and individual—as Paul A. M. Dirac's *The Principles of Quantum Mechanics*, published in 1930. But while Dirac's book was preceded by a series of the author's papers, Gibbs's appeared without any preparatory works that might have foreshadowed it. It must have been a genuine surprise to the scientific community, particularly since it would have seemed to some to represent not just a change of direction on Gibbs's part, but an actual reversal.

By this time Gibbs was widely known as one of the grand masters of thermodynamics. His work exemplified Maxwell's definition of thermodynamics as "the investigation of the dynamical and thermal properties of bodies, deduced entirely from what are called the First and Second Laws of Thermodynamics, without any hypothesis as to the molecular constitution of bodies."[11] The last phrase is the crucial one: Gibbs's work stood as the prime example of pure thermodynamics, unmarred by any appeal to hypotheses about the molecular structure of matter. The energeticists, who opposed and even scorned the use of molecular concepts, and especially the attempt to reduce thermodynamics to molecular mechanics, particularly valued Gibbs for having avoided these ideas in his work. Yet here he was in his new book discussing the principles of statistical mechanics and claiming, as his subtitle implied, that these principles could supply the "rational foundation of thermodynamics." Not only that, but according to the new Gibbs the laws of thermodynamics were only an "incomplete expression" of these more general principles. Did this apparent desertion of the thermodynamic camp really represent a sudden shift in Gibbs's basic outlook?

The answer to this question is certainly negative. Gibbs had not built molecular concepts into his papers on thermodynamics because he had "no need of that hypothesis," as Pierre-Simon Laplace is said to have remarked in a very different context. But when Gibbs thought he could clarify his discussion by referring to molecular behavior,

under the algebraic shell that concealed them?"

Gibbs's spare and abstract style and his unwillingness to include a variety of examples and applications to particular experimental situations made his work very difficult for potential readers. As a consequence, the literature of the late 19th century contains many rediscoveries of results already published by Gibbs. Such major figures as Helmholtz and Max Planck independently developed their own thermodynamic methods for treating chemical problems, quite unaware of the treasures concealed in the third volume of the *Transactions of the Connecticut Academy*.

Gibbs wrote no other major works on thermodynamics, thinking that he had said all he needed to say—perhaps all that in principle could be said—on the subject. He limited himself to a few short papers elaborating several points in his long memoir. He shifted his attention to other issues in physics and mathematics, and he rejected all suggestions that he write a treatise on thermodynamics that would make his work more accessible. In 1892 Rayleigh wrote to Gibbs urging him to expand on his ideas, saying that the original memoir was "too condensed and too difficult for most, I might say all, readers." Gibbs's answer strikes an unexpected note: He now thought that his memoir seemed "too *long*" and showed a lack of a "sense of the value of time, of my own or others, when I wrote it."

Shortly before his death, however, Gibbs did agree to a republication of his writings on thermodynamics, to which he planned to add some new material in the form of supplementary chapters. Among the matters that Gibbs listed for inclusion in this supplement, two seem particularly appealing to me. Unfortunately we have only the bare titles, "On Similarity in Thermodynamics" and "On Entropy as Mixed-up-ness," and we shall never know what Gibbs planned to write on either subject. (One cannot help wondering if "entropy as mixed-up-ness" is related to the mixing process that Gibbs used in chapter XII of his *Statistical Mechanics* to discuss the irreversible increase of entropy.)

Statistical mechanics: Molecular concepts

I shall not even try to comment on the various subjects to which Gibbs devoted his attention in the decade after

he had not hesitated to do so. These references are infrequent, but they do occur—in his treatment of the reaction carrying NO_2 into N_2O_4, for example, and in his mention of the "sphere of molecular action" when he introduced his analysis of the thermodynamics of capillarity. The discussion of the Gibbs paradox is the best known of the passages in "On the Equilibrium of Heterogeneous Substances" where molecules are mentioned. When two gases are allowed to mix at constant temperature and pressure, there is an increase of entropy. The amount of this increase is "independent of the degree of similarity or dissimilarity" between the two gases, unless they are identical—the same gas—in which case there is no entropy increase at all. Gibbs's discussion of this situation is explicitly molecular, and leaves no doubt that he had none of the objections to molecular arguments that some of his thermodynamic followers attributed to him.

If we try to trace the development of Gibbs's ideas leading up to his book on statistical mechanics, we meet with only very limited success. The first sign of a statistical approach to the second law of thermodynamics in Gibbs's writings comes in the remarkable last sentence of his analysis of the Gibbs paradox. Because it is possible, in principle, for the ordinary molecular motions in a gas mixture to produce unmixing, with the corresponding decrease in entropy, one cannot exclude the possibility that this process will occur, unlikely as it may be. "In other words," Gibbs wrote, "the impossibility of an uncompensated decrease of entropy seems to be reduced to improbability." This remark, so different in character from everything else in Gibbs's memoir, was apparently not developed at all.

In 1884 Gibbs made one of his rare appearances at a meeting away from New Haven. He presented a paper in Philadelphia to the American Association for the Advancement of Science, "On the Fundamental Formula of Statistical Mechanics with Applications to Astronomy and Thermodynamics." Only the brief published abstract survives of this talk, but one can nevertheless draw some conclusions from it about the state of his thinking on the subject in at that time.

Gibbs had evidently studied the papers of Maxwell and Ludwig Boltzmann closely enough to decide that they had created a subject deserving a new name that would separate it from the limiting connotations of "the kinetic theory of gases." The new discipline treated bodies of arbitrary complexity moving according to the laws of mechanics, but it investigated their motions statistically. It considered, in Maxwell's words, "a large number of systems similar to each other in all respects except in the initial circumstances of the motion, which are supposed to vary from system to system," and it asked about "the *number* of these systems which at a given time are in a phase such that the variables which define it lie within given limits."[11] Gibbs coined the name "statistical mechanics" in his Philadelphia paper to denote the discipline based on such investigations.

The fundamental formula mentioned in his title described the evolution in time of the distribution function or phase density of a population of similar systems. Boltzmann had been discussing other versions of this equation since 1868, and had recently stressed its importance, as had Maxwell.[15] This equation is now generally known as Liouville's theorem, but Gibbs never called it that.

He must have gone on thinking about this statistical mechanics, for he listed it in 1888 among several possible topics for papers he was considering writing but nothing appeared in print until his 1902 treatise. He was already lecturing on it at Yale, where his courses on the subject bore various titles: "On the *A Priori* Deduction of Thermodynamic Principles from the Theory of Probabilities," "Theoretical Thermodynamics" and "Dynamics and Thermodynamics." From what we know about Gibbs's teaching, the very fact that he gave such courses implies that he had already carefully thought through the subject

and organized it to his own satisfaction. His student Henry Bumstead wrote that Gibbs "seldom, if ever, spoke of what he was doing until it was practically in its final and complete form."[16]

We find evidence of Gibbs's mastery of statistical mechanics in an unlikely place—the obituary notice for Clausius that he wrote in 1889. He compared Clausius's contributions to "molecular science" with those of Maxwell and Boltzmann: "In reading Clausius, we seem to be reading mechanics; in reading Maxwell, and in much of Boltzmann's most valuable work, we seem rather to be reading in the theory of probabilities." Gibbs recognized that "the larger manner in which Maxwell and Boltzmann proposed the problems of molecular science" allowed them to go further than Clausius. He nevertheless praised Clausius's "remarkable insight" and noted that his "hypothesis" about disgregation, which suffered from "the acknowledged want of a rigorous demonstration," had later found its justification in Boltzmann's work. No careful reader of Gibbs's obituary of Clausius could have doubted his active interest in the molecular approach to thermodynamics. There is, however, no indication that such careful readers actually existed.

By 1892 Gibbs was devoting much of his energy to writing up the results of his years of work. In the letter to Rayleigh referred to above, Gibbs wrote: "Just now I am trying to get ready for publication something on thermodynamics from the *a priori* point of view, or rather on 'Statistical Mechanics,' of which the principal interest would be in its application to thermodynamics—in the line therefore of the work of Maxwell and Boltzmann. I do not know that I shall have anything particularly new in substance, but shall be contented if I can so choose my standpoint (as seems to me possible) as to get a simpler view of the subject." Judging from Gibbs's extensive notes on statistical mechanics, some of which bear dates from about this time, he was working on a lengthy manuscript of more or less the size, structure and contents of the work

that appeared in 1902. Lecture notes taken by a student in Gibbs's course for the academic year 1894–95 confirm that he had already worked out his own way of developing statistical mechanics and that he did not change it significantly over the years.

Equipartition and the irreversibility problem

One of the most remarkable things about Gibbs's book is that it reflects so little of the state of the subject at the time it was written. The two problems discussed with most urgency in the 1890s were the failures of the equipartition theorem and the difficulties and obscurities in Boltzmann's statistical explanation of irreversibility. The failure of the equipartition theorem to account for the specific heats of all the common gases had been recognized since Maxwell's first paper on the kinetic theory in 1860, and no real progress had been made despite many efforts to understand the difficulty.[17] Gibbs had no answer to this major unsolved problem either, but he was quite aware that it posed a serious threat to the status of the whole subject. His reaction was to set forth his statistical mechanics "as a branch of rational mechanics" and to give up "the attempt to frame hypotheses concerning the constitution of material bodies." "Difficulties of this kind," he wrote in his preface, "have deterred the author from attempting to explain the mysteries of nature, and have forced him to be contented with the more modest aim of deducing some of the more obvious propositions relating to the statistical branch of mechanics." In giving up the attempt to deal with the failures of equipartition, Gibbs may have sensed that the limits of that theorem could not be established within a theory that was "a branch of rational mechanics."

Gibbs could not avoid the question of the origin of irreversibility if he was going to supply the rational foundation for thermodynamics that the subtitle of his book promised. But he apparently decided at an early stage of his work that Boltzmann's methods for dealing with this question were not what he wanted. None of his research notes mention either Boltzmann's kinetic method, which made molecular collisions the basis of his *H*-theorem or the combinatorial method Boltzmann introduced in 1877. Nor did Gibbs comment on Boltzmann's papers of the 1890s, which elaborated and clarified his views on irreversibility. We are not even certain[17] that he read Boltzmann's *Lectures on Gas Theory* when they appeared in 1895 and 1898.

His own discussion of the irreversible approach to equilibrium is set forth the chapter XII entitled "On the Motion of Systems and Ensembles of Systems Through Long Periods of Time." Gibbs treated it as a mixing process in phase space, but his guarded language and the absence of equations in this chapter suggest that he may not have been completely satisfied with this analysis of irreversibility. At several places in his notes Gibbs asks himself questions like "What is the entropy of a system not in equilibrium?" Perhaps his uncertainty on this and related matters led to the delay in publishing the manuscript that he was working on in 1892. It is not impossible that the pressure to contribute to the Yale Bicentennial series propelled Gibbs into completing a book that he might otherwise have considered not quite ready to publish.

Serenity: Love of abstract truth

One sometimes still hears the story that Gibbs never received proper recognition during his lifetime. It is patently untrue. His name and work may have been quite unknown to the general educated public in this country and abroad, but how many knew of Maxwell or Carl Friedrich Gauss? Gibbs did receive almost every honor that the learned world could grant—election to the great academies of many lands, honorary degrees from universities in three countries, several prizes. He also received in good measure what he valued most, the respect and admiration of those colleagues everywhere whose own work put them in a position to appreciate his. In 1901 the Royal Society of London, which had elected Gibbs a foreign member in 1897, awarded him its Copley Medal, perhaps the highest distinction a scientist could receive at that time.

Among the personal qualities that impressed Gibbs's family and associates—his approachability, his lively sense of humor, his kindliness and sympathetic approach to students and colleagues alike, his readiness to give counsel and help—one seems to stand out, his serenity. He appears to have been a man utterly at home in his surroundings, a happy man; a friend even called him "the happiest man she ever knew."[1] While some of his notable contemporaries—Henry James, Charles Sanders Peirce, Henry Adams—found it impossible to come to comfortable terms with their American environment, Gibbs had no such problem. Yale and New Haven offered him what he needed.

In a famous section of his *Democracy in America*, Alexis de Tocqueville analyzed the reasons which he saw for the lack of attention paid to theoretical science in this country[18]: "Nothing is more necessary to the culture of the higher sciences or of the more elevated departments of science than meditation; and nothing is less suited to meditation than the structure of democratic society." However valid this generalization may or may not have been, it surely did not impose limits on Gibbs. He did find that needed opportunity for sustained thought, "that calm . . . necessary for the deeper combinations of the intellect," in de Tocqueville's phrase.

Nevertheless Gibbs might not have disagreed with de Tocqueville's generalization, and he actually formulated a similar one of his own. In writing about his late friend Hubert A. Newton, who had started as a mathematician but then turned to observational astronomy, Gibbs commented that this change of direction was probably due to "the influence of his environment." Gibbs thought it was to be expected "that the questions which nature forces on us are likely to get more attention in a new country and a bustling age, than those which a reflective mind puts to itself, and that the love of abstract truth which prompts to the construction of a system of doctrine, and the refined taste which is a critic of methods of demonstration, are matters of slow growth." The last part of Gibbs's sentence could be taken as a description of characteristic features of his own way of doing science, which was "part of the never-ending meditation."[19]

* * *

The final text of this paper was completed while I was visiting professor of the history of science at Harvard University. The complete version of this article appears in Proceedings of the Gibbs Symposium, Yale University, May 15–17, 1989, *edited by D. G. Caldi and G. D. Mostow (1990); the reader is referred to that paper for a more detailed bibliography. This abridged version appears with the permission of the American Mathematical Society. I am grateful to the Yale University Library for permission to quote from the Gibbs Collection at the Beinecke Rare Book and Manuscript Library—the source for many of Gibbs's unpublished writings.*

References

1. L. P. Wheeler, *Josiah Willard Gibbs: The History of a Great Mind*, 2nd ed., Yale U. P., New Haven (1952). This is the authorized biography of Gibbs and, unless otherwise noted, is my source for biographical information.

2. *The Early Work of Willard Gibbs in Applied Mechanics*, L. P. Wheeler, E. O. Waters, and S. W. Dudley, eds., Schuman, New York (1974), p. 43.

3. J. W. Gibbs, *The Scientific Papers of J. Willard Gibbs*, H. A. Bumstead, R. G. Van Name, eds., Longmans, Green and Co, London (1906), reprinted 1961, Dover, New York. Citations of Gibbs's papers are from this edition.

4. For further discussion, see M. J. Klein, in *Springs of Scientific Creativity: Essays on Founders of Modern Science*, R. Aris, H. T. Davis, R. H. Stuewer, eds., U. Minnesota P., Minneapolis (1983), p. 142.

5. R. Clausius, Philos. Mag. **43**, 106 (1872).

6. J. C. Maxwell, *Theory of Heat*, Longmans, London (1871), p. 186.

7. T. Andrews, *Philos. Trans. R. Soc.* London **159** (1869).

8. E. B. Wilson, Sci. Monthly **32**, 210 (1931).

9. J. Thomson, *Proc. Roy. Soc.* **20**, (1871), pp. 1–8.

10. See ref. 6 and J. C. Maxwell, letter to P. G. Tait, 1 December 1873, in C. G. Knott, *Life and Scientific Work of Peter Guthrie Tait*, Cambridge U. P., Cambridge, England (1911), p. 115.

11. For selected writings by Maxwell, see *The Scientific Papers of James Clerk Maxwell*, W. D. Niven, ed., Cambridge U. P., Cambridge, England (1890).

12. R. G. Osterweis, Trans. Conn. Acad. Arts Sci. **38** 103, (1949).

13. P. Duhem, *Josiah-Willard Gibbs à propos de la Publication de ses Mémoires scientifiques*, A. Hermann, Paris (1908).

14. O. Knudsen, in *The Michelson Era in American Science, 1870–1930*, S. Goldberg, R. H. Stuewer, eds., AIP, New York (1988) p. 224.

15. L. Boltzmann, *Lectures on Gas Theory*, S. G. Brush, trans., U. Calif. P. (1964), pp. 274–90. See also ref. 11.

16. H. A. Bumstead, in ref. 3, p. *xxiv*.

17. M. J. Klein, in *The Boltzmann Equation: Theory and Application*, E. G. D. Cohen, W. Thirring, eds., Springer-Verlag, New York (1973), p. 53.

18. A. de Tocqueville, *Democracy in America*, vol. II, Random House, New York (1945). See also N. Reingold, *Nature* **262**, 9 (1976)

19. W. Stevens, *Trans. Conn. Acad. Arts Sci.* **38**, 161 (1949). ∎

Wallace Clement Sabine and acoustics

PHYSICS TODAY/FEBRUARY 1985

Wallace Clement Sabine was the world's first and most celebrated architectual acoustical scientist. He received his graduate training in the physics department at Harvard University and remained on Harvard's faculty until his death, in 1919. His reputation is based not only on his contributions to acoustics, but also on his teaching and administrative contributions to the physics curricula at Harvard and his services to the US during World War I.

The Jefferson Physical Laboratory has represented for many scientists an opportunity to learn, to exchange ideas with colleagues worldwide, and to explore the boundaries of science. (See the article by Gerald Holton in this volume.) Sabine, who joined the Jefferson Laboratory two years after it opened, was one of the first beneficiaries of these resources.

Early years

Sabine's early education and his experiences at Harvard foreshadowed the lives of many other scientists who have studied there.[1] He came from the Midwest—Columbus, Ohio. Strong pressure from his mother, who said, "A farmer's life is not to be the portion of my son," motivated him to develop his aptitude for physics. He attended a college near his home, Ohio State University, where he came under the influence of an inspiring physics teacher, Thomas Corwin Mendenhall, who became his mentor and lifelong friend.

By chance, John Trowbridge, then the director of Harvard's Jefferson Physical Laboratory, had met Sabine two years before his graduation at a meeting of the American Association for the Advancement of Science in

Philadelphia, where Trowbridge was chairman of a session. Sabine was so excited by that session that he was bold enough to tell Trowbridge of his determination to go to Harvard for his graduate studies. Trowbridge later said that his view of the value of large public scientific gatherings was broadened by the experience of meeting young men like Sabine.

Sabine received his Bachelor of Arts degree from Ohio State in June 1886, at the age of 18. After passing Harvard's qualifying examination, he began graduate study in the fall of 1886. Trowbridge welcomed him heartily to his own course; that year Sabine also studied with Benjamin Osgood Peirce, Edwin H. Hall, James Mills Peirce and William E. Byerly. At the end of his first year at Harvard, he was awarded a two-year Morgan Fellowship, which greatly relieved the financial burden on his parents of keeping two children in graduate school—Sabine at Harvard and a sister at MIT. During the next two summers he supplemented his fellowship stipend with employment at the Bell Telephone Laboratories.

During his second year, he assisted Trowbridge in developing laboratory experiments in electricity, electric lighting and photography. Trowbridge listed Sabine as coauthor on three papers published in 1888, two in the *Proceedings* of the American Academy of Arts and Sciences on the spectral absorption of ultraviolet light by metals, and the third in the *American Journal of Science* on the use of steam in spectrum analysis.

Sabine received a Master of Arts degree in 1888, at the age of 20, and in the fall continued to work at the Jefferson Lab with Trowbridge. A year later he was an appointed assistant with a stipend of $500. This position led to another joint paper with Trowbridge, on electrical oscillations in air, which appeared in 1889, also in the *Proceedings* of the American Academy.

In the summer of 1890, he was put in charge of the Summer School of Physics, at a stipend of $400. For reasons that are not quite clear, Sabine never worked toward a doctorate, but Harvard awarded him an honorary DSc in 1914.

That same summer of 1890, Sabine received an appointment as instructor. He devoted much of his energy and talents to building up the department's laboratory courses and designing and making apparatus for instruction. In 1893, Sabine published his only book, *A Student's Manual of a Laboratory Course in Physical Measurements*. The manual contained outlines for 70 experiments in mechanics, sound, heat, light, magnetism and electricity. In June 1895, at age 27, Sabine was made assistant professor of physics.

About this time, Sabine began the independent research that made him, within a few years, the world's first and most celebrated acoustical scientist. This dual commitment—to research as well as to teaching—is directly attributable to the encouragement he received from both Harvard President Charles W. Eliot and Trowbridge, and to the superior resources of the new Jefferson Physical Laboratory.

Sabine's loyalty to teaching remained high. Years later, responding to a request for advice, he wrote: "My principal hope is that the [Carnegie] Institute may not serve to separate the research and the educational functions of scientific men. . . . The instructor who does not engage in research. . . soon teaches Science as isolated facts rather than as groups of problems. . . . If he be engaged in research work, the spirit of it inevitably enters into his teaching. On the other hand, an investigator who does not teach serves a diminished constituency. Next to its direct results, the value of scientific work lies in its stimulating influence on every activity, not of the individual but of the country at large, and this can

Leo L. Beranek is director of Bolt, Beranek and Newman, in Boston, Massachusetts. He did research on acoustics in the constant temperature room at the Jefferson Physical Laboratory from 1938 to 1947.

Using the resources of the Jefferson Physical Laboratory, Sabine changed architectural acoustics from an obscure body of knowledge to an experimental science.

Leo L. Beranek

Wallace C. Sabine, who founded the scientific discipline of architectural acoustics with investigations that began in Harvard University's Jefferson Laboratory and Fogg Art Museum. (Photo courtesy AIP Niels Bohr Library.)

best be attained through its connection with the universities."

A new science

Sabine's acoustical contributions had their beginning when the auditorium in the Fogg Art Museum, which was opened in 1895 (later renamed Hunt Hall when the Fogg museum moved to a new building) was found to be unsuitable for lectures. The drawings of figure 1 show the plan and the center-line cross section of the lecture room that Sabine included in his paper.[2,3]

In an effort to salvage the disastrous lecture hall, Eliot turned to Trowbridge for advice and, on receiving his recommendation, asked the newly appointed assistant professor Sabine for help. After some thought and consultation with his mentor at Ohio State, Mendenhall, Sabine accepted the assignment to try to correct the poor acoustics.

Although Eliot sought only to correct a defect in a single auditorium, Sabine was determined to discover by scientific experiment the secrets of an age-old baffling subject: architectural acoustics. His colleagues warned him that the problem was so complex as to preclude satisfactory solution. They even referred to his new assignment as "a grim joke." Sabine realized that success would have to come from his own efforts and, courageously he ignored the amused incredulity of his colleagues.

At the time of Sabine's investigation, the literature of architectural acoustics showed no consensus on the dimensions for an acoustically acceptable room, or the materials of which to build it, or the means to correct existing defects. Sabine recognized that, broadly speaking, only the size and shape and the interior materials, including both the furnishings and the audience, constituted the prime variables in the acoustics of rooms. In most repair work, only the materials could be varied. It was obvious that the reverberation of the

Lecture room in Hunt Hall (then called the
Fogg Art Museum) at Harvard University: **a**
floor plan, **b** cross section. **Figure 1.**

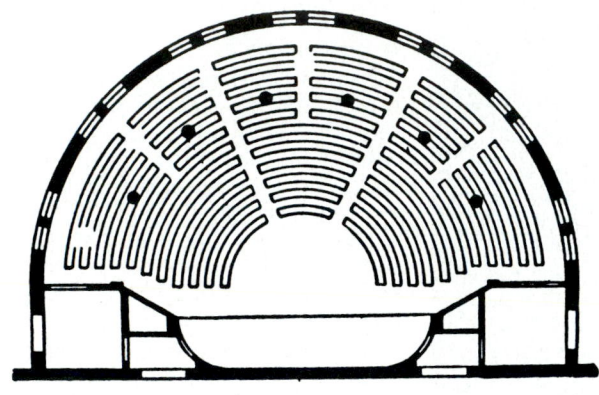

a

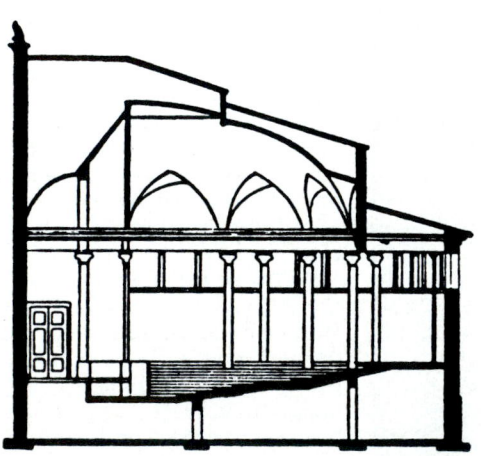

b

Fogg lecture room was too long for
spoken words to come through clearly,
and that sound-absorbing materials
would have to be added. Sabine concluded that his first task would be to
determine the relative absorbing power
of the substances that might be used in
correction. To do this, he needed to be
able to measure the rate of decay of
sound in the room as he introduced
various kinds and amounts of materials.

Sabine groped for a suitable way to
measure the decay of sound intensity in
a room and concluded that the two
known optical methods for measuring
the intensity of sound, that is, observing
a sensitive manometric gas flame either
by a micrometer telescope or by
photography, were unsatisfactory. He
then chose the measuring setup shown
in Figure 2. The sound source was a 512-
Hz organ pipe, blown from a double
tank, water-sealed and noiseless. The
air supply was turned on and off by an
electro-pneumatic valve. The electric
current required for this purpose also
produced a mark on the cylinder of a
chronograph that measured intervals
of time between half a second and 10
secondswith an accuracy of about one
hundredth of a second. As sound
detector, Sabine used the ears of a
human observer whose only duties were
to turn the air supply to the organ pipe
on and off and to squeeze a hand bulb to
produce a mark on the cylinder of the
chronograph   whenever   the
reverberation in the room became
inaudible.

At the outset, Sabine tested many
features of his apparatus in both the
unfinished lecture room of the Boston
Public Library and the reverberant
lobby of the old Fogg Art Museum. In
particular, he determined how long a
time the organ pipes had to sound
before the sound field reached a steady
state. He also determined that the
viscosity of the air in the rooms had a
negligible effect on the reverberation
at 512 Hz. Other observers have since
found that at higher frequencies vis-
cous sound absorption is in fact significant.

Sabine's early measurements yielded

some interesting results; for example:
▶ Measurements in Steinert Hall (Boston) showed that the duration of audibility is nearly the same in all parts of an auditorium
▶ In the large lecture hall of Jefferson, he found that the duration of audibility is independent of the position of the source
▶ In the Fogg lecture room, the efficiency of an absorber proved to be largely independent of its position
▶ Also in Fogg, Sabine found that over a period of time different experienced observers recorded nearly identical durations of audibility for a given stimulus and condition.

With these preliminaries out of the way, Sabine began his research in earnest in the Fogg lecture room during the late fall of 1895. He obtained permission to transport more than 400 seat cushions (filled with hair, covered with canvas and light damask) from Sanders Theater across the wide street to the lobby of the Fogg museum, provided only that they were returned to Sanders within a few days. Introducing these seat cushions into the lecture room a few at a time, he measured the duration of the sound decay at 500 Hz as a function of linear meters of seat cushions. To avoid interference from street noises, he made his measurements after midnight. Next he made similar measurements for curtains, cloth, canvas, hair-felt and oriental rugs. He also made extensive measurements of the same type in Sanders Theater, where moving the cushions from the lobby to the hall was easier.

Over the next five years, the Sanders seat cushions were lugged repeatedly to and from the Fogg and various other rooms, generally after classes were done at the end of the day, and returned to Sanders before classes assembled again the next morning. With two laboratory assistants, Sabine moved the cushions and made his measurements between midnight and 5 am, on either three consecutive or three alternate nights per week. The amount of work involved was prodigious, but, according to his biographer, W. D.

Orcutt, Sabine scarcely missed a daytime class except when illness confined him to bed.

Sabine made another significant measurement, essential to his understanding of the lecture room in the summer of 1897. One day, at the close of a lecture, he determined the duration of the residual sound before and immediately after half the audience had left, and then after all of it had gone, thus achieving a measure of "audience absorption"—both per person and per square meter of floor space occupied. He repeated this experiment under better controlled conditions in the large lecture room at Jefferson three years later.

During the first two years after Sabine undertook his research at the Fogg lecture room, Eliot encouraged the acoustical experiments. But then he became exasperated with the delay and issued a reprimand that Sabine could not ignore. In early 1898, Sabine prepared a paper that contained a drawing and specifications for recommended treatment of the Fogg lecture room. That paper served as a guide for the renovation and became the first of 13 articles that Sabine published on architectural acoustics. It describes the placement of hair-felt blankets, 0.75 and 1.05 inch in thickness, on 21 surfaces of the room, and the covering of the platform with a thick carpet. The lecture room was returned to use in October 1898, and although its circular shape prevented it from ever being fully satisfactory, it remained in use for 75 years, until the building was torn down to make room for Canaday Hall. The University paid Sabine an honorarium of $500 for his efforts and, in addition, covered the costs of the experimentation, including the two assistants.

In about 1898, Sabine began to use the constant temperature room in the basement at the west end of Jefferson as a reverberation chamber. (See the figure on page 177.) This room appealed to him for two reasons: first, because it and the rooms extending above it to the roof were separated structurally from the

rest of the building, thus normal activities elsewhere in Jefferson could not be heard; and, second, because its walls were constructed of brick and cement, making it highly resistant to excitation by sound. It was here that Sabine first measured the sound absorpton at 512 Hz of brick and concrete, both painted and unpainted.

The new Boston Symphony Hall

Sometime early in 1898, Major Henry Lee Higginson, the founder and benefactor of the Boston Symphony Orchestra, having learned that the City of Boston was determined to destroy the old Music Hall to make way for a street, revived plans for a new hall for his orchestra. Apparently Higginson discussed those plans and also some apprehensions he had about the acoustics of the new hall with Eliot. Eliot, who had been pleased with Sabine's recommendations for the Fogg lecture room, convinced Higginson that the architects for the new hall, McKim, Mead and White, could benefit from Sabine's experience. With Higginson's permission, Eliot asked Sabine whether he would consider consulting with the architects. Sabine hesitated before responding. What troubled him was that he had not yet determined a generalized formula that would allow him to extrapolate from his measured results to the plans for an unbuilt structure.

Believing that there must be such a generalized formula, Sabine devoted the next fortnight to a feverish restudy of the seat-cushion data at 512 Hz that he had taken in a dozen or so rooms of various sizes and reverberances around Harvard. On Saturday evening, 29 October 1898, the nswer came to him. He turned to his mother who was nearby and said, "I have found it at last! It's a hyperbola!" (See Figure 4.) The very next day he wrote to Eliot: "You may be interested to know that the curve in which the duration of the residual sound is plotted against the absorbing material is a rectangular

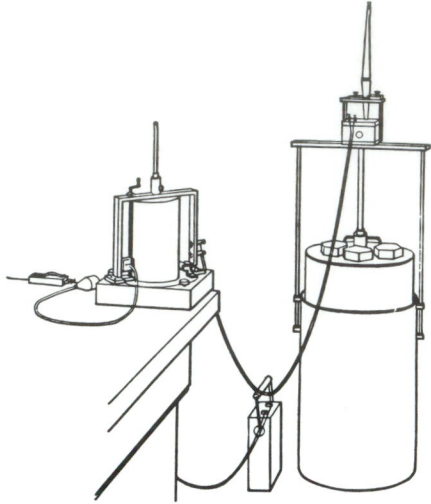

Acoustical setup for controlling the sounding of an organ pipe and accurately measuring intervals of time.   **Figure 2.**

hyperbola with displaced origin; that the displacement of the origin is the absorbing power of the walls of the room; and that the parameter of the hyperbola is very nearly a linear function of the volume of the room...It is only necessary to collect further data in order to predict the character of any room that may be planned, at least as respects reverberation." The formula, known as the Sabine formula, is:

$$T = KV/S_t\bar{\alpha}$$

$$\bar{\alpha} = (S_1\alpha_1 + S_2\alpha_2 + \ldots)/S_t$$

where T is the time in seconds it takes for sound to decay 60 decibels in the hall; K is nearly a constant, inversely proportional to the speed of sound; V is the room volume; S_t is the total area of all surfaces in the room; and $\bar{\alpha}$ is the average sound absorption coefficient of the surfaces in the room, each of which has area S_n and sound absorption coefficient α_n.

Eliot, reinforced by Sabine's new confidence, wrote forthwith to Higginson to tell him of the kind of help he thought Sabine could give the architects. The encouragement and resources of Harvard during Sabine's previous three years of study now bore fruit, and his findings were about to be subjected to practical test.

Over the next eight months Sabine frantically collected the new data he thought necessary. He studied the effect of organ-pipe intensity on the accuracy of his data. He assembled measurements comparing the absorbing power of carpets, cheesecloth, cretonne cloth, cork, open windows, and seats of several kinds. He conducted tests in various university rooms, including the lecture hall, the constant temperature room and room 41 in Jefferson; the faculty room, clerk's room and dean's room in Harvard's University Hall; and five rooms at the Harvard Botanic Gardens. He also

returned to the two rooms he had experimented in at the Boston Public Library.

Following the initiation of the Boston Symphony Hall project, McKim, Mead and White had worked with Higginson's building committee to choose a suitable shape and size for the hall. Acting primarily on the advice of musicians and a few lay observers who had heard music in a number of European halls, they decided to model the design after the Neues Gewandhaus in Leipzig, Germany, but to increase the seating capacity by about 70% over the Gewandhaus's 1560 seats. Sabine later wrote, "At this stage calculation was first applied . . . which, proportions being preserved, would have doubled the volume . . . resulting in a [calculated] reverberation time of 3.02 seconds." That time is half again as much as he had calculated for the Gewandhaus. Sabine continued, "This would have differed from the chosen result by an amount that would have been very noticeable." Today we know that the proposed hall would have been completely unacceptable.

Sabine and the architects agreed that the hall should not be significantly greater in length than the Gewandhaus. Attention then turned to the existing Boston Music Hall, which Boston audiences found quite satisfactory. The Music Hall was rectangular in shape, of nearly the same length, and had a calculated reverberation time only slightly longer than in the Gewandhaus. It accommodated the desired larger audience by having two balconies instead of one, whose floors lay parallel to the sloping main floor. Sabine wrote that, "the real discussion was based on only two buildings—the present Boston Music Hall and the Leipzig Gewandhaus; one was familiar to all and immediately accessible, the

other familiar to a number of those in consultation. . . . It should, perhaps, be added immediately that neither hall served as a model architecturally, but that both were used rather as definitions and starting points on the acoustical side of the discussion."

In addition to reverberation time, Sabine also advised the architects on other points, described after the fact in a letter to Higginson in February 1899. These included such recommendations as eliminating the balcony seating along the sides of the stage, narrowing the stage and splaying its sides. The effect of this, he said, will be to "increase the loudness or volume of the sound, and at the same time . . . better the 'attack.' The reflected sound and the sound coming directly [to the listeners] will more accurately unite. . . . In respect to loudness, I do not think that the new hall will, on the whole, be at a disadvantage in comparison with the old. . . . In respect to reverberation, the two halls will be very nearly the same. . . ."

Although Sabine makes no mention of it, the ornate interior of Boston Symphony Hall, with its coffered ceiling and sidewall niches and statues, contributes significantly to its good acoustics. Sabine paid particular attention to eliminating ventilation noise, writing: "Transmission of disturbing noises through ventilation ducts . . . is practically a legitimate and necessary part of the subject. . . . [Ventilation] is best secured by a system . . . known as 'distributed floor outlets.' It has the additional merit of being, perhaps, the most efficient system of ventilation."

Sabine was to suffer bitter disappointment following the opening of Symphony Hall on 15 October 1900. Musicians and critics from around the world who heard the hall in its early

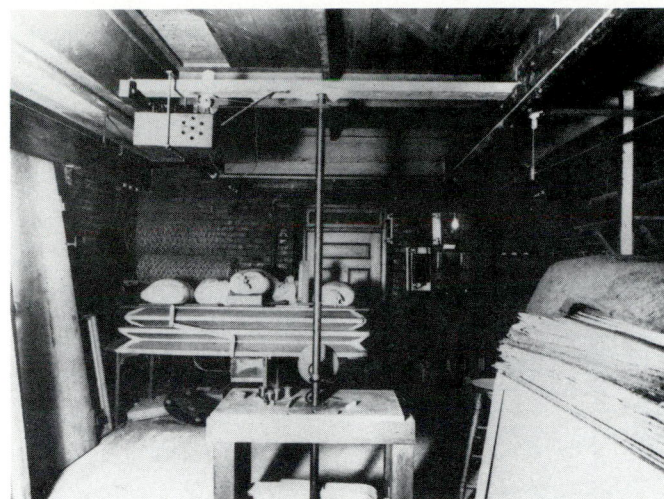

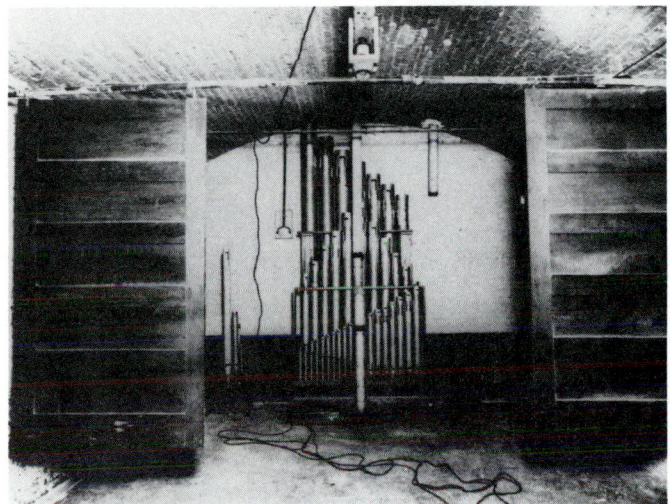

Constant temperature room in the Jefferson Physical Laboratory. On the left of the cutaway view is shown the control room with the wind chest feeding the organ pipes and the motor-driven rotating shaft connected to the sound-diffusing panels in the room below. Below left is a view of the organ pipes and the rotating sound-diffusing panels. The rotating panels greatly reduced the fluctuations in intensity of the decaying reverberating sound, making it possible for the observer to determine more accurately when the reverberation became inaudible. Below right shows the box in which the observer was enclosed to eliminate the sound-absorbing effect of his clothing. The photos were taken around 1919. (Photos courtesy of Riverbank Acoustical Laboratories, Geneva, Illinois.) **Figure 3.**

years compared it unfavorably with the better halls of Europe. Two years after it opened, an article appeared in the 31 December 1902 edition[4,5] of the prestigious *Boston Evening Transcript*, entitled "Boston Symphony Hall—a Scientific Analysis of its Acoustics—The Hall Said to Possess Wonderful Adaption to the Transmission of Pure Notes to All Parts of the House—The Dissident Judgement of a Music Critic." The article, signed by Frank Waldo, PhD, praises the hall and quotes from articles by a Dr J. B. Upham and Sabine. Appended to the article was a note, in square brackets, by the *Transcript*'s respected music critic, William Foster Apthorp:

This is all very well; but, like many essays on musical subjects by scientists, it arrives at conclusions with which most musicians find it difficult to agree. To begin with, neither the late Dr. Upham nor Mr. Sabine can be rightly deemed

competent to express a musical opinion of any weight whatever; both come musically in the amateur class. And, to conclude with, we have not yet met the musician who did not call Symphony Hall a bad hall for music. Expert condemnations of the Hall differ, as far as we have been able to discover, only in degrees of violence.
—W.F.A.

Boston Symphony Hall, unchanged today, is rated the greatest hall in the western hemisphere and, for its size, one of the greatest in the world. One explanation for its poor initial reception may have been its size; it seats 2625 persons, almost twice the 1400 seats typical of European halls. Thus, with an orchestra of 90 or so—the usual number of musicians in 1900—the music sounded thin in Symphony Hall. With today's complement of 104 musicians, the sound is fuller and louder.

A second disappointment came to

Sabine when the reverberation time of the completed Symphony Hall was measured at half a second shorter than he predicted in his published paper of 1900 (calculated as 2.3 seconds versus 1.8 seconds actual). The reason for the discrepancy was that he had understated the absorbing power of the audience by calculating it on a "per person" rather than a "per square meter" basis. His earlier measured data were published both ways, but his choice of the per-person prediction of the reverberation time was unfortunate because the area per person was greater in Symphony Hall than in the Jefferson lecture hall, where the earlier measurements were conducted. So unhappy was Sabine that in no paper after 1900 did he mention Symphony Hall, and when questioned about it he would only say that the real test of his work must come with actual use of the hall for the exact purpose of its erection.

I hasten to add that the effect of

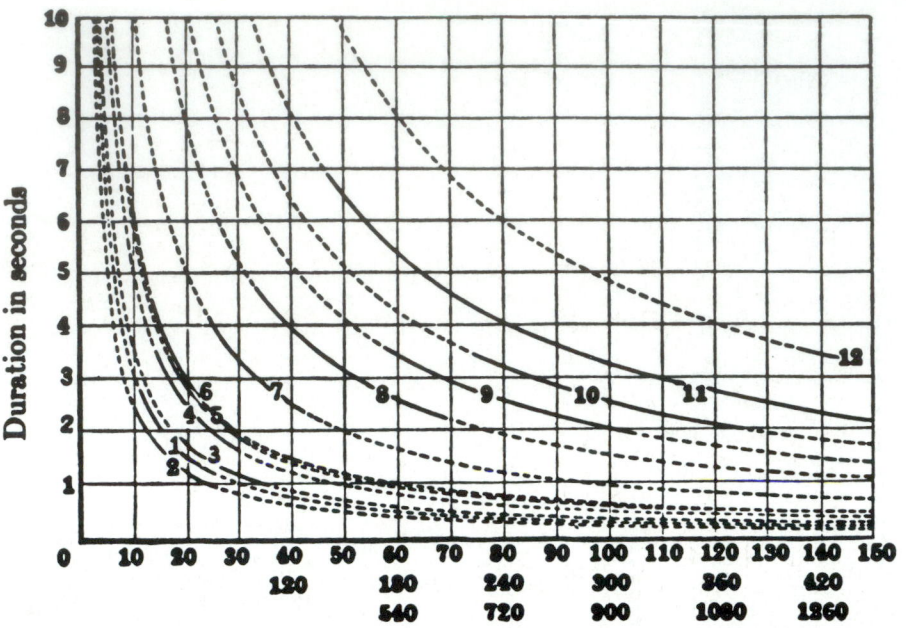

Plot of the audible duration of sound at 512 Hz as determined by a listener from the instant the organ pipe was cut off. The graph, reproduced from Sabine's original paper, shows curves for 12 rooms. The solid portion of each curve shows the measured data as a function of the number of square feet of Sanders Theater seat cushions introduced into the room. Room No. 11 is the Fogg lecture room and No. 12 is Sanders Theater. The dashed portions of the curves are hyperbolas fitted by Sabine to the measured (solid line) portions. Figure 4.

Total absorbing material

Sabine's choice of predictions in the design of the volume and shape of the new hall was of no consequence. His primary concern was to compare his calculations for the new Symphony Hall with those for the Boston Music Hall. Because the volumes and areas per seat were nearly equal, his calculations for both halls were about equally high. The Boston Music Hall, now called the Orpheum Theater, is still standing, although renovations have completely changed the interior.

Back to teaching

One can imagine that those five years of intensive nightly measurements would have aroused questions and comments among his teaching colleagues. Perhaps they got the impression that he was apathetic toward his academic work, because his only opportunity to sleep was during the day. Another irritant to his colleagues may have been his preemption for his project of laboratory assistants, who were, consequently, of little help to other faculty members because they too had to sleep during the day. Still remembered in 1900 were the stories that circulated about Sabine and his assistants dragging hundreds of seat cushions from Sanders Theater to various Harvard buildings late in the day and dragging them back in the early morning. Whether he was impelled by this collegial reaction or by other disappointments, without apparent reason Sabine threw himself wholly into teaching in the academic years 1900 to

1904.

In 1900 Sabine asked to be assigned direct responsibility for "Physics C," an introductory experimental physics course, as well as for three other courses covering light, heat, thermodynamics and physical optics—courses that were open to both graduates and undergraduates. But it was the undergraduate Physics C that interested him most. Eliot had noted in his annual report for 1899–1900 that enrollment was much higher in chemistry courses than in physics—a fact he thought most curious. After all, he commented, physics was broader in scope than chemistry, and there was considerable demand for competent teachers trained in heat, light and electricity. Eliot's prescience was rewarded: In the three years following Sabine's embarking on Physics C, the attendance jumped from 69 to 236, prompting Trowbridge to write in his report of 1902–03: "The marked success of this elective is due to Professor Sabine." A year later Trowbridge added: "Physics C now occupies three large laboratories in Jefferson. The success is due to the constant effort of Professor Sabine to increase its instrumental appliances and to raise its intellectual level."

Strangely, Sabine went without promotion for ten years, despite the high quality of both his teaching and research. Theodore Lyman once intimated to me that Sabine may have been neglected in the wake of the adverse criticism of Symphony Hall. Furthermore, Higginson paid him no consult-

ing fee, even though Eliot had set a precedent for doing so at the Fogg.

Recognition

In 1904, Sabine began to accept commissions from architects to advise on the acoustics of churches, cathedrals, auditoriums and theaters by the dozens, and the Boston Opera House (now demolished to make room for a parking lot). His need for reverberation data over a wide range of frequencies on diverse building materials sent him back to his laboratory in Jefferson's constant temperature room in successive bursts of intense all-night, three-nights-a-week activity in 1904–05, 1908–09, and again in 1914–15. In 1906, he wrote, "Each problem has been taken up as it has been brought to the writer's attention by an architect in consultation either over plans or in regard to a completed building. This method is slow, but it has the advantage of making the work practical. . . ." His papers, the last of which was published in 1915, and his consultations established his fame.

Sabine worried constantly about the propriety of accepting fees or royalties for his work in acoustics. Early in his Harvard career he raised the question with Eliot, who later wrote, "I repeatedly pointed out to him that Harvard University gained much whenever its professors contributed to the public welfare, . . . and, further, that no sharp and fixed line can be drawn between Pure and Applied Science, since what is pure today may easily have valuable

Interior of the Boston Symphony Hall. (Courtesy of Boston Symphony Orchestra.) **Figure 5.**

applications tomorrow."

Finally, in 1905, Sabine's promotion came—a double advancement to full professor, which his biographer says caused Sabine real embarrassment and mystified his family and friends. Eliot later commented on this delayed promotion, but offered no explanation: "Sabine served five years as instructor in physics and ten years as assistant professor at Harvard University before he reached the position of full professor, in spite of the fact that he quickly proved himself an admirable teacher and a highly successful investigator." As a full professor, his salary was raised to the standard professorial $4000 and it remained the same for the next 10 years.[6]

Soon after his promotion, Sabine again surprised his colleagues when, after many years of not participating in faculty meetings, he proposed, early in 1906, a radical reorganization of graduate study in applied science at Harvard. This proposal, and the sudden death of Dean Nathaniel S. Shaler, elevated Sabine to the post of dean of the graduate school of applied science, which he held for nine years with an annual supplement to his professorial salary of $500. His biographer says that he took the position of dean reluctantly, and only after persuasion by Eliot, because his primary interest continued to be research and teaching.

William F. Osgood has commented on Sabine's success at discovering and attracting to Harvard promising young scientists—among them Percy W.

Bridgman, Edwin C. Kemble, Francis Wheeler Loomis, and Theodore Lyman. In 1976, Kemble, the last link between Sabine and ourselves, wrote[4] to me, "Sabine was personally responsible for my coming to Harvard for graduate work. My undergraduate patron, Dayton C. Miller, who was also an acoustics man, wrote to Sabine and Harvard offered me a fellowship which Sabine paid for out of his personal pocket. I took Sabine's Optics the first year and liked him exceedingly. He had me at his Marlborough Street home for dinner and I remember once lunching with him to talk over a personal problem. But in the spring of 1917 he was off to war . . . two years later he was gone."

In July 1915, Sabine was appointed "Hollis Professor of Mathematics and Natural Philosophy," though he hardly had an opportunity to enjoy his new status. He was completely absorbed with the dismal prospect of World War I and, taking a leave of absence, he served the allied war effort. First he prepared in Switzerland a planning report for the International Tuberculosis Commission. Then, in 1917, he visited the battlefield twice on scientific missions, returning to the US in September. Thereafter, until the war ended in November 1918, he spent four days of each week in Washington advising on aircraft design and production and from Tuesday to Thursday he met his classes in Cambridge. It might be said that de facto he was the first chief scientist of the US Air Force.

On 7 January 1919, he underwent an

operation to remove a malignant kidney. His death followed three days later. The obituary "minute" prepared by his colleagues for presentation to the Harvard faculty contains a fitting conclusion, "He succeeded by reason of a combination of qualities among which were unending patience and untiring energy."

* * *

I am deeply grateful to Geraldine Stevens for her invaluable editorial help in the preparation of this article. The article is based on a talk given on 4 May 1984, at the centennial celebration of the Jefferson Physical Laboratory.

References

1. W. D. Orcutt, *Wallace Clement Sabine: A Biography*, Plimpton, Norwood, Mass. (1933).

2. W. C. Sabine, *Collected Papers on Acoustics*, T. Lyman, ed., Harvard U. P. (1922), Dover, New York (1964). Reference 3 and ten other papers are included.

3. W. C. Sabine, "Architectural Acoustics: Reverberation," (published in seven parts), Am. Arch. Build. News (1900), April 7 and 21, May 5, 12 and 26 and June 9 and 16.

4. L. L. Beranek, "The Notebooks of Wallace C. Sabine," J. Acoust. Soc. Am. **61**, 629 (1977).

5. L. L. Beranek, J. W. Kopec, "Wallace C. Sabine, Acoustical Consultant," J. Acoust. Soc. Am. **69**, 1 (1981).

6. Details about Sabine's appointments at Harvard were supplied by Robert Shenton, secretary of the Harvard Corporation. □

HEINRICH HERTZ AND THE DEVELOPMENT OF PHYSICS

In addition to confirming Maxwell's electromagnetic theory, Hertz's experimental and theoretical work one hundred years ago helped lay the foundation for quantum theory and relativity.

Joseph F. Mulligan

PHYSICS TODAY/MARCH 1989

The discovery that electromagnetic radiation in the microwave and radio regions of the spectrum displays the same basic behavior as visible light—reflection, refraction, diffraction, interference, polarization—was made in Karlsruhe in 1888 by Heinrich Rudolf Hertz. Some special events last year marked the centennial of Hertz's momentous discovery. The Technical University in Karlsruhe, at which Hertz did his electromagnetic research, held a symposium on Hertz and the consequences of his work, and devoted one complete volume of its publication *Fridericiana* to him.[1] In the US, the Microwave Theory and Techniques Society of the Institute of Electrical and Electronics Engineers held a symposium in New York City to celebrate Hertz's achievements. On display at this meeting, from the Science Museum in London, was a refurbished set of replicas of Hertz's original apparatus.[2] These replicas were exhibited at the MIT Museum in Cambridge, Massachusetts, before returning to London late last year.

These tributes to Hertz concentrated in great part on his life, his experiments at Karlsruhe and the significance of his work for microwave and radio-frequency technology. While this article will look at these more well-known aspects of Hertz's work, it also will cover his lesser-known work, including his theoretical contributions to electro-

magnetism and his work in fields related to what we now call modern physics.

Hertz's life

Hertz was born in 1857 into a well-to-do German family in Hamburg. A very bright student, he did equally well in the humanities and the sciences, and showed great manual skill in drawing, sculpting, woodworking, metal fabrication and the design and construction of scientific apparatus such as sensitive galvanometers and voltaic cells. These abilities were to stand him in good stead throughout his scientific career. Hertz started out studying engineering, but soon gave it up for physics. In 1878 he enrolled at the University of Berlin to work under Hermann von Helmholtz and Gustav Robert Kirchhoff.[3]

Hertz quickly established himself as a promising student by solving a prize problem, posed by Helmholtz, on whether the electric charge moving in a conductor has inertial mass. Helmholtz then proposed in 1879, through the Berlin Academy of Sciences, another prize for anyone who was able to "establish experimentally any relation between electromagnetic forces and the dielectric polarization of insulators." Encouraged by Helmholtz, Hertz considered taking this on as his doctoral dissertation subject, but a careful analysis led him to the conclusion that "any decided effect could scarcely be hoped for, but only an action lying just within the limits of observation." He therefore put the problem aside, only to return to it— and solve it—nine years later, in 1888, as part of his classic experimental work on electromagnetic waves.

Instead of tackling this difficult prize problem, Hertz

Joseph Mulligan is a professor of physics at the University of Maryland, Baltimore County, in Catonsville, Maryland.

Apparatus that Heinrich Hertz used to study the polarization and refraction of electromagnetic waves. The wire frame, almost 2 m high and strung with parallel wires, acted as an analyzer for the polarized radiation produced by Hertz's spark transmitter. The 30° prism, when filled with hard pitch, weighed more than 1000 lbs. (Courtesy of the Deutsches Museum, Munich.)

Figure 1.

chose a theoretical dissertation on inductive effects in charged rotating spheres, and he successfully completed it under Helmholtz's direction in January 1880. His dissertation contains some powerful and (for that time) sophisticated mathematics, and was completed in less than a year—clear evidence that Hertz was a first-rate mathematical physicist in addition to being a remarkable experimentalist. He took his doctoral examination in February 1880 after only three semesters at Berlin, and received his degree *magna cum laude*, a rare distinction in Helmholtz's institute.

Hertz stayed on in Berlin for three years as Helmholtz's assistant at the Physical Institute, during which time he published 13 scientific papers on a variety of subjects. He was appointed *Privatdozent* at the University of Kiel in 1883, was called to Karlsruhe as professor of physics in 1885 and spent his final, difficult years from 1889 to 1893 as successor to Rudolf Clausius in Bonn. Hertz never had robust health and suffered from recurring problems with his teeth, jaw, nose and eyes. He died of chronic blood poisoning on 1 January 1894, just a few months before his 37th birthday. His mentor Helmholtz, whom he revered and loved, died just eight months after Hertz, at age 73.

Theoretical electromagnetism

Hertz learned electromagnetic theory from Helmholtz, one of the very few physicists on the continent to appreciate the importance of the work of James Clerk Maxwell. Most German physicists accepted the action-at-a-distance theories of Franz Neumann and Wilhelm Weber, which maintained that electromagnetic forces propagated from one body to another at infinite speed, as was then assumed also to be the case for gravitational forces. Helmholtz had developed a compromise theory that attempted to reconcile these ideas with Maxwell's theory of the electromagnetic field. His compromise was to accept the action-at-a-distance theory for free space, but to adopt Maxwell's theory in the case of dielectrics, in which electromagnetic disturbances were assumed to propagate by means of the polarization of the medium.

Hertz's first year in Kiel was not a happy one. He was lonely, longed for the stimulating scientific atmosphere of Berlin and lacked the equipment necessary to continue the experimental research on electromagnetism he had begun there. It seemed natural, then, for him to turn to Maxwell's theory (as modified by Helmholtz) for guidance on the crucial experiments needed to decide once and for all among the competing theories of electromagnetism.

It did not take Hertz long to see the inconsistency of his mentor's ideas, however, and he proceeded in a very important, and often overlooked, paper of 1884 to derive Maxwell's equations from first principles by a new method.[4] This avoided both the mechanical analogies Maxwell had originally used and the explicit idea of the displacement current he had introduced. Both of these approaches were out of favor in Germany, and Hertz wrote his paper in an attempt to win his German colleagues over from the theories of Neumann and Weber to that of Maxwell. In this paper Hertz obtained Maxwell's fundamental equations in the symmetric form we use today, free of scalar and vector potentials. The fundamental equations of the electromagnetic field in this form were always referred to at the turn of the century as "Maxwell's equations in the Hertz–Heaviside form," in recognition of the contributions made by Hertz and by Oliver Heaviside, the self-taught British physicist and electrical engineer who made many important contributions to electrical theory. Hertz always insisted that it was Heaviside who deserved credit for this form of Maxwell's equations, but it was the clarity of Hertz's presentation that won over physicists to this formulation.

In 1890, after his experiments on electromagnetic waves had demonstrated his experimental skill, Hertz wrote two other important theoretical papers on electromagnetism. The first of these, "On the Fundamental Equations of Electromagnetism for Bodies at Rest,"[5] helped overcome the repugnance most German physicists retained for Maxwell's theory even after Hertz's convincing experimental verification of its predictions. Arnold Sommerfeld recalled that on reading this paper of Hertz, "the shades fell from his eyes," and he understood electromagnetic theory for the first time.[6] The once-neglected equations of Maxwell, stripped by Hertz of the mechanical models on which Maxwell had based them, soon became the basis for all research in electromagnetism and optics in Germany, as they were throughout the rest of the scientific world.

Experiments on electromagnetism

Hertz's 1884 paper on Maxwell's theory of the electromagnetic field convinced him that the crucial proof needed to decide between Maxwell's ideas and those of Neumann, Weber and Helmholtz was not to establish the relationship between electromagnetic forces and dielectric polarization, as set forth by Helmholtz in the prize problem of 1879. Rather it was to determine whether empty space behaves like all other dielectrics. To prove that it does he set out to measure the speed of long-wavelength electromagnetic waves in air—the best available approximation to free space—to see if they propagated at the speed of light, as Maxwell's theory predicted.

One reason that Hertz accepted a professorship at Karlsruhe, despite the lack of advanced students there to test and develop his ideas, was the presence of a considerable amount of equipment that he could use in his research. Among this apparatus he found a pair of Riess spirals, which he used for a lecture demonstration on electromagnetic induction. Impressed by the distances over which changes in the current in the primary coil produced detectable electrical changes in the secondary coil, Hertz set out to modify this arrangement until any direct inductive effects could be ruled out. The only explanation of observed currents in the secondary coil would then be the propagation of electromagnetic radiation through space. Hertz's many modifications of this apparatus finally produced the experimental equipment he needed to verify Maxwell's theory.

He began with a primary circuit, or transmitter, consisting of a piece of copper wire forming a closed loop except for a small air gap between its two ends, and driven by the spark discharge from an induction coil; and a secondary passive circuit, or receiver, consisting of a similar coil with an air gap between its ends. Hertz could adjust the length of this secondary gap with a micrometer screw and obtain an estimate of the signal

Hertz in military uniform. This photograph was probably taken in 1877, when Hertz was 20 years old and serving his year of military service with the First Railway Guards Regiment in Berlin. In the fall semester of 1877 he began his physics studies in Munich but a year later transferred to the University of Berlin to study under Hermann von Helmholtz and Gustav Kirchhoff. (Courtesy of the Deutsches Museum, Munich.) **Figure 2.**

strength at the position of the receiver from the distance the spark could jump. Hertz obtained all of his monumental results in this way.

In September 1889 in Heidelberg, Hertz delivered the keynote address at the annual meeting of the German Association for the Advancement of Natural Science and Medicine. In this talk, "On the Relations Between Light and Electricity," Hertz pointed out that the thin thread on which the success of his research in Karlsruhe in 1885–89 hung was the fact that sparks about 0.01 mm in length and lasting about a microsecond can be seen by a dark-adapted human eye in a perfectly dark room. He pointed out that "the success of a workman depends upon whether he properly understands his tools."[7] Even today, physicists who have tried to repeat Hertz's experiments with his Karlsruhe apparatus have come away with a new respect for his experimental abilities and with amazement at what he was able to accomplish with such primitive equipment.

Hertz later improved on the sensitivity of his original apparatus by adjusting the dimensions of his receiver until it was in resonance with the transmitter. The tuning required was not very precise, because the transmitter produced a highly damped oscillating spark that radiated a rather broad band of frequencies. When Hertz tuned the detector to any frequency within that band, he achieved resonant enhancement of the signal. He also found that the distance over which signals could be detected increased considerably when he reduced the inductance and capacitance of the transmitting coil and thus increased the frequency. (As we know today, the energy radiated varies as the fourth power of the frequency.) The reduction in inductance and capacitance also sufficiently shortened the wavelengths to allow Hertz to carry out standing-wave measurements in the limited laboratory space he had available.

With this apparatus Hertz set out to verify the predictions of Maxwell's theory. By using the variation in the length of the sparks across the adjustable secondary air gap to locate the nodes and antinodes in standing-wave patterns, he was able to obtain the wavelength of the radiation, which was about 9 meters. He obtained the frequency f of his transmitter from the equation $f = 1/(2\pi\sqrt{LC})$, using calculated values for the inductance L and capacitance C of his circuit. He then obtained the speed of light from the equation $c = f\lambda$. His results were not very accurate because of uncertainties in his calculated value for the frequency and because diffraction effects and spurious reflections perturbed his wavelength measurements. Despite these problems, Hertz was able to show in early 1888 that these long-wavelength electromagnetic waves did not travel at an infinite speed, but at a speed of the same order of magnitude as that of visible light. In this way, he verified one of the major predictions of Maxwell's theory.

Hertz then proceeded, in his classic paper of 1888, to show that 66-cm waves—microwaves—travel in straight lines and can be reflected, refracted and polarized in the same way light waves can.[5] His source in these experiments was a linear dipole positioned on the focal line of a cylindrical parabolic reflector so as to produce a linearly polarized plane wave. His detector coil was symmetrically positioned on the focal line of an identical parabolic reflector. Again, sparking across an air gap was both the source and the means of detecting the radiation.

To study refraction Hertz used a wooden container in the shape of a prism 1.5 m long (see the figure on page 181). This box, whose cross section was an isosceles triangle 1.2 m on a side with an apex angle of about 30°, was filled with asphalt or hard pitch; the resulting prisms weighed over 1000 lbs. To study polarization Hertz used an octagonal frame almost 2 m high strung with parallel conducting wires. These wires could be aligned at various angles with respect to the axis of the transmitter by rotating the frame. By making use of this property of the device Hertz was able to show that the radiation was indeed linearly polarized and that the amount of radiation transmitted depended on the orientation of the wires with respect to the axis of the transmitter. In this way he established the transverse nature of radiation.

By painstaking experiments such as these, Hertz confirmed fully the predictions of Maxwell's theory and eliminated from consideration the theories of Neumann, Weber and Helmholtz. As he stated at the end of his Heidelberg lecture:[8]

The connection between light and electricity, of which there were hints and suspicions and even predictions in the theory, is now established.... Optics is no longer restricted to minute ether waves, a small fraction of a millimeter in length; its domain is extended to waves that are measured in decimeters, meters and kilometers. And in spite of this extension, it appears merely ... as a small appendage of the great domain of electricity. We see that this latter has

become a mighty kingdom.

Hertz's measurement of the speed of unguided electromagnetic waves at radio frequencies was the first such measurement to use a standing-wave technique. This extension of measurements of the speed c of electromagnetic waves from time-of-flight techniques in the visible to standing-wave techniques in the radio and microwave regions of the spectrum has an interesting parallel in the research in the 1970s that extended modern standing-wave techniques for measuring c from the radio and microwave regions back into the visible.[9] In fact Hertz's description of what had to be done to measure the speed of radio waves sounds much like Charles Townes's outline of the best way to measure the speed of light, namely by wavelength and frequency measurements on stabilized lasers.[10]

Beyond verifying Maxwell's theory

While working to verify Maxwell's theory, Hertz made a number of other discoveries, any one of which could have established a physicist's scientific reputation. For example, Hertz appears to have been the first to record resonance curves for the tuning of a radio receiver to a transmitter. The figure below, which is taken directly from one of Hertz's papers, shows the spark length he observed in the air gap of his receiver as a function of the total length of the wire in the rectangular coil of his detector. One can see a clear resonance at about 450 cm, indicating that this length of wire had a distributed inductance that led to resonance at the frequency of the transmitting circuit and so produced stronger and longer sparks in the receiver's air gap.

Hertz's 1889 paper, "On the Propagation of Electric Waves by Means of Wires," contains a very detailed

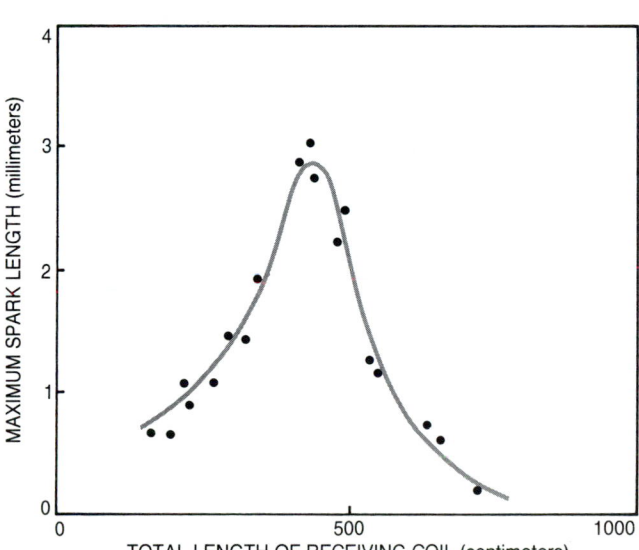

Resonance curve for tuning Hertz's receiving coil to the frequency of a transmitter. The receiving coil was a rectangle, two of whose opposite sides were fixed in length at about 100 cm and whose other two sides varied from 10 cm to 250 cm. (Based on a drawing published by Hertz, ref. 5.) **Figure 3.**

discussion of the "skin effect."[5] He was able to show conclusively that at the frequencies he used in his research the current was confined to the exterior skin of the wires, as Heaviside and J. H. Poynting had predicted on the basis of Maxwell's equations.

The ingenuity with which Hertz confirmed this prediction shows a great experimentalist at work. He studied the effect on a high-frequency current through a central conductor of adding, one by one, 24 additional conductors arranged in a circle parallel to and surrounding the first conductor. He found that the current in the central conductor decreased as the other conductors were added and finally went to zero when the other conductors completely surrounded it, even though the resistance of the inner wire was much smaller than the combined resistance of all the outer wires. Hertz extended this work to a convincing study of the shielding of electrical equipment from high-frequency radiation. He showed that a conducting sheet only $\frac{1}{20}$ mm thick (or even a wire gauze) completely surrounding a spark gap was sufficient to shield the spark gap completely from strong, high-frequency radiation in the 100-MHz range.

Hertz carried out many of his experiments with electromagnetic waves of 66-cm wavelength, and thus deserves to be called the discoverer of microwaves. He constructed the first coaxial transmission line and used it to carry 50-MHz, or 6-m, radiation. He made a coaxial line 5 m long and 30 cm in diameter, with a wire running down the center through insulating supports.[11] The outer conductor consisted of 24 copper wires set along the outer perimeters of seven equidistant circular rings of heavy wire. With this coaxial line Hertz was able to study the electric field configurations in the space between the inner and outer conductors.

Hertz also designed a resonant detector to explore this electric field. It consisted of 125 turns of 1-mm copper wire wound in a tight helix 1 cm in diameter and then bent into a toroid 12 cm in diameter, with a small spark gap interrupting the windings at one point in the toroid. This detector was small enough to be introduced between the central wire and the outer conducting shell of the coaxial line. (As has been pointed out by J. H. Bryant,[11] this was the first "slotted line," although the term came into use only with the development of radar just before and during the Second World War.) Hertz was able to detect standing waves in the line and show that the wavelength inside the coaxial line was the same as in free space—in this case, 6 m.

The most important practical consequence of Hertz's experiments was, of course, our present worldwide system of radio and microwave communications, which was developed through work on Hertzian waves by Karl Braun, Gugliemo Marconi, Oliver Lodge, Lee de Forest and their successors. Because Hertz's interest was focused exclusively on fundamental physics, he never realized that his experiments might have remarkable consequences for communications technology.

But Hertz's experimental work was not confined to electromagnetic waves. He also published papers on meteorology, mechanics, the hardness of materials, the photoelectric effect and cathode rays. The latter two areas of his research are particularly important for the history of modern physics.

The photoelectric effect

The year 1987 was the centenary of Hertz's discovery of the photoelectric effect. (See Giorgio Margaritondo's article in PHYSICS TODAY, this volume, page 188.) Although Hertz discovered this effect by accident in the course of his work on electromagnetic waves, he recognized its importance immediately, because the only other effect of light on electrical phenomena known at that time was the change in the electrical resistance of selenium when it was exposed to light. Hertz saw that an interaction between light and electricity of the kind he had observed could clarify the nature of both light and electricity. He therefore interrupted his electromagnetic wave research for six months and explored "this new and very puzzling phenomenon."

In a 16-page paper in the *Annalen der Physik* in 1887 Hertz described his observations of the photoelectric effect.[5] In the course of his experiments with spark transmitters and receivers he had noticed that the maximum length of the spark in the gap of his detecting coil was decidedly shorter when he enclosed the receiving spark gap in a dark case to make it easier to see the sparks. He suspected at once that light from the transmitting spark was perturbing the receiving spark. As we know today, this light falling on the pole pieces of the secondary spark gap produced electron emission from the poles and thus facilitated sparking across the gap.

Figure 4 shows the original photoelectric apparatus Hertz used to study the effect of light on an electric discharge. To clarify how light enhanced the sparking, Hertz used poles of copper, brass, aluminum, iron, tin, zinc and lead; next he varied the light source, using sunlight, gas flames and electric arcs, among other sources; and lastly he interposed various gases, liquids and solids between the light source and the receiving spark gap. The interposed materials included metal and glass sheets of various thicknesses, paraffin, shellac, mica, gate, wood, cardboard, paper, rock salt and more esoteric substances such as ivory, feathers and the skins of animals.

Hertz's observations suggested to him that the radiation that was effective in increasing the observed spark length lay near the short-wavelength limit of the visible spectrum. He then used a quartz prism to disperse the radiation from his light sources, and found that the wavelengths enhancing the sparks were indeed in the ultraviolet. This was, of course, the case for the metals he used as his pole pieces because they all had work functions close to 4 eV, which puts their threshold wavelength below 310 nm, in the ultraviolet.

After six months of intense effort on the photoelectric effect, Hertz put it aside "so as to direct my attention once more to the main question," which in this highly productive period of his life was the nature and properties of radiofrequency waves. In 1888 Wilhelm Hallwachs, a *Privatdozent* at Leipzig, published a paper in which he extended Hertz's work on the photoelectric effect. In his diary entry of 29 January 1888, Hertz commented, "My consolation must be that I expected and accepted this happening; moreover, I have reached the goal that I wanted to reach, the goal for which I set aside those other experiments."

The six months Hertz stole from his electromagnetic wave research to straighten out the physics of the photoelectric effect were well spent. His results, although obviously incomplete, provided a firm experimental foundation for the work of Hallwachs, Julius Elster and Hans Geitel, Philipp Lenard and Robert Millikan, which led to today's full picture of the photoelectric effect. The experimental data of these researchers, in turn, eventually led to the verification of Einstein's photoelectric equation, which played such an important role in the development of quantum physics.

Cathode rays

While serving as an assistant to Helmholtz in Berlin, Hertz was a friend of Eugen Goldstein, who was doing experiments under Helmholtz's direction on the conduc-

Photoelectric apparatus.
Hertz used this equipment in Karlsruhe in 1887 to study the effect of light on the spark jumping a gap between two electrodes in an evacuated bell jar. (Courtesy of the Deutsches Museum, Munich.)
Figure 4.

tion of electricity in gases. Goldstein, who gave the name "cathode rays" to the emissions from the negative electrode in evacuated Geissler tubes, passed on to Hertz his enthusiasm for this area of research. Hertz published some of his Berlin work on the glow discharge in evacuated tubes in a paper entitled "Experiments on the Cathode Discharge,"[8] and he used this work as his *Habilitationsschrift*—the postdoctoral thesis required to become a university lecturer—when he was appointed *Privatdozent* at Kiel in 1883.

At that time it was unclear whether cathode rays were waves or particles. Hertz succeeded in deflecting them with magnetic fields, but failed to do so with electric fields. This was because he was able to achieve only a poor vacuum, so that ions from the residual gases built up static charges that tended to nullify the applied electric field. Hertz was also unable to observe the magnetic field of the cathode-ray beam outside the discharge tube and came to the incorrect conclusion that the cathode rays were not particles but waves or some new kind of disturbance in the ether.

To see if the cathode rays were indeed some kind of wave, Hertz tried to diffract them by sending a thin beam of cathode rays through a diffraction grating, but he obtained negative results.

When Hertz arrived in Bonn in 1889, he completed one last paper on the propagation of radio waves on wires, and then began to look around for an exciting field of research. When Lenard became his assistant in April 1891, Hertz was stimulated to return to the study of electric discharges in gases. He noted that he "was rewarded by the immediate and unexpected discovery" that cathode rays could pass through thin sheets of metal. Lenard took hold of this idea and in December 1892 produced a beam of cathode rays outside the evacuated discharge tube by using windows made of thin metallic foil of the kind Hertz had employed.

This work of Hertz and Lenard made up for Hertz's failure to deflect cathode rays in electrostatic fields, for it made possible unambiguous experiments on cathode rays

that were outside the discharge tube and therefore were not perturbed by static charges within. Such experiments were performed by Lenard, J. J. Thomson and others, and led to Lenard's Nobel Prize in 1905 "for his work on cathode rays" and to Thomson's Nobel Prize in 1906 "for his theoretical and experimental investigations on the conduction of electricity by gases" and especially for his measurement of the charge-to-mass ratio of the electron. If Hertz had lived, it is likely that he would have shared in one of these prizes. As Hendrik Lorentz recalled in his Nobel lecture in 1902, "Immediately after Maxwell I named [as one of the founders of Maxwell's electromagnetic theory] Hertz, that great German physicist, who, if he had not been snatched from us too soon, would certainly have been among the very first of those whom your academy would have considered in fulfilling your annual task."[12] Health problems forced Hertz to abandon all experimental work at the end of 1892, and he died just one year later.

The work of Hertz and Lenard was important for Wilhelm Roentgen's dicovery of x rays in late 1895. Roentgen was using a cathode-ray tube to study the effect of cathode rays on the luminescence of certain chemicals when he first observed his mysterious x radiation. Hertz and Lenard had clearly produced x rays in some of their earlier experiments, but they were never able to distinguish those rays from cathode rays because at that time they were confused as to the exact nature of cathode rays. In his fundamental 1896 paper on x rays Roentgen refers to the work of Hertz and Lenard, pointing out that they had observed the deflection of cathode rays in a magnetic field.[13] This led Roentgen to conclude that x rays were not the same as cathode rays, because he had found that x rays could not be deflected magnetically. Rather, Roentgen thought x rays were produced bythe collision of cathode rays with glass or metal surfaces inside the cathode-ray tube, although he was not able to explain the exact mechanism. It was only later that others clarified the distinction between the processes producing the continuous and sicrete x-ray spectra. The short-wavelength cutoff of the continuous x-ray spectrum then became a crucial piece of evidence in the development of Max Planck's quantum theory of radiation.

Impact on relativity

Hertz did other work that turned out to be important for the development of modern physics. For example, Planck used Hertz's dipole oscillators in his treatment of black-body radiation because Hertz had already calculated how such oscillators emit and absorb radiation; Planck ac-

Wilhelm Roentgen, who discovered x rays in 1895. Roentgen was helped in his discovery by his knowledge of the properties of cathode rays as determined by many physicists, including Hertz and Philipp Lenard, Hertz's assistant in Bonn. (Courtesy of AIP Niels Bohr Library, Landé Collection.) **Figure 5.**

knowledged this in his 1918 Nobel Prize address.[12]

Hertz's contributions to the development of relativity theory were less direct, but two aspects of his influence are worth considering. First, Hertz's experiments on electromagnetic waves eliminated all action-at-a-distance theories of electromagnetism, and this rejection carried over to gravitational interactions. As Max von Laue wrote, "Albert Einstein's greatest contribution, a theory of gravitation according to which it propagates with the velocity of light, thus traces back directly to Heinrich Hertz."[14] Hertz himself raised this question about gravitation in his Heidelberg address: "We are at once confronted with the question of direct actions at a distance. Are there such? Of the many in which we once believed there now remains but one—gravitation. Is this too a deception? The law according to which it acts makes us suspicious."

Second, just as the first of Hertz's 1890 theoretical papers on electromagnetism was instrumental in establishing Maxwell's equations as the basis for all research on electromagnetism and light in Germany, so his second paper, "On the Fundamental Equations of Electromagnetism for Bodies in Motion,"[5] advanced ideas that eventually led to Einstein's special theory of relativity. In this paper Hertz postulated that the ether contained in a body moved with the body, even though he realized that this postulate was a temporary one that would one day have to be replaced by a more complete theory. Both Lorentz and Einstein had high regard for Hertz's paper. Although Lorentz made the opposite assumption—that the ether was always at rest and the body moved through it—he liked many of Hertz's ideas and modified them to his own purposes. Einstein wrote of Hertz's paper, "A study of Heinrich Hertz's investigation into the electrodynamics of moving bodies will give the reader a clear insight into the conception, prevalent at that time, concerning the electrodynamics of Maxwell."[15]

These varied contributions of Hertz show him to be the last great classical physicist and the precursor of a new generation of modern physicists. The use of his work by the founders of both quantum theory and relativity in their struggle to break with classical physics shows his ideas to be part of the groundwork for the two most important physical theories of the 20th century.

Place in history

Hertz's German colleagues had profound respect for him both as a physicist and as a man, and expected him to take Helmholtz's place as "Reichs-Chancellor of German physics." In a letter to Helmholtz written on 6 January 1894, a few days after Hertz's death, Ludwig Boltzmann paid tribute to Hertz's work by pointing out its impact on the direction of physics research: "One should emphasize the extraordinary import of Hertz's discoveries in relation to our whole concept of Nature, and the fact that beyond a doubt they have pointed out the only true direction that investigation can take for many years to come."[16]

Einstein, in an unpublished manuscript written in 1895, refers to "the wonderful experiments of Hertz,"[17] and elsewhere writes about "the great revolution forever linked with the names Faraday, Maxwell and Hertz"[18]—a clear indication of his appreciation of the importance of Hertz's work for the development of physics.

Bust of Hertz in the entrance courtyard at the Technical University in Karlsruhe. The inscription reads, "In this place Heinrich Hertz discovered electromagnetic waves in the years 1885–1889." (Photograph courtesy of IBM.) **Figure 6.**

The founder of quantum theory lavished similar praise on Hertz. In 1931 Planck summarized the importance of Hertz's research on electromagnetic waves[19]:

Thus gradually the universal significance of Maxwell's ideas became to be more and more recognized on all sides, till at last the crucial experiments of Heinrich Hertz with very rapid electrical oscillations were crowned with an unexampled success, by the production of electrical waves of a few centimeters wavelength. Through this discovery, which produced the greatest sensation in all the scientific world, the speculations of Maxwell were translated into fact and a new epoch of experimental and theoretical physics was begun.

Hertz's brief life came to an end just when his ability and experience put him in a position to make even greater contributions to physics. After Hertz's death, Planck paid a final tribute to his colleague in an address to the Berlin Academy of Sciences[20]:

Spoken or unspoken, the name of Hertz will be among the first of this generation, as long as men pay attention to electric waves. But we, the association of physicists, we will sun ourselves in the glow of this name, yes, we will share in its glory, for he was truly one of us.

References

1. Fridericiana: Z. Univ. Karlsruhe **41** (1988).
2. For an excellent catalog of this exhibit of Hertz's apparatus, see J. H. Bryant, *Heinrich Hertz: The Beginning of Microwaves*, IEEE, New York (1988).
3. The best English-language sources on Hertz's life are R. McCormmach, in *Dictionary of Scientific Biography*, C. C. Gillispie, ed., Scribner's, New York (1970–78); J. Hertz, M. Hertz, C. Susskind, eds., *Heinrich Hertz: Memoirs, Letters, Diaries*, San Francisco P., San Francisco (1977). This volume, arranged and edited by Hertz's two daughters, consists of facing pages in German and English, with an excellent biographical introduction by M. von Laue. On the influence of Helmholtz on Hertz's contributions to physics, see J. F. Mulligan, Am. J. Phys. **55**, 711 (1987).
4. H. Hertz, *Miscellaneous Papers*, Macmillan, London (1986), p. 273. On the importance of this paper see S. d'Agostino, Hist. Stud. Phys. Sci. **6**, 261 (1975).
5. H. Hertz, *Electric Waves*, Dover, New York (1962). This volume contains all of Hertz's experimental papers on electromagnetic radiation.
6. A. Sommerfeld, *Elektrodynamik*, Akademische Verlagsgesellschaft Geest und Portig KG, Leipzig (1954), p. 2.
7. H. Hertz, *Miscellaneous Papers*, Macmillan, London (1896), p. 313; reprinted with an introduction by E. C. Watson in Am. J. Phys. **25**, 335 (1957).
8. H. Hertz, *Miscellaneous Papers*, Macmillan, London (1896).
9. J. F. Mulligan, Am. J. Phys. **44**, 960 (1976).
10. C. H. Townes, in *Advances in Quantum Electronics*, J. R. Singer, ed., Columbia U. P., New York (1961), p. 5.
11. For a good photograph of this coaxial line and of the detector Hertz used inside this line, see J. H. Bryant, *Heinrich Hertz: The Beginning of Microwaves*, IEEE, New York (1988), p. 33.
12. *Nobel Lectures in Physics, 1901–1921*, Elsevier, New York (1967).
13. W. C. Roentgen, translated in H. A. Boorse, L. Motz, eds., *The World of the Atom*, Basic, New York (1966), p. 395.
14. M. von Laue, in *Heinrich Hertz: Memoirs, Letters, Diaries*, J. Hertz, M. Hertz, C. Susskind, eds., San Francisco P., San Francisco (1977), p. xxxv.
15. A. Einstein, quoted in E. Segrè, *From Falling Bodies to Radio Waves*, Freeman, New York (1984), p. 182.
16. L. Boltzmann, quoted in L. Koenigsberger, *Hermann von Helmholtz*, English edition, Dover, New York (1965), p. 421.
17. A. Einstein, "On the Examination of the State of the Aether in a Magnetic Field," unpublished 1895 manuscript referred to in A. Pais, *'Subtle is the Lord . . . ,'* Oxford U. P., New York (1982), p. 130.
18. A. Einstein, in *James Clerk Maxwell: A Commemoration Volume 1831–1931*, Cambridge U. P., Cambridge, England (1931), p. 70.
19. M. Planck, in *James Clerk Maxwell: A Commemoration Volume 1831–1931*, Cambridge U. P., Cambridge, England (1931), p. 62.
20. M. Planck, *Physikalische Abhandlungen und Vorträge*, vol. 3, F. Vieweg und Sohn, Braunschweig (1958), p. 288. ∎

100 YEARS OF PHOTOEMISSION

A century of scientific struggle has given us powerful
photoemission spectroscopies for probing electronic structure
and has brought us to the threshold of great advances
based on new, high-brightness sources of synchroton radiation.

Giorgio Margaritondo

PHYSICS TODAY/APRIL 1988

*In a series of experiments on the effects of resonance
between very rapid electric oscillations that I carried out
and recently published, two electric sparks were produced
by the same discharge of an induction coil, and therefore simultaneously. One of these, spark A, was the discharge
spark of the induction coil, and served to excite the primary
oscillation. The second, spark B, belonged to the induced
or secondary oscillation. I occasionally enclosed spark B in
a dark case so as to make observations more easily, and in
so doing I observed that the maximum spark length became
decidedly smaller inside the case than it was before.*

With these words, Heinrich Hertz announced in 1887 the
discovery of the photoelectric effect.[1] Although he realized that the phenomenon was important, he certainly
could not have imagined how fundamental its role in
physics was to be over the next 100 years. Even now,
many physicists do not completely understand that role.
For example, the effect is often, but incorrectly, credited
with leading Albert Einstein to the quantization of the
electromagnetic field. Photoemission experiments actually made their most important contributions to field
quantization *after* Einstein had formulated the theory, by
demonstrating its validity.

The photoelectric effect is crucial to today's science
and technology. Photoemission spectroscopy, for example,
is one of the most advanced and productive areas of
experimental physics, serving as a leading probe of the
electronic structure of atoms, molecules, solids and solid
interfaces. (See figure 1.) In this article I trace the
development of photoemission techniques from Hertz's
discovery to the present, when we stand at the threshold of
further great advances stimulated by the advent of new,
high-brightness sources of synchroton radiation. In reviewing this century of struggle with technical and
conceptual problems, I concentrate on landmark results
such as those of Hertz, Philipp Lenard, Einstein and Kai
Siegbahn. I should emphasize, however, that these
breakthroughs were made possible by a long series of
contributions by scientists who are often forgotten. The
centennial of the photoelectric effect is a good time to
celebrate all the scientists, major and minor alike, who
contributed to the development of modern photoemission.[2]

An accidental discovery

In 1879 the Berlin Academy of Sciences offered a prize for
research "to establish experimentally any relation
between electromagnetic forces and the dielectric polarization of insulators." Hermann Helmholtz called Hertz's
attention to this problem, and stimulated him to initiate
his historic experiments on the existence of James Clerk
Maxwell's electromagnetic waves. Hertz performed these
experiments at the Technische Hochschule in Karlsruhe,
where he was a professor of experimental physics. His
apparatus consisted basically of an oscillating circuit
containing a spark gap, which generated the waves, and a
second, suitably tuned circuit, which received them. Their
reception was established by observation of a spark across
a gap in the second circuit. As Hertz's words cited above

Giorgio Margaritondo is a professor of physics at the University of
Wisconsin, Madison, and is associate director of the university's Synchrotron Radiation Center.

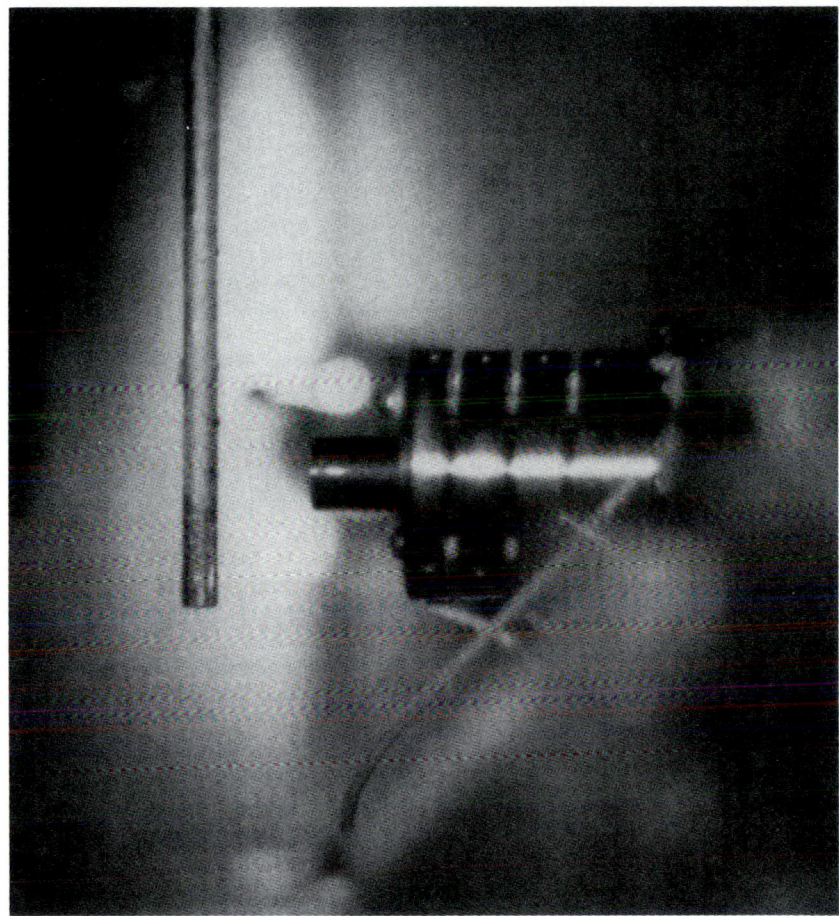

Photoelectron analyzer facing the first cleaved single crystal of the high-temperature superconductor $YBa_2Cu_3O_x$. The tiny specimen, grown by Patricia Morris and William Bonner, is mounted on the vertical rod, together with similar crystals. In this experiment at the Wisconsin Synchrotron Radiation Center, a Bellcore–Wisconsin team headed by Ned Stoffel, Marshall Onellion and the author probes the surface of the crystal using an angle-resolved photoelectron energy analyzer. Helen Farrell of Bellcore designed the photoemission system. **Figure 1**

indicate, the weakness of the induced spark prompted him to enclose the secondary spark gap in a dark case to make the observations easier. This, however, revealed an unexpected interaction between the two sparks. As we now know, the secondary spark was facilitated by the light-induced emission of electrons from the electrodes. The photoelectrons in the secondary spark gap were generated by ultraviolet photons emitted by the primary spark. The case absorbed the ultraviolet radiation, making it more difficult to produce the secondary spark.

These events annoyed Hertz because they interfered with his main line of research: "I had no intention of allowing this phenomenon to distract my attention from the main object I had in view, but it occurred in such a definite and perplexing way that I could not altogether neglect it."[1] Eventually, his appreciation of the importance of the new phenomenon prevailed, and he interrupted his main experiments to study the effect in detail. Figure 2 illustrates the experimental system that produced the best results. Two sparks were generated at the gaps d and f by the Ruhmkorff coils a and e, which were powered by the

same series of Bunsen cells, b, through the mercury circuit breaker c. Hertz studied the interference between the two sparks by placing a plate, p, between them.

Hertz used plates made of an impressive variety of materials: metals (both thick and thin sheets), paraffin, shellac, resin, ebonite, various kinds of glass, porcelain, earthenware, mica, agate, wood, pasteboard, paper and also, *ad abundantiam*, ivory, horn, animal hides and feathers. He performed his most significant experiments with crystals. Some of these, notably rock salt, had minimal effects on the interference between the sparks. Hertz's observations, corroborated by refraction experiments, led him to identify ultraviolet light as the immediate cause of the phenomenon: "After what has now been stated, it will be agreed that the light of the active spark must be regarded as the prime cause of the action. And if the observed phenomenon is an effect of light at all it must be solely an effect of the ultraviolet light."

Hertz's experiments could not identify electrons as the other main ingredient of the phenomenon. The problem, of course, was that the electron itself had not yet been discovered. Soon after Hertz's announcement, it became clear that ultraviolet radiation caused the emission of negatively charged particles from solids. In 1888 Wilhelm Hallwachs found that a negatively charged, insulated zinc plate would lose its charge when exposed to ultraviolet radiation, while a neutral plate would become positively charged.[2] In 1897 J. J. Thomson announced the discovery of the electron, and within two years he had demonstrated that the negative particles emitted in the photoelectric effects were electrons: "The following paper contains an account of measurements of m/e and e for the negative electrification discharged by ultra-violet light. The value of m/e in the case of ultra-violet light is the same as for the cathode rays."[3]

Lenard obtained the same result independently,[4] using the apparatus shown in figure 3. This apparatus is a precursor of the modern electron analyzers used for photoemission spectroscopy. Ultraviolet radiation produced by a spark bombarded a cathode, U, in an evacuated glass tube. The cathode could be biased with an external dc voltage supply. The anode E consisted of a screen at ground potential, with a small hole. The electrodes α and β, which were connected to electrometers, detected

photoelectrons passing through the hole in the anode. The trajectory of these electrons could be modified by inducing a magnetic field with a pair of Helmholtz coils, represented in the figure by the dotted circle. Without the field, electrode α detected the electrons. When the coils were operated, electrode β detected the electrons for certain combinations of cathode bias voltage and magnetic field strength. From these data, Lenard computed the electron charge-to-mass ratio e/m. By using various cathode bias voltages, he made another fundamental discovery: For each cathode there was a maximum value of the photoelectron kinetic energy. The intensity of the ultraviolet light did not affect this maximum kinetic energy, although it did determine the number of photoelectrons emitted per unit time. These results remained unexplained until Einstein's quantum theory of the photoelectric effect.

Discovery of the photon

In 1905 Einstein published an article titled "On a Heuristic Point of View about the Creation and Conversion of Light," in which he suggests that electromagnetic energy is quantized.[5] This article includes the quantum theory of the photoelectric effect, for which Einstein received the Nobel Prize in 1921. A common misconception is that Einstein derived the concept of the photon from the results of experiments on the photoelectric effect. This would not have been possible with the data available in 1905, although Einstein acknowledged that Lenard's experiments on photoemission were one of his inspirations: "The production of cathode rays by ultraviolet light can be better understood on the assumption that the energy of light is distributed discontinuously in space." Einstein's derivation of the concept of photons was actually based entirely on statistical mechanics.

The revolutionary character of this derivation cannot be overestimated. At the time, the overwhelming evidence appeared to favor Maxwell's theory of the electromagnetic field. Lenard's results on the photoelectric effect seemed not to be a serious challenge to this theory. Five years earlier, Max Planck had rather unwillingly initiated the quantum revolution, but he had stopped short of questioning Maxwell's picture of the radiation inside his version of the blackbody. Against this formidable *status quo*, Einstein proposed a simple but powerful argument. He considered the entropy in a volume V_1 due to radiation in a frequency interval ν to $\nu + d\nu$. He assumed that Wilhelm Wien's radiation law is valid in the spectral region containing this interval, and demonstrated that the change in entropy caused by an isothermal change in volume from V_1 to V_2 is $(k_B\, dE/h\nu) \ln(V_1/V_2)$, where dE is the total energy of the radiation in the frequency interval ν to $\nu + d\nu$. This relation is equivalent to the expression for an ideal gas as long as the number of particles in the gas is $dE/h\nu$. Thus the radiation itself consists of "particles" of energy $h\nu$.

In the last part of the article, Einstein used this

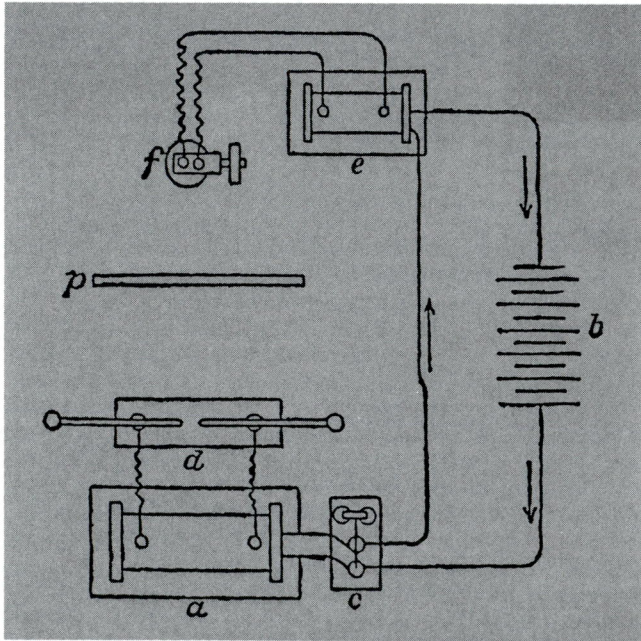

Schematic diagram of the experimental system that Heinrich Hertz used in 1887 to study the newly discovered photoelectric effect. (From H. Hertz, *Electric Waves*, McMillan, London, 1900.) **Figure 2**

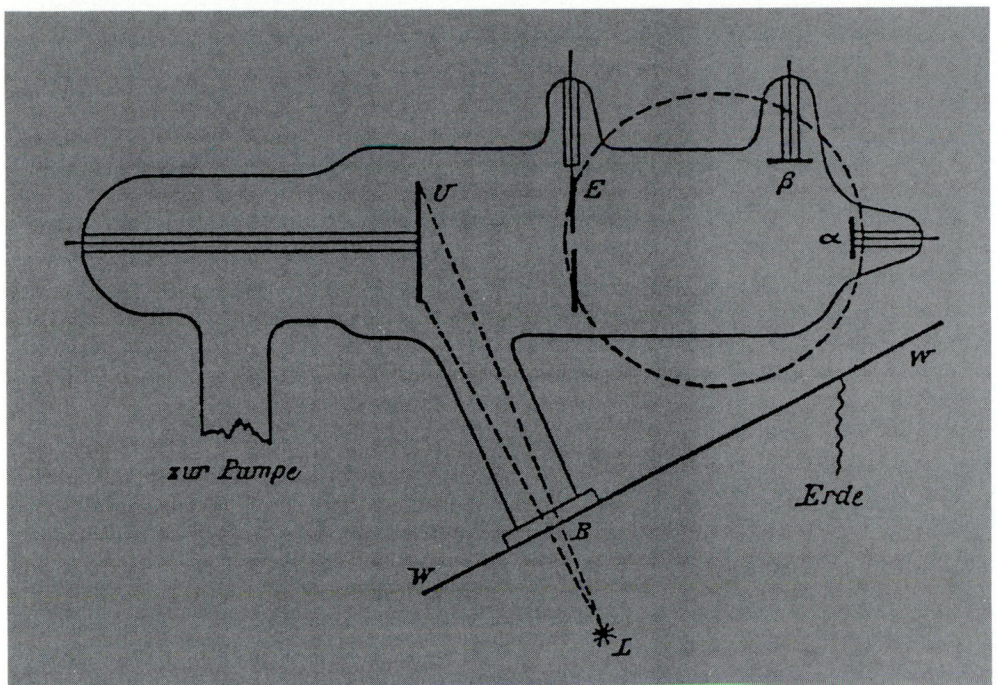

Apparatus used by Philipp Lenard to demonstrate that the particles emitted in a photoelectric process are electrons. The instrument shown in this diagram is a precursor of modern photoelectron spectrometers. (From P. Lenard, *Ann. Phys. (Leipzig)* **2**, 359, 1900.) **Figure 3**

elegant result to formulate the quantum theory of the photoelectric effect. In this theory he predicted, based on the transfer of energy from photons to electrons, that the energy of the photoelectron would be $h\nu - P$, where P is "the amount of work that the electron must produce on leaving the body." For photoelectrons emitted with the maximum kinetic energy from a metal, P coincides with the work function. Einstein's linear frequency law was consistent with Lenard's results, but could not be tested with the data available in 1905. The first experiments on the frequency effects after 1905 produced somewhat ambiguous results. Some experimenters claimed that the maximum photoelectron velocity, rather than the maximum energy, depended linearly on the frequency. This incorrect conclusion was the product of insufficient and inaccurate data.

In 1912 the classic experiments by Arthur Llewelyn Hughes and by Owen Williams Richardson and Karl Taylor Compton clearly demonstrated the validity of Einstein's frequency law.[6] Figure 4 shows some of Hughes's data. Experiments in 1916 by Robert A. Millikan made the evidence in favor of Einstein's model complete.[7] These 1912 and 1916 experiments were the most important contributions of the photoelectric effect to the development of quantum physics. They definitively established the quantization of the electromagnetic field. Together with Niels Bohr's theory of the hydrogen atom, published in 1913, these results made irreversible the revolutionary process started by Planck.

Photoemission spectroscopy evolves

In the late 1910s, three decades after the discovery of the photoelectric effect, conditions appeared to be good for the development of a new spectroscopy based on photoelectrons. A strong theory linked the energy distribution of photoelectrons to the ground state distribution of electrons in the emitting system. In 1914 H. Robinson

and W. F. Rawlinson began using x-ray lines to excite photoelectrons, opening the way for detailed analysis of the energy distribution of photoelectrons.[8] However, two problems delayed the actual birth of modern photoemission spectroscopy by more than three decades: poor energy resolution and insufficient quality of photoemitting surfaces.

The resolution was determined by the bandwidth of the photon source and by the resolution of the electron analyzer. Robinson and Rawlinson's use of x-ray lines improved the photon resolution, but their pioneering experiments, as well as those of Maurice de Broglie[9] and others, were severely limited by poor analyzer resolution.

Surface quality dramatically affects photoelectric emission in solids because the excited electrons have a very short mean free path. In a typical experiment, the kinetic energy of the photoelectrons ranges from a few electron volts to a few hundred electron volts. At these energies, electrons can travel only a few angstroms or a few tens of angstroms inside the solid before being inelastically scattered. Thus only electrons that absorb photons in a region very close to the surface can travel to the surface, escape and become photoelectrons. This situation is complicated by surface contamination, which dramatically affects phenomena in the thin escape region.

Starting in the early 1900s several scientists explored the short "escape depth" for photoelectrons. In 1919 Compton and L. W. Ross treated the experimental and theoretical problems encountered in measuring the escape depth.[10] However, their work had severe deficiencies. For example, it did not find evidence that the mean free path depends on the electron energy. It did, however, clearly establish that the escape depth is on the order of angstroms or tens of angstroms. Unfortunately, this fundamental fact was largely ignored in subsequent decades, and surface contamination rendered most of the corresponding data almost useless. The problem was

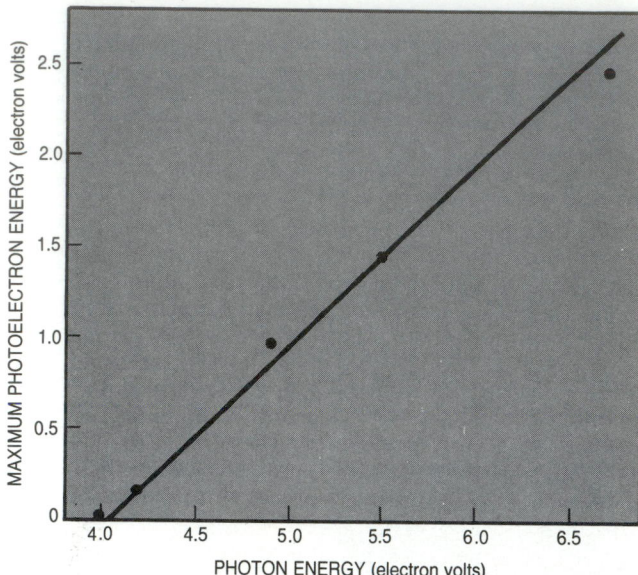

Maximum kinetic energy of electrons dislodged from cadmium, as a function of photon energy. These experimental data, taken by Arthur L. Hughes in 1912, demonstrated the linear relation predicted by Albert Einstein's model.[6]   **Figure 4**

solved completely only with the advent of commercial ultrahigh-vacuum systems capable of routinely delivering pressures of 10^{-10}–10^{-11} torr. At these pressures, a clean surface, prepared, for example, by cleaving a crystal, becomes contaminated only after several hours or days, giving one time to perform photoemission experiments on surfaces essentially free of contamination.

In the years between the two world wars, photoemission spectroscopy went into a deep depression as experimenters shifted their attention to the competing techniques of x-ray absorption spectroscopy and x-ray emission spectroscopy, which could achieve better resolution. Photoemission research did, however, produce fundamental results outside of spectroscopy. In 1923 Kenneth Kingdon and Irving Langmuir discovered that a cesium coating lowers the work functions of surfaces.[11] This result indirectly influenced the discovery of the first photocathode with a good quantum yield, the Ag–O–Cs photocathode.[12] The subsequent development of many kinds of efficient photocathodes had an important impact on optical research and technology. Another fundamental event of the 1920s was Arnold Sommerfeld's formulation of his theory of metals. In the context of this theory, the work function, which is the most fundamental parameter for the photoelectric effect, enters into several other phenomena such as thermionic emission and field emission.

Modern photoemission spectroscopy

In the 1950s and 1960s a series of experimental breakthroughs made photoemission spectroscopy a leading probe of the electronic structures of atoms, molecules and solids. The two branches of the field—x-ray photoemission spectroscopy and ultraviolet photoemission spectroscopy—are quite different and evolved rather independently. In XPS the photons have enough energy to extract photoelectrons from deep, atomic-like core levels. In UPS the photoelectrons are extracted mainly from valence states.

The development of XPS was primarily the result of an extensive experimental program at the University of Uppsala, Sweden, by Kai Siegbahn and his coworkers.[13] This work, which produced a second Nobel Prize both for photoemission and for the Siegbahn family, began in the early 1950s. Siegbahn had considerable experience in the design and construction of high-resolution electron spectrometers for β-ray spectroscopy. When he turned his attention to the photoelectrons created by x rays he estimated that the new instruments could achieve resolutions competitive with those obtained in x-ray emission and absorption spectroscopy. His first attempts, however, were frustrated by sample contamination. In the late 1950s Siegbahn described the solution to this problem in a personal letter to John G. Jenkin, R. C. G. Leckey and J. Liesegang, physicists at La Trobe University in Bundoora, Australia: "I recall that my students on the project and I had been unable to record any photoelectron spectrum at all from any sources we tried for several months, in spite of the fact that the instrument was running very well for radioactive sources using ThB. We then late one night tried a newly split NaCl crystal and suddenly we recorded our first photoelectron spectrum with extremely sharp lines and with the expected intensities."[2]

The "extremely sharp lines" are produced by the excitation of electrons from the core levels of the atoms in the solid. They are clearly visible in figure 5, which shows a spectrum for sodium chloride recorded by Siegbahn's group several years after the initial success. One could use these lines to analyze the chemical composition of the specimen. Furthermore, in the early 1960s Siegbahn and his coworkers demonstrated that one can use the energy positions of the lines to extract detailed information on the chemical bonds of each atomic species. The charge distribution of the valence electrons involved in bonding modifies the core energy levels with respect to the free-atom values, and therefore modifies the kinetic energies of the corresponding photoelectrons. Measurements of these "chemical shifts," which were already known in x-ray absorption and emission spectroscopy, immediately became a primary goal of x-ray photoemission spectroscopy. The excellent signal levels produced by Siegbahn's instruments made it possible to extend XPS to molecules in the gas phase. Three decades of research, initiated by Siegbahn's work, have transformed XPS into an exceedingly sophisticated technique. Virtually all aspects of the core-level photoemission process have been analyzed and exploited to extract information, including a variety of many-body effects and differences due to chemical shifts between bulk and surface atoms in solids.

The birth of UPS, like that of XPS, was prompted in part by advances in instrumentation. Important examples are the development of high-intensity ultraviolet lamps such as the helium line source, the use of windowless capillary connections between source and sample, and the construction of high-resolution electron energy analyzers based on electrostatic deflection. Unlike that of XPS, however, the development of UPS cannot be attributed to a single group. Early experiments revealed clear connections between the ultraviolet photoemission spectra and the band structures of solids. William Spicer's formulation of the three-step model provided a simple conceptual framework for the interpretation of the UPS curves.[14] This model divides the solid-state photoemission process into three independent steps: optical excitation, transport to the surface and emission into the vacuum. This approximation made it possible to analyze the many factors that affect the photoemission spectra, and emphasized the contribution of the ground state electronic structure, which dominates the first step. Similar work has established a clear connection between the ground state electronic structures of molecules in the gas phase and their UPS curves.

Several articles in PHYSICS TODAY have described the evolution of XPS and UPS in recent years.[15] Investigators can now control or scan virtually all the parameters relevant to a photoemission experiment: photon energy, direction and polarization; photoelectron energy, direction and spin polarization; and various characteristics of the sample. One gets better measurements as one controls and scans more variables—and this flexibility and effectiveness is itself the best measure of the status of modern XPS and UPS.

Spectacular advances in the quality of instrumentation over the past two decades are largely responsible for today's advanced state of photoemission spectroscopy. For example, the extensive use of efficient electron detectors such as the channeltron is crucial in modern angle-resolved photoemission experiments. The channel plate, an area-sensitive detector, has been used in sophisticated electrostatic analyzers that display the spatial distribution of the photoelectron intensity.

The major advance in instrumentation was the development of synchrotron radiation as a tunable, intense, polarized source of ultraviolet and x-ray photons. The complete control that synchrotron radiation gives the experimenter over the photon's parameters has had an impact on almost every area of photoemission spectroscopy. For example, by tuning the photon energy one can control the escape depth of the photoelectrons and distinguish between the contributions from bulk and surface electronic states. Experimenters studying surface chemical and physical phenomena such as corrosion, catalysis and the formation of interfaces make extensive use of this technique.[15] The intensity and brightness of synchrotron radiation are important when one studies the angular distribution of photoelectrons with small-area detection schemes. An example is the use of angle-resolved photoemission to measure the energies and **k** vectors of valence electrons in crystalline solids and to map their band structures. Figure 6 shows the results of an early application of this technique.[16] Modern XPS and

UPS techniques go beyond the measurement of electron energies, the classic domain of photoemission, to explore the states of the electrons—that is, their wavefunctions—giving us a complete understanding of the electronic structure.

The future

One of the most amazing aspects of this centennial of photoemission is that the field is at the threshold of another rejuvenation. The construction of a new generation of synchrotron radiation sources based on magnetic undulators is now under way. These sources will have unprecedented intensities and brightnesses. The undulators at the Advanced Light Source at Berkeley and at the low-emittance ring at Trieste, for example, will be orders of magnitude brighter than sources at any existing soft-x-ray facility. These new sources will be a boon to those photoemission techniques that are now severely restricted by low signal levels. Furthermore, they will make possible novel photoemission experiments that are not feasible at present signal levels.

The study of the spin polarization of photoelectrons is one area of research that is now exceedingly difficult due to low signal levels. Measurements of the spin polarization of photoelectrons by Maurice Campagna and his coworkers at KFA (Jülich, West Germany) and by other

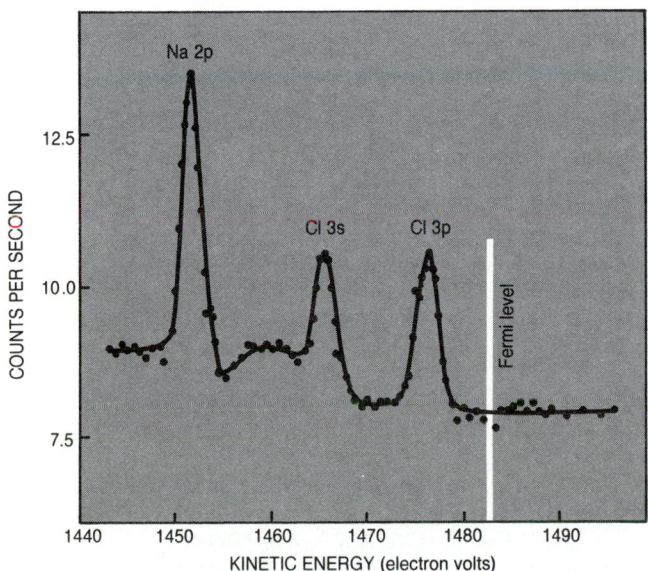

X-ray photoemission spectrum of sodium chloride recorded at the University of Uppsala by Kai Siegbahn and his coworkers. The spectrum exhibits sharp lines caused by the excitation of photoelectrons from atomic-like core levels. **Figure 5**

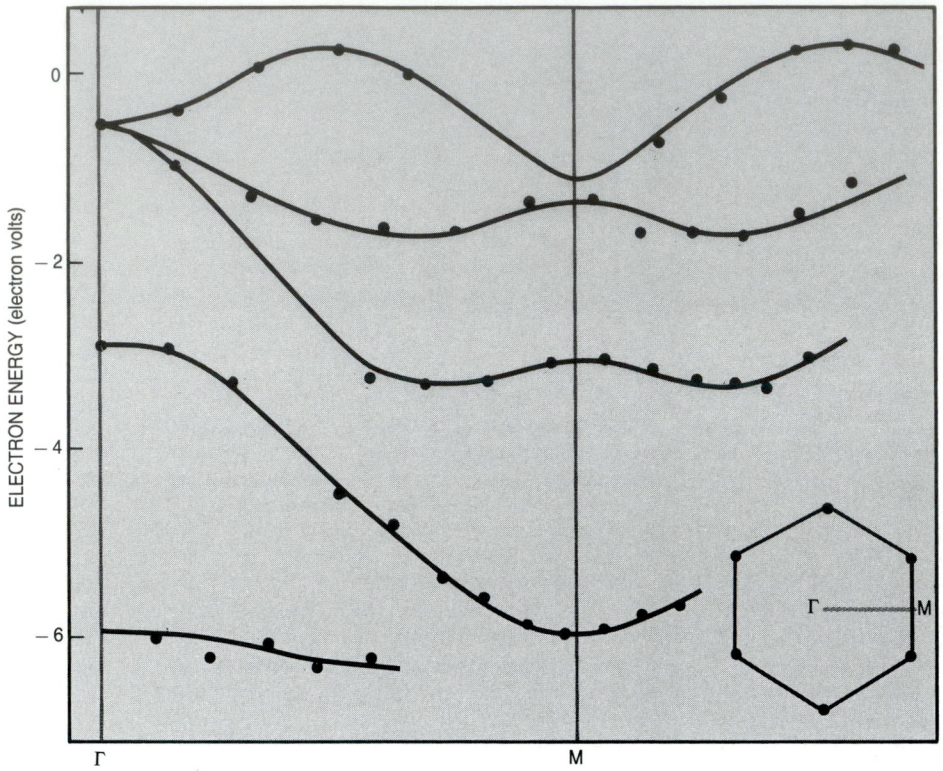

ELECTRON MOMENTUM **k**

Band structure of gallium selenide as measured by angle-resolved photoemission. Since the mid-1970s, photoemission data have yielded plots of electron energy as a function of the **k** vector. This figure shows one of the first such plots, obtained in 1976 by Bell Laboratories scientists at the Wisconsin Synchrotron Radiation Center. The band structure is mapped for the Γ–M direction in the Brillouin zone, as shown in the inset.[16]   **Figure 6**

groups have already yielded fundamental information on the magnetic properties of solids.

Photoemission microscopy and experiments resolved in time are just two novel photoemission techniques that stand to benefit from the new sources. Past attempts to implement photoemission microscopy had to trade off energy resolution against space resolution. The new sources will make it possible to achieve good space and energy resolution at the same time. Instrumentation based on multilayer coatings with high reflectivities for x rays will have energy resolutions on the order of a tenth of an electron volt and spatial resolutions on the order of a few hundred angstroms. These instruments will produce micrographs that show the chemical situation of each element in a specimen—information that is of crucial importance in molecular biology, microelectronics and other areas. The excitement generated by these research opportunities is the best celebration of Hertz's discovery, and the best way to begin the second century of photoemission research.

References

1. H. Hertz, Ann. Phys. (Leipzig) **31**, 983 (1887). Translated in H. Hertz, *Electric Waves*, McMillan, London (1900).

2. Interesting historical reviews of the photoelectric effect, including many references to lesser known contributors, can be found in the following publications: F. K. Richtmyer, E. H. Kennard, T. Lauritsen, *Introduction to Modern Physics*, McGraw–Hill–Kogakusha, New York (1955); M. Cardona, L. Ley, *Photoemission in Solids*, vol. I, Springer-Verlag, Berlin (1978); J. G. Jenkin, R. C. G. Leckey, J. Liesegang, J. Electron Spectrosc. **12**, 1 (1977).

3. J. J. Thomson, Philos. Mag. **48**, 547 (1899).

4. P. Lenard, Ann. Phys. (Leipzig) **2**, 359 (1900); **8**, 149 (1902). P. Lenard, Wien. Ber. **108**, 1649 (1899).

5. A. Einstein, Ann. Phys. (Leipzig) **17**, 132 (1905).

6. A. L. Hughes, Philos. Trans. R. Soc. London, Ser. A **212**, 205 (1912). O. W. Richardson, K. T. Compton, Philos. Mag. **24**, 575 (1912).

7. R. A. Millikan, Phys. Rev. **7**, 355 (1916).

8. H. Robinson, W. F. Rawlinson, Philos. Mag. **28**, 277 (1914).

9. M. de Broglie, C. R. Acad. Sci. **172**, 274 (1921).

10. K. T. Compton, L. W. Ross, Phys. Rev. **13**, 374 (1919).

11. K. H. Kingdon, I. Langmuir, Phys. Rev. **21**, 380 (1923).

12. L. R. Koller, Phys. Rev. B **36**, 1639 (1930). N. R. Campbell, Philos. Mag. **12**, 173 (1931).

13. K. Siegbahn, C. Nordling, A. Fahlman, R. Nordberg, K. Hamrin, J. Hedman, G. Johansson, T. Bergmark, S.-E. Karlsson, I. Lindgrenf, B. Lindberg, *ESCA: Atomic, Molecular and Solid State Structure Studied by Means of Electron Spectroscopy*, Almqvist & Wiksells, Uppsala (1967). K. Siegbahn, C. Nordling, G. Johansson, J. Hedman, P. F. Heden, K. Hamrin, U. Gelius, T. Bergmark, L. O. Werme, R. Manne, Y. Baer, *ESCA Applied to Free Molecules*, North Holland, Amsterdam (1969).

14. See, for example, W. E. Spicer, Phys. Rev. **112**, 114 (1968).

15. F. J. Himpsel, N. V. Smith, PHYSICS TODAY, December 1985, p. 60. J. H. Weaver, PHYSICS TODAY, January 1986, p. 24. R. S. Bauer, G. Margaritondo, PHYSICS TODAY, January 1987, p. 26.

16. P. K. Larsen, G. Margaritondo, J. E. Rowe, M. Schluter, N. V. Smith, Phys. Lett. A **58**, 623 (1976).

Michelson and measurement

A boy who grew up in Western mining camps became the preeminent master of precision measurement by light waves, largely through his playful work with interferometers and his quest to know the 'innards' of his instruments.

Loyd S. Swenson Jr

PHYSICS TODAY/MAY 1987

Measurement, as much as if not more than mathematics, is the essence of modern natural philosophy and artful invention. To measure is to begin to understand and to manipulate. Yet modern scientific instrumentation is so complex and so thoroughly prepackaged that it is difficult to know enough about what goes on inside the "black boxes" that we use to measure. Only a hundred years ago it was still possible for some people to at least appreciate and for a few to manipulate nearly every instrument in specialized usage. One such person in physical optics was Albert Abraham Michelson, a US naval officer who became a world-renowned experimenter and preeminent master of measurement by light waves largely through his playful work with interferometers and interferometry. Measurement was his mistress and muse, and by virtue of his quest to understand the "innards" of all his instruments, Michelson's name became practically synonymous with extremely precise measurement.

During the half-century from 1880 to 1930, Michelson perfected his skills as an optical scientist and gained the reputation of being one of the best experimental manipulators of optical apparatus in the world. He did this largely by designing and developing ever better, more precise and more accurate instruments and techniques for measuring extremely small and extremely large lengths, widths, angles

and distances. Central to his quest to understand visible nature was a device, eventually called the "interferometer," that he constantly worked at improving or adapting to novel uses. He did not create this device *de novo*, nor did he patent or market it after its principles of operation came to bear his name.

Yet Michelson's interferometer, first custom-made in Berlin in 1880 as a one-of-a-kind device by the scientific instrument makers Schmidt und Haensch, became a standard laboratory and "high tech" industrial tool by the time of the First World War. It had already been used in various permutations for a host of applications, as Michelson himself listed in 1902: measuring indices of refraction, thicknesses of soap films, coefficients of expansion and the gravitational constant; testing screws for uniformity of pitch; measuring light waves themselves; analyzing spectral lines; determining standards of length; testing the Zeeman effect; assisting astronomical measurements; and checking astronomical aberration.

These and other uses of Michelson's interferometer are far less famous than its supposedly central role in intellectual history as the primary instrument that "disproved" the existence of a luminiferous, or light-bearing, ether. This most celebrated story is often garbled, and many still think, mistakenly, that the Michelson-Morley experiment directly inspired young Albert Einstein to declare in 1905 that the electromagnetic ether was superfluous and to invent the first special, or restricted, theory of

relativity. (See the article by John Stachel on page 209.)

Spectacular as the interaction between experimental optics and eletromagnetic theory building was, the mainline contributions of Michelson and his interferometric methods to classical optics far outstripped in both breadth and depth the indirect impact his work had on the scientific revolutions of the 20th century. Here it is important to emphasize that at the turn of the century invisible lights of all sorts were being extended or discovered—ultraviolets, infrareds, radio waves, x rays, emissions from radioactive materials and even microwaves. In all these discoveries interferometric methods sooner or later played a part. The electromagnetic spectrum of radiant energy transfer was being filled. On 10 December 1907 the Swedish Royal Academy of Sciences awarded Michelson its seventh Nobel Prize in Physics—the first ever to be given to an American in science. It was no accident and it was quite proper at the time that the citation read, "for his precision optical instruments and the spectroscopic and metrological investigations conducted therewith." Michelson had arrived as a world-renowned experimental physicist well before Max Planck and Einstein made theoretical physics a profession.

This article will attempt a cameo portrait of Michelson's life and work, trying to show historically how he conceived and designed an instrument that should rank with the telescope, microscope, barometer and thermom-

Loyd Swenson Jr is a professor of history at the University of Houston, in Houston, Texas.

Albert Abraham Michelson (1852–1931) sketching himself while a cadet midshipman at the US Naval Academy in Annapolis, Maryland. Michelson was a member of the class of 1873. (Michelson Collection, Nimitz Library, US Naval Academy; courtesy of AIP Niels Bohr Library.)

eter in the pantheon of scientific tools. My focus is on Michelson's role in creating an apparatus, then a series of devices, through which to see and measure aspects of the cosmos never so well seen before. My purpose is to capture the essential features, somewhat exaggerated because of the need for economy, of Michelson's interferometry in words that epitomize his achievements.

Optics before Michelson

Michelson was born in Strzelno, Prussian Poland, on 19 December 1852, and was brought by his Jewish merchant parents to New York City late in 1855; from there the family traveled onward via Panama to San Francisco in 1856. While the Civil War was raging back East, young Albert spent his boyhood days in California and Nevada mining camps and grew into a promising student. After President Lincoln's assassination Albert assumed the middle name of Abraham. From 1866 to 1869 he attended Boys' High School in San Francisco, where the principal, Theodore Bradley, took care to raise the promising boy into an accomplished youth with "great aptitude for scientific pursuits." After a summer spent back home in contests for an appointment to the US Naval Academy from the territory of Nevada, the ambitious young man boarded one of the first West-to-East transcontinental railroad trains to plead his case in person at the White House. Amazingly, he succeeded in getting an overquota appointment to the academy from President Ulysses S. Grant. Thus, Michelson became a midshipman, persevered and graduated from Annapolis in 1873, ninth in rank in a class of 29 but first in optics and second in other physical science subjects.

After a furlough and then two years at sea under sail and with auxiliary steam, Ensign Michelson was assigned instructional duties in physics and chemistry back at the Naval Academy. There, under the direction of Commander William T. Sampson, Michelson began his career in optics with a lecture demonstration in which he tried to determine the velocity of light using a rotating mirror, Foucault's method. Meanwhile, he had married well—Sampson's wife's niece, Margaret Heminway—and his first son had been born. During the academic year 1877–78 Michelson studied the published accounts of the acknowledged French leaders in optical research. Soon he found several ways to improve upon their efforts. His frontier boyhood, his schooling in San Francisco, his Naval Academy training and his maritime experiences were beginning to pay dividends.

At mid-century the lingering Newtonian–Huyghensian controversy over the ultimate nature of light had seemingly been settled in favor of the undulatory, vibratory, wave or ether theory rather than the corpuscular, ballistic, particle or emission theory. The contest appeared dramatically resolved by overwhelming experimental evidence before Michelson got involved. There were many reasons for this, not least of them the fascinating cooperation, then competition, between Armand Hippolyte Louis Fizeau and Jean Bernard Léon Foucault to be first to measure the velocity of light, then to prove that its velocity in water is less than in air—which result would and did favor waves over particles.

Thomas Young in London and Augustin Jean Fresnel in Paris had done much to advance the wave theory of light through studies of interference and transverse wave motion, respectively. Jean Baptiste Joseph Fourier and Dominique-François-Jean Arago likewise had been influential in clarifying conduction, radiant heat and polarization phenomena. Josef Fraunhofer, by developing spectroscopy in the 1820s; Michael Faraday, by developing the basic ideas of electromagnetic field theory in the 1840s; and Charles Wheatstone, by developing experimental acoustics, optics and electrical measuring devices in the 1850s, also helped set the stage. When James Clerk Maxwell's grand synthesis of light, electricity and magnetism finally appeared in treatise form in 1873, it failed to gain immediate and widespread acceptance, although the next 15 years of experimental work proved its majestic significance.

Most of these scientific heroes in Michelson's chosen field were dead by the time he appeared to take up their challenges. He admired the combined genius of Wheatstone, Arago, Foucault and Fizeau above all. They were the kind of experimenters that he most trusted. Another generation of leaders in his profession, only a few years older

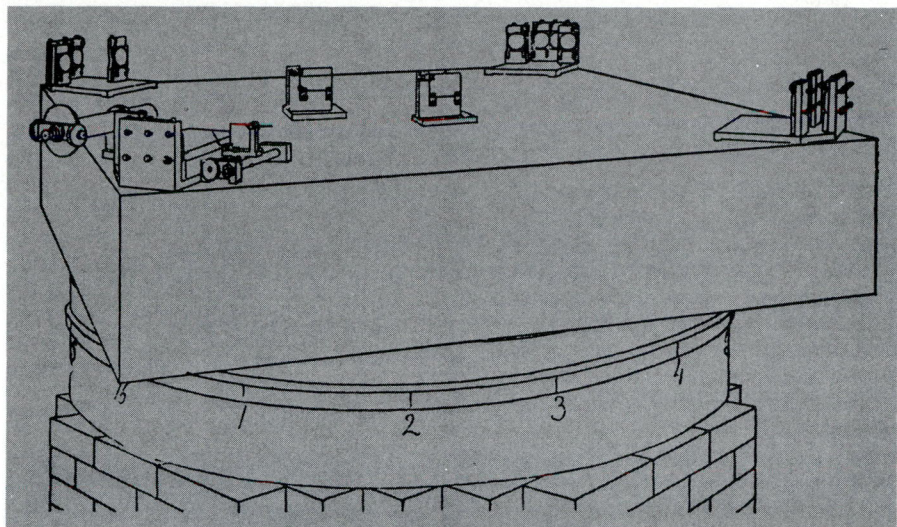

than Michelson, were impressed by his potential. Simon Newcomb, for example, the Canadian-American professor of mathematics and astronomy at the US Naval Observatory, and Hermann von Helmholtz, the German physician turned physiologist and physicist, were both solicitous and encouraging mentors for young Michelson. Maxwell had just died, but William Thomson (later Lord Kelvin) and Lord Rayleigh were lively supporters of Michelson's plans for the uses of light-wave measurements.

During his 1880–82 postgraduate studies in Europe, Michelson met experts who were most helpful because their work was more specifically spectroscopic. In Berlin he met Gustav R. Kirchoff and Hermann Karl Vogel; in Heidelberg, Robert W. Bunsen and Georg H. Quincke; and in Paris, Jules C. Jamin, Éleuthère-Élie-Nicolas Mascart, M. Alfred Cornu, Alfred Potier and Gabriel Lippmann, who would win the next Nobel Prize for Physics after Michelson's. Hendrik A. Lorentz in Leiden became Michelson's chief friendly critic.

Developing the interferometer

Having found several relatively cheap and facile ways to improve on Foucault's revolving mirror for the determination of the velocity of light, Michelson confided to his patron Newcomb a plan for answering Maxwell's despair over ever finding a way to measure second-order effects—effects on the order of 1 in 100 000 000, for example, the square of the ratio of the velocity v of the Earth in its orbit to the velocity V of light. (Only later was the upper case symbol V for the velocity of light replaced by the lower case c.)

Already in the late summer of 1880 Michelson must have conceived his hope of comparing the paths of light at right angles as a method for finding the

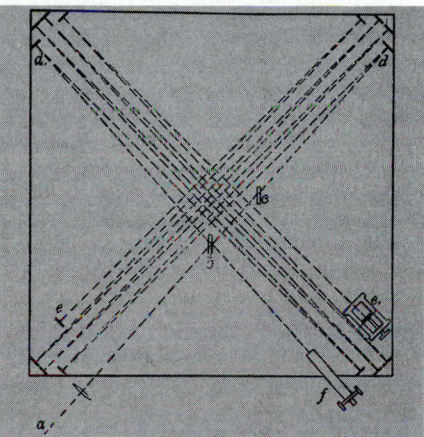

Ether drift apparatus and light path. The optical bench is atop a massive stone slab that floats on mercury (top drawing). The multiple mirror configuration (bottom drawing) increases the path length by a factor of about ten. (From Michelson's 1903 book, *Light Waves and Their Uses*.)

motion of the Earth relative to the ether, for he wrote to Newcomb from Berlin on 22 November 1880 saying that Helmholtz "could see no objection to it, except the difficulty of keeping a constant temperature." That objection, it turned out in the long run over the next half-century, was to be a major one indeed. Throughout the few years (1883–89) that Michelson collaborated with his neighboring chemist Edward W. Morley, temperature control plagued their experiments. And until the end of Dayton C. Miller's complementary work in the 1930s, temperature control was a perpetual problem in interferometry.

Serendipity was never more evident in experimental science than here, for Michelson was in the process of discovering, at least for himself, that one could convert any microscope or any telescope into an interferometer simply

Ruling engine for making diffraction gratings. Michelson built the device at the University of Chicago. (Courtesy of AIP Niels Bohr Library.)

by using only diametrically opposed portions of its lenses. After the disappointments of the ether drift tests in 1881 and 1887, Michelson created at least two dozen different designs for interferometers to illustrate "Light waves and their uses." That was the title of his Lowell Lectures at Harvard in 1899 and of his first book, published by the University of Chicago Press in 1903. Thus Michelson turned his lemon into lemonade.

By 1930 it was common to explain all interferometers either as devices for dividing wavefronts or separating different frequencies from point or line sources, or as instruments for dividing amplitudes by partial reflection from two or more multiple-path beams. Michelson's original interferometer of 1881 was considered the exemplar of the latter category. His stellar interferometer, conceived in 1890 and applied to satellites of Jupiter (but not to stars until the early 1920s), was considered an exemplar of the former category. The technical heart of most earlier amplitude dividers was some sort of beam splitter, usually a half-silvered mirror. Fizeau, Jamin and Mascart had been the pioneer inventors of such devices. Then in the 1890s Michelson, Rayleigh, Ernst Mach, Otto R. Lummer, J.-M. Rene Benoit, and Ernst Pringsheim, among others, picked up the pace of development.

At first Michelson called his inven-

tion an "interferential refractometer," in honor of Jamin's apparatus for measuring indices of refraction of various gases and liquids. However, when he began to think of using interference methods to measure the standard meter bar in Paris, he called his modified device an "interferential comparator." As its versatility increased and became more obvious, others began to call it simply the Michelson interferometer. Although Michelson himself had too many versions in mind and in operation to glorify any one of them, he did admit that one clone, commercially produced by Hilger and Watts, had "proved most generally useful." Later modifications by others produced the Fabry–Perot interferometer after 1896 and the Lummer–Gehrcke multiple-beam interferometer after 1903. The Twyman–Green (1919), Meggers–Peters (1918) and Gehrcke–Lau (1927) modifications later served as optical machines to help make other, better, more precise optical machines.

Precision length measurements

Perhaps the most significant achievement of Michelson's interferometry in the context of his times was not the ether drift tests but rather the measurement of the platinum–iridium standard meter bar in terms of immaterial light waves. Since at least 1827, scientists had realized that a wavelength definition of the meter (suppo-

sedly one forty-millionth of the Earth's circumference) would be preferable to defining it as the distance between two lines inscribed on a bar made of an alloy and housed in some bureau somewhere. Not only would a wavelength definition for the metric system be far more precise, it might be inherently free from secular variations with time and it would be easily available in any laboratory anywhere. If the fundamental standard were lost or destroyed, Michelson's method could replace it with duplicates, he bragged, "which could not be distinguished from the originals."

Morley and Michelson in 1887 and 1889 published essays on the feasibility of establishing a light wave as the "ultimate standard of length," and then Michelson, with other assistants, proceeded to do the job. Using his basic interferometric scheme of counting fringes past a fiducial mark in his eyepiece, Michelson was able to devise a decimal series of intermediate standards, "etalons," that allowed him to perform this delicate and difficult task. He eventually settled on the homogeneous red radiation of cadmium vapor as his source (vacuum tubes had just been invented!). He discovered some ingenious ways of arranging his beam splitter, mirrors, mirror carriages, spectroscope and microscope, and created an elaborate apparatus for temperature control and internal manipulation. Michelson and Benjamin A. Gould used American instrument makers to fabricate many of the accessories for this instrument. The final result? "The number of light waves in a standard meter was found to be, for the red radiation of cadmium 1 553 163.5, for the green 1 966 249.7, for the blue 2 083 372.1—all in air at 15 °C, and at normal atmospheric pressure."

This value, checked and rechecked through the years, stood the test of time until 1960, when a spectral line of krypton was chosen to replace the cadmium red. No wonder the world of science and technology was so impressed by Michelson's expertise.

Why go to all this trouble when the most extreme variation of the standard meter was known to be only about one-thousandth of a millimeter in any of its copies? Michelson wrote in 1899:

The answer is that the requirements of scientific measurement are growing more and more rigorous every year. A hundred years ago a measurement made to within one thousandth of an inch was considered rather phenomenal. Now it is one of the modern requirements in the most accurate machine work. At present a few measurements are relied upon to within one ten-thousandth of an inch. There are cases in which an accuracy of one-millionth of an inch has to be attained and it is even possible to detect differences of one five-millionth of an inch. Past experience indicates that we are merely anticipating the requirements of the not-too-distant future in producing means for the determination of such small quantities.

Analogy with sound

Michelson was a violinist, Morley was a pianist and Miller was a flautist—all amateurs to be sure, but musicians nonetheless, who thought continuously of sound and light wave motions. The musical talents of these three protagonists of the classical ether drift experiments were considerable, and their commitment to acoustics as well as optics was formidable. Thinking analagously about waves (even though these are longitudinal for sound and transverse for light), could hardly have been avoided without split-brain surgery. Acoustical interference studies, especially researches on reverberation in architectural acoustics, and the beginnings of anechoic chambers for exploring noise, "beats" in music and sound shadows must have had much influence on the trio.

In fact, Michelson began his first Lowell Lecture in 1899, "Wave motion and interference," by predicting that someday soon there would be "a *color music*, in which the performer, seated before a literally chromatic scale, can play the colors of the spectrum in any succession or combination . . . producing at will the most delicate and subtle modulations of light and color, or the most gorgeous and startling contrasts and color chords." Michelson thus foresaw the gigantic sound and light shows that are so popular now throughout Europe, Asia and the United States, especially at historical monuments.

Acoustical interferometers appear to have been used before optical ones were common, as evidenced by the apparatuses of Karl Koenig and Georg Hermann Quincke, but apparently they were not so-called.

Null result

Michelson's interferometry grew out of his embarrassment in failing to find "the relative motion of the Earth and the luminiferous ether." With his 1881 experiment he hoped to discover the resultant velocity of all the Earth's motions through the ether of space. Its null results he interpreted as falsifying the hypothesis of a stationary ether.

After Michelson became a civilian professor of physics at the Case School of Applied Science in Cleveland and began his collaboration with the older chemist Morley at Western Reserve University, the two in 1886 corroborated Fizeau's work on Fresnel's drag coefficient and published "Influence of motion on the velocity of light." They reported that light traveling within a stream of flowing water moves fractionally faster (by $7/16$) with the current than outside it or against it. Conclusion? Oddly enough, "that the luminiferous ether is entirely unaffected by the motion of the matter which it permeates."

Then in the classic Michelson–Morley paper of 1887, "On the relative motion of the Earth and the luminiferous ether," they set a more modest goal for their improved ether drift experiment. Couching their argument in terms of corpuscular versus undulatory theories of astronomical aberration, their report begins by stating the rationale for testing the hypothesis of a stationary ether through which the Earth moves. Seeking only Earth's orbital velocity (its speed and direction), they designed the apparatus to increase the path length of the cross of light to about ten times its 1881 value. The influences of Potier and Lorentz were here evident. Null results, once again presented in tabular and graphic form, reinforced the fame of their failure, within significant figures. They had expected a change corresponding to about 0.4 wavelength. They observed a change of less than 0.04 wavelength. Less than 10% of their expectation meant a *conclusive* null result. Yet in their supplement to that classic paper, Michelson and Morley offered at least seven ideas—four possibilities for laboratories and three for observatories—for attacking all over again the problem of the motion of the whole Solar System through space!

In his final Lowell Lecture, "The ether," Michelson confessed:

> The experiment is to me historically interesting, because it was for the solution of this problem that the interferometer was devised. I think it will be admitted that the problem, by leading to the invention of the interferometer, more than compensated for the fact that this particular experiment gave a negative result.

Into the 20th century

In 1889 Michelson split from Morley as he left Cleveland for the first chair in physics at the brand new Clark University in Worcester, Massachusetts. Before the break, however, he and Morley had discussed several new applications for interferometry that should give positively stunning results. One was an experiment to measure the angular diameter of heavenly bodies; Michelson tested the feasibility of such a measurement by focusing on a moon of Jupiter.

Michelson was again lured away to a new home when the University of Chicago was established along with the Columbian Exposition in 1892-93 (which marked the 400th year since 1492). Incredibly busy organizing a new physics department and supervising the completion of the new Ryerson Physics Laboratory, Michelson nonetheless began to build an engine to rule the tiny grooves for diffraction gratings. He also built and tested a large vertical interferometer beside his building, and he laid plans for the construction of two more major inventions: an echelon spectroscope and, with Samuel Wesley Stratton, a large mechanical harmonic analyzer. All this plus personnel and personal matters put a strain on Michelson's marriage, and a painful divorce ensued in 1898. (See Albert Moyer's article on page 202.) The very next year he delivered his series of eight Lowell Lectures at Harvard, was elected president of The American Physical Society and remarried, this time to a socialite Chicago student 28 years of age named Edna Stanton. Michelson's first family became lost in the shuffle as he started his second. While his science meant most to him, it clearly was not an all-consuming passion.

Michelson's professional life, unlike his personal life, appears not to have slowed down after the turn of the century. He was long obsessed by the effort to get his ruling engine for the production of big diffraction gratings working better than Henry Augustus Rowland's engine at Johns Hopkins University. He failed in this. Michelson also became involved—somewhat reluctantly and marginally, but inevitably—in science policy issues, local, regional and national. As his honors accumulated and "little science" grew into bigger if not yet "big" science, Michelson was forced to change his style a bit—in teaching, in research and in service. Demonstrations, grantsmanship and graduate students all demanded more time. He answered the call of duty back to the Navy for optical research on rangefinders in World War I. He encouraged younger colleagues such as George Ellery Hale in their work in astrophysics and in the entrepreneurship required to build bigger telescopes. He ordered Robert A. Millikan to take over all his graduate student duties but cheered him on in his oil-drop experiments on the quantization of charge and in his mobilization

Stellar interferometer that Michelson used in 1920 to make the first measurement of a star's diameter. (Courtesy of William Osborn, Case Western Reserve University.)

efforts for the Great War. In a bid to win the directorship of the newly established National Bureau of Standards, Michelson asked Stratton in 1902 to plead his case in Washington, DC. But Stratton himself got the position. Michelson also supported Henry Gordon Gale's big 1914 project to use interferometry to measure the rigidity of the Earth and thus to "smell oil."

Immediately after the war Michelson began the project of using his stellar interferometer to measure the angular diameter of the red giant star Betelgeuse in the shoulder of Orion. Results were formally announced in December 1920: Betelgeuse must be almost as large as the orbit of Mars. Newspaper publicity on that discovery rivaled the noise from Arthur Eddington's 1919 eclipse expeditions to test Einstein's general relativity and prove the bending of starlight by the curvature of space.

In the mid-1920s Michelson supervised several major projects: a new measurement of the velocity of light between two mountains 22 miles apart above Los Angeles; an attempt at a large field in Clearing, Illinois, to measure the effect of the Earth's rota-

tion on the velocity of light; completing original designs for ether drift tests in ongoing attempts to respond to the challenges of Miller's experiments; and as president of the National Academy of Sciences in 1923, to carry through a major fund drive that he inaugurated. In 1927, the year before he retired from the University of Chicago and moved permanently to Pasadena, California, Michelson published his second and last book, *Studies in Optics*. Therein, he summed up his life's work, explained his conservative attitude toward the ether (an attitude he held in spite of the recent achievements of relativity and quantum mechanical theories) and tried to advise physical scientists to pay more attention to better measurements than to mathematical physics.

Michelson, the master of measurement through interferometry, was a rival to Rowland for domination of science in America in the last decade of the 19th century. By the second decade of the 20th century Einstein had appeared to assume that mantle worldwide. But Einstein's theories could not have been accepted without Michelson's measurements.

When Michelson became the first American Nobel laureate in science (President Theodore Roosevelt had won the Nobel Peace Prize the year before), it marked the moment in history that world-class status in scientific research began to be given to the United States. Michelson's inventions and innovations, as well as the technologies and techniques that ensued from his work, were always of such fundamental importance to the progress of physical science "toward the next decimal place" that it is quite fitting to call his flourishing years the Michelson era in American science.

Michelson had important precursors, colleagues, competitors and successors, but he was preeminent, as his daughter's biography, *The Master of Light*, portrays him. Over the years, as the Nobel Prize matured in prestige and remuneration, Michelson, a mere professor but a militant perfectionist, also matured. He played the perfect role model of a meticulous authority and yet he remained a continuously productive artist in physics. Until the day he died, 9 May 1931, Michelson pursued the elusive goal of measuring the velocity of light, perhaps the most funda-

Michelson in his laboratory at the University of Chicago sometime in the period 1900–10. (Courtesy of AIP Niels Bohr Library.)

mental constant in nature, under even better conditions and to ever higher degrees of accuracy. He was indeed, as many of his heirs have said, a virtuoso without a peer in measurement. Perhaps his best judgment of that fundamental constant was his 1926 value, $299\,796 \pm 4$ km/sec.

Michelson's last book was published the year before the Optical Society of America dedicated its annual meeting to him on the 50th anniversary of the beginning of his scientific career. Michelson had used, as he titled one of his last papers, "Light waves as measuring rods for sounding the infinite and the infinitesimal." When he died he was hardly less a believer in the wave theory of light and its concomitant ether. Although he acquiesced in Einstein's relativty theories (with a few reservations), he died confident in the knowledge that he had indeed sounded the nature of light and measured its

velocity as precisely as was then possible.

* * *

I thank my colleagues James H. Cooke and D. T. McAllister and my student David Payne for their insightful critiques.

Bibliography

- The centennial of the Michelson–Morley experiments properly begins with the centennial of the first Michelson interferometer, which was celebrated in Potsdam and Caputh, East Germany, in 1981. The proceedings appear in *Astron. Nachr.* **303** (1982); my contribution begins on p. 39.
- Michelson's two books are *Light Waves and Their Uses* (U. of Chicago P., Chicago, 1903) and *Studies in Optics* (U. of Chicago P., Chicago, 1927). The second book appeared the peak year of the quantum mechanical revolution.
- Other works on interferometry that I recommend are S. Tolansky, *Interference Microscopy for the Biologist* (Charles C. Thomas, Springfield, Ill., 1968); A. C. Candler, *Modern Interferometers* (Hilger

and Watts, London, 1951); W. E. Williams, *Application of Interferometry* (Wiley, New York, 1948); and W. H. Steel, *Interferometry*, 2nd ed. (Cambridge U. P., New York, 1984). See also D. C. Miller, Rev. Mod. Phys. **5**, 203 (1933).
- For bibliographies, see my book *The Ethereal Aether: A History of the Michelson–Morley–Miller Aether Drift Experiments, 1880–1930* (U. of Texas P., Austin, 1972) and Max Iklé's extensive bibliography in his German translation of Michelson's first book, *Lichtwellen und ihre Anwendungen* (J. A. Barth, Leipzig, 1911).
- Other relevant works include the biography by Michelson's daughter, D. M. Livingston, *The Master of Light* (U. of Chicago P., Chicago, 1979); my book on the social history of physics, *Genesis of Relativity: Einstein in Context* (Burt Franklin, New York, 1979); G. Holton, Y. Elkana, eds., *Albert Einstein, Historical and Cultural Perspectives* (Princeton U. P., Princeton, N. J., 1984); and A. Pais, *Subtle Is the Lord: The Science and Life of Albert Einstein* (Oxford U. P., New York, 1982). □

Michelson in 1887

The year of the famous ether drift experiment and the use of light waves as a standard of length was also a year for personal problems, including a damaging scandal.

Albert E. Moyer

Albert Abraham Michelson in 1887. (Photograph courtesy of Clark University Archives.)

In July 1907, Bernhard Hasselberg, a member of the Nobel Prize committee, confided to George Ellery Hale that Albert Michelson was his choice for the year's physics award: "In *my* opinion he is in every way, and absolutely without comparison, the most meritorious of all the gentlemen now proposed to us."[1] After the Swedish committee chose Michelson, making him the first American ever picked for a Nobel science prize, Hasselberg was even more complimentary. He wrote to Hale that Michelson's selection "is the *best* of all which have been made up to this date. Our earlier laureates Röntgen, Lorentz, Zeeman, Becquerel, Curie, Rayleigh, Lenard and J. J. Thomson are indeed men of eminent scientific merits, but for my part I consider the work of Michelson as more fundamental and also by far more delicate." Even if we allow for the idiosyncrasies of the Nobel selection process and for Hasselberg's personal biases, we still must grant that by 1907 Michelson had earned the respect of the international

Albert E. Moyer is an associate professor of the history of science at Virginia Polytechnic Institute and State University, in Blacksburg, Virginia. His special interest is the development of American physics.

physics community.

Exactly 20 years earlier, in 1887, Michelson had begun the investigation with Edward Morley that eventually provided one of the main justifications for his Nobel Prize. What was this work? It was not the "Michelson–Morley experiment," the famous interferometer test for Earth's drift through the supposed ether. Rather, it was another collaboration with Morley using the same interferometer—an investigation to determine whether wavelengths of light could provide a standard unit of length. Whereas the Nobel committee stressed the importance of the latter research along with other precision measurements, it failed even to mention the ether drift test. This omission does not mean that scientists in 1907 were unaware of the test. During the prior two decades, Hendrik Lorentz, George Fitzgerald, Henri Poincaré and Joseph Larmor had been grappling with its confusing result. Even the royalty of British physical science, Lords Kelvin and Rayleigh, had been keeping tabs on "Michelson's celebrated experiment" and its "well-known negative result."[2] Awareness of an experiment, however, differs from appreciation. Widespread appreciation of the test's significance would

come only after Michelson's Nobel Prize, when the physics community grasped the full implications of Albert Einstein's special theory of relativity.

In this article, I return to the crucial year of 1887—the year in which Michelson, assisted by Morley, undertook the two research projects that independently contributed to Michelson's Nobel Prize and to his post-Einsteinian fame. Although I discuss both the standard-of-length and the ether drift project, I do not analyze them in detail; such studies already exist.[3] Rather, I aim to show how Michelson, in his 1887 investigations, fit into the larger picture of the day's physics, including both conceptual and institutional aspects, especially in the United States. Because Michelson's personal circumstances influenced his professional affairs, I also provide a glimpse of his private life.

Joining a network of physicists

Physicists in the United States in 1887 felt cut off from the scientific capitals of Europe—Berlin, Paris and London. Michelson felt especially isolated. A professor since 1882 at the fledgling Case School of Applied Science in Cleveland, he felt separated from even his American colleagues,

Edward W. Morley around 1885. (Photograph courtesy of Case Western Reserve University Archives.)

such as Henry Rowland at the decade-old Johns Hopkins University, John Trowbridge at Harvard's new Jefferson Physical Laboratory and Willard Gibbs at Yale. In a letter to Rowland, Michelson complained of "the inconvenience of being so far removed from scientific centers," and added that "I would much prefer a position farther East."[4] Other scientists shared this perception of Cleveland's isolation. When the American Association for the Advancement of Science met there in the summer of 1888, the permanent secretary excused the disappointingly low attendance by pointing to the "natural aversion to going into the interior of the country in August."[5]

The founders of Case had sought to bring science and its practical applications to the Midwestern industrial city. But the five-year old program suffered a setback in the fall of 1886 when a fire destroyed the main academic building, including much of the physics apparatus that Michelson had painstakingly assembled. As Michelson began the year 1887, damaged facilities compounded the problem of isolation.

However, Michelson, like many of his more enterprising colleagues throughout the United States, found ways to compensate. In response to the immedi-

ate problems caused by the fire, he relocated his laboratory and salvable apparatus to a building of nearby Western Reserve University, the home institution of chemist Morley, Michelson's research partner since 1885. As for the more basic problem of isolation, Michelson followed the lead of other active researchers by participating in an informal informational network. The network, which the scientist sustained through letters and travel, was both national and international in scope.

Michelson had become well connected within the domestic branch of the network. At the beginning of his career, while investigating the speed of light as a student and instructor at the US Naval Academy, he had formed a close professional bond with Simon Newcomb. Newcomb, the influential head of the government's Nautical Almanac Office, opened many doors for Michelson at home and abroad. Michelson also established close ties with two of the nation's most esteemed academic physicists, Rowland and Gibbs. For example, around 1884, when Michelson began to plan for a definitive ether drift experiment—a refined version of the experiment he originally performed in Germany in

1881—he consulted with Gibbs. Through conversation and correspondence, he questioned the Yale theorist about "the feasibility of the experiment."[6] Other members of Michelson's American network included physicists George Barker and Alfred Mayer, both of whom in earlier years had tried to get Michelson a job outside the Navy. Barker, of the University of Pennsylvania, finally arranged for Michelson's appointment as a professor through an influential friend at Case. "I am sure," Barker wrote to his friend, "the Case School will never regret this step."[7]

Michelson also cultivated relationships with prominent Europeans, the anchors of the physics network. These ties dated back to his days as a postgraduate student in Germany and France. Like other Americans faced with the shortcomings of American graduate schools, Michelson was particularly attracted to the University of Berlin, the site of Hermann von Helmholtz's famous laboratory. Beginning in the fall of 1880, Michelson attended Helmholtz's lectures on theoretical physics. In addition, he turned to Helmholtz for advice on his first ether drift test and his newly devised interferometer. While in Germany, Michelson also studied optics and spectroscopy under the masters in these fields. In the fall of 1881, he went on to Paris, expanding his list of scientific contacts to include the leaders in French optics. (See the article by Loyd Swenson in this volume.)

Although he visited London too, Michelson became close to top British physicists only after returning to the United States. In 1884 Michelson attended the Montreal meeting of the British Association for the Advancement of Science, chaired by Rayleigh, and heard the subsequent "Baltimore lectures," delivered at Johns Hopkins by William Thomson (later Lord Kelvin). Rayleigh and Thomson quickly became two of the most consequential members of Michelson's network, providing him with perspective and encouragement on his research. While in North America, they advised him to repeat his German interferometer experiment, but only after first checking

Armand Fizeau's earlier result on the speed of light in moving water. Michelson accepted this advice, reporting to Thomson in March 1886 that he and Morley had confirmed Fizeau's result.[8] A letter from Rayleigh then provided the incentive for Michelson and Morley to push on to the next stage of their ether research, a repeat of the interferometer test. Michelson wrote back to Rayleigh in March 1887 that he had been distressed by the lack of interest of his "scientific friends" and the "slight attention" the German test received. He continued, "Your letter has however once more fired my enthusiasm and it has decided me to begin the work at once."

Experimenters and measurers

European physicists in the period around 1887 made a distinction that helps us understand the relation between Michelson's research and the physics of his day. The Europeans, particularly the Germans, distinguished between experiment and measurement.[9] Whereas "experimental physicists" typically investigated new phenomena related to unsettled theories, the "measuring physicists" typically devised precision instruments to refine the quantities specified in established theories or to update recognized physical constants. Michelson was primarily a measuring physicist, as evidenced by his attempts to make precise determinations of Earth's motion through the ether and of a standard of length using wavelengths of light.

Michelson fit in well with other Americans who favored using their laboratories for measurement.[10] Rowland, for example, showed this propensity in his measurements of both the ohm and the mechanical equivalent of heat and in his manufacture of diffraction gratings for use in his spectral studies. Most Americans, however, emphasized the experimental rather than the measuring side of laboratory practice. Well-known experimental work of the day included Mayer's study of the configurations of floating magnets, Trowbridge's electrical research on the "superficial energy" between alloys and Edwin Hall's systematic investigation of conductors, which led to the disovery of the "Hall effect." Rowland also complemented his measurements with experiments, including an inquiry into the magnetic effect of moving electric charges. As in Europe, very few scientists specialized in theoretical physics

or treated it as a distinct area of teaching and research. Gibbs proved to be the exception, with contributions to thermodynamics, statistical mechanics and the electromagnetic theory of light. The dearth of theorists reflected in part the lack of advanced mathematical training in the United States. Michelson himself once admitted to Mayer that he made "no pretense" of being a mathematician.[11]

To make their exacting measurements, Michelson and the other Americans needed precision instruments. For this they did not need to turn to their European colleagues for help. Rather, they drew on an indigenous tradition of first-rate engineering and manufacture. Two of the world's foremost fabricators of optical apparatus were in the United States, John Brashear in Pittsburgh and Alvan Clark in Cambridge, Massachusetts. Michelson and Morley relied on Brashear's shop for the delicate optical components required in their ether drift and standard-of-length investigations. And for all its disadvantages, living in Cleveland allowed them to gain an edge on fellow Americans because they had immediate access to the local firm of Warner and Swasey.[12] This precision manufacturing company, for which Morley served as a consultant, had won the contract to furnish all the mechanical and structural portions of the telescope for the Lick Observatory, the largest refracting telescope in the world. Shortly after Warner and Swasey completed the bulk of the Lick project in October 1887, Michelson and Morley began working with them to devise a mechanically advanced instrument for the next phse of their standard-of-length investigation.

Conceptual framework

Whether they were experimenters or measurers, most American physicists around 1887 worked within the same general conceptual framework. Echoing many of their European colleagues, they assumed that all physical phenomena could be represented in terms of underlying atoms, molecules and ethers that obey the laws of classical mechanics. Although they shared the basic precepts of this traditional outlook as they pertained to gases, heat, electricity, magnetism, light and even gravity, the physicists were quite unsettled about specific mechanical theor-

ies and models. Michelson in particular coupled an orthodox trust in the existence of the luminiferous ether to an equally orthodox mechanical view of the atomic processes that caused light waves. Throughout an address to the 1888 Cleveland meeting of the AAAS, he confidently spoke of "the forces and motions of vibrating atoms and of the ether which transmits these vibrations in the form of light."[13] Although convinced of the existence of the ether, Michelson was unsure of Augustin Fresnel's specific hypothesis that the ether was essentially at rest relative to the moving Earth. Thus, he set out with Morley to make a definitive measurement of the relative Earth–ether motion.

The measurement's well-known null result threatened Fresnel's overall theory of the ether but not the notion of an ether itself. In fact, Michelson and Morley candidly reported their perverse result and then attempted to account for it with alternative explanations. First they considered the ether theories of George Stokes and Lorentz, concluding, however, that both of those options had their own weaknesses. Then they rationalized that the Earth's irregular surface perhaps had trapped portions of the ether, thus obscuring the "ether wind" they aimed to detect. Michelson and Morley never questioned the basic concept of the ether, just its particular character.

Michelson and Morley began working in earnest on the ether project soon after Rayleigh's encouraging letter. In April 1887, Morley reported to his father: "Michelson and I have begun a new experiment. It is to see if light travels with the same velocity in all directions. We have not got the apparatus done yet, and shall not be likely to get it done for a month or two."[14] Curiously, the measurement of ether drift was not to be the first application of their new interferometer—an ingenious arrangement of mirrors mounted on a sandstone megalith floating on mercury. They first employed the new device in June to explore the possibility of using the wavelength of one of the spectral lines emitted by thallium, lithium, hydrogen or especially sodium as a standard of length. In the prior half century, especially in France, the quest for an unequivocal and invariant measure of a meter had come to preoccupy many researchers.

They initially addressed the standard-of-length question for a simple

Interferometer built by Michelson and Morley in 1887 for their famous ether drift experiment. (Michelson Collection, US Naval Academy.)

reason. While preparing the interferometer for the ether test, they first had to ensure that the perpendicular paths of the two light beams were of approximately equal lengths; they did this by adjusting the position of one of the plane mirrors at the end of one of the paths. They realized that if the mirror "moved parallel with itself a measured distance by means of the micrometer screw," then the distance could be correlated with an exact number of wavelengths; they could determine this number simply by counting alterations in the circular interference fringes as the mirror was moved.[15] Although Michelson had thought of the method "several years ago," they now confirmed that the method "seemed likely to furnish results much more accurate" than any prior determinations of length.

Not until July 1887 did Michelson and Morley finally set the adjusted interferometer into slow rotation and look for shifts in the interference fringes indicating the relative motion of the Earth and ether. A month later Michelson wrote to Rayleigh that the tests "have been completed and the result is decidedly negative."[16] In the ensuing months and years, Michelson came to value the ether drift test not for its null finding but for its role in the creation of a refined interferometer applicable to more pressing investigations, especially the determination of a standard of length. The Nobel Prize committee would agree.

Institutional resources

To announce their findings, Michelson and Morley traveled in early August to the meeting of the AAAS at Columbia College in New York City. Samuel Langley, president of that large, open-door organization, captured in his welcoming remarks the sense of community that the association provided to isolated researchers.[17] Specifically, he reminded his colleagues that many of them had come to their first meetings "as solitary workers in some subject for which they had met at home only indifference." But they discovered at the meetings other scientists "caring for what they cared for, and found among strangers a truer fellowship of spirit then their own familiar friends had afforded." Michelson found his "fellowship of spirit" in Section B, Physics, one of eight disciplinary divisions available to the 729 members in attendance. Section B, established in 1882, constituted the only national coalition of physicists.

During the week-long meeting, the association elected Michelson to be the new vice president in charge of the section. For the next 12 months, Michelson, at age 34, served as the spokesman for the loose-jointed but aspiring American physics community.

Michelson and Morley jointly presented two of the 38 papers read in Section B. One dealt with the relative velocity of the Earth and the ether, and the other covered the method for establishing a standard of length. A few days after the close of the meeting, *Science* reported that "the most important paper" of Section B was Michelson and Morley's discussion of a standard of length.[18] The magazine devoted a full paragraph to characterizing the results. As for their ether paper, *Science* labeled it "a second paper of great interest" and summarized it in only a few lines.

The opinion of AAAS members on the relative value of the two research projects was apparently shared by the leadership of the National Academy of Sciences—the other main, but much smaller and more exclusive, scientific society in the United States. In early November 1887, the Academy invited Michelson and Morley to the annual "Scientific session" to present their latest thoughts on establishing a standard of length.[19] Of the 22 papers read at this meeting, held again at Columbia College, theirs was the only one by nonmembers. The audience was obviously impressed: Michelson was elected a member the next year.

In late December, Michelson and Morley presented their standard-of-length paper to one other appropriate audience, the Civil Engineers' Club of Cleveland. Both men were members of this regionally active group, which was certain to appreciate the practical implications of the standard-of-length research and the interferometer itself. Morley actually read the paper during the evening meeting, but Michelson participated in the discussion afterwards. "Permit me to state," Michelson commented, "that Professor Morley has given me more than my due in attributing so large a share of this work to me. Without his assistance, our present results would never have been attained."

The AAAS and the NAS did more than provide national forums for airing results. The two societies also financed a limited amount of new research. Faced with the expense of precision instruments, Michelson had emerged as one of the nation's most active fund raisers for science.[20] Dollars from the NAS's Bache Fund helped build the Cleveland interferometer. The Bache board of directors matched $500 they had awarded Michelson in March 1887 with an equal amount in late July. These were sizable sums in a time when a person could have his horse shod or buy a bushel of potatoes for about $1. Michelson and Morley supplemented the latter grant with one of the first appropriations ever made by the AAAS. During the August meeting the council of the association allotted the two men $175 "to aid them in the establishment of a standard of length."

Michelson and Morley published their interferometer findings on both the ether drift and the standard of length in the *American Journal of Science*. This multidisciplinary journal, published in New Haven, Connecticut, and edited by James and Edward Dana, provided publishing physicists with their only serious domestic outlet; the *Physical Review* was still six years in the future. Through the efforts of two associate editors, physicists Trowbridge and Barker, the *Journal* managed to carry the best products of American physics. In addition, it kept the community posted on international publications. Throughout 1887, for example, Trowbridge and Barker presented abstracts of articles from foreign journals such as *Annalen der Physik und Chemie, Comptes Rendus*, the

Philosophical Magazine and *Nature*. The *Journal*'s section on "Scientific intelligence" added reports of domestic and foreign meetings as well as reviews of physics books.

Few Europeans read the *American Journal of Science*, so to ensure that their findings reached Europe, many American physicists published or reprinted their articles in foreign journals. Their favorite outlets were British—the *Philosophical Magazine* and *Nature*. Michelson and Morley arranged for each of their two articles to appear simultaneously in the *American Journal of Science* and the *Philosophical Magazine*. The ether drift paper initially appeared in the two journals in November 1887, while the article on the standard of length came out in December.

Personal problems

The December publication on the standard of length would seem to be an appropriate conclusion to this year in the life of Albert Michelson. The article appears to crown 12 months in which Michelson assumed the top office in the physics section of the AAAS, embarked on research that would lead to a Nobel Prize, and completed the ether drift test that would bring him fame in later times. A seldom mentioned scandal in the fall of 1887, however, marred this otherwise storybook year. It damaged his marriage and disrupted his research. In light of Michelson's personal problems, it is perhaps remarkable that he was able to accomplish so much in 1887.

Dorothy Livingston, Michelson's youngest daughter by his second wife, traces his private life in her biography of him.[21] She explains that tensions riddled his first marriage, to Margaret Heminway. The two had quite different backgrounds, Margaret coming from a wealthy New York family and Michelson born in Prussian Poland but raised in mining towns in California and Nevada. Each was strong willed, Margaret wanting to control those about her and Michelson obsessive about his research. The tensions contributed to, and in turn were heightened by, Michelson's psychological illness that surfaced in the fall of 1885. Morley speculated that the illness was precipitated by overwork on the Fizeau experiment—by "the ruthless discipline with which he drove himself to a task he felt must be done with such perfection that it could never again be called into question." Whatever its cause, this "softening of the brain," as Morley characterized the illness, meant that Michelson had to quit teaching at Case and move to New York for full-time medical care. Mor-

ley guessed that Michelson would never work again; Margaret apparently agreed, attempting to have him committed permanently to an asylum. Within a few months, however, he recovered and returned to Case. According to Livingston, he returned a cynical man completely alienated from his wife.

Well aware of the long hours that Michelson was putting into the interferometer during the spring and early summer of 1887, Margaret feared a relapse of his illness. Consequently, she arranged for him to combine the business of the August AAAS meeting with the pleasure of a family vacation in the East. (The family included two sons and a daughter.) When they returned to Cleveland, Livingston points out, they were appalled to discover that the maid had stolen some family valuables and fled. Although she was arrested and most of the stolen items were recovered, this event set the stage for the scandal soon to engulf Michelson.

Scandal. The Michelsons, needing a new maid, hired 19-year-old Ruth Whitfield. After several weeks in the family's employ, Ruth, along with her aunt Emma Whitfield, confronted Michelson at home with a demand: Pay $100 to Ruth and $25 to Emma if he wanted them to keep silent about "improper liberties" he had taken with Ruth. This confrontation occurred on 8 October 1887, a Saturday night. Michelson arranged for the women to meet him at his Case laboratory on Monday morning. With his lawyer and Morley at his side—and with a police detective hidden in an adjoining room—he asked the women to repeat their allegation and demand. When they did, Michelson had them arrested and charged with blackmail.

The *Cleveland Plain Dealer* and *The Leader and Herald* reported on the arrest and the subsequent preliminary hearing.[22] According to their accounts, Michelson's lawyer testified about the Monday morning confrontation, recalling that when he asked what complaint Ruth had with Michelson, Emma replied: "He hugged her, and kissed her, and asked her to go to his room." When he asked what would happen if Michelson did not pay the money, Emma answered, "I'll expose him and make trouble for his family." The women's attorney also had an opportunity to cross-examine Michelson, who denied Ruth's allegation. One newspaper reported a portion of Michelson's testimony:

"When this occurrence is alleged to have taken place between you and Ruth, was your wife at home?" asked Attorney Kaiser, the defen-

Margaret Heminway, Michelson's first wife. (Photograph courtesy of Dorothy Michelson Livingston and the Michelson Collection, US Naval Academy.)

dants' attorney.

"I believe she was not at home."

"Was not one of their objects in calling on you Saturday night to get Ruth's trunk?"

"They asked for the trunk after making their demands."

"While they were at your house did your wife come to the door?"

"Yes."

"Did you say to her, 'Please retire, I have some private business with these ladies?'"

"I did."

"Where did the conversation occur?"

"In the kitchen."

"Did you caution these women not to speak so loudly lest your wife might hear?"

"I did."

"Did you promise to pay them something?"

"I did."

"How much?"

"One hundred dollars."

"Why did you not pay it at the time?"

"In the first place, I had not the money, and in the second place I had no intention of paying it."

On Wednesday evening, the day after the hearing, a police constable arrested Michelson. According to the newspapers, Ruth Whitfield "swore to an affidavit... charging Mr. Michelson with having committed assault and battery upon her on October 5, 6, and 7." At this point, the story dropped from the newspapers. Court records, however, continue the account. The records reveal that the justice of the peace who ordered Michelson's arrest set his bail at $100. Summoned before the same justice three days later along with Ruth Whitfield, Michelson "waived examination," whereupon the justice transferred the case to police court.

Documents from police court disclose that in November the case against Michelson "was called up and dismissed." Likewise, a grand jury report from November indicates that Ruth and Emma Whitfield were not indicted for blackmail. Very likely, the opposing lawyers arranged an exchange whereby Michelson dropped his charge and Ruth Whitfield dropped her countercharge.

The timing of these events proved inopportune for Michelson. The Civil Engineers' Club had scheduled Morley, and presumably Michelson, to present their standard-of-length paper on the evening of 11 October, the day after the Whitfields' arrest. The club's minutes state that the "paper had been expected" that evening. Instead, Morley and Michelson rescheduled the paper for late December. Also, during the very week that the story was breaking in the newspapers, Michelson's mentor Newcomb was in town along with other dignitaries for the official unveiling of the Lick telescope.

Repercussions. Whereas the details of the episode remain clouded, the aftermath is clearer. The scandal contributed to the collapse of the Michelsons' already foundering marriage. Another decade passed, however, before Margaret obtained a divorce, with the judge finding that Michelson had been "guilty of extreme and repeated cruelty to the complainant."[23]

The episode also disrupted Michelson's interferometer research. Michelson and Morley had planned originally to repeat the ether drift test every few months to guarantee that their initial result was not a quirk. In particular, they wanted to be sure that the negative result did not reflect a chance occurrence of Earth's various motions canceling each other and thus producing a negligible resultant velocity relative to the stationary ether. Actually, there were several reasons for not repeating the test that autumn: Michelson and Morley were moving their laboratory to a permanent site from the temporary quarters they had been using since the fire; they were busy with their classes at the beginning of the academic year; they had begun to suspect, as they expressed it in a "supplement" to their ether drift article, that it was "hopeless" to detect an effect with the interferometer located at the surface of the Earth; and they had shifted their attention to the more tractable standard-of-length investigation. The scandal added to all this and helped scuttle their plans for an immediate repetition of the ether test.

Finally, the scandal seems to have soured the Case administration's attitude toward Michelson. When Case first hired Michelson, Barker asserted that the school would never regret that step. However, by the time Michelson moved on in 1889 to an attractive professorship at Clark University, Morley could report to his father that the administrators "are glad to have him go from the Case School."[24] Personally disappointed, Morley added, "They certainly lose one of the first two physicists in the country in losing Michelson."

Morley realized at an early date what Hasselberg and the Nobel committee came to appreciate in 1907: that Mi-

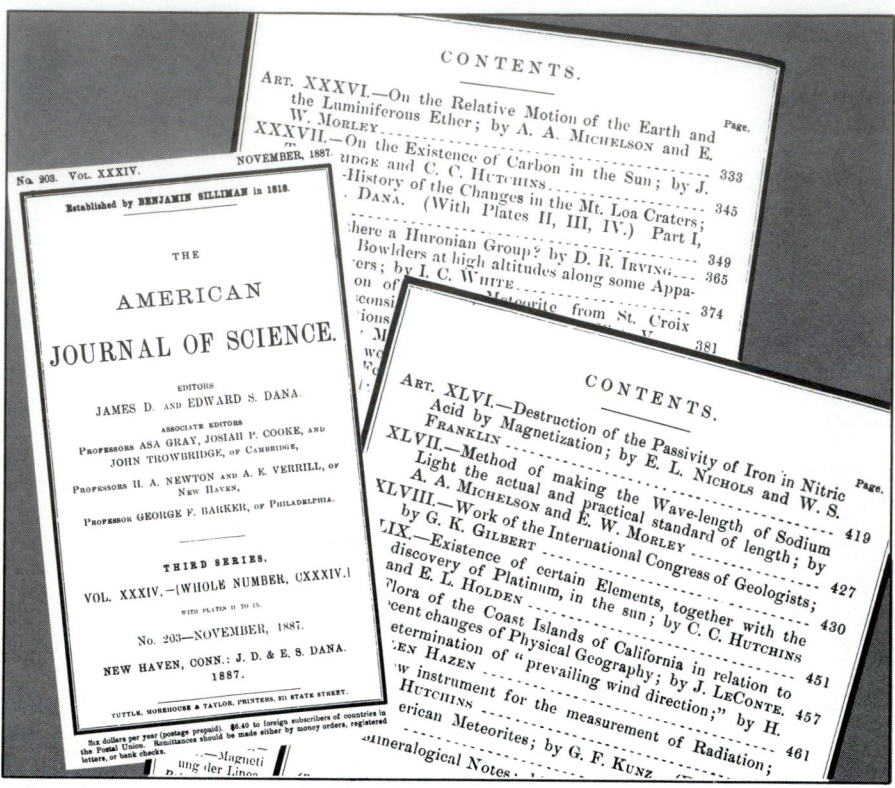

Journal pages. The November 1887 issue of the *American Journal of Science* contains Michelson and Morley's report on their ether drift experiment. Their article on the use of light as a standard of length is in the December issue.

chelson excelled as a physicist. Indeed, already in 1887 Michelson came near to being the "complete" physicist. That is, he was involved as deeply and creatively as possible for an American of his day with all aspects of the physics enterprise—with the individuals, ideas and institutions of late 19th-century physics. Becoming a consummate physicist, however, exacted a price. Michelson's personal life reeled from psychological distress and marital discord. One wonders what Michelson in later years remembered most about 1887, his achievements in physics or his personal problems. Both contributed to making 1887 a pivotal year for Michelson.

* * *

For their advice in preparing this article, I thank Dorothy Michelson Livingston, Theodore McAllister and Loyd Swenson Jr. For their assistance in obtaining documents, I thank Mary Catalfamo of the Michelson Collection at the US Naval Academy, Judith Cetina of the Cuyahoga County Archives, Dennis Harrison of the Case Western Reserve University Archives and Ann Sindelar of the Western Reserve Historical Society.

References

1. B. Hasselberg, letters to G. E. Hale, 5 July 1907 and 29 December 1907, Michelson Collection, Nimitz Library, US Naval Academy, Annapolis, Md.

2. W. Thomson (Lord Kelvin), letter to J. W. Strutt (Lord Rayleigh), 23 January 1907, in S. P. Thompson, *The Life of Lord Kelvin*, vol. 2, Chelsea, New York (1976), p. 1194. J. W. Strutt (Lord Rayleigh), in *Scientific Papers by Lord Rayleigh*, vol. 5, Dover, New York (1964), p. 678. See also R. S. Shankland, Phys. Teach. **15**, 19 (1977); G. Holton, Isis **60**, 133 (1969); A. Pais, *Subtle is the Lord: The Science and Life of Albert Einstein*, Oxford U. P., New York, (1982), chs. 6,8.

3. The following fundamental studies also contain good bibliographies: L. S. Swenson, *The Ethereal Aether*, U. Texas P., Austin (1972); D. M. Livingston, *The Master of Light*, Scribner's, New York (1973); R. S. Shankland, Am. J. Phys. **32**, 16 (1964); J. M. Bennett, D. T. McAllister, G. M. Cabe, Appl. Opt. **12**, 2253 (1973). I am grateful to these authors for much of the background information in this article.

4. A. A. Michelson, letter to H. Rowland, 6 November 1885, in *Science in Nineteenth Century America*, N. Reingold, ed., Hill and Wang, New York (1964), p. 311.

5. F. W. Putnam, in *Proceedings of the AAAS: Thirty-seventh Meeting, Held at Cleveland, August, 1888*, AAAS, Salem, Mass. (1889), p. 416.

6. A. A. Michelson, letter to J. W. Gibbs, 15 December 1884, in *Science in Nineteenth Century America*, N. Reingold, ed., Hill and Wang, New York (1964), p. 307.

7. G. Barker, letter to J. Stockwell, 5 May 1881, in R. S. Shankland, Am. J. Phys. **32**, 22 (1964).

8. A. A. Michelson, letter to W. Thomson, 27 March 1886, and letter to J. W. Strutt (Lord Rayleigh), 6 March 1887, in D. M. Livingston, *The Master of Light*, Scribner's, New York (1973), pp. 117, 123.

9. C. Jungnickel, R. McCormmach, *Intellectual Mastery of Nature*, vol. 2, U. Chi-

10. D. J. Kevles, *The Physicists*, Knopf, New York (1978), ch. 3. A. E. Moyer, *American Physics in Transition*, Second printing, Tomash/AIP, New York (1986). Angeles (1983).

11. A. A. Michelson, letter to A. Mayer, 26 June 1880, in *Science in Nineteenth Century America*, N. Reingold, ed., Hill and Wang, New York (1964), p. 286.

12. The Leader and Herald (Cleveland), 17 October 1887, p. 8. Cleveland Plain

13. A. A. Michelson, quoted in A. E. Moyer, *American Physics in Transition*, Second printing, Tomash/AIP, New York (1986).

14. E. W. Morley, letter to S. B. Morley, 17 April 1887, in *Science in Nineteenth Century America*, N. Reingold, ed., Hill and Wang, New York (1964), p. 312.

15. A. A. Michelson, E. W. Morley, Am. J. Sci. **34**, 427 (1887). See also A. A. Michelson, E. W. Morley, Am. J. Sci. **34**, 333 (1887); A. A. Michelson, E. W. Morley, J. Assoc. Eng. Soc. **7**, 110, 153 (1888).

16. A. A. Michelson, letter to J. W. Strutt (Lord Rayleigh), 17 August 1887, in R. S. Shankland, Am. J. Phys. **32**, 32 (1964).

17. S. Langley, in *Proceedings of the AAAS: Thirty-sixth Meeting, Held at New York, August, 1887*, AAAS, Salem, Mass. (1888), p. 342; see also p. 345.

18. Science **10**, 86 (1887).

19. *Report of the NAS for the Year 1887*, US Government Printing Office (1888), pp. 10, 28.

20. *Report of the NAS for the Year 1889*, US Government Printing Office (1891), p. 41. See also A. A. Michelson, letter to J. Billings, 15 May 1889, Michelson Collection, Nimitz Library, US Naval Academy, Annapolis, Md.

21. D. M. Livingston, *The Master of Light*, Scribner's, New York (1973), esp. pp. 111–136.

22. The Leader and Herald (Cleveland), 11 October 1887, p. 5; 12 October 1887, p. 5; 13 October 1887, p. 8; 16 October 1887, p. 6. Cleveland Plain Dealer, 11 October 1887, p. 8; 12 October 1887, p. 8; 14 October 1887, p. 5; 16 October 1887, p. 15. See also the record from the Justice's Court for Cleveland Township, 12 October 1887, and the Court of Common Pleas "Journal," 17 November 1887, Cuyahoga County Archives, Cleveland, Ohio; "Police Court case records," 22 November 1887, Western Reserve Historical Society, Cleveland, Ohio.

23. *Decree of Divorce of Margaret H. Michelson from Albert A. Michelson*, Michelson Collection, Nimitz Library, US Naval Academy, Annapolis, Md.

24. E. W. Morley, letter to S. B. Morley, 2 June 1889, in *Science in Nineteenth Century America*, N. Reingold, ed., Hill and Wang, New York (1964), p. 313.   □

Einstein and ether drift experiments

**Recently discovered letters, written
at the turn of the century to his fiancée, shed new light
on the origin of the special theory of relativity.**

PHYSICS TODAY/MAY 1987

John Stachel

Volume 1 of *The Collected Papers of Albert Einstein*,[1] published in 1987, contains a number of previously unpublished lecture notes, examination papers and letters by Einstein. Among the most notable new items are 42 letters written between 1898 and 1902 to his fiancée Mileva Marić, whom he met while they were fellow students of physics at the Swiss Polytechnical School in Zurich, which both entered in 1896.

These letters confirm Einstein's later recollection that he had begun to work on the electrodynamics of moving bodies many years before submitting his epochal 1905 paper on special relativity to *Annalen der Physik*.[2] They also record Einstein's continued interest, in the years between 1899 and 1901, in designing an optical experiment to test the putative motion of the Earth through the ether—which should have been detectable according to the then prevalent interpretation of Maxwell's theory.

While there is no mention of Albert A. Michelson in any of the letters in volume 1, which covers the period from Einstein's birth until he got an appointment as patent clerk at the Swiss Patent Office, there is strong indirect evidence[3] that he must have known of the Michelson–Morley experiment by 1899. Here I will review briefly the new evidence of Einstein's early theoretical and experimental work on the electrodynamics of moving bodies.

Einstein's first comments on the subject, which appear in a remarkable letter that has been dated to August 1899, were inspired by a rereading of Heinrich Hertz's basic papers on Maxwell's electrodynamics.

I am more and more convinced that the electrodynamics of moving bodies, as currently presented, is not correct, and that it should be possible to present it in a simpler way. The introduction of the term "ether" into theories of electricity leads to the notion of a medium of whose motion one can speak without, I believe, being able to associate any physical meaning with such a statement.

Einstein is clearly skeptical about the concept of a movable ether, a concept that was basic to Hertz's theory of the electrodynamics of moving bodies. Whether this skepticism already extended to the concept of the ether itself, as was certainly the case by 1905, is more doubtful. On the whole, Einstein's views in this letter seem similar in many ways to those of Hendrik A. Lorentz, who postulated a universal but immobile ether. But there is no mention of Lorentz in any surviving letter of Einstein's until December 1901, when he states that he intends to study what Lorentz and Paul Drude have written on the subject. So it is entirely possible that Einstein arrived at his views in 1899 independently of Lorentz.

Einstein refers in this letter to the need for "radiation experiments" to decide between various views of electrodynamics. In September of 1899, he writes:

A good way of investigating how a body's relative motion with respect to the luminiferous ether affects the velocity of propagation of light in transparent bodies occurred to me in Aarau [a Swiss town Einstein had recently visited]. I have also thought of a theory on this subject that seems to me to be very plausible. But enough of this!

Einstein goes on to commiserate with Marić, who was preparing to take the intermediate examinations at the Polytechnical School, a set of examinations that Einstein had already passed. There is no further evidence in the letters about the nature of Einstein's experiment or of his theoretical ideas.

A couple of weeks later, he informs Marić that

I also wrote to Professor [Wilhelm] Wien in Aachen about the work on the relative motion of the luminiferous ether with respect to ponderable matter, which "the boss" [Heinrich Friedrich Weber, Einstein's physics professor at the Polytechnical School] treated in such a stepmotherly fashion.

This remark partially confirms the narrative that Rudolf Kayser, Einstein's son-in-law, gives in his 1930 biography of Einstein[4] (written with Einstein's cooperation and approval):

He encountered at once, in his second year of college [1897–98], the problem of light, ether and the Earth's movement. This problem never left him. He wanted to construct an apparatus which would accurately measure the Earth's movement against the ether. That his intention was that of other important theorists, Einstein did not yet know. He was at

John Stachel is professor of physics at Boston University and editor of *The Collected Papers of Albert Einstein*.

that time unacquainted with the positive contributions, of some years back, of the great Dutch physicist Hendrik Lorentz, and with the subsequently famous attempt of Michelson. [Michelson first performed his experiment in 1881, and repeated it in 1887, partly in response to a criticism by Lorentz, before Lorentz's first major work on the electrodynamics of moving bodies in 1892.] He wanted to proceed quite empirically, to suit his scientific feeling of the time, and believed that an apparatus such as he sought would lead him to the solution of a problem, whose far-reaching perspectives he already sensed.

But there was no chance to build this apparatus. The skepticism of his teachers was too great, the spirit of enterprise too small. Kayser's account still does not offer any clues to what Einstein's experimental design could have been. The only evidence known to me on this question is the record of Einstein's 1922 lecture at Kyoto University, "How I created the theory of relativity," kept by the physicist Jun Ishiwara.[5] He was the first Japanese to publish on the theory of relativity, had visited Einstein on a trip to the West, was instrumental in getting Einstein to come to Japan and acted as translator of his lectures. Ishiwara's record, in Japanese, of Einstein's lecture includes Einstein's description of an experiment that occurred to him while he was a student:

So I wanted to demonstrate by some means this motion of the Earth relative to the ether.... At the time when I posed this problem to myself I never doubted the existence of the ether and the motion of the Earth. Thus, I predicted that if light from a source is reflected by a mirror, it should have different energies depending on whether it is propagated parallel or antiparallel to the direction of motion of the Earth; and I proposed verifying this with two thermocouples, by measuring the difference in the heat produced in each.

This may well be a description of the idea that Einstein had in Aarau, although his reference to "the velocity of

propagation of light in transparent bodies" suggests that he may have had in mind some variant of Armand Fizeau's well-known experiment on this subject. Indeed, it is curious to note that in 1854 Fizeau had proposed an experiment on the difference in energy between light rays moving in opposite directions, which was actually performed in 1902 by Nordmeyer.[6]

In the "Aarau" letter of September 1899, Einstein explains why he turned to Wien for support of his ideas: "I read a very interesting paper from the year 1898 by this man [Wien] on the same topic." The paper[7] was the text of Wien's report to the Society of German Scientists and Physicians, "On questions relating to the translatory motion of the luminiferous ether." Here Wien discusses both Hertz's concept of a moving ether and Lorentz's concept of an immobile ether, and he briefly considers 13 experiments bearing on the question. The last one he mentions is the Michelson–Morley experiment. It is reasonable to conjecture that Einstein read this account in 1899, and that it thus represents the minimum information he had about that experiment by then. Here is Wien's account:

The Michelson–Morley experiment. If the ether is at rest, then the time a light ray needs to travel back and forth between two glass plates must change if the plates are moving. The change depends on the quantity v^2A^2 [v is the velocity of the plates; A is the reciprocal of the speed of light], but should be observable by the application of interferometry.

The negative result is incompatible with the assumption of an ether at rest. This assumption can only be maintained by means of the hypothesis that the linear dimensions of rigid bodies are altered by motion through the resting ether in the same ratio, so as to compensate for the lengthening of the path of the light ray.

Wien does not make it clear that interference between two perpendicular rays is the basis of the Michelson–Morley experiment. It is possible that Einstein did not see a more detailed account of the experiment until he read Lorentz's 1895 monograph[8] or Drude's book on optics, which contains a summary[9] of Lorentz's theory; both books

include detailed discussions of the Michelson–Morley experiment. Just when Einstein read Lorentz and Drude is not clear, although as noted above, in a letter from December 1901 he states his intention to study their work.

Einstein's next comment on relative motion occurs in a letter to Marić written in March 1901: "How happy and proud I will be when the two of us together will have brought our work on relative motion to a successful conclusion." This comment raises the intriguing question of the nature of Marić's role in their collaboration. Her letters to Einstein (only ten from the period of the first volume have been found) contain no substantial references at all to physics. His letters to her contain references to joint study of books, requests for her to look up data, and one or two other mentions of joint work; but these letters give no indication of any ideas she contributed to their work.

Writing to his friend and former fellow-student Marcel Grossman in September 1901, Einstein returns to the subject of ether drift experiments:

On the investigation of the relative motion of matter with respect to the luminiferous ether, a considerably simpler method has occurred to me, which is based on customary interference experiments. If only relentless fate would give me the necessary time and peace! When we see each other, I will tell you about it.

The reference to "customary" [*gewöhnlich*] interference experiments in this letter is intriguing but puzzling. Any suggestion that Einstein had in mind nothing more than a repeat of the Michelson–Morley experiment seems to be ruled out by Einstein's report, in a subsequent letter, that Alfred Kleiner of the University of Zurich was enthusiastic about his experimental proposal. Kleiner was a well-informed experimenter, who later wrote a number of surveys of the then current state of physics. It is hard to believe that he would not have known enough about the Michelson–Morley experiment to recognize a description of it.

Einstein was also developing his theoretical ideas on electrodynamics during this period. During the same month, he wrote Marić:

I am now working very eagerly on

Einstein's 1899 letter on the ether

The following is part of a letter Albert Einstein wrote to Mileva Marić, in August 1899. It is reprinted from The Collected Papers of Albert Einstein *with the permission of Princeton University Press (© 1987 by the Hebrew University of Jerusalem).*

Ich hab den Band Helmholtz zurückgetragen & studiere gegenwärtig noch einmal aufs Genaueste Hertz' Ausbreitung der elektrischen Kraft. Der Anlaß dazu war, daß Helmholtz' Abhandlung über das Prinzip der kleinsten Wirkung in der Elektrodynamik nicht verstand. Es wird mir immer mehr zur Überzeugung, daß die Elektrodynamik bewegter Körper, wie sie sich gegenwärtig darstellt, nicht der Wirklichkeit entspricht, sondern sich einfacher wird darstellen lassen. Die Einführung des Namens, "Äther" in die elektrischen Theorien hat zur Vorstellung eines Mediums geführt, von dessen Bewegung man sprechen könne, ohne daß man wie ich glaube, mit dieser Aussage einen physikalischen Sinn verbinden kann. Ich glaube, daß elektrische Kräfte nur für den leeren Raum direkt definierbar seien, von Herz auch betont. Ferner werden elektrische Ströme nicht als "Verschwinden elektrischer Polarisation in der Zeit" sondern als Bewegung wahrer elektrischer Massen aufzufassen sein, deren physikalische Realität die elektrochemischen Äquivalente zu beweisen scheinen. Mathematisch sind sie dann immer in der Form $\partial X/\partial x + \cdot + \cdot$ aufzufassen. Die Elektrodynamik wäre dann die Lehre von den Bewegungen bewegter Elektrizitäten & Magnetismen sein [sic] im leeren Raum: Welche von beiden Anschauungen gewählt werden muß, werden ja die Strahlungsversuche ergeben müssen. —Bis jetzt hab ich übrigens von Rektor Wüst keine Nachricht. Ich werde ihm nächstens schreiben.

Hier im Paradies ist es fortgesetzt sehr schön, zumal wir wunderbares Wetter haben. Doch haben wir immer unangenehme Besuche von Mamas Bekannten, deren stumpfsinnigem Geschwätze ich durch die Flucht zu entrinnen pflege wenn nicht grade gegessen wird.

---

an electrodynamics of moving bodies, which promises to become a capital paper. I wrote you that I doubted the correctness of the ideas about relative motion [that letter has not been found]. But my doubts were based solely on a simple mathematical error. Now I believe in it more than ever!

This passage suggests that Einstein had already adopted some version of the relativity principle—which is not to say that he had yet disentangled his ideas on relative motion from their electrodynamical background, let alone given them the kinematical foundation that proved essential to the formulation of the special theory of relativity. But the passage does suggest that he may have fully expected the outcome of his experiment to be negative.

In December, as mentioned above, Einstein wrote that he had

spent the whole afternoon with Kleiner in Zurich and explained my ideas on the electrodynamics of moving bodies to him.... He advised me to publish my ideas about the electromagnetic theory of light for moving bodies together with the experimental method. He found the experimental method proposed by me to be the simplest and most appropriate one conceivable.... I shall most certainly write the paper in the coming weeks.

A few days later in December, he wrote Marić:

I now want to buckle down to work and study what Lorentz and Drude have written on the electrodynamics of moving bodies. [Jakob] Ehrat [a friend and former fellow Polytechnical School student, who was now an Assistant there] must get the literature for me.

Whatever reading and writing he may have done at this time, Einstein published nothing on the subject for $3\frac{1}{2}$ years. Surviving correspondence sheds very little light on what happened. Perhaps a reading of Lorentz's work temporarily shook his faith in the relativity principle; perhaps he saw that the problems involved in upholding it were greater than he had anticipated. I have speculated elsewhere[10] on the question of what happened between 1902 and 1905, but there are unfortunately no relevant new letters from this period.

In summary, the newly discovered correspondence with Marić proves that Einstein was concerned with the theoretical and experimental aspects of the electrodynamics of moving bodies from at least 1899 on. He was very much interested in ether drift experiments, and appears to have designed at least two, which he hoped to carry out himself. While he was almost certainly aware in a general way of the existence of the Michelson–Morley experiment from late 1899 on, it is not mentioned at all in his surviving letters from that period. The new evidence thus serves to confirm, at least for the period 1899–1902, Gerald Holton's conclusion that the experiment did not play a significant role in Einstein's work.[3] But ideas about ether drift experiments did form an important strand in his thinking about the complex of problems that ultimately led him to develop the special theory of relativity.

References

1. J. Stachel *et al.*, eds., *The Collected Papers of Albert Einstein: The Early Years (1879–1902)*, Princeton U. P., Princeton, N. J. (1987). All Einstein quotations are translated from this volume with the kind permission of the Hebrew University of Jerusalem.

2. A. Einstein, Ann. Phys. (Leipzig) **17**, 891 (1905).

3. For studies of the relationship of the Michelson–Morley experiment to Einstein's work, see the fundamental article by G. Holton, reprinted in G. Holton, *Thematic Origins of Scientific Thought*, Harvard U. P., Cambridge, Mass. (1973), p. 261. See also J. Stachel, Astron. Nachr. **303**, 47 (1982).

4. R. Kayser [under the pseudonym A. Reiser], *Albert Einstein: A Biographical Portrait*, Boni, New York (1930), p. 52.

5. J. Ishiwara, *Einstein Kyôzyu-Kôen-roku* [The Record of Professor Einstein's Lectures], Kabushika Kaisha, Tokyo (1971), p. 79. Widely differing English translations of the relevant passages on the origins of special relativity have appeared. [See, for example, PHYSICS TODAY, August 1982, p. 45, and the letter by Arthur Miller on page 9 of this issue.] Fortunately, they all agree more or less closely on the passage cited. (I have also consulted a German translation prepared by H. J. Haubold and E. Yasui, whom I thank for making it available to me.) For the translation used here, see J. Stachel, Astron. Nachr. **303**, 47 (1982).

6. See J. Stachel, Astron. Nachr. **303**, 47 (1982) for references and details.

7. W. Wien, Ann. Phys. (Leipzig) **65**(3), Beilage (1898), p. i.

8. H. A. Lorentz, *Versuch einer Theorie der elektrischen und optischen Erscheinungen in bewegten Körpern*, Brill, Leiden (1895). Einstein later recalled that this was the only work by Lorentz he read before writing his paper on special relativity.

9. P. Drude, *Lehrbuch der Optik*, Hirzel, Leipzig (1900). Chapter VIII of section 2 of the part on physical optics is entitled "Bewegte Körper."

10. See J. Stachel, Astron. Nachr. **303**, 47 (1982); and my article to appear in *Atti del Convegno Internazionale: L'Opera di Einstein.* □

The impact of special relativity on theoretical physics

Rudely uniting absolute space and universal time into a single, changeable 'space–time,' the theory has become a fact of life, part and parcel of our view of nature; this essay illustrates its impact by means of some early and late examples of its use.

J. David Jackson

PHYSICS TODAY/MAY 1987

As one of my colleagues put it, "Asking about the impact of special relativity on theoretical physics is like asking about the impact of Shakespeare on the English language." An impossibly large, even senseless, task. Special relativity is a fact of life, part and parcel of the way nature is. If its impact on everyday life is slight, its impact on physicists' thinking is profound. Space and universal, inexorable time were rudely united into space–time by Albert Einstein's discovery of special relativity. Explorations of the properties of space–time, of the covariance of physical laws, and of the physical invariants of nature, together with quantum mechanics, led to the formulation of relativistic quantum field theories. The external symmetries of space–time were augmented by abstract spaces for internal symmetries corresponding to invariant (conserved) quantities such as isospin, strangeness and charm. The whole worldview of modern theoretical physics can be traced back to the fundamental postulate or idea that physical phenomena do not change just because you happen to be moving by, instead of standing still, when observing them.

Rotations and other transformations in internal spaces have replaced "moving by," but the idea is the same.

The idea that the laws of nature, at least on a scale that does not encompass the whole universe, are independent of translational motion of the system under study is very old—certainly it was part of Galileo's knowledge. In special relativity, the idea is elevated to be the first postulate:

▶ There exists a triply infinite set of equivalent Euclidean frames of reference, moving with constant velocities in rectilinear paths relative to one another, in which all physical phenomena occur in an identical manner.

The second postulate is the joker, simplicity itself:

▶ There exists in nature a limiting, invariant speed.

The existence of a limiting, invariant speed is a crucial part of special relativity; it happens to be the speed of light, but it is shared by all massless particles and approached very closely by cosmic rays and by ordinary particles in accelerators.

For many physicists at the turn of the century special relativity was intimately intertwined with electromagnetism and light—not surprisingly, since swiftly moving light and slowly moving matter seem so distinct, and since the historical roots of relativity lay in attempts to reconcile electromagnetic experimental results with a mechanistic ether. Einstein did not help by using as his second postulate "light is

aways propagated in empty space with a definite velocity c, independent of the state of motion of the emitting body," or by his use of light signals to synchronize his clocks, or by titling his 1905 paper "On the electrodynamics of moving bodies." That the laws of relativity are very general was recognized by Wolfgang Pauli[1] in 1919 and by others even earlier.

From the two postulates, plus some reasonable assumptions about the isotropy of space–time and such, flow the consequences of special relativity, with its demotion of time to just another coordinate and its blurring of the distinction among before, now and after. Although there are many seemingly more bizarre consequences of special relativity, it was the destruction of the preferred position of time that made relativity difficult for so many in the early days and even now. All of this is old, familiar stuff. The broad sweep from 1905 to the present day is also a well-trodden path.[2]

Einstein was apparently familiar with the Michelson-Morley experiment, published 18 years earlier, when he wrote his paper on special relativity, even though he does not refer to it explicitly. (See the article by John Stachel in this volume.) Whether Einstein acknowledged it or not, the Michelson-Morley experiment was a very important component of the general milieu of theoretical physics in 1905. It certainly played a central role in the efforts of Hendrik A. Lorentz and others to devel-

J. David Jackson is professor of physics at the University of California, Berkeley. He is a particle theorist and the author of a well-known text on classical electromagnetism. He has recently been working on the SSC.

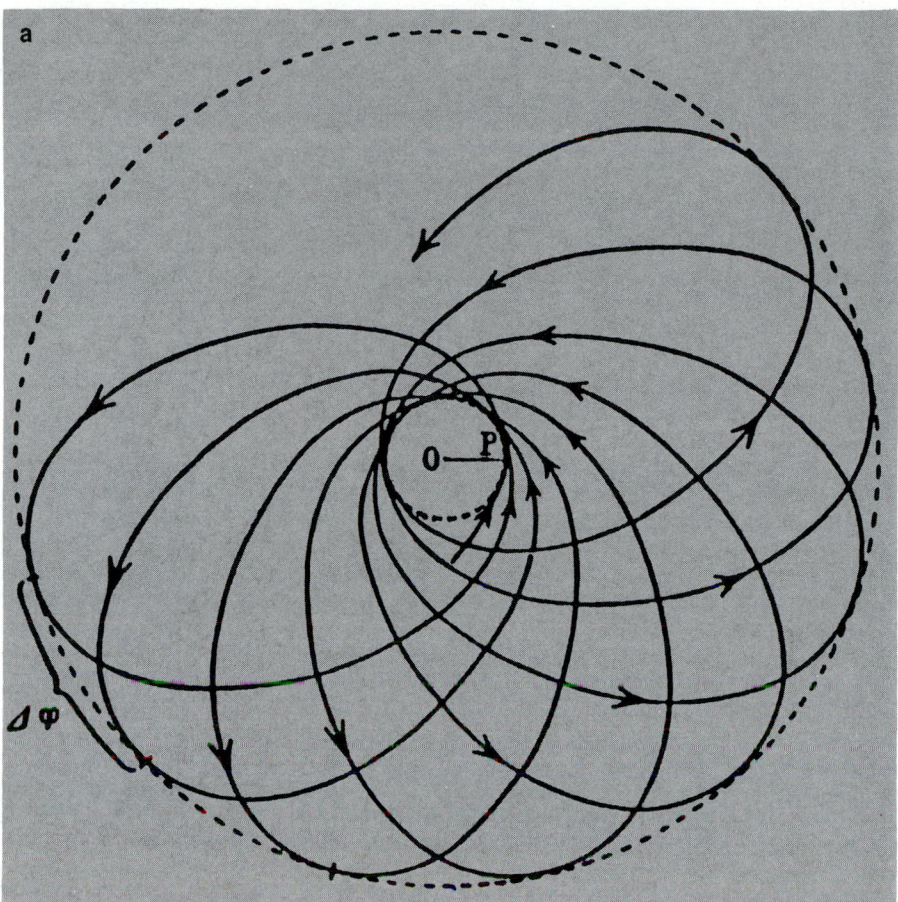

op a consistent electrodynamic theory of matter within the framework of the Maxwell equations and an ether. But with his 1905 paper Einstein changed the rules of the game. In one stroke, he made the ether meaningless, generalized classical mechanics to relativistic speeds and comfortably ensconced an etherless Maxwell's theory within his framework. Small wonder that his peers balked for a few moments at swallowing such a large pill!

This informal essay is in the nature of a lecture written down, with the impact of special relativity on theoretical physics illustrated—rather superficially, I fear—by examples of its use. Some are well known, others perhaps less familiar. The examples are taken largely from the first 30 years after Einstein's discovery. In recent times, the theory has so permeated theoretical particle physics that it is pointless to describe its effect. Instead, I give a few illustrations of the efficacy of viewing a physical phenomenon in different reference frames, thus emphasizing that there is no preferred inertial frame and returning the discussion to the theme of this issue, the Michelson–Morley experiment.

Electrons and atoms

One test of special relativity is to verify the expressions for energy and momentum as functions of speed for a particle, such as the electron—recently discovered when Einstein devised his theory. By 1905, this test was already

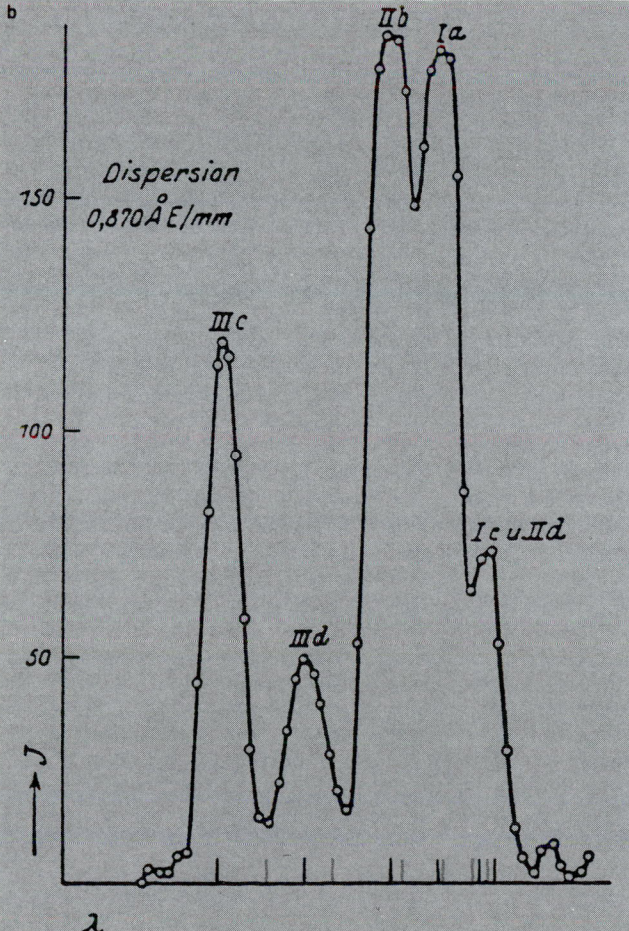

Hydrogen atom. Above is Arnold Sommerfeld's diagram for the orbit of an electron in a hydrogenlike atom. Relativistic corrections perturb the perfect Keplerian orbits of the Bohr model, leading to splittings between energy levels seen as fine structure in the spectra of hydrogenlike ions. (From reference 4.) The graph at left shows the fine structure of the He$^+$ spectrum obtained by Fritz Paschen (from *Ann. Phys. Leipzig*, **50**, 901, 1916). The data are photometer readings of a photographed spectrum in the vicinity of the 4686-Å line ($n = 4 \rightarrow n = 3$); the scan covers 0.6 Å. The tick marks on the abscissa are the predictions of Sommerfeld's model, in perfect agreement with the data; we have added dotted lines indicating the predictions of Max Abraham's "absolute theory."

under way, thanks to the electrodynamic models of the electron of Max Abraham, Lorentz and others. Abraham's model was a rigid sphere of charge; for a moving electron it gave the relations

$$E = E_0\left[\frac{1}{\beta}\ln\left(\frac{1+\beta}{1-\beta}\right) - 1\right]$$
$$cp = E_0\left[\frac{1+\beta^2}{2\beta^2}\ln\left(\frac{1+\beta}{1-\beta}\right) - \frac{1}{\beta}\right]$$

where β is v/c and E_0 is the energy of the sphere at rest, $e^2/2a$. Experiments done in 1901–02 by Wilhelm Kaufmann agreed well with Abraham's expressions. By 1904, Lorentz had proposed a model in which the electron's charge is subject to the FitzGerald–Lorentz contraction. His formulas were

$$E = E_0\gamma(1 + \beta^2/3)$$
$$cp = E_0'\gamma\beta$$

where γ is the familiar Lorentz factor of special relativity, $(1 - \beta^2)^{-1/2}$. The factors E_0 and E_0' are different. Lorentz concluded that Kaufmann's data were as consistent with his formulas as with Abraham's.

Meanwhile, Jules Henri Poincaré, building on Lorentz's work but removing, at least formally, certain inconsistencies (electrostatic instability, for example), arrived in 1905, and more fully in 1906, at the expressions

$$E = \gamma mc^2$$
$$p = \gamma mv$$

Einstein obtained the same relations, at the same time, on purely kinematic grounds. These are the well-tested and familiar expressions of today.

But model builders die hard, particularly if there are experimenters aiding and abetting. Abraham was in battle against Lorentz and Poincaré. In early 1906 Kaufmann announced a new set of measurements and declared that the Poincaré–Einstein formulas were definitely ruled out by experiment. A. H. Bucherer entered the fray with his own model and his own observations. By 1908, Bucherer, on the basis of new data, had abandoned his model and accepted the standard expressions, but the dust did not fully settle until almost 1920—with all the theorists (except Einstein, who floated aloof above it all) agonizing as the experimental sands shifted around them.[3] It seems that then, as now, the scars of such battles where reputations are on the line color the participants' attitudes about broader issues for all time.

A brilliant example of the full exploitation of special relativity in the early years is Arnold Sommerfeld's calculation of the fine structure of hydrogenlike atoms in the old quantum theory. In late 1915, W. Wilson and Sommer-

Llewellyn Hilleth Thomas showed that a precessional effect in special relativity produces an additional energy of interaction of the spin of an orbiting particle. Born in London in 1903 and educated at Cambridge, Thomas taught at Ohio State University and Columbia University. During World War II he worked at the Aberdeen Proving Ground in Maryland. He is now retired from North Carolina State University. (Photo courtesy of AIP Niels Bohr library.)

feld independently generalized Niels Bohr's quantization condition to systems with more than one degree of freedom. The extension is expressed in terms of generalized momenta p_i and coordinates q_i:

$$\oint p_i\,dq_i = 2\pi n_i\hbar$$

where n_i is an integer and the integration is over a full cycle of the motion of the ith degree of freedom.

Sommerfeld applied the extended formalism to the nonrelativistic hydrogen atom (elliptical orbits), introducing radial (n_r) and angular (k) quantum numbers for the two degrees of freedom in the plane and showing that the Bohr energy formula still applied, provided Bohr's quantum number n is replaced by $n_r + k$. The presence of only the sum of the two quantum numbers, not n_r or k separately, results in the degeneracy (in a nonrelativistic description) of the energy levels with respect to orbital angular momentum, a result that is preserved in quantum mechanics proper. Sommerfeld then moved on to a relativistic description of the atom. He noted that the familiar kinetic energy $p^2/2m$ is only an approx-

imation to the special relativistic expression

$$T = c\sqrt{p^2 + m^2c^2} - mc^2$$

where m is the rest mass of the particle. The Hamiltonian for a hydrogenlike atom then is

$$H = c\sqrt{p^2 + m^2c^2} - mc^2 - \frac{Ze^2}{r}$$

Sommerfeld saw that angular momentum, suitably defined (in terms of polar coordinates) as $\gamma mr^2\dot{\theta}$, is conserved and so is still quantized as an integral (the quantum number k) multiple of $\hbar$. Separation of p^2 into angular and radial parts led to

$$H = c\left(p_r{}^2 + \frac{k^2\hbar^2}{r^2} + m^2c^2\right)^{1/2} - mc^2 - \frac{Ze^2}{r}$$

For constant energy E (the quantity sought for), the expression

$$H = E$$

is a function of r alone and can be solved for p_r in terms of r and constants. Sommerfeld then used the quantization rule to find the energy eigenvalues. In this way, he found the

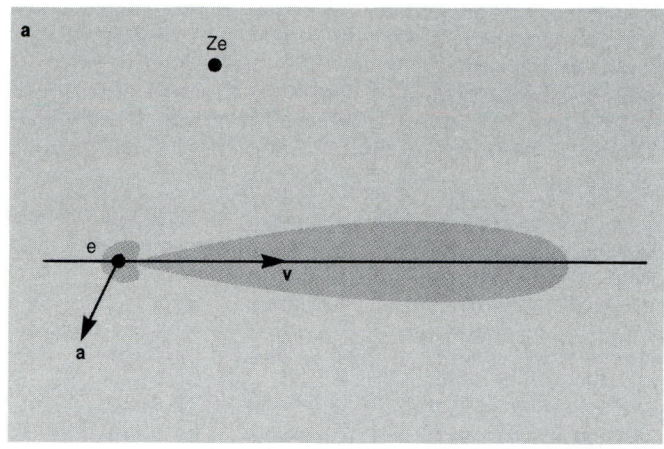

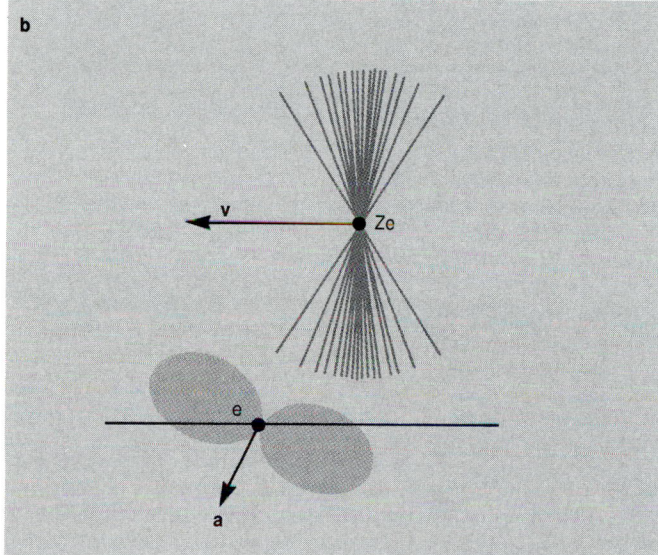

Bremsstrahlung as seen in two reference frames. **a:** In the laboratory frame, the heavy nucleus (charge *Ze*) is at rest while the swift, light particle (charge *e*) shoots by, emitting a spray of radiation in the forward direction. **b:** In the rest frame of the light particle, the electromagnetic field of the nucleus appears as a pulse of radiation that is scattered as dipole radiation by the light particle.

rather complicated expression[4]

$$E = mc^2 \left[1 + \frac{\alpha^2 Z^2}{(n_r + \sqrt{k^2 - \alpha^2 Z^2})^2} \right]^{-1/2} - mc^2$$

where α is Sommerfeld's fine structure constant, $e^2/\hbar c$, approximately $1/137$.

The relativistic corrections for the hydrogen atom ($Z = 1$) are tiny. Because α is so small, the square root in the denominator above is nearly equal to k. By expanding the overall square root in powers of α^2, one recovers the Bohr formula

$$E = -\frac{\alpha^2 mc^2}{2(n_r + k)^2}$$

But the amazing thing is that Sommerfeld's result, with integer k and n_r, agreed to fantastic precision with Friedrich Paschen's careful observations of hydrogenlike atoms such as He$^+$ (see the figure on page 213). Einstein and

many others were impressed. With reduced-mass corrections included, the equation withstood the assaults of greater and greater precision until 1947, when Willis E. Lamb and R. C. Retherford observed the minute splitting between the $2s_{1/2}$ and $2p_{1/2}$ states in hydrogen caused by radiative corrections.

Some of you knowledgeable about quantum mechanics and spin will say, "But fine structure is caused by spin–orbit coupling, not just the relativistic increase in mass! We treated those separately in our quantum mechanics classes." Others will say, "But Sommerfeld got the degeneracies wrong. His k quantum number is really $j + \frac{1}{2}$ and states with the same j but different orbital angular momenta are still degenerate." My answer is that spin–orbit coupling is a relativistic effect and the fact that Sommerfeld obtained precisely the energy formula found 12

years later with the Dirac equation, even if the most detailed correspondences are not quite right, shows that there is a deep connection between special relativity and the concept of spin. I prefer to think that Sommerfeld was somehow prescient, not just lucky. His relativistic treatment within the old quantum theory captured the essence of the spinning electron without knowledge of that degree of freedom!

An interesting sidelight of Sommerfeld's work on fine structure is how he used the experimental data and his quantization methods in support of Einstein's theory.[4] He noted that the relativistic kinetic energy of the electron in the various electrodynamic models and special relativity could be written to first order in the common form

$$T = (p^2/2m)(1 + 3a\beta^2/4 + \cdots)$$

where a is 1 for special relativity and 4/5 for Abraham's model, for example. He then calculated the first-order correction to the energy levels and arrived at the same expression as given above, except that on the left E is multiplied by a, as are both factors of $\alpha^2 Z^2$ on the right. Evidently, the factors of a are so positioned that in the limit of small $\alpha^2 Z^2$ were recover the Bohr formula, but the fine-structure spacings are proportional to a. Sommerfeld then compared the formula with the experimental spacings and concluded that the data conclusively rule out a value of a of 4/5 but are in complete agreement with a = 1. (See the figure on page 213.) Sommerfeld felt, and I do too, that disproof of the "absolute theory" (as he called it) and support of special relativity precise spectroscopic measurements was a nice touch.

Waves and particles

My next examples of the impact of special relativity on theoretical physics—really on physics—are two sides of a coin: the particulate nature of light (Compton effect, 1922) and the wave nature of matter (the de Broglie hypothesis, 1923). Both involved special relativity intimately, and had profound impact on physical thought. They can, I suppose, be described as second-order effects of special relativity. They are also so well known as to require only brief mention.

Recall that Arthur Holly Compton reported seeing in the scattering of x rays a secondary line at longer wavelengths than the incident radiation, a line whose wavelength was a function of angle. The explanation lies in the conservation laws of relativistic energy and momentum for a collision between a massless photon (x ray) of energy E and momentum E/c and a stationary electron of mass m_e. The scattering act transfers some energy and momentum to the recoiling electron. The energy E' of the scattered photon is hence less than the energy of the incident photon, so its wavelength is longer. Direct application of the formulas of special relativity leads to the Compton relation

$$\Delta\lambda = \frac{2\pi\hbar}{m_e c}(1 - \cos\theta)$$

It is striking that the Compton effect, which firmly established the *particle* nature of light quanta (objects carrying both energy and momentum and behaving as particles in relativistic kinematics), has a formula displaying the *wave* nature of the electron through $\hbar/m_e c$. We even call it the Compton *wavelength*!

In his PhD thesis, Louis de Broglie argued that if a light wave has a particulate nature that can be described by an energy–momentum $(E/c, \mathbf{p})$, related to the frequency–wave-vector $(\omega/c, \mathbf{k})$ by

$$E = \hbar\omega$$
$$\mathbf{p} = \hbar\mathbf{k}$$

then material particles should have associated with them waves and wave-like behavior with the same connections. Thus he predicted that particles of momentum p should show diffraction phenomena of wavelength $2\pi\hbar/p$, a prediction triumphantly confirmed by experiment in 1926. De Broglie made his whole discussion explicitly relativistic to describe light and matter waves in the same framework. He showed that requiring the phase advance of a wave following a semiclassical closed path to be an integral multiple of 2π yielded the relativistic Wilson–Sommerfeld quantization rules. He thus had relativistic wave mechanics almost within his grasp. The relevant point for us is his exploitation of special relativity to press the particle–wave duality for matter as much as for light.

As an aside, I note a bit of a puzzle: Einstein, the inventor of special relativity and of the light quantum for the photoelectric effect, took about 10 years to admit in print that a light quantum has directed momentum as well as energy associated with it.[5] It seems amazing that someone who was aware of the four-vector character of both $(\omega/c, \mathbf{k})$ and $(E/c, \mathbf{p})$ and who had invented a bundle of light energy with $E = \hbar\omega$ did not get around to accepting $\mathbf{p} = \hbar\mathbf{k}$ for 11 years and did not explain the Compton effect in advance. Perhaps it is a measure of how ingrained the continuous wave theory of James Clerk Maxwell was, and also of how small a photon's momentum E/c is for a given energy as compared with that

of a nonrelativistic particle, whose momentum times c is the geometric mean of the kinetic energy and the rest energy, $(2mc^2T)^{1/2}$. Only with energetic photons (x rays and above) does one easily see the transfer of momentum.

Spin and Thomas precession

My next example of how special relativity spread in influence concerns Llewllyn Hilleth Thomas, of the Thomas precession and the sector-focusing cyclotron. The story of the alkali doublets, the anomalous Zeeman effect and the introduction of the spin of the electron by George Uhlenbeck and Samuel Goudsmit in 1925 is familiar to all.[6] But the assumed g-factor of 2 for the electron, necessary for the Zeeman effect, was an embarrassment in the fine structure, where a straightforward calculation of the spin–orbit interaction gave splittings a factor of 2 too large. The location of the error and its elimination to yield the correct fine structure are due to Thomas. It is unfortunate that most physicists are aware only of the short letter[7] Thomas published in *Nature* in early 1926, if of that. It baffled even the greats like Pauli for a time. There is also a longer paper[8] by Thomas that is a marvel to behold. His contribution concerning the fine structure is sophisticated and profound; but not only does the detailed paper contain the Thomas precession, it has the complete development of the relativistically covariant equations for the motion of a spin with any g-factor in an arbitrary external electromagnetic field.

The effect discovered by Thomas is purely kinematic and results from the lack of commutativity of successive Lorentz transformations as a particle moves in a curved path. He showed that the coordinate axes in the particle's rest frame precess with respect to the laboratory axes with an angular velocity

$$\boldsymbol{\omega}_\mathrm{T} = \frac{\gamma^2}{\gamma+1}\frac{\mathbf{a}\times\mathbf{v}}{c^2}$$

where γ is the usual Lorentz factor and $\mathbf{a}$ is the acceleration. For a particle of spin $\mathbf{s}$, this Thomas precession adds a term to the potential energy equal to $\mathbf{s}\cdot\boldsymbol{\omega}_\mathrm{T}$. For a nonrelativistic electron moving in a central electrostatic potential energy $V(r)$, the simple spin–orbit energy (caused by the electric field appearing partially as a magnetic field in the electron's rest frame) and the Thomas precession energy neatly combine to give

$$U = \frac{g-1}{2mc^2}\mathbf{s}\cdot\mathbf{L}\frac{1}{r}\frac{\mathrm{d}V(r)}{\mathrm{d}r}$$

The -1 is from the Thomas preces-

sion. With $g = 2$, there is a reduction by a factor of 2, as needed by experiment. One sometimes speaks of a Thomas factor of $\frac{1}{2}$, but more properly it would be $(g-1)/g$. The form of U shows why a subtle and delicate consequence of special relativity, the Thomas precession, can have a gross effect on the energies: The Thomas precession energy is of order $1/c^2$, but so is the spin–orbit energy.

In his longer paper,[8] Thomas describes spin relativistically in terms of a second-rank antisymmetric tensor $S^{\alpha\beta}$ or alternatively an axial four-vector S^α that reduces to a three-vector in the particle's rest frame. Possible covariant equations for spin motion in external fields are greatly restricted by the assumptions that the equation is linear in the spin vector and in the electromagnetic field strengths and that it involves the particle's four-velocity u^α and at most its first time derivative. If the mechanical motion of the particle is described by the Lorentz force equation, the relativistic generalization of the torque equation for a spin magnetic moment in a magnetic field is

$$\frac{\mathrm{d}S^\alpha}{\mathrm{d}\tau} = \frac{e}{mc}\left[\frac{g}{2}F^{\alpha\beta}S_\beta + \frac{g-2}{2c^2}u^\alpha S_\lambda F^{\lambda\mu}u_\mu\right]$$

Here τ is the proper time, $F^{\alpha\beta}$ is the electromagnetic field tensor, and summation over repeated indices is implied. An equivalent equation, somewhat longer but physically more transparent, can be deduced by considering the time derivative with respect to the *laboratory time* of the *rest-frame* spin vector $\mathbf{s}$:

$$\frac{\mathrm{d}\mathbf{s}}{\mathrm{d}t} = \frac{e}{mc}\mathbf{s}\times\left[\left(\frac{g-2}{2} + \frac{1}{\gamma}\right)\mathbf{B}\right.$$
$$-\frac{g-2}{2}\frac{\gamma}{\gamma+1}(\boldsymbol{\beta}\cdot\mathbf{B})\boldsymbol{\beta}$$
$$\left.-\left(\frac{g}{2} - \frac{\gamma}{\gamma+1}\right)\boldsymbol{\beta}\times\mathbf{E}\right]$$

For small velocities (that is, β near 0, γ near 1), one recovers the expected torque on a magnetic moment of $g e\mathbf{s}/2mc$, but at high speeds there are major modifications. The last term (involving $\mathbf{E}$) results from the combined spin–orbit interaction and Thomas precession. The first term describes the precession of the magnetic moment in the magnetic field. Note that if the g-factor were exactly 2, the precession would be precisely at the cyclotron frequency $eB/\gamma mc$. The small departure of the electron's and muon's g-factors from 2 causes a precession more rapid by a factor of $(g-2)\gamma/2$; this difference has formed the basis of high-precision measurements of $g-2$ for the electron and muon. The Thomas

Evan James Williams analyzed bremsstrahlung from the electron's point of view, greatly clarifying the physical understanding. Williams initially worked as an experimenter in Ernest Rutherford's lab and then as a theorist specializing in collision phenomena. Born in Wales in 1903 and educated at Manchester and Cambridge, Williams taught at Manchester University, Liverpool University and University College of Wales, in Aberystwyth. During World War II he worked on antisubmarine warfare; he died, back at Aberystwyth, in 1945. (Photo courtesy of Ian Callaghan, Manchester University)

equation forms the basis of the discussion of spin motion in present-day high-energy particle accelerators.[9]

Two comments. The first of the spin-motion equations illustrates explicitly an important aspect of the use of special relativity, the covariance of physical laws. The first postulate, in effect, demands that physical laws be relationships among quantities having the same Lorentz transformation properties. The left-hand side of the spin-motion equation is the proper-time derivative of a four-vector and hence is also a four-vector, because proper time is an invariant quantity. The right-hand side involves two four-vectors and a second-rank four-tensor combined in such a way as to give a four-vector. The mere requirement of covariance or invariance of physical laws is a powerful tool in delimiting the possibilities. Often, as in this example with the assumptions stated above, the result is unique.

The second comment is that 32 years after the fact, Thomas's equations were rediscovered by a group at Princeton who were unaware of the previous publication.[10] Thomas's paper in the *Philosophical Magazine* stands as a *tour de force* with one immediate application, but which had to wait 40 years for broader use (and see the credit go to others).

The Dirac equation

The next example of special relativistic thinking, the Dirac equation,[11] is so celebrated that it might be passed over. Yet it is worth a brief discussion because it shows how an *idée fixe* can have brilliant consequences, even though the driving requirement is later found not to be really necessary. Paul A. M. Dirac's aim in late 1927 was a relativistic generalization of the Schrödinger equation,

$$i\hbar\frac{\partial\psi}{\partial t} = H\psi$$

Because Dirac required the relativistic

wavefunction to be interpreted as a probability amplitude, he rejected equations—such as those familiar from electromagnetism and already discussed in quantum mechanics by Oskar Klein, Erwin Schrödinger and others—involving second derivatives with respect to time. The continuity equation emerging from such a theory involves a probability density (if it is interpreted as such) that is not positive definite. For Dirac, this was unacceptable. He therefore returned to the structure of the Schrödinger equation with its first-order time derivative. He then sought an equation that treated energy and momentum on the same (relativistic) footing, with the usual wave mechanical equivalence, according to which $(E, c\mathbf{p})$ is replaced by $(i\hbar\partial/\partial t, -ic\hbar\nabla)$. He was led in a then mysterious, but now familiar, way to an equation involving four-component wavefunctions and 4×4 noncommuting matrices, and first order in space–time derivatives. For physicists the Dirac equation is a thing of beauty and a joy to behold. Further, it contains a seeming miracle, the electron spin, and with the correct g-factor! The demand of linearity in E and $\mathbf{p}$ forced the four-component wavefunction on Dirac. The added degrees of freedom turned out to be the two signs of the spin direction and two signs of the energy (which led to positrons).

The Dirac equation describes quarks and leptons of all sorts, not just electrons. It contains the previously *ad hoc* nonrelativistic spin of Pauli in the appropriate limit (and with its spin–orbit interaction correctly including Thomas precession). Its two signs of the energy emerged later to be the description of particles of spin ½ and their antiparticles. All of these from Dirac's demand for a wave equation first order in the time derivative and so also first order in spatial derivatives—what bounty to be obtained from a requirement that was later found unnecessary!

To clarify briefly the last remark, let me say that particles occur in nature as fermions (with intrinsic angular momenta that are odd half-integral multiples of $\hbar$) or bosons (with intrinsic angular momenta that are integral multiples of $\hbar$). Spin-½ fermions are described by the Dirac equation, but bosons are described by other wave equations. The simplest is the second-order equation rejected by Dirac. When multiparticle descriptions ("second quantization" techniques) were developed, it was found that the "probability density" that was not positive definite is actually a charge density, with its sign depending on the preponderance of particles or antiparticles. One can even develop single-particle descriptions of bosons, analogous to the Dirac equation, based on a relativistic second-order differential equation.

Before leaving Dirac, we must return to the Sommerfeld formula for the energy levels of hydrogenlike atoms. Solution of the Dirac equation for an electron in a fixed Coulomb potential leads to the same energy levels, with the Sommerfeld angular momentum integer k interpreted as $j + \frac{1}{2}$, with j equal to $\frac{1}{2}$, $\frac{3}{2}$, ... The orbital angular momentum quantum number l does not appear explicitly, only the total angular momentum. Indeed, the idea of separately conserved spin and orbital angular momentum quantum numbers, absent in Sommerfeld's model, but accepted after the work of Pauli and of Uhlenbeck and Goudsmit, is also absent in Dirac's description. Spin *is* an important concept and Sommerfeld did not have it, but its separation from orbital motion is only valid nonrelativistically. Strictly speaking, there is only total angular momentum. For me, this throws some light on the success of Sommerfeld's calculation and the close connection of spin with relativity that Dirac made manifest.

Chronologically, the next major development with special relativity as a key ingredient was quantum field theory. With Dirac's work on quantization of the radiation field pointing the way, a fully general formulation of the quantum theory of fields, with continuous degrees of freedom, was developed at the hands of Werner Heisenberg, Pauli and others. Continuous systems in classical physics, apart from electromagnetism, were known, of course. Euler–Lagrange equations of motion can be derived from the principle of least action, based on a Lagrangian density $\mathscr{L}$ that is a function of the fields and their derivatives. Since quantum mechanics appears most naturally as a generalization of Hamilton's equations of motion for canonically conjugate coordinates, the Hamiltonian played a central role in quantum field theory, with time singled out through, for example, the use of equal-time commutation (or anticommutation) relations for the field operators. Creation and annihilation operators for particles in specific states of motion emerge from normal-mode expansions and the quantization of the simple harmonic oscillator Hamiltonians in the standard way.

While the Hamiltonian is important, the Lagrangian density and its space–time integral, which is the action integral, are extremely valuable constructs. Invariances of the Lagrangian density under various transformations (translations, rotations, internal "rotations" and so forth) lead to conserva-

tion laws. The first postulate of relativity imposes the requirement of Lorentz invariance on the action integral and on the Lagrangian density. The Euler–Lagrange equations of motion emerge naturally in covariant form and the condition that $\mathscr{L}$ be a Lorentz scalar restricts mightily the forms of hypothetical interactions among the fields.

The story of the emergence of the Lagrangian, the principle of least action and the use of path integrals as a direct approach to quantum mechanics and quantum field theory is not my task here. It is identified with the name of Richard Feynman above all others, but the idea goes back to Dirac, who in 1933 stressed the superiority of the approach via the Lagrangian because of its manifest invariance with respect to Lorentz transformations.[12] The various approaches to relativistic quantum field theory can be found in standard texts.[13]

In our discussion of random examples of the roles and uses of special relativity in theoretical physics we have come rather far from the practicalities of atomic spectra and Sommerfeld's atom. In part, that is because special relativity permeates the work place of the physicist, if not the layman, from the lowest energies in atomic systems (where the precision is so high that the tiny relativistic effects must be included) to the highest laboratory energies in the giant particle accelerators (where relativistic effects are gross and must enter even the crudest considerations).[14] It is, as I said at the beginning, a pointless exercise to try to count the ways. Instead, I close with examples that illustrate the truth of what Albert Michelson and Edward Morley found: There is no preferred frame of reference.

Frames of reference

Anyone considering swiftly moving charged particles learns very soon to examine the physical processes in more than one inertial reference frame. Consider bremsstrahlung, that is, the emission of radiation by a light, charged particle on collision with an atomic nucleus. The process was first calculated by Hans Bethe and Walter Heitler, Sommerfeld and others on the basis of second-order perturbation theory in quantum mechanics. But as was shown brilliantly by Carl Friedrich von Weizsäcker and especially Evan James Williams, the physical understanding of bremsstrahlung and many other processes benefits greatly when they are viewed in appropriate inertial frames.

Conventionally, bremsstrahlung is described in the laboratory as the acceleration of a swiftly moving, light,

charged particle by the Coulomb field of an essentially stationary heavy nucleus, accompanied by the emission of radiation (see the figure on page 215). In the Weizsäcker-Williams method of virtual quanta,[15] one instead views the process in the rest frame of the incident particle. In this frame the heavy nucleus is incident at high speed upon the stationary light particle. The nucleus's Coulomb field appears, because of the Lorentz transformation, very much like a pulse of electromagnetic radiation with transverse electric and magnetic fields having a broad frequency spectrum with a number of virtual quanta per unit frequency interval inversely proportional to frequency.[16] These virtual photons, of frequencies ω', are scattered by the stationary light particle into a wide angular distribution—proportional to $\frac{1}{2}(1 + \cos^2\theta')$, at least at low frequencies. If the motion is highly relativistic, these photons, which are predominantly at low frequencies and large angles, are caused by the relativistic Doppler shift to appear in the laboratory at moderate to high frequencies $\gamma\omega'$ and small angles on the order of $1/\gamma$. Bremsstrahlung thus has an alternative description as the scattering of virtual quanta in the rest frame of the incident particle.

There is, in fact, an apparent dilemma, and a neat explanation due to Weizsäcker. The spectrum of virtual quanta of the heavy nucleus contains energies $\hbar\omega$ far above the rest energy of the light particle. Such quanta, when scattered and transformed to the laboratory frame, would have energies on the order of $\gamma\hbar\omega'$, far greater than γmc^2, violating conservation of energy—the largest possible photon energy is $(\gamma - 1)mc^2$. As Weizsäcker showed in detail, however, this violation does not occur, because the scattering of quanta with energies $\hbar\omega'$ greater than mc^2 in the incident particle's rest frame is described by the Compton cross section, not the simple Thomson (recoilless) cross section. The Compton scattering cross section falls off with increasing photon energy, the scattered photons have lower energies because of recoil, and the angular distribution changes. All of these features combine to conserve energy back in the laboratory and give all the details of the less transparent calculations.

A wide variety of inelastic collisions of charged particles can profitably be described by the Weizsäcker–Williams method, whereby the fields of a "projectile" are replaced by an equivalent spectrum of photons interacting with a "target." Usually, but not always, the target is at rest in a reference frame other than the laboratory.

Another example of the usefulness of

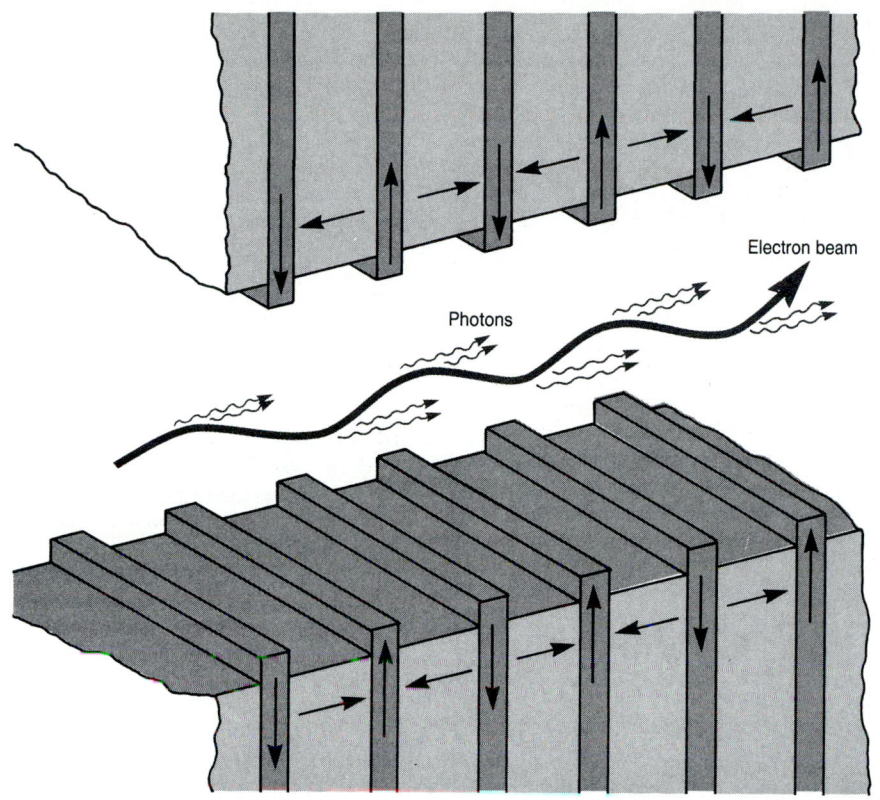

Undulator magnet for enhancing the synchrotron radiation from a beam of electrons. Such devices are also the active parts of free-electron lasers. Steel pole pieces are shown by verticle arrows; $SmCo_5$ permanent magnets, assembled with alternating polarities, are shown by horizontal arrows.

shifting from one reference frame to another is in understanding the frequency of radiation created by electron undulators, devices in which a relativistic electron beam is caused to make very small sideways oscillations by a periodic array of magnets of alternating polarity.[17] The array can be realized by two sets of samarium–cobalt permanent magnets, alternating in polarity, separated by iron pole pieces; the set looks rather like the black and white keys on a piano, with a corresponding keyboard above (see the figure above).[18] The beam passes through the middle and experiences in succession magnetic fields directed upward and downward with spatial wavelength λ_u. The magnetic force causes the beam to undulate from side to side with the same basic period.

Because the electrons are accelerated, they radiate. But at what frequency? To determine it, we first make a Lorentz transformation to the approximate rest frame of the beam. The electrons are almost at rest, so the magnetic force in that frame is negligible, but there is an electric field, equal to $\gamma\beta\times\mathbf{B}$, which causes the electrons to oscillate (because $\mathbf{B}$ is periodic). The oscillating electrons emit dipole radiation in the well-known way. The wavelength is governed by the periodicity of the magnetic field, as seen in this reference frame. Because of the Fitz-Gerald–Lorentz contraction, the magnet structure rushing by has a wavelength λ'_u of λ_u/γ. This is the wavelength of the radiation emitted in the electrons' rest frame. If we now return to the laboratory, we can ask for the wavelength of the radiation there. The relativistic Doppler shift (the Lorentz transformation for light), in the limit of large γ, takes the form

$$\lambda \simeq (\lambda'/2\gamma)(1 + \gamma^2\theta^2)$$

where λ' is the wavelength in the moving frame and θ is the angle of the radiation in the laboratory relative to the direction of motion. With the dipole wavelength from the rest frame, the result for the wavelength of the radiation from an undulator is

$$\lambda \simeq (\lambda_u/2\gamma^2)(1 + \gamma^2\theta^2)$$

The factor γ^2 means that the radiation can have extremely short wavelengths. For example, an undulator wavelength λ_u of 4 cm is realistic, as is a γ of 3000 (for electrons of 1.5 GeV). The undulator radiation would then have a wavelength of roughly 22 Å (an energy of 560 eV): soft x rays. The motion in the rest frame of the beam is not completely nonrelativistic and sinusoidal. As a result, a number of harmonics of the fundamental are radiated; the relative intensities depend on the details of the undulator. For an undulator of N periods, the radiation peaks have a height proportional to N^2, a fractional width $\Delta\lambda/\lambda$ of order $1/N$ and an angular spread of order $1/\gamma\sqrt{N}$. Undulators provide intense, coherent, tunable sources of soft x rays and vacuum ultraviolet for condensed matter and biological research; they also serve as the "engines" that drive free-electron lasers.

The phenomenon of channeling radiation is my last example of how one can understand essential features by viewing a situation in different inertial frames. If high-energy positrons are incident on a crystal in a direction closely parallel to a crystal plane, they are channeled between the planes like a stream in a straight Alpine valley with mountain ranges on either side. In general, there is motion from side to side as the positron tests the slopes. This transverse motion implies acceleration and hence radiation, just as in the undulator. The frequencies emitted are governed by the basic parameters of the crystal lattice and the relativistic parameter γ of the incident positrons.

The average transverse electrostatic potential experienced by a positron passing between crystal planes is approximately quadratic. (The period of transverse oscillation is long compared with the transit time for one lattice spacing. The longitudinal periodicity is thus washed out.) The "spring constant" $V''(0)$ for the effective oscillator can be estimated by dimensional arguments to be roughly 13.6 eV/a^2, where a is the distance between lattice planes. The transverse electric field in the laboratory, $\mathbf{E}_\perp$, is $V''(0)\mathbf{x}_\perp$. If the positron were nearly at rest in the laboratory, its transverse frequency of oscillation ω_0 would simply be $(V''(0)/m_e)^{1/2}$, corresponding to optical photon energies of a few electron volts.

Let us go to the moving coordinate frame in which the positron is almost at rest, the frame moving with the velocity of the incident beam. There the positron experiences the transformed electrostatic force of the lattice. This has electric and magnetic components, but since the positron's motion is nonrelativistic in this frame, only the electric force is important. The transformation law of the transverse electrostatic field is

$$\begin{aligned}\mathbf{E}'_\perp &= \gamma\mathbf{E}_\perp \\ &= \gamma V''(0)\mathbf{x}_\perp \\ &= \gamma V''(0)\mathbf{x}'_\perp\end{aligned}$$

(The transverse coordinates are, of course, unchanged by the Lorentz transformation.) The "spring" is thus stiffer by a factor of γ in the positron's average rest frame, and the frequency

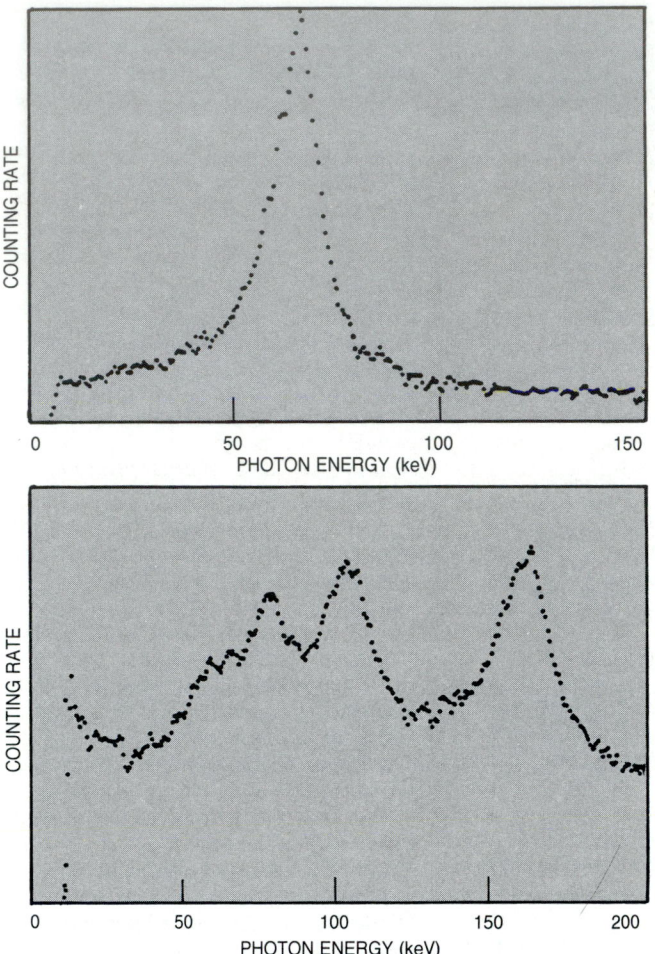

Channeling radiation spectra for 55-MeV positrons (**a**) and electrons (**b**).[21] The particle beams are parallel to the (110) planes of diamond crystals. Note that the positron spectrum exhibits a single peak, while the electron spectrum is highly structured.

of its transverse motion is $\omega_0\sqrt{\gamma}$. The dipole radiation from the oscillation appears in the laboratory in the usual narrow forward cone with angular width of order $1/\gamma$ and with a frequency ω given approximately by $2\gamma^{3/2}\omega_0/(1 + \gamma^2\theta^2)$. For 50-MeV positrons, γ is around 100 and the photon energies are 2000 times optical, or tens of keV.[19] For 10-GeV positrons, the energies are 10^7 times optical—around 50 MeV![20]

To the extent that the potential is truly quadratic, even a quantum mechanical treatment gives only the fundamental line. Observations always show a strong peak at the fundamental, sometimes with smaller peaks near multiples of the basic energy. For high-energy electrons the emission occurs in the same general energy range, but the electron spectra can exhibit a number of prominent peaks.[19,21] (See the graphs above.) The difference arises because the electrons experience a funnel-shaped electrostatic interaction, not a simple harmonic oscillator potential. The frequency of classical transverse motion then depends on the

amplitude, and for a given amplitude the periodic motion corresponds to a superposition of many harmonics of the basic frequency. (In quantum language, the energy levels of transverse motion are not equally spaced and transitions are not restricted to those between adjacent levels.)

These examples are but illustrative of the value of viewing physical phenomena in different inertial frames. The absence of a preferred frame of reference, established by the Michelson–Morley experiment 100 years ago and bothersome to many at the time, has been turned with Einstein's guidance into a very positive virtue. I have tried to document how a few practitioners used special relativity to good effect in its early days, and how it still gets used to simplify and clarify our understanding.[22] Equivalent inertial frames are everywhere. Use them for profit and enjoyment!

References

1. W. Pauli, *Relativitätstheorie* (*Enzyklopädie der Mathematischen Wissenschaften*, vol. 19), Teubner, Leipzig (1921).

2. A. Pais, *Inward Bound*, Oxford U. P., New York (1986).

3. For a succinct account, see A. Pais, *Subtle is the Lord: The Science and Life of Albert Einstein*, Oxford U. P., New York (1982), ch. 7. The details are described by A. I. Miller, in *Some Strangeness in the Proportion*, H. Woolf, ed., Addison–Wesley, Reading, Mass. (1980), p. 66.

4. Sommerfeld's original paper is not easily accessible. In English, an authoritative source is the translation of the third edition of his book *Atombau und Spektrallinien*: A. Sommerfeld, *Atomic Structure and Spectral Lines*, H. L. Brose, transl., Methuen, London (1923), ch. VIII, sect. 2 and mathematical note 16; see also sect. 7. A number of texts from the 1930s present his derivation.

5. A. Pais, *Subtle is the Lord: The Science and Life of Albert Einstein*, Oxford U. P., New York (1982), p. 408.

6. E. Segré, *From X-Rays to Quarks*, Freeman, San Francisco (1980), ch. 8. A. Pais, *Inward Bound*, Oxford U. P., New York (1986), ch. 13.

7. L. H. Thomas, Nature **117**, 514 (1926).

8. L. H. Thomas, Philos. Mag. **3**, 1 (1927).

9. B. W. Montague, Physics Reports **113**, November 1984, page 1.

10. V. Bargmann, L. Michel, V. L. Telegdi, Phys, Rev. Lett. **2**, 435 (1959).

11. P. A. M. Dirac, Proc. R. Soc. London, Ser. A **117**, 610 (1928).

12. P. A. M. Dirac, Phys. Z. Sowjetunion **3**, 64 (1933); reprinted in *Selected Papers on Quantum Electrodynamics*, J. Schwinger, ed., Dover, New York (1958), p. 312.

13. See, for example, C. Itzykson, J.-B. Zuber, *Quantum Field Theory*, McGraw–Hill, New York (1980).

14. W. K. H. Panofsky and E. M. Purcell address aspects of relativistic engineering in *Some Strangeness in the Proportion*, H. Woolf, ed., Addison–Wesley, Reading, Mass. (1980), p. 66.

15. C. F. von Weizsäcker, Z. Phys. **88**, 612 (1934). E. J. Williams, Det. Kgl. Danske Viden. Selskab., Math-fys. Medd. **13**(4) (1935). See also E. J. Williams, Proc. R. Soc. London **139**, 163 (1933).

16. The spectrum and some of the applications are discussed in a number of texts. One is J. D. Jackson, *Classical Electrodynamics*, 2nd ed., Wiley, New York (1975), sect. 15.4.

17. H. Motz, J. Appl. Phys. **22**, 527 (1951).

18. D. Attwood, K. Halbach, H.-J. Kim, Science **228**, 1265 (1985).

19. M. J. Alguard *et al.*, Phys. Rev. Lett. **42**, 1148 (1979). R. L. Swent *et al.*, Phys. Rev. Lett. **43**, 1723 (1979).

20. J. Bak *et al.*, Nucl. Phys. B **254**, 491 (1985).

21. H. Park *et al.*, J. Appl. Phys. **55**, 358 (1984).

22. The effectiveness of the use of different inertial frames is illustrated masterfully in E. M. Purcell, *Electricity and Magnetism* (*Berkeley Physics Course*, vol. 2), 2nd ed., McGraw–Hill, New York (1985). □

letters

Relativity and quantum theory

Because my article "The impact of special relativity on theoretical physics," pages 212–220, was prepared in haste and subjected to limited review by others, I overlooked the insightful work of my old friend and office mate in graduate school, Lawrence Biedenharn. In a paper entitled "The 'Sommerfeld puzzle' revisited and resolved," Biedenharn establishes an intimate connection between the relativistic old quantum theory of Arnold Sommerfeld and the relativistic quantum mechanics of Paul A. M. Dirac for the Kepler problem (with its special O(4) symmetry).[1] He shows that the correct correspondence for the nonrelativistic Sommerfeld problem is not the spinless Schrödinger equation, but rather the Schrödinger equation with spin as a dynamically independent variable.

My comment that Sommerfeld's "treatment within the old quantum theory captured the essence of the spinning electron without knowledge of that degree of freedom" is given explicit meaning in Biedenharn's paper, to which I commend you all. I thank Pekka Pyykkö of the University of Helsinki for drawing Biedenharn's work to my attention.

On a separate point, Lee C. Pittenger has kindly pointed out that in discussing wigglers I may have misled the unthinking reader in identifying the wavelength of the periodic magnetic structure rushing by the electron in its rest frame directly with the wavelength of the dipole electromagnetic radiation emitted in that frame. It is of course the frequencies that are the same. The correct statement is that the wavelength of dipole radiation in that frame is $\lambda' = \lambda_{u}/\gamma\beta$, while the wavelength of the magnetic structure is $\lambda_{u}' = \lambda_{u}/\gamma$. In my article, I treated the two as identical, which is legitimate only for γ much larger than 1.

Reference

1 L. C. Biedenharn, Found. Phys. **13**, 13 (1983).

J. DAVID JACKSON
*University of California
Berkeley, California*

7/87

Werner Heisenberg and the beginning of nuclear physics

The advent of quantum mechanics caused a greater transformation in the understanding of physical reality of microscopic phenomena than the change in the understanding of macroscopic phenomena brought about by relativity.

Arthur I. Miller

PHYSICS TODAY/NOVEMBER 1985

Great advances in science alter our view of the world. Galileo Galilei's theory of motion, Albert Einstein's theory of special relativity and Werner Heisenberg's invention of quantum mechanics, with its subsequent interpretation by Niels Bohr and Heisenberg himself, spring immediately to mind (figure 1). In exploring these episodes we must recognize that historical narrative without investigation of conceptual transformation is just chronology.

An excellent case study to illustrate the shift in points of view is that of the roots and ramifications of Heisenberg's ground-breaking papers in nuclear theory, "On the Structure of the Atomic Nucleus," published[1] in 1932–33. Investigating the roots of these papers brings into bold relief a fascinating aspect of the development of atomic physics, namely, that for Bohr, Heisenberg and Erwin Schrödinger, among others, conceptual problems were often more critical than considerations of empirical data.

During 1926–27 the most critical fundamental problem of physics was the nature of physical reality on the atomic level. The continuing struggle of physicists such as Bohr and Heisenberg to understand nature at a level beyond sense perceptions, forced them to grapple with problems that cut to the very core of how we construct knowledge: How thoroughly connected are visualization and intuition? How is the intuition that leads to one sort of visualization replaced by another? How did the imagery of subatomic phenomena represented in an analogous fashion to atomic transitions (figure 2) or Yukawa's nuclear force (figure 3) arise? By what process was the first representation replaced by the second, a precursor of Feynman-type diagrams? Probing such problems offers insight into the transformation of the imagery used to describe the atom—from models of the atom viewed as a tiny solar system (figure 4), to the diagrams of energy levels (figure 2), and then to Feynman diagrams (figure 3). The ramifications of replies by Bohr and, particularly, Heisenberg to these problems have set the course of subnuclear physics and altered dramatically our understanding of physical reality in ways that are

still not completely understood.

I will proceed as follows: First, I will show that the roots of Heisenberg's nuclear exchange force are found in his June 1926 discovery of the exchange energy in atomic processes. The discussion in early 1927 of the Bohr–Heisenberg interpretation of quantum mechanics completed setting the stage for Heisenberg's introduction, in June 1932, of the nuclear exchange force. Next I will analyze two early ramifications of Heisenberg's exchange force: Enrico Fermi's theory of beta decay (January 1934) and Hideki Yukawa's meson theory of nuclear forces (November 1934). The epilogue connects these developments in nuclear physics with the modern notion of forces transmitted by particles.

New atomic physics

From June 1925 through early 1927 certain longstanding problems in atomic physics, such as the anomalous Zeeman effect and the spectrum of the helium atom, were solved with the new quantum mechanics Heisenberg had invented in June 1925. In this heady period so many stunning results appeared every month, it was as if years were passing. However, the physical

---

Arthur I. Miller is professor of history and philosophy of science at University College London, in London, England.

Werner Heisenberg in 1926, one year after he had invented the matrix, or quantum, mechanics. Figure 1

meaning of the intermediate manipulations was unclear, although they produced results that could be compared with experiment: Quantum mechanics lacked correspondence rules for the physical meaning of its symbols. In other words, the mathematical symbols and operations of quantum mechanics (its syntax) did not yet possess unambiguous meanings (semantics). When Bohr and Heisenberg were struggling in Copenhagen to interpret quantum mechanics, Heisenberg wrote[4] to Wolfgang Pauli on 23 November 1926: "What the words 'wave' and 'corpuscle' mean we know not any more."

Heisenberg provided[5] an excellent glimpse into their struggles in his November 1926 paper, "Quantum mechanics," where he emphasized that our "customary intuition" (gewöhnliche Anschauung) cannot be extended into the atomic domain. This term appears often in the German-language scientific literature of 1923–27. The German usage of this phrase is loaded with subtleties because the notion of intuition (Anschauung) as it was used by Bohr, Max Born, Heisenberg, Pauli and Schrödinger is rooted in Kantian philosophy. By "customary intuition" these physicists meant the intuition or

mental imagery that is constructed from objects or phenomena that we actually observe in the world of sense perceptions. Examples include the mental imagery of a solar-system atom, with its positively charged nucleus and satellite electrons; light as strictly a wave phenomenon whose properties are abstracted from those of water waves; and electrons that behave like billiard balls. Customary intuition is the mental imagery that is imposed on classical physics. Heisenberg went on to describe how one by one these customary intuitions had to be given up: Data from dispersion had undermined the picture of the solar-system atom, the experiments of Arthur Compton, Hans Geiger and Walther Bothe had forced physicists to consider the light quantum seriously, and according to Einstein's results on the ideal quantum gas, free electrons had lost their individuality. In fact, Heisenberg had been forced to abandon customary intuition to invent the matrix, or quantum, mechanics. In the original formulation of quantum mechanics, Heisenberg characterized the atom with measurable quantities such as its spectral lines. Yet in key scientific papers, Heisenberg and Born, among others, lamented

the loss of a directly "intuitive" interpretation of the new atomic physics.

Erwin Schrödinger thought he knew how to bring back customary intuitions—somewhat altered, of course. In a paper of March 1926 on wave mechanics, Schrödinger explained[6] in a highly subjective tone that he had been driven to formulate his theory because he had been "repelled by the methods of transcendental algebra, which appeared very difficult to me and by the lack of visualizability [Anschaulichkeit]" of Heisenberg's quantum mechanics. Classical realists, such as Einstein and the venerable Hendrik A. Lorentz, had nothing but praise for Schrödinger's wave mechanics, in which the bound electron was represented as a charged wave surrounding the nucleus.

Heisenberg thought otherwise. On 8 June 1926 he wrote[4] to Pauli that "what Schrödinger writes on the visualizability of his theory . . . I consider trash. The great accomplishment of Schrödinger's theory is the calculation of matrix elements." And it was to this end that Heisenberg exploited[7] what he insisted was only the mathematical equivalence of the two theories in a remarkable paper entitled, "Many-

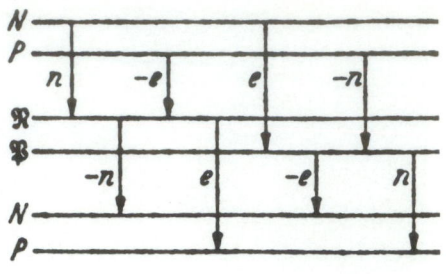

Beta decay depicted as a "cascade-like" process analogous to the transition between atomic energy levels. N and P denote a neutron and proton, and $\mathfrak{N}(\mathfrak{P})$ a "hypothetical stationary state" of a neutron and proton, n and — n a neutrino and antineutrino, and e and — e an electron and positron. (From G. Wentzel, *Zeitschrift für Physik* **104**, 34, 1936.)   Figure 2

Gregor Wentzel's "schemata" for depicting beta decay as a "two-stage process" that is intermediated by a meson. (From reference 22, German edition.)   Figure 3

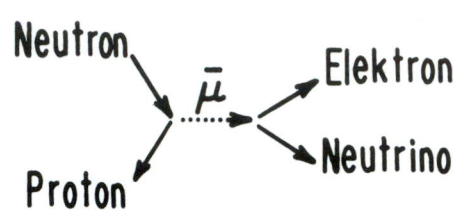

body problem and resonance in quantum mechanics" completed in June 1926. Heisenberg warned that Schrödinger's "intuitive pictures" should not be imposed on quantum mechanics, whose invention had required their denial in the first place, and that Schrödinger's intuitive pictures should be especially avoided in the many-body problem. In March 1926, Schrödinger had suggested that many-body problems should temporarily be set aside in both wave and matrix mechanics. These theories use classical potential functions for point particles that are actually interpenetrating states of vibration; this causes severe conceptual difficulties for the many-body problem. Heisenberg's direct rebuttal to Schrödinger's pessimistic assessment came in June 1926, in a paper in which Heisenberg *treated* the many-body problem—he writes that we must set limitations "on the discussion of the intuition problem."

Heisenberg's desire to demonstrate that matrix mechanics could deal with many-body systems was the catalyst for his June 1926 paper. But the results in this paper drew him ever deeper into the "intuition problem." An even more fundamental problem for him was the connection of matrix mechanics with Bose–Einstein statistics and Pauli's "prohibition against equivalent orbits." At first sight, in either matrix or wave mechanics, atoms should obey classical statistics. So Bose–Einstein statistics seemed to contradict these theories. Yet, for example, only the ortho and para states of the two-electron helium atom are possible. Also Pauli's prohibition against equivalent orbits seemed not to fit into the mathematics of matrix or wave mechanics.

It suffices here to sketch how Heisenberg set the stage for the many-body

calculations that followed. He analyzed the spectrum of two identical systems coupled through an interaction that is symmetrical in the coordinates of both systems. The total energy of the unperturbed total system is degenerate. Using first-order perturbation theory Heisenberg demonstrated that the symmetrical interaction breaks the degeneracy. The resulting energy spectrum contains two series of energy levels, one higher than the other. Each energy level contains a contribution that can be interpreted as the result of two systems exchanging places, that is, as a resonance, or exchange, energy.

Applying these results to the helium atom, and including the electron's spin for which there was then no mathematical formulation, Heisenberg reached the following conclusions: The reduction of the statistical weight to one, that is, the number of possible states of the two-electron system to only the totally antisymmetric state, was an example of Bose–Einstein counting. This choice satisfied automatically "Pauli's prohibition against equivalent orbits." Consequently, concluded Heisenberg, Bose–Einstein statistics and Pauli's prohibition against equivalent orbits are consistent with quantum mechanics. Owing to the indistinguishability of electrons (which Heisenberg assumed to be a consequence of Bose–Einstein statistics), any depiction of the exchange energy "makes no sense physically." To Heisenberg these results augured further restrictions on the "reality of corpuscles," and hence on their depiction. Heisenberg went on to extend his results to a system of N identical spin-½ particles. He recognized that the statistics of spin-½ particles go "beyond the Bose–Einstein" statistics for reasons that may reside in the "phase relations between

the component systems or particles." As he wrote in his sequel paper "Quantum mechanics," of November 1926, one would like to determine to "what extent Pauli's prohibition of equivalent orbits makes necessary a modification of the Einstein statistics."

Independent of Heisenberg, P. A. M. Dirac was also investigating[8] the symmetry properties of systems composed of identical particles. Using a more general approach, Dirac concluded that of all possible wavefunctions for systems of identical particles, only the symmetric and antisymmetric combinations are acceptable. He showed that the symmetric combination is associated with Bose–Einstein statistics and that the antisymmetric is associated with spin-½ particles. The vanishing of the antisymmetric wavefunction when two electrons are in the same state "is just Pauli's exclusion principle." Up to this point Heisenberg's and Dirac's results are the same, as Dirac himself wrote in a footnote mentioning that Born had just notified him of Heisenberg's "many-body" paper. But Dirac went on to use an ideal gas to prove that antisymmetric wavefunctions are associated with a new statistics *intrinsic* to spin-½ particles.

Heisenberg recalled[9] in an interview (25 February 1963) that the many-body paper was "full of excitement for me because so many new things came up [that] for quite a considerable time I did mix up Bose statistics with Fermi–Dirac statistics." So, on the way to solving the matrix mechanics for the spectrum of helium, Heisenberg found it necessary to invent new concepts such as the exchange energy and, independently, the Fermi–Dirac statistics. Courage to persevere in solving a problem by overcoming obstacles through inventing new concepts, and an intuitive feel for which obstacles

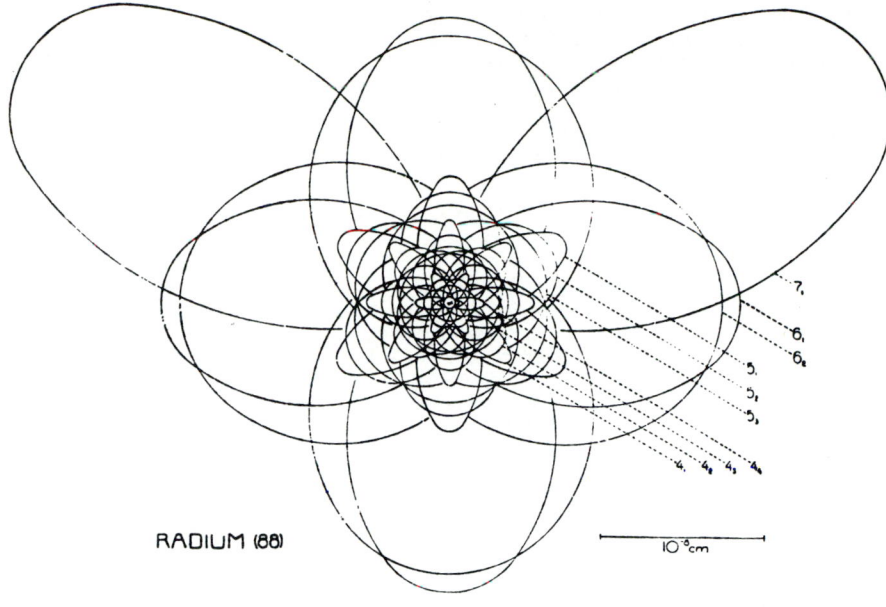

RADIUM (88)

10^{-8} cm

Bohr's atomic theory. This rendering of the radium atom from Bohr's theory depicts what Max Born wrote of in 1923 as "a remarkable and alluring result of Bohr's atomic theory [that] the atom is a small planetary system . . . the thought that the laws of the macrocosmos in the small reflect the terrestrial world exercises a great magic on mankind's mind."[2] (Figure from reference 3.) Figure 4

should be bypassed for the moment, are trademarks of Heisenberg's style of research. We recall that in the 1925 quantum-mechanics paper, he was undeterred by his lack of understanding why xy does not equal yx, that is, by his ignorance of matrix algebra. This style would stand him well while working with Bohr during 1926–27, and again during the 1930s, when he invented the exchange force in nuclear physics despite having to deal with "Bose" electrons. We next turn to this work.

Restricted metaphors

After the heroic struggles of late 1926 through early 1927 in Copenhagen, Bohr and Heisenberg arrived independently at apparently separate resolutions for the paradoxes posed by the wave–particle duality of light and matter.

In the 1927 paper on the uncertainty principle, Heisenberg further explored[10] the concept of intuition and, in fact, entitled the paper "On the intuitive [*anschaulich*] content of the quantum-theoretical kinematics and mechanics." Focusing exclusively on his own quantum mechanics, with its unvisualizable particles, and encouraged by the redefinitions of macroscopic physical reality required by special and general relativity, Heisenberg proposed that in the atomic domain a revision of our usual kinematical and mechanical concepts "appears to follow directly from the fundamental equations of the quantum mechanics." In this way Heisenberg boldly demarcated the notion of "to be understood intuitively" from the visualization of atomic processes. He permitted the syntax of quantum mechanics to determine the restrictions on such perception-laden symbols as position and momentum. These restrictions are the uncertainty relations. Consequently, Hei-

senberg redefined the concept of intuition through the theory's mathematics and separated intuition from visualization. The paper on the uncertainty principle was, therefore, a turning point in Heisenberg's view of the nature of physical reality. In later years Heisenberg recalled Pauli's response to the uncertainty principle paper as, "*Es wird Tag in der Quantentheorie*" ("Day is breaking over the quantum theory").

Bohr disagreed with Heisenberg. The dilemma, in Bohr's view, was that whereas images constructed from our customary intuitions had to be separated from the laws of physics, we are forced to phrase these laws in a language tempered by sense perceptions because it is the only language we have. By September 1927 Bohr found[11] that he could grasp both horns of the dilemma with the principle of complementarity, which he took to be rooted in the wave–particle duality of matter and radiation. In the atomic domain this principle separates intuitive pictures from the actual development of atomic systems in space and time: The development is governed by physical laws that require a "departure from visualization in the usual sense."

Heisenberg agreed that the complementarity principle placed restrictions on metaphors from the macroscopic world, that is, from ordinary perceptions. But he remained wary of these metaphors, owing to their previous disservices. In fact, in a letter of 16 May 1927 to Pauli, Heisenberg wrote[4] that there are "at present between Bohr and myself differences of opinion on the word 'intuitive'."

To summarize thus far: By 1929 Heisenberg and most other physicists realized that in the atomic domain visualization and visualizability are mutually exclusive. Visualization re-

sults from the cognitive apparatus of our mind acting on sense data to produce, somehow or other, the sort of mental imagery that until 1923 physicists had imposed cavalierly on all physical theories. So visualization is what Heisenberg referred to in his 1926 paper, "Quantum mechanics," as the "customary intuition" that could not be extended into the atomic domain. Visualizability, on the other hand, concerns the intrinsic characteristics of atomic entities that are not open to our perceptions—the electron spin, for example. In classical physics, visualization and visualizability are synonymous. In the 1927 paper on the uncertainty principle, Heisenberg permitted the mathematics of quantum mechanics to decide the meaning of the theory's symbols, that is, the intrinsic properties of subatomic particles are revealed through the mathematics of quantum theory. But he still resisted any imagery of atomic phenomena. Thus, for Heisenberg, as of 1927, in the atomic domain visualizability and intuition did not yet possess a depictive or visual component. (In Kant's philosophy *Anschaulichkeit*—visualizability—is less abstract than *Anschauung*—intuition. For example, a typical definition of *Anschaulichkeit* that may be found in a German-language philosophical dictionary is: "the immediately given . . . the readily graspable in the *Anschauung*." In 1927 Heisenberg found it necessary to invert the Kantian notions of *Anschauung* and *Anschaulichkeit*, no mean feat.)

The nuclear exchange force

In 1932 Heisenberg introduced the depictive component of visualizability in another virtuoso performance, Part I of his nuclear physics papers published in the *Zeitschrift für Physik*. In these papers we can glimpse—as perhaps

1933 Solvay Conference. Some of the *dramatis personae* discussed here appear among the participants sitting (left to right): E. Schrödinger, I. Joliot, N. Bohr, A. Joffe, M. Curie, P. Langevin, O. W. Richardson, E. Rutherford, T. De Donder, M. de Broglie, L. de Broglie, L. Meitner and J. Chadwick. Standing (left to right): E. Henriot, F. Perrin, F. Joliot, W. Heisenberg, H. A. Kramers, E. Stahel, E. Fermi, E. T. S. Walton, P. A. M. Dirac, P. Debye, N. F. Mott, B. Cabrera, G. Gamow, W.Bothe, P. M. S. Blackett (at back), M. S. Rosenblum, J. Errera, E. Bauer, W. Pauli, J. E. Verschaffelt, M. Cosyns (at back), E. Herzen, J. D. Cockcroft, C. D. Ellis, R. Peierls, A. Piccard, E. O. Lawrence and L. Rosenfeld. (AIP Niels Bohr Library)   Figure 5

nowhere else in 20th-century physics—the struggles of a scientist trying to frame a theory in a situation where basic laws of physics, as well as current theory, seem to be violated. As was his style, Heisenberg meant his research in nuclear physics to cover more than this one discipline. Consequently it is helpful to survey[12] the state of fundamental physics in 1932.

The euphoria that followed the successes of both quantum theory and Dirac's relativistic theory of the electron, formulated in 1928, was short-lived. Several interconnected problems—for example, the interpretation of the electron's negative energy states and the explanation of the continuous beta-ray spectrum from the decay of nuclei—still remained unresolved. Furthermore, because protons and electrons were assumed to constitute nuclei, nitrogen-14 had the wrong statistics. Thus physicists suggested that nuclear electrons may not obey quantum theory (Dirac's equation included) because their spins would be suppressed, among other reasons. Some physicists, principally Bohr, suggested that energy conservation may not hold in beta decay because the energy spectrum of the beta rays for a supposedly two-body final state is continuous.

In 1932 James Chadwick's discovery of the neutron led, as Heisenberg wrote[1] in Part I of his paper, to an "extraordinary simplification for the theory of the atomic nucleus." In addition to settling the problem of statistics, the neutron suggested to Heisenberg a way to relate the problems of beta decay to the form of the attractive force that binds the nucleus. As Heisenberg wrote[13] to Bohr on 20 June 1932, "The basic idea is to shift the blame for all principal difficulties [of fundamental physics] onto the neutron and to refine quantum mechanics in the nucleus."

To explain the statistics of nitrogen-14, Heisenberg assumed that the neutron is a particle with spin ½. (However, in Part III of the paper he expressed[1] reservations even on this point.) Next Heisenberg had to decide whether the neutron is a composite particle, consisting of a proton and electron, or a fundamental particle. The conception of the neutron as a bound state, or as a collapsed hydrogen atom, had been proposed by Ernest Rutherford in 1920 and was advocated by Chadwick. However, a non-elementary neutron required the nuclear electron to obey incorrect statistics. So Heisenberg wrote, "it does not seem appropriate to elaborate further on such a picture." Yet, Heisenberg proposed, if under the proper circumstances a fundamental neutron decays into a proton and electron, then the "conservation laws of energy and momentum are probably no longer applicable." At this time Heisenberg had not yet accepted Pauli's hypothesis of a neutrino and he straddled the fence regarding Bohr's proposal that energy conservation may be violated in beta decay. In fact, Heisenberg continued[1] in Part I to use energy conservation for discussing the energetics of beta decay. In Part III he wrote that pushing the "energy law beyond its validity is

Feynman diagrams

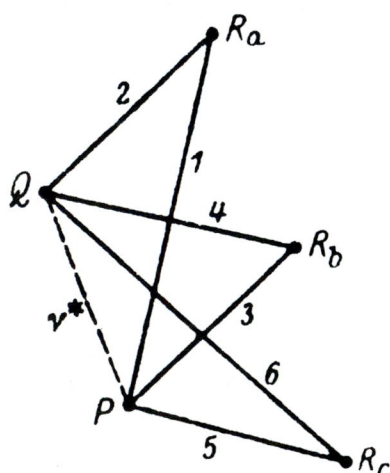 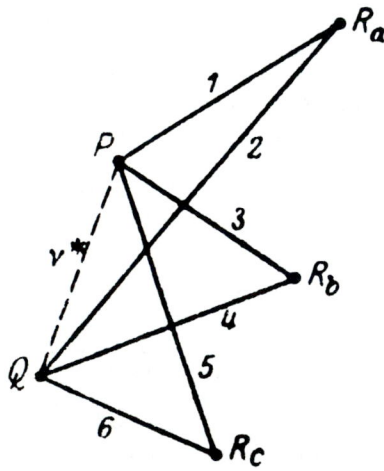

Heisenberg approved of Feynman diagrams as "intuitive" in accordance with the new meaning of "visualizability" (reference 23). He recalled (reference 9) on 13 February that the diagrams in the Kramers–Heisenberg paper of 1925 (*Zeitschrift für Physik* **31**, p. 681) that resulted from photon–atom scattering were suggested by the mathematics of this late version of atomic physics, and so were like Feynman diagrams: "We drew these pictures. First you go from this level to that and then from that level to here. Then one could easily see that in going from here to here you can go by different ways. That's like Feynman diagrams nowadays." In this representative term diagram from the Kramers–Heisenberg paper, R_a, R_b, R_c, Q and P are levels in an atom from which light is scattered. The incident light causes the atom to make transitions from a state P to a state Q via the intermediate states R. Energy need not be conserved in the intermediate transitions. (Figure from W. Heisenberg, H. Kramers, *Z. Phys.* **31**, 681, 1925.)   Figure 6.

indeed logically possible throughout (e.g., the strong validity of the classical selection rules in quantum mechanics)." Although Heisenberg considered that the idea of a fundamental neutron was far more satisfactory, owing principally to the problem of where the electrons originated in beta decay, he dealt also with a composite neutron. In fact, in his 1932–33 papers on nuclear physics, Heisenberg tried in every way to include electrons in the nucleus.

The force between a charged proton and a neutral neutron cannot be of the same sort as that between two charged particles, and any proposed form for the nuclear force must yield the property of saturation. Thus Heisenberg drew the "analogy" of the attractive force between a proton and neutron and the exchange force that is the dominant factor for the stability of H_2 and H_2^+ molecules. The failure of the old Bohr theory to account for the stability of the H_2^+ ion had been taken as an indicator of fundamental problems. In 1927 Fritz London and Walter Heitler extended the concept of exchange energy (and thereby quantum mechanics too) into chemistry and formulated the theory of the homopolar bond. They discovered that the ion's stability is due to the exchange energy. No visualization of the exchange energy. No visualization of the exchange force is possible because it cannot be developed from intuitions constructed from the world of sense perceptions (see box, Figure 7, a and f). Rather, the visualizability for the exchange energy is given by the quantum mechanics, although it defies depiction, thus frame f is empty. In the solution to the problem of the H_2^+ ion, a strictly quantum-mechanical contribution to the ion's energy can be described metaphorically as the two protons sharing the single electron. The exchange contribution to the helium atom and the H_2 molecule arises through the indistin-

guishability of the electrons, and so, as Heisenberg himself cautioned[7] in his many-body paper of 1926, any depiction of this force "makes no sense physically." In 1928 Heitler had written[15] that it is as "yet incomprehensible what exchange in reality means."

As if in response to Heitler, in 1932 Heisenberg extended the concept of exchange energy to the nucleus in a way that offered a depictive component of visualizability through the mathematics of the theory, and thus, in this case, to an understanding of what the exchange energy "in reality means."

Modern nuclear physics begins with this passage[1] in Heisenberg's "On the structure of the atomic nucleus I":

> Suppose we bring the neutron and proton to a separation comparable to nuclear dimensions; then in analogy to the H_2^+ ion, the negative charge will undergo a migration [*Platzwechsel*], whose frequency is given by a function $J(r)/h$ of the separation r between the two particles. The quantity $J(r)$ corresponds to the exchange [*Austausch*], or more correctly, migration integral [*Platzwechselintegral*], of molecular theory. The migration can again be made more intuitive by the picture of electrons that have no spin and follow the rules of Bose statistics. But it is surely more correct to regard the migration integral $J(r)$ as a fundamental property of the neutron–proton pair, without intending to reduce it to electron motions.

Yet as we shall see, Heisenberg went on to discuss nuclear electrons that "follow the rules of Bose statistics." This was necessary for him to relate beta decay to the neutron–proton force. Heisenberg played all ends against the middle in this theoretical free-for-all, which offered him the "arbitrariness [in seeking] a formulation that will not lead sooner or later to internal difficulties."

Heisenberg's change of terminology

from "exchange" to "migration" signals a new concept to follow. With this switch from exchange to migration, Heisenberg's visualizability of the migration of the electron in the neutron-proton force is, however unintentionally, the ingredient required to draw g in Figure 7. In this figure the quantity $J(r)$ is the attractive force between a fundamental nuclear proton and a composite nuclear neutron. The attractive force operates through the "migration" of charge from the neutron to the proton, which, capturing the Bose electron, becomes a neutron. In Heisenberg's nuclear exchange force the neutron and proton do not merely change places. The metaphor of motion is of the essence here, and that is why I rendered *"Platzwechsel"* as "migration."

The mathematics of Heisenberg's nuclear theory offered a new intuition, or mental imagery, for the atomic and subatomic domains. The reason is that the graph in frame g, and as we shall discuss in a moment, frames h, i and j as well, in the box, Figure 7 could not have been conceived without quantum mechanics. Needless to say, owing to our cognitive apparatus all graphs must be drawn with the usual figure and ground arrangement—that is, with continuity and discontinuity side by side.

In the subsequent literature most physicists referred to the quantity $J(r)$ as an "exchange force." The exchange force was a center of attention at the 1933 Solvay Conference on nuclear physics (figure 5). To a first approximation, Heisenberg assumed that the exchange forces between neutrons and protons are static and central. Because $J(r)$ is a short-range force, the expressions most frequently used for it were

$$J(r) = ae^{-br}$$

$$J(r) = ae^{-(br)^2}$$

where a and b are parameters that can be adjusted to fit, for example, the binding energies of nuclei.

Visualization and visualizability

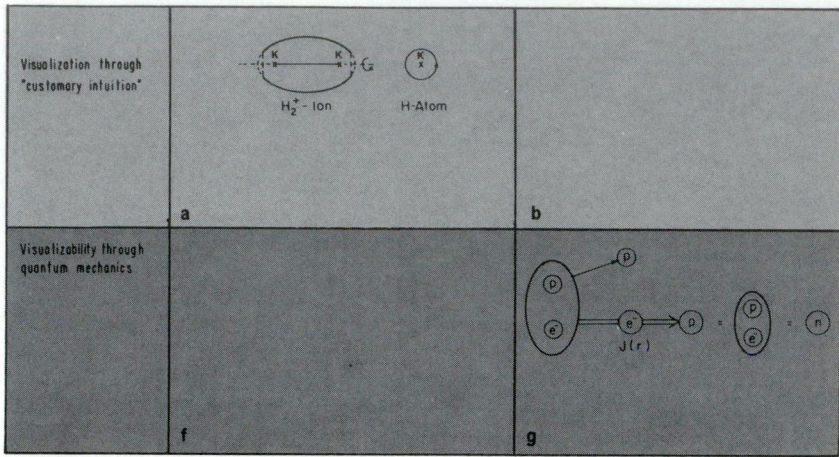

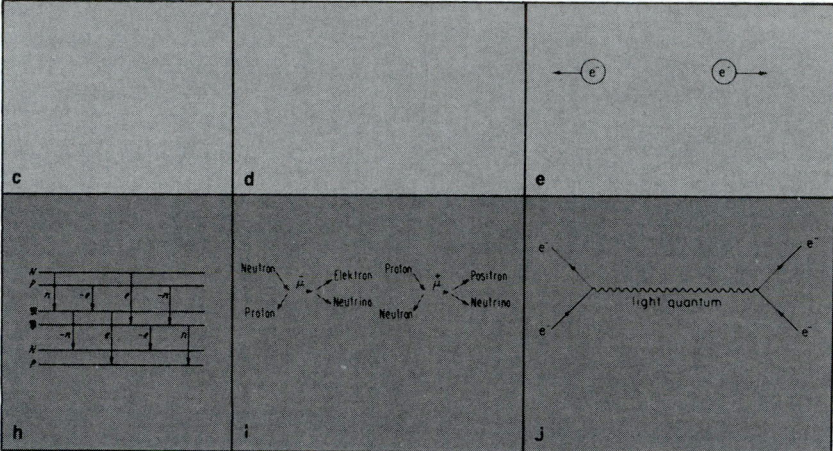

The changing notions of physical reality that began with Heisenberg's extension of the exchange concept from molecular physics to nuclear physics are in frames **g** through **j**. Frames **a**, **b**, **c**, **d** and **e** contain visualizations according to pictures constructed from objects actually perceived—that is, customary intuition. Until 1923 customary intuition was imposed on all theories. Frame **a** is from Pauli's unsuccessful attempt in 1922 to deduce a stable H_2^+-molecule ion from Bohr's theory of the atom with its solar-system imagery of atoms and molecules; K denotes a "positive center," or proton. Frames **f**, **g**, **h**, **i** and **j** depict visualizability according to quantum mechanics. Thus frames **b**, **c**, and **d** are empty. By a new mode of intuition is meant that the graphs in frames **g**, **h**, **i** and **j** could not have been conceived of without quantum mechanics. Frame **f** is empty. Frame **e** is the anthropomorphic, or intuitive, representation for the Coulomb repulsion between two electrons. Frame **j** is a Feynman diagram for the new visualizability, or intuition, for the repulsive interaction between two electrons in which they exchange a light quantum. Today the words exchange and migration are synonymous. (Frame **a** is from reference 13.)

Figure 7.

Ramifications

Other sorts of exchange forces were proposed that agreed with available data better than Heisenberg's. These forces did not require inappropriate migratory electrons. But Heisenberg's visualizability of a nuclear exchange force with something exchanged turned out to be fruitful. It was instrumental in Enrico Fermi's thinking towards his 1934 theory of beta decay,[16] whose success, wrote[17] Heisenberg in 1935, was "proof of the existence of exchange forces" in nuclei. Fermi's theory was referred to as an "intuitive theory"—in the new interpretation of the term intuitive—because the theory's mathematics permitted an analogy of the electromagnetic transition in a two-level atomic system to the redefined intuition or mental image. Here a neutrino and an electron are emitted instead of the photon in the transformation of a neutron into a proton when an atom decays (compare figures c and h in the box above).

In 1934 Igor Tamm elaborated[18] and further generalized Fermi's intuitive theory of beta decay. Tamm assumed that the neutron and proton result from the splitting of a degenerate state by the exchange energy that originates in the neutron's emission of the electron and neutrino, or inversely, the proton's emission of a positron and neutrino. The splitting of the energy levels is analogous to what occurs in molecules. So far Tamm offered nothing really new. But next he generalized the mental imagery in Fermi's theory of beta decay as follows: "The role of light particles (ψ-field) [electrons and neutrinos] providing an interaction between heavy particles corresponds exactly to the role of the photon (electromagnetic field), providing an interaction between electrons...." Using suitably modified methods of quantum electrodynamics Tamm went on to show that the interaction energy from Fermi's beta-decay theory was too small to explain the binding energies of neutrons and protons in nuclei. Thus, contrary to the hopes of Heisenberg and Fermi, a theory of beta decay could not cover the neutron–proton force as well.

Directly following on Tamm's paper, and similar in thrust, is a paper[19] by D. Iwanenko. Like Tamm, Iwanenko compared Heisenberg's "interaction *exchange* energy ... between proton and neutron [to] the birth and absorption of a photon in the case of two electrons."

It is noteworthy that Iwanenko used the term exchange even though he compared the mechanism of the neutron–proton force to the Coulomb interaction mediated by a photon. It is tempting to suggest here that Iwanenko, like most other physicists at that time, missed the purpose of Heisenberg's change of terminology from exchange to migration. Hideki Yukawa did not. In November 1934 Yukawa considered[20] the exchange force between the neutron and proton to be a "migration force describable with a *Platzwechselintegral* [migration integral]."

Yukawa set out to modify the theory of Heisenberg and Fermi by changing the analogy of the field of "light particles" transmitting the neutron–proton force to the exchange of a single, new "heavy particle." The "interaction between the elementary particles," continued Yukawa, "can be described by means of a field of force, just as the interaction between charged particles is described by the electromagnetic field. ... This field should be accompanied by a new sort of quantum just as the electromagnetic field is accompanied by the photon." Yukawa found that he could write a Hamiltonian for his theory that was equivalent to Heisenberg's Hamiltonian (reference 1, Part I) if one takes for the migration integral

$$J(r) = g \frac{e^{-ur}}{r}$$

for describing the exchange of a legitimate Bose particle, where g is a coupling constant and u is the inverse of the range of the nuclear force associated with the exchanged particle. Yukawa's choice for $J(r)$ stands in stark contrast with earlier choices, which were almost arbitrary exponentials chosen for expediency in comparing theory with empirical data. We may conjecture with confidence that Yukawa's explicit use of the German word *Platzwechselintegral* in his English-language paper was to signal that he meant not a metaphorical exchange, or *Austausch*, but a real "migration." And thus did Heisenberg's metaphor of motion become physical reality.

Epilogue

The scientific literature supports Gregor Wentzel's recollection[21] that Yukawa's "ingenious idea . . . was not received, wherever it became known, with immediate consent or sympathy." This situation changed in 1937 when physicists believed Yukawa's meson had been discovered in cosmic-ray phenomena. So it would seem reasonable to expect in the 1937 literature some sort of diagrammatic representation of particles transmitting forces. After all, the proper verbal descriptions abounded in the literature and, with hindsight, such suggestive mathematical formalisms as those proposed by E. C. G. Stueckelberg. But no diagrams of exchange particles can be found from that time. In fact a conceptualization from Dirac's 1927 quantization of the electromagnetic field of the Coulomb force transmitted by a photon played little, if any, role in subsequent work on electromagnetic processes such as electron–electron scattering. Nor have I found in the correspondence I have studied thus far diagrams depicting the transmission of forces by particles. In the published literature there are depictions of beta decay like the one in figure 2. But figure 2 was, after all, the sort of diagram that was the basis of Fermi's intuitive theory of beta decay.

An interesting development occurred in 1943, when Wentzel proposed a "didactic" device for depicting Yukawa's treatment of beta decay as an "indirect" process that is mediated by a virtual meson (compare figures d and j in the box of Figure 7). In Heisenberg's original conception, the exchange force was transmitted by a real electron with incorrect properties. These "schemata," wrote[22] Wentzel, depict the "beta-decay process as a two stage transition." Wentzel's figure (see figure 3) depicts the nuclear ex-

change force that had been first proposed by Heisenberg with no analogy to the photon as an intermediary of the interaction between two electrons. We recall that Iwanenko and Tamm did make the photon analogy, although without any diagram. Yukawa's more successful theory of the nuclear exchange force was formulated directly along the lines proposed by Heisenberg, namely, by introducing a "migration integral" for describing the nuclear force. It was Yukawa's ingenious idea to propose a "migration integral" that described the actual "migration" of a proper particle, that is, a proper boson and not a Bose electron. But Yukawa offered no depiction of this process.

In the 1949 preface to the English edition of his book, Wentzel wrote that extensive changes were necessary only in the section where his figure (figure 3) appeared, and "here it was easy to modernize the text and to adapt it to the present state of knowledge." Wentzel's book became a valuable stop-gap until the publication of up-to-date books such as *Mesons and Fields* (Row, Peterson, Evanston, Illinois, 1955) by Hans Bethe and Frederic de Hoffmann. But the transition from Wentzel's schematics of 1943 to the Feynman diagrams of 1949 required further transformations of what constitutes physical reality and its accompanying mental imagery.

By 1943 the mental imagery of most physicists had already undergone several transformations, for is it not true that figure 3 is another way of "seeing" the energy levels in figure 2? In turn, figure 2 is a descendant of the energy-level representations in atoms that replaced the customary intuition of the solar-system atom that the intuitive (in the pre-1927 meaning of this word) language of classical physics had imposed on the Bohr theory (figure 4). Analysis of these transformations of mental imagery, which resulted in Feynman diagrams (compare figures e and j in the box, Figure 7), is beyond the scope of this article.[23]

Clearly, as we have seen, fundamental progress in science is accompanied by transformations of mental imagery. Heisenberg's nuclear exchange force was another step on a path that reached back to scientific work that he had accomplished in 1926. This path reaches even further back, because it is defined by a culturally based mode of mental imagery—customary intuition—that had been of importance to philosopher–scientists such as Ludwig Boltzmann, Albert Einstein and Hermann von Helmholtz, who emphasized this mode of visual thinking. Despite the advances that resulted from their scientific research, their mental imagery remained based in customary

intuition. On the other hand, Heisenberg's scientific research forced him to liberate his mental imagery from customary intuition. His results set the course of subnuclear physics.

By and large, research on the frontiers of physics continues on the path taken by Boltzmann, Einstein, Helmholtz and Heisenberg. A hallmark of a well-developed and fruitful scientific theory in the 20th century is some sort of mental imagery. As Heisenberg recalled[9] on 25 February 1963 of his style of research: "The picture changes over and over again, and it's so nice to see how such pictures change."

References

1. W. Heisenberg, Z. Phys. **77**, 1 (1932), Part I; **78**, 156 (1932), Part II; **80**, 587 (1933), Part III.

2. M. Born, Naturwissenschaften **27**, 537 (1923).

3. H. Kramers, H. Holst, *The Atom and the Bohr Theory of Its Structure*, Gyldendal, London (1923).

4. W. Pauli, *Wissenschaftlicher Briefwechsel mit Bohr, Einstein, Heisenberg, u. A.*, A. Hermann, K. v. Meyenn, V. F. Weisskopf, eds., Springer-Verlag, Berlin (1979).

5. W. Heisenberg, Naturwissenschaften **14**, 501 (1926).

6. E. Schrödinger, Ann. Phys. (Leipzig) **70**, 734 (1926).

7. W. Heisenberg, Z. Phys. **38**, 411 (1926).

8. P. A. M. Dirac, Proc. Roy. Soc. (A) **112**, 692 (1926).

9. Archives for the History of Quantum Physics, AIP, New York.

10. W. Heisenberg, Z. Phys. **43**, 172 (1927).

11. N. Bohr, Nature (Supplement) 580 (1928).

12. See also, J. Bromberg, Hist. Stud. Phys. Sci., **5**, 307 (1971); L. Brown, L. Hoddeson, PHYSICS TODAY, April 1982, p. 36; R. Stuewer, in *Otto Hahn and the Rise of Nuclear Physics*, W. Shea, ed., Science History, New York (1984).

13. Letter on deposit at the AIP Center for History of Physics.

14. W. Pauli, Ann. Phys. (Leipzig) **68**, 177 (1922).

15. W. Heitler, Z. Phys. **46**, 47 (1928).

16. E. Fermi, Z. Phys. **88**, 161 (1934).

17. W. Heisenberg, in *Pieter Zeeman*, Nijhoff, The Hague (1935), p. 108.

18. I. Tamm, Nature **133**, 981 (1934).

19. D. Iwanenko, Nature **133**, 981 (1934).

20. H. Yukawa, Proc. Phys.–Math. Soc. Japan **17**, 48 (1935).

21. G. Wentzel, in *The Physicist's Conception of Nature*, J. Mehra, ed., Reidel, Dordrecht (1973), p. 380.

22. G. Wentzel, *Einführung in die Quantentheorie der Wellenfelder*, Deuticke, Vienna (1943), translated by C. Houtermans, J. M. Jauch, as G. Wentzel, *Quantum Theory of Fields*, Interscience, New York (1949).

23. See A. I. Miller, *Imagery in Scientific Thought: Creating 20th-Century Physics*, Birkhäuser, Boston (1984). □

Hideki Yukawa and the meson theory

Some 50 years ago, in his first research contribution, a young Japanese theoretical physicist explained the strong, short-range force between neutrons and protons as due to an exchange of "heavy quanta."

Laurie M. Brown

PHYSICS TODAY/DECEMBER 1986

A little over 50 years ago, Hideki Yukawa, a young Japanese theoretical physicist at the University of Osaka, proposed a fundamental theory of nuclear forces involving the exchange of massive charged particles between neutrons and protons. He called the exchanged particles "heavy quanta" to distinguish them from the light quanta of the electromagnetic field, but now they are known as pions, the lightest members of a large family of particles called mesons. These days, the meson theory seems to be a straightforward application of quantum field theory to the nuclear forces, but it could not have appeared so 50 years ago. Otherwise it would not have been invented by an obscure, unpublished Japanese physi-

cist who had never even traveled abroad, but instead by one of the many Western physicists, some of them of great reputation, who were working on the theory of nuclear forces. Furthermore, Yukawa's paper was totally neglected for more than two years, although it was written in lucid English and published[1] in a respected journal of rather wide circulation. The meson theory has turned out to be an important paradigm for the theory of elementary particles, as seminal as Ernest O. Lawrence's cyclotron has been for its experimental practice. In this article I will trace the intellectual trail Yukawa followed to arrive at his theory, using unpublished documents that have recently been discovered among

Yukawa's papers at Kyoto University. (These documents have been organized and cataloged in the Yukawa Hall Archival Library, and I am grateful to Rokuo Kawabe and Michiji Konuma for translations and comments.)

The problem of nuclear forces

The meson theory was the result of a powerful creative act. It incorporated a number of ideas that are common-

Hideki Yukawa, at right, with (from left) his father-in-law Genyo Yukawa, his mother-in-law Michi and his wife Sumi at their home in Osaka in 1932. (Photos illustrating this article courtesy Michiji Konuma; from reference 9, reproduced with permission of Sumi Yukawa.)

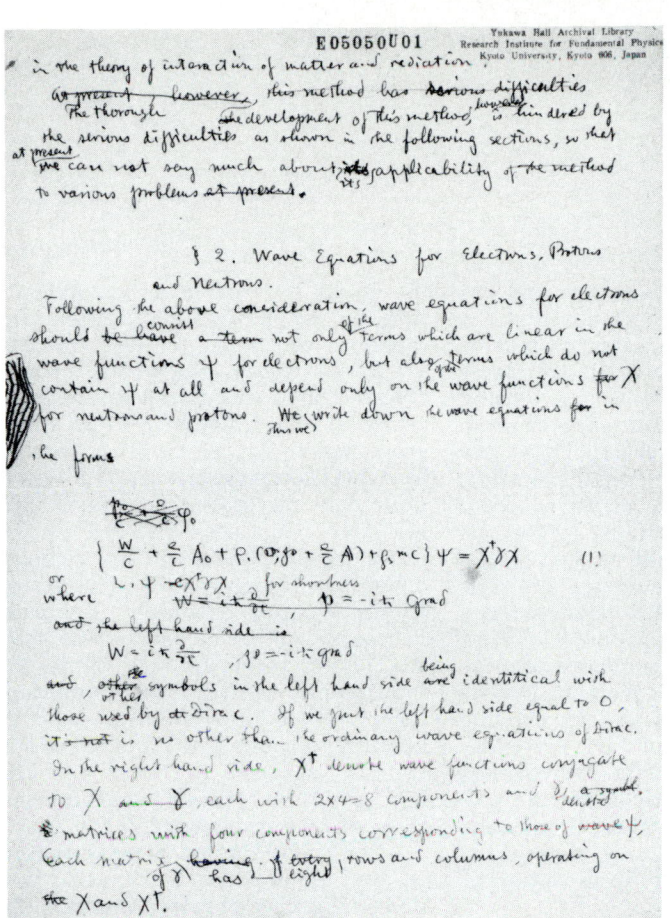

Manuscript page containing Yukawa's wave equation, based on the Dirac equation, for describing the intranuclear electron field having the nucleon-exchange current as a source. Equation 1 in the manuscript corresponds to equation 3 in the this article. (The documents illustrating this article come from the Yukawa Hall Archival Library, Kyoto University, courtesy of Ziro Maki.)

place today, but which were novel and surprising 50 years ago. These are some of the original features proposed by Yukawa:

▶ The nuclear binding force is transmitted by the exchange of massive charged particles (the heavy quanta).

▶ The range of the force is inversely proportional to the mass of the quantum.

▶ There are *two* nuclear forces, one strong and one weak, and hence two coupling constants.

▶ The weak interaction is also mediated by the exchange of the heavy quanta (charged intermediate bosons).

▶ The heavy quanta are unstable, decaying via the weak interaction.

It was a great achievement to propose such a bold solution for the nuclear-force problem, for it was difficult at that time even to grasp and to formulate the dimensions of the problem. At a symposium held in Minneapolis in 1977 on the subject of nuclear physics in the 1930s, Eugene Wigner remarked:[2]

> The disorientation in physics which existed . . . [around 1932] is hard to imagine. We can remember it hardly more than we can

remember the time when man's chief purpose was to hunt some animals, when he did not have chairs to sit in, did not have beds, did not have covers, did not have a jacket, did not have anything, not to mention that he did not have a radio to speak into. It is hard to remember that.

The physicist's view of matter before the discovery of the neutron was that everything, including the nucleus, consisted of protons and electrons and was held together by electric and magnetic forces.[3] In that picture, the nucleus of mass number A contained A positive protons and $A - Z$ negative electrons; the neutral atom had, of course, Z extranuclear electrons. Thus, matter was said to be made of pure electricity. Within this framework, quantum mechanics was scoring impressive successes in explaining phenomena that involved the extranuclear electrons, in fields such as spectroscopy, chemistry and magnetism, but the behavior of the intranuclear electrons was quite another matter. The mere presence of electrons in the nucleus seemed to violate important principles of physics. One problem, apparently insurmountable within the conventional framework of quantum mechanics, was that an electron, having low mass and confined within a space as small as a nucleus, would according to Werner Heisenberg's uncertainty principle have a kinetic energy so large that it

would rapidly escape the nucleus. Another difficulty was that certain nuclei, those with an odd number of intranuclear electrons, should have had magnetic fields about a thousand times larger than that inferred from the hyperfine structure of atomic spectra. One such nucleus—nitrogen-14, with 14 protons and 7 electrons—should have had half-integral spin and obeyed Fermi–Dirac statistics, but was known to have integral spin and to obey Bose–Einstein statistics.

As early as 1920, Ernest Rutherford had suggested that there might exist a neutral nuclear constituent that he called the "neutron," consisting of a proton and an electron much more tightly bound than in the hydrogen atom. James Chadwick, working in the Cavendish Laboratory in Cambridge in 1932, assumed, when he discovered the neutron, that he had found Rutherford's composite neutron, which had been sought at the Cavendish for a decade. Chadwick did consider the possibility that his neutron was a new elementary particle, but said[4] that the idea had "little to recommend it."

Within a few weeks after Chadwick's announcement, Heisenberg wrote the first part of a three-part paper presenting[5] a theory of nuclear structure in which protons and neutrons were the principal constituents (*Bausteine*) of the nucleus, instead of the protons and electrons (and sometimes alpha particles) of the earlier models. (See the article by Arthur Miller, on Heisenberg, this volume, page 222.) Heisenberg did not challenge the idea that the neutron was composed of a proton and an electron, but by suppressing the electron degrees of freedom in his nuclear Hamiltonian he was able to construct the first genuinely quantum-mechanical nuclear theory. (Dmitri Iwanenko in Leningrad, at about the same time, suggested[6] treating the neutron as an elementary particle in the nucleus, but he does not seem to have developed a detailed theory of this type.) So far as nuclear systematics was concerned, Heisenberg treated the proton and the neutron as the charged and neutral states of a single entity, our present-day nucleon, and he introduced operators that changed the charge of the nucleon, turning neutrons into protons or vice versa. The force that he introduced between the proton and neutron had charge-exchange character, analogous to that in

Laurie Brown is professor of physics and astronomy at Northwestern University, Evanston, Illinois. This article is based upon a paper presented at an international symposium on the golden jubilee of the meson theory, Kyoto, Japan, 15–17 August 1986.

the hydrogen molecular ion H_2^+; the force between neutrons was analogous to the exchange force in the hydrogen molecule, but between protons, which Heisenberg regarded as elementary particles, there was only ordinary Coulomb repulsion. Thus in assigning forces Heisenberg clearly differentiated the "elementary" proton from the "composite" neutron. Nevertheless, despite its unsymmetric treatment of what we now call the nucleon doublet, Heisenberg's theory, especially as modified during the following year by Wigner and Ettore Majorana, is correctly regarded as the foundation of the modern phenomenology of nuclear structure.

While Heisenberg had seized the opportunity presented by the discovery of the neutron to formulate a phenomenological theory of nuclear forces, he was not the sort of physicist who would sidestep the issues raised by the presence of composite neutrons (and hence, implicitly, electrons) in the nucleus. After all, to make a spin-$\frac{1}{2}$ neutron out of the proton and electron required the suppression of a half-quantum of spin. Furthermore, although Heisenberg was as sensitive as anyone could be to the further difficulties that arose from a model nucleus containing additional "loose" electrons, that is, those not bound in neutrons, his conception of beta-decay radioactivity and of the large radiative effects observed in cosmic-ray and certain laboratory gamma-ray experiments demanded[7] loose electrons in the nucleus. In this connection he said,[5] in part III of his three-part paper, that "the α particles in the nucleus must be built of protons and electrons (not protons and neutrons)" and that those electrons must add their contribution to radiative scattering to that of the "free nuclear electrons." Thus, about half of the contents of Heisenberg's three-part paper of 1932 was concerned with problems raised by the nuclear electrons and with the theory of the neutron itself as a proton–electron compound.

To summarize: In working out the systematics of nuclear structure, Hei-

senberg regarded the proton and the neutron as sisters, so similar that the exchange currents in terms of which he expressed his charge-exchange forces were formulated in terms of raising and lowering operators adapted from the Pauli spin matrices. Those are precisely the modern isospin operators, although Heisenberg employed them without the accompanying idea of the charge independence of nuclear forces. Nevertheless, he insisted that electrons of both the bound and loose varieties were present in the nucleus. This view so characterized Heisenberg's nuclear model that in 1933 Wigner, listing[8] "three different assumptions possible concerning the elementary particles" in the nucleus, began his list: "(a) The only elementary particles are the proton and the electron. This point of view has been emphasized by Heisenberg and treated by him in a series of papers."

Yukawa's first research

Yukawa graduated from Kyoto Imperial University in 1929, the first year of the Great Depression and a difficult time in which to find a job. He therefore continued to live with his family in Kyoto, where his father, Takuji Ogawa, was a professor of geography at the university. For three years, together with Sinitiro Tomonaga—his classmate, the son of a Kyoto philosophy professor, and a future fellow Nobel laureate—Yukawa worked as an unpaid assistant in the theoretical-physics research room of Professor Kajuro Tamaki. Intensely ambitious, he was afraid that he had arrived upon the physics scene too late to make a fundamental contribution to quantum theory. Much later, Yukawa wrote[9] in his autobiography, "As I was desperately trying to reach the front line, the new quantum physics kept moving forward at a great pace."

After his graduation Yukawa reflected that there were two outstanding problems facing physics. One was the structure of the nucleus. The other consisted of the questions of principle raised by Paul A. M. Dirac's relativistic electron theory, especially the puzzle

Yukawa's Green's function solution of the wave equation (equation 3 in the text) for the hypothetical Bose electron of mass m within the nucleus. (From document F01 010 U01.)

that the theory required negative-energy states, for which there appeared to be no meaning in the theory of relativity. With these problems in mind, Yukawa began to look into hyperfine structure, a small but observable splitting of atomic spectral lines. The effect was believed to be due to the action of the nuclear magnetic field on the extranuclear atomic electrons, the main influence being on those electrons that penetrate nearest to the nucleus and thus feel the strongest magnetic field. At the same time, because of the strong attractive electric field near the nucleus, those electrons would attain velocities near that of light. Thus the problems associated with electrons in or near the nucleus and the problems of Dirac's electron theory would be present at the same time, and Yukawa hoped they would shed light upon each other.

As it happened, Enrico Fermi in Rome was thinking along the same lines. He was an experienced and established theorist and, moreover, did not have to deal with Tamaki. When Yukawa handed his manuscript to Tamaki, the professor stored it in a safe place and said that he would study it when he had the time. Meanwhile, Fermi published[10] his treatment of the problem, discouraging Yukawa from pursuing it further or from considering what it might mean for nuclear structure, as he had originally planned. Instead, his attention was drawn to the appearance[11] of the first part of a monumental paper by Heisenberg and Wolfgang Pauli, which set forth a relativistic quantum theory of the electromagnetic field in interaction with Dirac electrons. That paper Yukawa later repeatedly called a partial "settling of accounts" and "in certain respects the final balance sheet of the quantum theory originated by Planck."

The account that needed settling was wave–particle duality. The quantum of action (Planck's constant) was introduced into physics by Max Planck in 1900 in his treatment of the spectrum of blackbody radiation. In 1905 Einstein showed that Planck's theory implied that electromagnetic radiation, whose behavior had been seen as fundamentally wavelike, must at times behave in ways that are particlelike. But waves and particles are conceptually incompatible, so Einstein's quantum theory gave rise to what was called the wave–particle paradox. After Niels Bohr's successful treatment of the hydrogen spectrum in 1913, Planck's

quantum of action became recognized as setting the scale of quantum effects in matter as well as radiation. The mystery was only deepened by Louis de Broglie's conjecture that electrons have a wavelike nature in addition to their particulate properties, which was soon experimentally confirmed. By 1929, after the new quantum mechanics of Heisenberg and Erwin Schrödinger had resulted in a broad understanding of atomic phenomena, Yukawa felt that the time had come to make a fresh attack on resolving the wave–particle paradox.

Dirac had already in 1927 introduced methods that took into account the quantum nature of the electromagnetic field when calculating, for example, the spontaneous decay rates of atomic states (determining the intensity of the lines in atomic spectra), but Heisenberg and Pauli's 1929 paper was the first fully relativistic quantum theory of the electromagnetic field. Dirac had quantized only the transverse degrees of freedom of the field, those corresponding to free radiation, and he had left the static Coulomb potential in its classical form. Heisenberg and Pauli found a way to quantize both transverse and longitudinal components of the electromagnetic vector potential, as required in a fundamental relativistic theory. But they were discouraged to find that the theory predicted an infinite electron mass, as well as other infinities that contradicted observation. Finding and curing the source of those infinities became a major task of theoretical physics in the 1930s, a task that was at least provisionally accomplished only in the 1940s by the so-called renormalization program, in which Tomonaga played a leading role.

Partly out of disappointment over losing his unwitting race with Fermi on the hyperfine-structure problem, Yukawa applied his energy to quantum field theory, hoping that he could complete the "settling of accounts" that Heisenberg and Pauli had begun. In *Tabibito* he vividly describes[9] the frustration of his attack on the infinities of quantum electrodynamics:

Each day I would destroy the ideas that I had created that day. By the time I crossed the Kamo River on my way home in the evening, I was in a state of desperation. Even the mountains of Kyoto, which usually consoled me, were melancholy in the evening sun. . . . Finally, I gave up that demon hunting and began to think that I

should search for an easier problem.

The demons, in fact, continued to haunt him throughout the rest of his career. Even after the provisional solution of renormalization in the 1940s, he continued to regard the forging of a finite quantum field theory as the prime objective of theoretical physics, and he continued to investigate fundamental theories (for example, nonlocal field theory) that were alternatives to the accepted renormalized theories.

The meson theory

For the time being, however, Yukawa resolved to think of other matters. He saw Chadwick's announcement of the neutron and soon afterward read part I of Heisenberg's paper on the stucture of atomic nuclei. This rekindled his interest in the nuclear-force problem, although he was not prepared to accept a purely phenomenological theory such as Heisenberg's. An unpublished document of early 1933 by Yukawa is entitled[12] "On the problem of nuclear electrons. I." It is in English, and begins, "The problems of the atomic nucleus, especially the problems of nuclear electrons, are so intimately related with the problems of the relativistic formulation of quantum mechanics that when they are solved, if they ever be solved at all, they will be solved together." Nevertheless, recognizing the importance of Heisenberg's contribution, he prepared a summary in Japanese of the two parts of Heisenberg's paper that had appeared; it became his first publication, although his first published *research* was his later meson paper.

In introducing his summary, Yukawa discussed the strengths and weaknesses of Heisenberg's model, placing particular stress on issues of principle. He wrote:[13]

In this paper Heisenberg ignored the difficult problems of electrons within the nucleus, and under the assumption that all nuclei consist of protons and neutrons only, considered what conclusions can be drawn from the present quantum mechanics. This essentially means that he transferred the problem of the electrons in the nucleus to the problem of the makeup of the neutron itself, but it is also true that the limit to which the present quantum mechanics can be applied to the atomic nucleus is widened by this approach. Though Heisenberg does not pres-

The Ogawa family. In front are Takuji, a professor of geography at Kyoto University, and his wife Koyuki; standing are (from left) Tamaki, Shigeki, Hideki and Masuki. Hideki adopted his wife's family name when he married Sumi Yukawa in 1932.

ent a definite view on whether neutrons should be seen as separate entities or as a combination of a proton and an electron, this problem, like the beta-decay problem . . . cannot be resolved with today's theory. And unless these problems are resolved, one cannot say whether the view that electrons have no independent existence in the nucleus is correct.

During 1933, Yukawa argued in several unpublished documents that Heisenberg's program could be carried one step further toward a fundamental theory of nuclear forces and beta decay. However, that step made it necessary to treat the neutron as a truly elementary particle because, he said,[14] any treatment of its structure would lie "outside the applicability of present quantum mechanics." He attempted to formulate the Heisenberg charge-exchange force by analogy to quantum electrodynamics, taking the exchange of an electron between a neutron and a proton in the theory of the nuclear force as analogous to photon exchange between charged particles in QED.

In classical electrodynamics, the electric and magnetic fields are derived from a relativistic vector potential A_μ that obeys the wave equation

$$\Box A_\mu = j_\mu \qquad (1)$$

Here, $\Box$ is the d'Alembertian operator $\nabla^2 - \partial^2/c^2\partial t^2$. The four-vector charge–current density j_μ is regarded as the "source" of A_μ. In a region free of charges and currents j_μ vanishes, and equation 1 describes the free electromagnetic field. The equation holds also in quantum electrodynamics, with A_μ as the quantized electromagnetic field and j_μ the appropriate quantized charge–current density, for example, that of the Dirac electron–positron field.

What Yukawa did was to start from the Dirac relativistic electron equation, regarded as the free-field equation of the classical electron field,

$$D\psi = 0 \qquad (2)$$

where D is the 4×4 matrix Dirac differential operator (including the electromagnetic potentials) and ψ is the four-component Dirac relativistic spinor wavefunction of the electron. He then modified this equation by introducing on its right-hand side a source term J depending on the neutron and proton (regarded as the neutral and charged states, respectively, of the nucleon):

$$D\psi = J \qquad (3)$$

While the source term J has a form somewhat similar to the electromagnetic current j_μ, it is significantly more complicated: It is constructed out of an eight-component nucleon spinor and its adjoint, and it contains Dirac matrices that act upon the spinor as well as matrices that change a neutron into a proton or vice versa. While j_μ (like A_μ) transforms as a relativistic four-vector, J (like ψ) must transform as a relativistic spinor. That was an essentially new and problematic aspect, as Yukawa realized. In QED the same j_μ that is responsible for the exchange of photons between charges, accounting for forces, is also the source of free photons. But if J were to account for beta decay as well as nuclear forces, then beta decay would necessarily be nonconserving of energy and angular momentum, as in any theory where the electron emerges unaccompanied from the nucleus. Of course, that was the case in Heisenberg's theory as well; Yukawa was trying to construct a theory that would justify Heisenberg's phenomenology, and not necessarily one with different practical consequences.

Yukawa used the known properties of the Green's function for the Dirac equation (with the electromagnetic four-potential set to zero) and wrote a solution to equation 3 in the following form: A Dirac operator acts upon an integral that for vanishing electron mass is entirely analogous to the retarded potential due to the source J. Explicitly, for zero electron mass, the integral is

$$\int d^3r' \, \frac{J(\mathbf{r}, t - |\mathbf{r}' - \mathbf{r}|/c)}{|\mathbf{r}' - \mathbf{r}|}$$

Here $\mathbf{r}'$ is the source point, $\mathbf{r}$ is the field point, t is the time and c is the velocity of light. However, for an electron of mass m there is an additional exponential factor, with oscillating character, in the integral. In his manuscript, Yukawa says:[14]

> Substituting this solution into [the Hamiltonian] H, we find the interaction energy that corresponds to Heisenberg's "*Platzwechsel*" interaction. Its exact form and the retardation effect can be seen, and a factor
>
> $$\exp(i\rho_3 mc|\mathbf{r}' - \mathbf{r}|/\hbar)$$
>
> [ρ_3 is a Dirac matrix] appears as a kind of phase factor, so that the term corresponding to Heisenberg's [*Platzwechselintegral*] $J(r)$ has a form like the Coulomb field and does not decrease sufficiently with distance.

He is referring here to the most striking difference between the nuclear force and electromagnetism or gravitation: its short range of influence. In the expression Yukawa derived from the Green's function, the small length $\hbar/mc$ does appear in the exponent and sets the scale of oscillation of the "phase factor," but it does not limit the range of the force.

However, for a time in preparing the quantized-field version of the theory based on equation 4, Yukawa thought he had the kind of solution he was seeking. On 3 April 1933, he gave his first talk to a meeting of the Physico-Mathematical Society of Japan; in the published abstract of that talk, he claims[15] to have obtained the nuclear charge-exchange force of range $\hbar/mc$, where the mass m is that of the electron. This gives a range that would still have been about 200 times too large. Nevertheless, had the result been mathematically correct, it would have been highly suggestive, for a heavier "electron" would have given a shorter range. Yukawa later recalled that when he gave the talk at Sendai, Yoshio Nishina (of Klein–Nishina fame) actually proposed to him that he introduce a "Bose electron." But in the manuscript that he read at Sendai, he withdrew the promising result, saying,

> The practical calculation does not yield the looked-for result that the interaction term decreases rapidly as the distance becomes larger than $\hbar/mc$, unlike what I wrote in the abstract of this talk.

On the blank reverse side of the previous page is written, "mistaken conjecture."

Yukawa was trying to derive Heisenberg's nuclear theory from deeper principles. However, that theory contained the questionable feature that the emission of an electron by a neutron violated the conservation of angular momentum, whether the neutron was treated variously as fundamental or composite. Pauli had proposed as early as 1930 that the electron emitted in beta decay might be accompanied by a light, neutral particle of spin $\frac{1}{2}$ (the neutrino), thus permitting the conservation of both linear and angular momentum, as well as energy. At the Solvay Conference of 1933, held in Brussels, he gave[16] his first public notice of the neutrino that he allowed to be published. Fermi also attended the conference, and upon returning to Rome made[17] a new, strikingly successful quantum field theory of beta decay that incorporated the neutrino.

At first Yukawa thought that Fermi had once again beaten him to the punch (as he had in the case of the hyperfine structure of spectral lines), for Yukawa had assumed with Heisenberg that the beta-decay interaction was the same as that responsible for nuclear forces. The success of Fermi's beta-decay theory suggested that the charge-exchange force might be implementable by the exchange of a *pair* of particles, namely an electron and a neutrino, rather than just a single electron. That would preserve all conservation laws. Yukawa began calculations along this line, as did Heisenberg and others. The first results using this approach were published[18] in independent letters to *Nature* by two Soviet physicists, Iwanenko and Igor Tamm. They both concluded that the Fermi field, as electron–neutrino exchange was called, could explain either the short range or the great strength of the nuclear binding force, but that the theory as formulated could not explain both simultaneously.

When Yukawa read the results of Tamm and Iwanenko in the fall of 1934, his explicit reasoning and subconscious intuition began to come together. As he recalled[9] in his autobiography:

> I was heartened by the negative result [of the Soviet scientists], and it opened my eyes, so that I thought: Let me not look for the particle that belongs to the field of nuclear force among the known particles, including the new neutrino.... When I began to think in this manner, I had almost reached my goal.... The crucial point came

to me one night in October.... My new insight was that [the range of the force] and the mass of the new particle that I was seeking are inversely related to each other. Why had I not noticed that before? The next morning I tackled the problem of the mass of the new particle and found it to be about two hundred times that of the electron.

Within weeks, Yukawa completed the formulation of his theory and reported on it at a branch meeting of PMSJ in Osaka. On 17 November 1934, he spoke on it at the annual PMSJ meeting in Tokyo. Meanwhile, he prepared[1] a paper in English on his theory—his first original research paper—and submitted it to the *Proceedings of the Physico-Mathematical Society of Japan*.

Mesons and 'mesotrons'

Yukawa's article, which appeared in February 1935, aimed at nothing less than a quantum field theory that would unify the nuclear binding force and the nuclear beta-decay interaction. That is, he set out, in the light of the insight gained by the analyses of Tamm and Iwanenko, to modify the theories of Heisenberg and Fermi in such a way that they could be combined. The model for such a theory was QED; following that model closely had important heuristic and pedagogical advantages. His paper appears transparent and appealing to us now, and it remains something of a puzzle why it was entirely ignored for two years in both Japan and the West. Yukawa's work began to be noticed only in 1937, when a particle whose mass closely fit the requirements of meson theory was detected in cosmic rays.

In his meson paper, Yukawa exploited the electromagnetic analogy from the outset, beginning with the "classical" theory. Generalizing d'Alembert's equation for the scalar potential of electromagnetism, which has the well-known static solution $1/r$ for a point source, Yukawa pointed out that a potential having a finite range, namely

$$U(r) = \pm g^2 \frac{e^{-\lambda r}}{r}$$

is a spherically symmetric static solution of the generalized wave equation

$$(\Box - \lambda^2)U = 0$$

He then introduced a source term J' on the right-hand side of this equation,

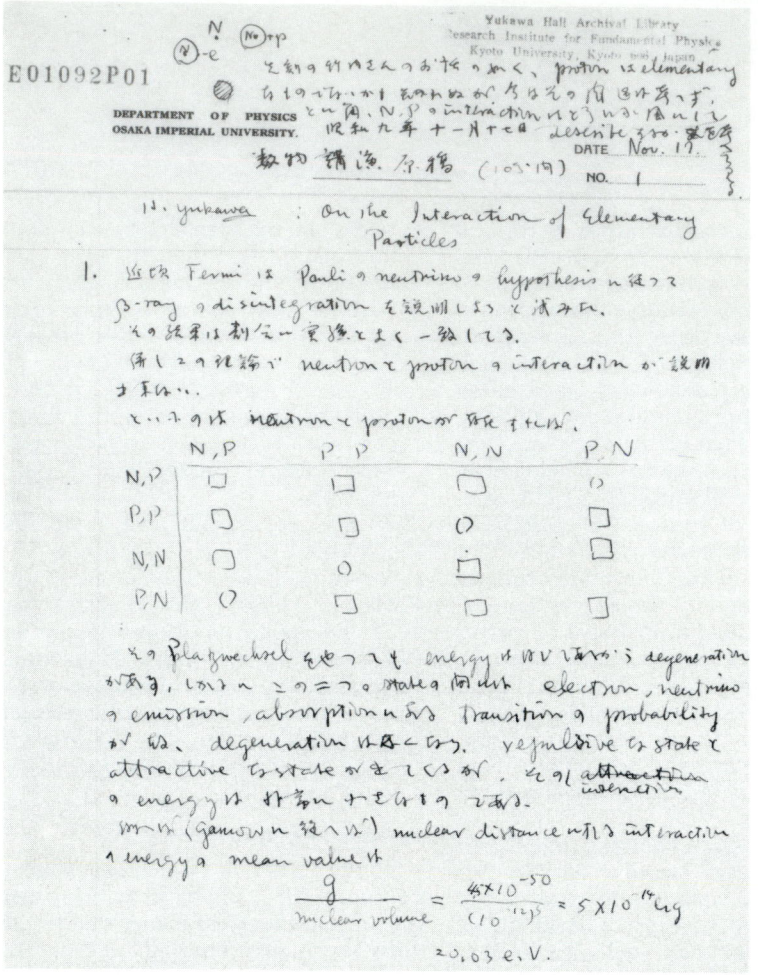

Notes for Yukawa's talk on meson theory at a meeting of the Physico-Mathematical Society of Japan on 17 November 1934. This talk introduced the meson theory of nuclear forces. (Document E01 092 P01.)

something like the source term J of equation 3. However, in this case the particle to be created has the transformation character of U—that is, it transforms like the electromagnetic scalar potential rather than a spinor. (I shall refer to Yukawa's U particle, which has no spin, as a "scalar" particle even though it does not transform as a Lorentz scalar.)

The new source term J' is constructed of nucleon spinors (as J had been) in such a way that the emission of a negatively charged particle corresponds to the transition neutron to proton, while the emission of a positively charged particle corresponds to the transition proton to neutron. Yukawa sought a solution analogous to that given in equation 4, and he obtained nearly the same result, but with an important difference: The exponential factor that appears when the mass m of the U particle is not zero here has the form

$$e^{-\mu|\mathbf{r}'-\mathbf{r}|}$$

The factor μ is determined from the change in energy $W_N - W_P$ in the transition of the heavy particle (what we now call a "nucleon") from neutron

to proton by emission of a positive U particle (a meson):

$$\mu = \sqrt{\lambda^2 - (W_N - W_P)^2/\hbar^2 c^2}$$

Thus, providing the energy difference is less than $\lambda c\hbar$, the exponential is a decreasing function of $|\mathbf{r}' - \mathbf{r}|$, and not the oscillating one found earlier in connection with the electron-emission theory. Upon quantizing the U field, Yukawa showed immediately that the quantity λ is $mc/\hbar$, so that the range of the force is inversely proportional to the mass of the U quantum. (That was the insight that crystallized the meson theory in Yukawa's mind that sleepless October night in 1934. While the range–mass relation is today so commonplace as to seem self-evident to most physicists, that was not the case in the 1930s. As late as 1938, the well-known Italian physicist Gian Carlo Wick took the trouble to present[19] an *anschaulich* explanation of Yukawa's result.)

Yukawa also assumed that the U field (or equivalently, the U quantum) was coupled to an additional charge-changing current constructed analogously to J' but with the electron and

neutrino replacing the proton and neutron. Thus he made the meson an "intermediate boson," carrying the weak as well as the strong interaction. The very notions of strong and weak nuclear interaction were not part of the general thinking before Yukawa made his meson theory. On the contrary, as we have seen, Heisenberg tried to make the beta decay and the nuclear charge-exchange forces the same; that is, he characterized them by a single coupling strength. After Fermi's beta-decay paper appeared, the same single-coupling idea was carried over into the so-called Fermi-field approach to nuclear forces. However, Yukawa introduced one large constant, g, and another small one, g', so that the decay amplitude of a neutron, say, into a proton, electron and neutrino was proportional to gg'. If one substituted for Fermi's coupling constant g_F the quantity $(4\pi/\lambda^2)gg'$, then, Yukawa pointed out, his theory of nuclear beta decay was effectively equivalent to Fermi's.

There was, however, an important additional consequence of Yukawa's assumption that the meson had a beta-decay interaction. It implied that the meson itself would decay radioactively if it were produced in a free state (that is, not bound in the nucleus). Curiously, Yukawa and his associates ignored this fact until 1938, when the Indian physicist Homi J. Bhabha pointed out[20] that important prediction of the theory. The short mean lifetime of the meson is the main reason why such particles are not found in abundance on the Earth.

The question of why mesons were not commonly observed was raised at Yukawa's first presentation of the theory, at one of the informal luncheon meetings of the nuclear-physics group at Osaka Imperial University, which Yukawa regularly attended. Yukawa responded that an energy sufficient to create a meson was required, that is, at least its rest energy of about 100 Mev, and such energies were in those days not available in the nuclear-physics laboratory. The only terrestrial anvil on which such a particle could be forged was the atmosphere, the hammer being the cosmic rays; Yukawa suggested to his experimenter colleagues at Osaka that they should look for U quanta in cosmic-ray cloud-chamber photographs.

In the *Physical Review* of 15 August 1936, Carl Anderson and Seth Neddermeyer published photographs of parti-

First page of Yukawa's rejected letter to *Nature* suggesting that particles observed in cosmic rays might be the mesons of his theory. The letter was sent on 18 January 1937. (From document E06 030 U02, Yukawa Hall Archival Library.)

cle tracks obtained in a cloud chamber operated in the field of a strong magnet atop Pike's Peak in Colorado. In some cases they could not positively identify the tracks as belonging to either electrons or protons. Yukawa wrote a letter to *Nature* pointing out that it was "not altogether impossible" that the ambiguous particles were his heavy quanta. That letter was rejected by the editor, but Yukawa succeeded in publishing[21] a "Short note" having the same contents in the *Proceedings of the Physico-Mathematical Society of Japan* in July 1937. By then, several experimental groups had reported[22] the observation of particles of both signs of electric charge and of mass between that of the electron and the proton, at roughly one-tenth of the proton mass.

The observations of the new cosmic-ray particles, which came to be called mesotrons, aroused[23] interest in Yukawa's theory. Yukawa and his students at Osaka had not been idle during the more than two years during which no one had paid attention to the heavy quantum. With Shoichi Sakata, Yukawa used the meson theory in a calculation of the K-capture process. That weak-interaction process had not been observed, but its existence had been suggested by the Austrian physicist Guido Beck, who visited Japan in 1935. By mid-1936, Yukawa had already begun to reformulate[24] the meson theory using the Pauli–Weisskopf relativistic theory of scalar particles.

(Pauli liked to call his theory "anti-Dirac" because it flouted some of the basic assumptions Dirac made in deriving his relativistic electron theory. Dirac's work had been interpreted to imply that fundamental particles had to have spin $1/2$.)

After the mesotron discovery, Yukawa and his students redoubled their efforts and soon became engaged in the development of the theory and its consequences, a process that so closely paralleled similar efforts in the West that one of the chief Western meson theorists, Nicholas Kemmer, later referred[25] to the Japanese and the Western efforts as a "joint enterprise." The mesotron turned out to be the second-generation heavy electron that we now call the muon, and the true meson of Yukawa, the pion, was found in cosmic rays only in 1947. That is another story, of course, one well worth the telling.

Fifty years later, the meson theory is still alive and well. In spite of its grandeur, today's standard model has not been able to calculate "low-energy" processes, such as meson–nucleon scattering, or the nuclear forces. But even when that becomes possible, Yukawa's meson will still be a shining star.

* * *

I wish to thank the US–Japan Cooperative Science Program and the Program in History and Philosophy of Science, both of the National Science Foundation, for research support.

References

1. H. Yukawa, Proc. Phys.-Math. Soc. Jpn. **17**, 48 (1935).
2. E. P. Wigner, in *Nuclear Physics in Retrospect*, R. Stuewer, ed., U. of Minnesota P., Minneapolis (1979), p. 160.
3. G. Gamow, *Constitution of Atomic Nuclei and Radioactivity*, Oxford U. P., Oxford (1931).
4. J. Chadwick, Proc. R. Soc. London, Ser. A **136**, 692 (1932).
5. W. Heisenberg, Z. Phys. **77**, 1 (1932); Z. Phys. **78**, 156 (1932); Z. Phys. **80**, 587 (1933).
6. D. Iwanenko, Nature **129**, 1312 (1932).
7. L. M. Brown, D. Moyer, Am. J. Phys. **52**, 130 (1984).
8. E. Wigner, Phys. Rev. **43**, 252 (1933).
9. H. Yukawa, *Tabibito* (The Traveler), L. M. Brown, R. Yoshida, trans., World Scientific, Singapore (1982).
10. E. Fermi, Nature **125**, 16 (1930); Z. Phys. **60**, 320 (1930).
11. W. Heisenberg, W. Pauli, Z. Phys. **47**, 1 (1929).
12. Document E05 030 U01, Yukawa Hall Archival Library, Kyoto University.
13. H. Yukawa, J. Phys.-Math. Soc. Jpn. **7**, 195 (1933).
14. Document E05 060 U01, Yukawa Hall Archival Library, Kyoto University.
15. H. Yukawa, Bull. Phys.-Math. Soc. Jpn. **7**, 131A (1933).
16. See, for example, L. M. Brown, PHYSICS TODAY, September 1978, p. 23.
17. E. Fermi, Ric. Sci. 4(2), 491 (1933); Nuovo Cimento 2, 1 (1934); Z. Phys. **88**, 161 (1933).
18. I. Tamm, Nature **133**, 981 (1934). D. Iwanenko, Nature **133**, 981.
19. G. C. Wick, Nature **142**, 994 (1938).
20. H. J. Bhabha, Nature **141**, 117 (1938).
21. H. Yukawa, Proc. Phys.-Math. Soc. Jpn. **19**, 712 (1937).
22. S. H. Neddermeyer, C. D. Anderson, Phys. Rev. **51**, 884 (1937). J. C. Street, E. C. Stevenson, Phys. Rev. **51**, 1005 (1937). Y. Nishina, M. Takeuchi, T. Ichimiya, Phys. Rev. **52**, 1198 (1937).
23. J. R. Oppenheimer, R. Serber, Phys. Rev. **51**, 1113 (1937). E. C. G. Stueckelberg, Phys. Rev. **52**, 41 (1937).
24. Document E02 130 P12, Yukawa Hall Archival Library, Kyoto University.
25. See N. Kemmer, Biog. Mem. Fellows R. Soc. **29**, 661 (1938). □

THE DISCOVERY OF NUCLEAR FISSION

Fermi's group bombarded uranium with neutrons in 1934, but it was almost five years before Hahn and Strassmann realized what these neutrons were actually doing. It required superb chemists to bring the comedy of errors to a close.

Emilio G. Segrè

PHYSICS TODAY/JULY 1989

Emilio Segrè was emeritus professor of physics at the University of California, Berkeley. He died suddenly on 22 April 1989 at the age of 84, three months after he presented the talk on which this article is based at the January meeting of the American Physical Society in San Francisco. He was awarded the Nobel Prize in physics in 1959.

Few modern discoveries have influenced mankind so rapidly and so profoundly as has nuclear fission, and few have had such an intricate history. Thus it is natural that the discovery of fission by Otto Hahn and Fritz Strassmann in December 1938 is remembered and commemorated in many places on its 50th anniversary. I participated in the early experiments in Rome, and later in the US, and I knew most of the principals well, except for Strassmann. I will try to present a brief outline of the discovery and its antecedents.

Transmutation by neutrons

I begin the story in 1934 with the first neutron bombardment of uranium.[1] Following the discovery of artificial radioactivity by Irène Curie and her husband Frédéric Joliot in Paris at the beginning of 1934, Enrico Fermi in Rome had started using neutrons from radon–beryllium sources, in lieu of alpha particles, to activate many common elements. Between March and April, with the help of Edoardo Amaldi, Oscar D'Agostino, Franco Rasetti and myself, Fermi had established[2] the reactions (n,p), (n,α) and (n,γ) or (n,2n). (The notation (A,B) means that a nucleus has been transformed by incident particle A, with the emission of particle B.) Our sources emitted about 10^7 neutrons per second.

Bombarding uranium with neutrons presented an especially interesting case, because we could expect to form element 93, the first transuranic element, by an (n,γ) reaction followed by subsequent beta decay. Indeed this does happen—but much more was in store. I remember that Rasetti was particularly eager to bombard uranium and thorium. The first communication on our results was dated 10 May 1934, about two months after the first neutron bombardment. For the sake of brevity I will omit most of the work on Th, which paralleled and often supplemented that on U.

The radioactivity produced by our neutron sources was not much greater than the natural radioactivity of uranium off the shelf. This caused severe technical problems. We could chemically remove some of the uranium's beta-active decay products before irradiating it. But they grew anew after just a few hours. They formed a large, confusing background of counts that had nothing to do with the neutron irradiation. The same trouble

Enrico Fermi's group in Rome, 1934. Left to right are Oscar D'Agostino, Emilio Segrè, Edoardo Amaldi, Franco Rasetti and Fermi.

affected all the European investigators and played a part in the errors we all made.

In Rome we immediately found that irradiated uranium showed a complex radioactivity with a mixture of several decay periods. We expected to find in U only the previously observed backgrounds, and so we started looking for an isotope of element 93 produced by an (n,γ) reaction on U^{238} followed by beta decay.[3]

A mistake in the chemistry

Here we made a mistake that may seem strange today. We anticipated that element 93 would resemble rhenium (element 75)—that it would be, in the language of Mendeleev, an eka-rhenium. Products of the bombardment are indeed similar to Re, but for a totally different and then unimagined reason: Some of the abundant *fission* products are isotopes of technetium, element 43. This element (so named because it was the first of the artificially produced elements) was discovered by Carlo Perrier and me three years later, in 1937, in molybdenum (element 42) bombarded with deuterons. It proved to be chemically very similar to Re. Thus it actually resembled what we erroneously expected in the 1934 uranium experiment.

Hahn and Lise Meitner, and Irène Curie, made the same error, presuming that 93 was an eka-Re. (Element 93, like all the now known transuranics, has a chemistry akin to that of the rare earths.) Even stranger is the fact that Niels Bohr did not object. Ten years earlier he had considered the filling of the 5f electron orbits and the formation of a new family of rare earths, although he did

not start it at exactly the right place. On the other hand, Aristid von Grosse at the University of Chicago pointed out in 1934 the possible analogy of element 93 to a rare earth.

Aluminum foil hides the prize

In Rome we also considered the possibility of the formation of short-lived alpha emitters in neutron-bombarded U. To test this hypothesis, we placed a uranium foil in front of an ionization chamber and irradiated it with slow neutrons. We thought that if alpha particles came from a short-lived substance created by the neutron bombardment, they would have a significantly longer range than the uranium background alphas. We therefore covered the uranium sample with a thin aluminum foil that would stop the U alpha particles. The results were negative, and the aluminum layer prevented us from seeing the big ionization pulses produced by the fission fragments! We did not publish this result, but it is in Amaldi's notebooks of the period.[5] I cannot say, however, that if we had seen the big pulses we would have understood their cause. A similar experiment was performed by Paul Scherrer and coworkers in Zurich and by Gottfried von Droste in Berlin. I've been told that the Swiss saw the big pulses but attributed them to a fault in the detector.

Another error was in not paying enough attention to a 1934 article by Ida Noddack in Berlin,[6] who criticized our chemistry and pointed to the possibility of fission. Much has been said of her prescience. Her article was certainly known to us in Rome, to Hahn and Meitner in Berlin and to Joliot and Curie in Paris. If any of us had really grasped its importance, it would have been easy to discover fission in 1935. It is equally astounding that Noddack did not try any experiment herself to check her ideas. It would have been quite easy for her too. Be that as it may, for more than three years all the investigators in the field considered only nuclear reactions that would lead to elements with atomic numbers near 92.

In Rome, we started by convincing ourselves that the radioactivity observed in bombarded uranium was not due to isotopes of elements between Pb and U, which turned out to be correct. We tried to establish some properties of substances extracted from the radioactive complex, and we found an ingredient that we thought behaved like an eka-Re. Its chief reaction was a precipitation with MnO_2. A few years ago, Franco Baroncelli in Italy repeated our old procedure and found that he could separate some isotopes of technetium (unknown in 1934) that simulate our results. Of course Tc behaves very much like Re, and it is a fission product! We should, however, have been suspicious of many things, and especially of the fact that our supposed element 93 could account for only a small fraction of the activity generated by the neutron bombardment.

Spurious transuranics

By June of 1934, the end of the academic year, we were confident enough of our imagined success at forming transuranic elements to publish this result. We also had a feeling, however, that the work was incomplete. We therefore refrained from naming the transuranic elements we thought we had found. Fermi was particularly upset when the press gave this work a lot of publicity.[7]

After the 1934 summer vacation, Bruno Pontecorvo joined our group. In October we suddenly discovered how to produce slow neutrons. This was a major find, and our group, depleted of its chemist by the departure of D'Agostino, suspended the work on transuranics and concentrated on studying slow neutrons. We did, however, look at the effect of slow neutrons on the U activity, to find out whether it involved neutron capture, and also to learn more about the substances produced in neutron bombardment. The work on transuranics in Rome stopped in the summer of 1935. At the end of the year Rasetti came to the US, Pontecorvo went to Paris and I became director of the Physics Institute in Palermo. Amaldi and Fermi, in Rome, concentrated their efforts on developing slow-neutron physics.

At this point Hahn and Meitner,[8] and Irène Curie,[9] entered the uranium fray. Both groups had a past history of great achievements in nuclear physics and chemistry.

In Berlin, Hahn and Meitner were the senior members of the group at the Kaiser Wilhelm Institute in Dahlem. Hahn was a renowned radiochemist; he had worked with Ernest Rutherford at Montreal 30 years earlier. At the beginning of his career he had discovered several new radioactive substances. Later, together with Meitner, he discovered the phenomenon of nuclear isomerism.

Meitner, an Austrian citizen born in Vienna, had been an assistant to Max Planck in Berlin; later she became the steady colleague of Hahn and worked in the same lab with him. She was a distinguished physicist and something of a rival to the Curies, both Marie and Irène. Meitner was of Jewish descent; Hahn was strongly anti-Nazi. The political situation made working conditions at Dahlem difficult and caused continuous anxiety.[10] Strassmann was the last to join the Dahlem group. He was a superior analytical chemist, and also resolutely anti-Nazi, a fact that hampered his career.

Irène Curie was the daughter of Marie, the discoverer

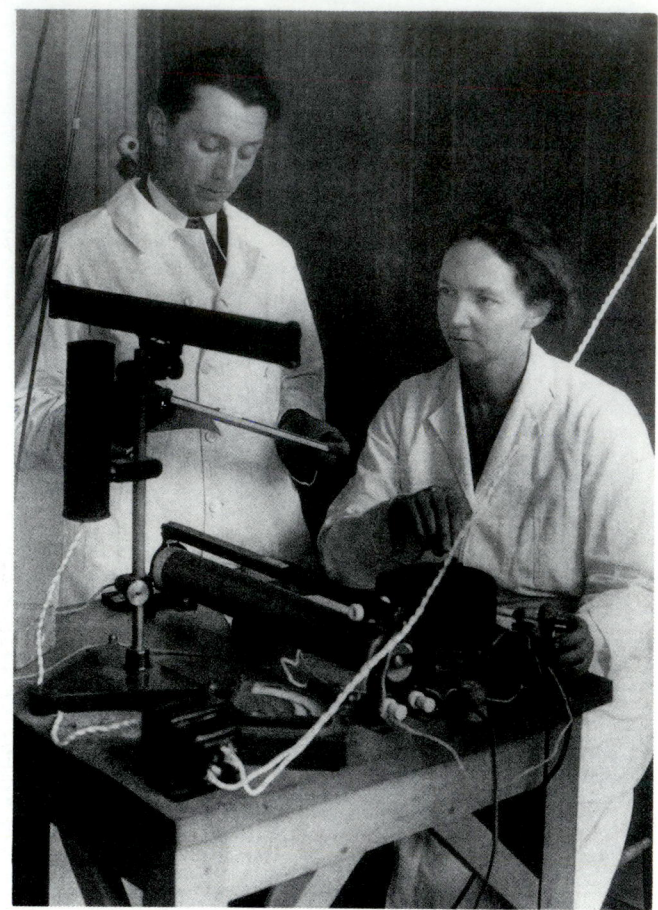

Frédéric Joliot and Irène Curie in their Paris laboratory, around 1934.

of radium, and she had learned radiochemistry and chemistry from her great mother. She was steeped in Marie's tradition, methods and techniques. She had married Joliot, and together they had performed important experiments on the positron and the neutron, though they had missed discovering these new particles. This was, however, soon compensated for by their memorable discovery of artificial radioactivity.

Berlin

Hahn and Meitner, working with a neutron source about as strong as the ones used in Rome and Paris, started by confirming our Rome results. This is somewhat surprising because they applied quite different chemistry. Their early papers are a mixture of error and truth as complicated as the mixture of fission products resulting from the bombardments. Such confusion was to remain for a long time a characteristic of much of the work on uranium. A firm and important result obtained by Hahn, Meitner and Strassmann was the identification of U^{239} as a beta emitter with a halflife of 23 minutes. For the rest, there is not much reason to follow the details. Their numerous papers, published mostly in *Naturwissenschaften*, record the ups and downs of these investigations that preceded the realization that fission was happening. At first the authors are Hahn and Meitner. Strassmann appears initially as collaborator and then, from July 1938, as a full-fledged coauthor. In 1937 the three published a longer summarizing paper in *Zeitschrift für Physik*[11] and a parallel paper in *Chemische Berichte*.[12] They mention 12 new radioactive isotopes attributed to elements of atomic number between 92 and 95, and a series of double isomeric states. (In fact, no transuranic element was properly discovered before 1940.) The supposed double isomers were particularly surprising. Double isomeric states were unknown then, and even today we know of less than a handful.

My own feeling at the time was that there was a mystery in uranium. In Palermo I could not work on it because I had no neutron sources. In the summer of 1936, when I visited Berkeley for the first time (and gratefully obtained radioactive material from Ernest Lawrence), I spoke at length with Philip Abelson, who was a student at that time, and pointed out to him the great uranium puzzle. I emphasized that the powerful neutron source offered by the cyclotron would make it easier to solve. A start could be made by irradiating U with fast or slow neutrons. Abelson made a few runs and gave me some of the resulting decay curves.

Paris

Irène Curie started working on neutron-bombarded thorium in collaboration with Hans von Halban and Preiswerk. As early as May 1935 they confirmed a 22-minute decay period we had found in Rome. This Th isotope is important as a precursor of U^{233}. More important, they soon found a 3.5-hour radioactive component, which chemically resembled lanthanum.[13] They did not realize, of course, that it was indeed La^{141}, a fission product! Unable to pin it down, Curie and her colleagues thought it an isotope of actinium.

In 1937 and 1938 Curie and Pavel Savitch concentrated their combined efforts on the study of the 3.5-hour substance. By July 1938 they arrived at the conclusion that the substance was not actinium, and that "all in all, the properties of [the 3.5-hour component] are those of La, from which up to now it can be separated only by fractional crystallization."[14] Had they been able to identify it as La, they would likely have discovered fission, as Hahn and Strassmann did a few months later by nailing down barium. Possibly their initial precipitate contained not only La, but also the chemically similar yttrium, also produced by fission, and it was these two substances that the fractional crystallization was separating.

In mid-May 1938, Hahn and Joliot met in Rome at the 10th International Chemistry Conference and discussed the Paris results. Hahn was convinced that there was something wrong in Curie's chemistry, and he decided to repeat some of her experiments.

Darkness and dawn

Two months earlier Hitler had annexed Austria to the Reich, and thus Meitner had lost the relative protection of her Austrian citizenship. She was now in imminent danger of arrest. In mid-July she fled the country in secrecy and haste, helped by Hahn and the Dutch physicist Dirk Coster. Hahn was greatly relieved when he heard that she had safely crossed into Holland. From there she proceeded to Copenhagen and then to Sweden. Hahn and Strassmann continued their work at Dahlem, and Hahn kept Meitner posted on their progress by frequent letters. He also made a point of disclosing to her the results he and Strassmann were obtaining, before publishing them or even mentioning them to anybody else, including his Berlin colleagues.

Hahn and Strassmann concentrated on what they thought were isotopes of radium and on the 3.5-hour product described by Curie and Savitch. They concluded that from the neutron bombardment of U^{238} they could

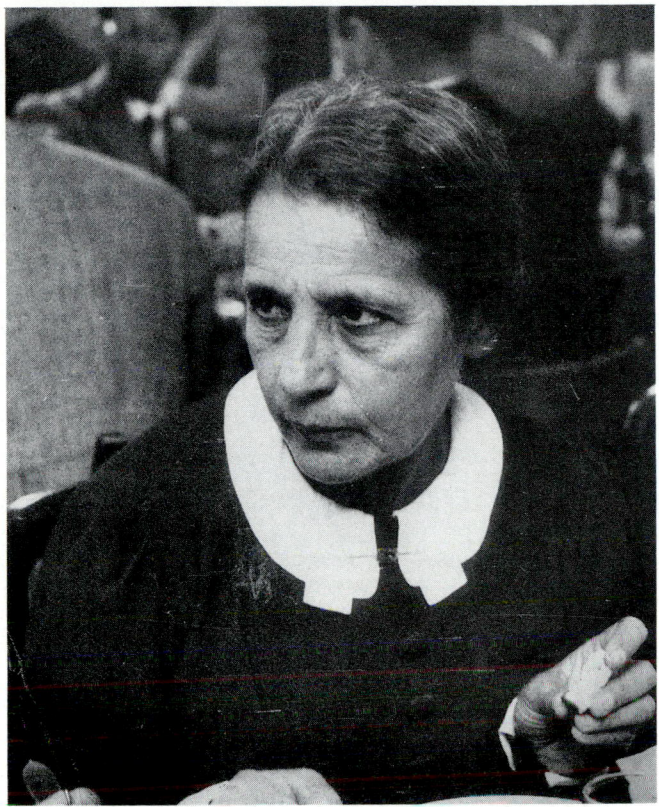

Lise Meitner in 1937, a year before she had to flee Berlin, where she had been collaborating with Otto Hahn for three decades.

Otto Hahn in Göttingen after World War II.

W. PRAGER

12 July 1938

$$\text{(1)} \quad _{92}U + n \rightarrow (_{92}U + n) \xrightarrow[\text{10 Sek.}]{\beta} {}_{93}\text{EkaRe} \xrightarrow[\text{2,2 Min.}]{\beta} {}_{94}\text{EkaOs} \rightarrow$$

$$\xrightarrow[\text{59 Min.}]{\beta} {}_{95}\text{EkaIr} \xrightarrow[\text{66 Std.}]{\beta} {}_{96}\text{EkaPt} \rightarrow$$

$$\xrightarrow[\text{2,5 Std.}]{\beta} {}_{97}\text{EkaAu ?}$$

$$\text{(2)} \quad _{92}U + n \rightarrow (_{92}U + n) \xrightarrow[\text{40 Sek.}]{\beta} {}_{93}\text{EkaRe} \xrightarrow[\text{16 Min.}]{\beta} {}_{94}\text{EkaOs} \rightarrow$$

$$\xrightarrow[\text{5,7 Std.}]{\beta} {}_{95}\text{EkaIr ?}$$

$$\text{(3)} \quad _{92}U + n \rightarrow (_{92}U + n) \xrightarrow[\text{23 Min.}]{\beta} {}_{93}\text{EkaRe ?}$$

8 November 1938

$$^{231}_{88}\text{Ra}_1 \xrightarrow[\infty \text{ 25 Min.}]{\beta} {}_{89}\text{Ac}_1 \xrightarrow[\infty \text{ 40 Min.}]{\beta} {}_{90}\text{Th ?}$$

$$_{88}\text{Ra}_2 \xrightarrow[\infty \text{110 Min.}]{\beta} {}_{89}\text{Ac}_2 \xrightarrow[\infty \text{ 4 Std.}]{\beta} {}_{90}\text{Th ?}$$

$$_{88}\text{Ra}_3 \xrightarrow[\text{mehrere Tage}]{\beta} {}_{89}\text{Ac}_3 \xrightarrow[\infty \text{60 Std.}]{\beta} {}_{90}\text{Th ?}$$

22 December 1938

$$\text{,,Ra I``?} \xrightarrow[< \text{1 Min.}]{\beta} \text{Ac I} \xrightarrow[< \text{30 Min.}]{\beta} \text{Th ?}$$

$$\text{,,Ra II``} \xrightarrow[\text{14} \pm \text{2 Min.}]{\beta} \text{Ac II} \xrightarrow[\sim \text{2,5 Std.}]{\beta} \text{Th ?}$$

$$\text{,,Ra III``} \xrightarrow[\text{86} \pm \text{6 Min.}]{\beta} \text{Ac III} \xrightarrow[\sim \text{mehrere Tage?}]{\beta} \text{Th ?}$$

$$\text{,,Ra IV``} \xrightarrow[\text{250—300 Std.}]{\beta} \text{Ac IV} \xrightarrow[< \text{40 Std.}]{\beta} \text{Th ?}$$

Some of the last steps in the discovery of fission, reproduced from 1938 papers of Hahn and coworkers. **Top:** In reference 15, from July 1938, Hahn, Meitner and Strassmann were still assuming that bombarding uranium with neutrons produced mostly transuranics, labelled "eka-gold," "eka-platinum" and so on, after the lighter elements they were presumed to resemble chemically. Hahn and his colleagues were forced to invoke three different imagined isomers of "eka-rhenium." Reaction 3 turns out to be correct, except that element 93 (now called neptunium) does not resemble rhenium at all. **Middle:** From Hahn and Strassmann, *Naturwissenschaften* **26**, 755 (1938), dated 8 November. Much of the activity observed after neutron bombardment was attributed to supposed isotopes of radium. **Bottom:** Finally, in reference 16, dated 22 December, Hahn and Strassmann have recognized fission. They have chemically identified "RaI" (now in quotes) as *barium*, which could only be a fission product.

obtain 16 nuclear species with atomic numbers ranging from 88 to 96, including a number of isomers.[15] The confusion was approaching its maximum. It was, however, the darkness before dawn, because the solution was not far off.

Early in December 1938 they thought they had established some decay chains in which the genetic relations appeared to be solidly known. Supposedly four isotopes of radium were decaying to Ac and then to Th. These putative chains were modifications of chains given in a previous paper (see the figure above). To make doubly sure, Hahn and Strassmann decided to identify the radium isotopes beyond any doubt. They submitted them to several stringent chemical tests using barium as a carrier and radium as a tracer. These superb experiments forced Hahn and Strassmann to conclude reluctantly that the hypothetical radium isotopes were in fact barium! In their historic 22 December paper[16] for *Naturwissenschaften* they wrote: "As chemists, in consequence of the experiments just described, we should change the schema given above and introduce the symbols of Ba, La, Ce in place of Ra, Ac, Th. As 'nuclear chemists,' working very close to the field of physics, we cannot yet bring ourselves to take such a drastic step, which goes against all previous experiences of nuclear physics." A few lines earlier, however, the authors had noted "that the sum of the mass numbers of Ba + Ma [technetium], for instance 138 + 101,

gives 239"—a clear sign that they were thinking of fission. This is the moment of the discovery of nuclear fission.

'What idiots we all have been!'

Hahn sent letters with these results, prior to their publication, to Meitner in Sweden. She showed them to her physicist nephew Otto Frisch, who was visiting from Copenhagen over the Christmas holidays.[10,17] Frisch and Meitner soon arrived at the idea of fission. A few days later Frisch returned to his lab in Copenhagen. In his words: "I was keen to submit our speculations—it wasn't really more at the time—to Bohr, who was just about to leave for the USA. He had only a few minutes for me, but I had hardly begun to tell him when he smote his forehead with his hand and exclaimed: 'Oh what idiots we all have been! Oh but this is wonderful! This is just as it must be! Have you and Lise Meitner written a paper about it?' "[17] Two weeks later their paper was received by *Nature*.[18]

The discovery of fission started a flood of investigations. The first and most obvious was the verification of the presence of fission fragments. One could calculate from the mass defects, or from the Coulomb repulsion of the fission fragments, that they had to be nuclei with kinetic energies of about 200 MeV; they would therefore be heavily ionizing. Such fragments were promptly observed by Frisch[19] and almost simultaneously by many others.

The chemical identification of the substances produced by neutron bombardment now took on a new aspect. Irène Curie, in 1938, had said, "It seems that uranium bombarded with neutrons gives an activity composed of almost every element." She was right, and now many rushed to disentangle the fission products.

While fission was being discovered, Fermi was in Stockholm collecting the Nobel Prize for "his demonstration of the existence of new radioactive elements produced by neutron irradiation, and for his related discovery of nuclear reactions brought about by slow neutrons." This citation has been variously interpreted, as far as the words "new radioactive elements" are concerned. If the word "isotopes" had been used instead of "elements," it would be clearer. Fermi emigrated directly from Sweden to the US, where he first heard the news about fission.

The discovery of fission was a sensation. The reaction in America can be seen by the spate of papers on fission that immediately appeared in the *Physical Review*. Luis Alvarez has vividly described the reaction at Berkeley.[20]

Verifying the transuranics

The transuranics still had some surprises in store. Joliot, trying to demonstrate fission, exposed a thin layer of

Fritz Strassmann (left) with Meitner and Hahn in Mainz, 1956.

uranium to neutrons and put next to it a sheet of bakelite in which he collected the fission fragments.[21] Edwin McMillan, independently, did the same experiment at Berkeley.[22] They both found a nonrecoiling activity remaining in the U layer. There was a 23-minute activity due to U^{239} formed by an (n,γ) reaction on U^{238}, already demonstrated by Hahn and Meitner. But there was an additional two-day activity.

Having come to Lawrence's Radiation Laboratory in 1938, I took advantage of the powers of the cyclotron as a neutron source to investigate this two-day component. I studied its chemical properties and concluded that they were those of a rare earth. I had expected that the two-day activity was due to a beta-decay product of U with atomic number 93 (as indeed it is), but I was expecting element 93 to have the chemistry of an eka-Re! My attempt to find a genetic relationship between the 23-minute and two-day activities failed because the beta rays from element 93 are uncommonly soft and led me astray. Thus another erroneous paper was added to the long list of blunders produced by irradiated uranium. Like many of the other blunders, however, it had some elements of truth. It showed that element 93 is similar to the rare earths.[23]

Finally, a few months later, McMillan and Abelson[24] chemically separated and recognized element 93 (neptunium), the daughter of U^{239}.

In the meantime Joliot, Fermi and many others had noted that the two fission fragments were particularly rich in neutrons. Most of the excess neutrons transformed into protons by beta decay, but it was conceivable that some neutrons were being set free. This opened the possibility of a chain reaction.

It was the beginning of 1939; war was threatening in Europe. Nuclear fission was becoming more than a scientific curiosity. I will not go into the subsequent story. It has been told many times.

The discovery of fission has an uncommonly complicated history; many errors beset it. Nature had, however, truly complicated the problem. One had to contend with the radioactivity of natural uranium and the presence of two long-lived isotopes—U^{235} and U^{238}. The heavier isotope, as is well known, does not undergo fission when bombarded by slow neutrons. The lighter isotope, which makes up only 0.7% of natural uranium, is responsible for all slow-neutron fission. This is a tricky setup. Above all, it seems to me that the human mind sees only what it expects.

* * *

The editors wish to thank Rosa Mines Segrè, the author's widow, for her gracious assistance in the preparation of this article.

References

For historical clarity, for some papers we give the date of receipt by the journal in square brackets.

1. E. Amaldi, O. D'Agostino, E. Fermi, F. Rasetti, E. Segrè, Ric. Sci. **5**, 452 (1934) [10 May 1934].
2. E. Fermi, Ric. Sci. **5**, 283 (1934) [25 March 1934]. E. Fermi, E. Amaldi, O. D'Agostino, F. Rasetti, E. Segrè, Proc. R. Soc. London **146**, 483 (1934) [25 July 1934].
3. E. Fermi, Nature **133**, 898 (1934).
4. A. von Grosse, J. Am. Chem. Soc. **57**, 440 (1934) [11 December 1934].
5. E. Amaldi, Phys. Rev. **111**, 1 (1984). See p. 278.
6. I. Noddack, Angew. Chem. **47**, 653 (1934) [10 September 1934].
7. L. Fermi, *Atoms in the Family*, AIP, New York (1988).
8. O. Hahn, L. Meitner, Naturwissenchaften **23**, 37 (1935) [22 December 1934]; **23**, 230 (1935) [2 March 1935].
9. I. Curie, H. von Halban, P. Preiswerk, C. R. Acad. Sci. (Paris) **200**, 1841 (1935) [27 May 1935].
10. F. Krafft, *Im Schatten der Sensation: Leben und Wirken von Fritz Strassmann*, Verlag Chemie, Weinheim (1981).
11. L. Meitner, O. Hahn, F. Strassmann, Z. Phys. **106**, 249 (1937) [14 May 1937].
12. O. Hahn, L. Meitner, F. Strassmann, Chem. Ber. **70**, 1374 (1937).
13. I. Curie, P. Savitch, J. Phys. Radium **8**, 385 (1937) [1 August 1937]; **9**, 355 (1938) [12 July 1938].
14. I. Curie, P. Savitch, C. R. Acad. Sci. (Paris) **206**, 906 (1938) [31 March 1938]; **206**, 1648 (1938) [30 May 1938].
15. O. Hahn, L. Meitner, F. Strassmann, Naturwissenchaften **26**, 475 (1938).
16. O. Hahn, F. Strassmann, Naturwissenchaften **27**, 11 (1939) [22 December 1938].
17. O. Frisch, *What Little I Remember*, Cambridge U. P., Cambridge (1979). O. Hahn, *Vom Radiothor zur Uranspaltung: Eine wissenschaftliche Selbstbiographie*, Vieweg, Braunschweig (1962).
18. L. Meitner, O. Frisch, Nature **143**, 39 (1939) [16 January 1939].
19. O. Frisch, Nature **143**, 276 (1939) [16 January 1939].
20. L. W. Alvarez, *Adventures of a Physicist*, Basic Books, New York (1987), ch. 4. L. Badash, E. Hodes, A. Tiddens, Proc. Am. Philos. Soc. **130**, 196 (1986).
21. F. Joliot, C. R. Acad. Sci. (Paris) **208**, 341 (1939) [30 January 1939].
22. E. McMillan, Phys. Rev. **55**, 510 (1939) [17 February 1939].
23. E. Segrè, Phys. Rev. **55**, 1104 (1939) [10 May 1939].
24. E. McMillan, P. H. Abelson, Phys. Rev. **57**, 1185 (1940) [27 May 1940]. ∎

Lise Meitner (1878–1968)

Lise Meitner, one of the pioneers of nuclear physics, was born in Austria on 7 Nov. 1878, the third of eight children of a Viennese lawyer. Being one of the first girls to enroll at the University of Vienna, she met occasional unpleasantness but was encouraged by Ludwig Boltzmann and in 1906 graduated with a PhD. She then studied with Max Planck in Berlin and later for a time became his assistant. But she also wanted to continue the work she had begun in Vienna on radioactivity and joined forces with Otto Hahn, who had studied radiochemistry with William Ramsay and Ernest Rutherford. That partnership lasted for over 30 years.

Her main interest was elucidation of radioactive transformations. With Hahn and Otto von Baeyer she developed accurate methods for the study of beta line spectra. The emission of so-called "Auger electrons" was first described and correctly interpreted by her, and in 1925 she showed that the beta lines were emitted after and not before the radioactive alpha transformation, a question much debated at the time. She also developed radioactive recoil, first recognized by Hahn, as a method for obtaining atomically thin layers of certain radioactive substances.

In one respect she backed the wrong hunch: She thought that the primary beta-rays (like alpha rays) were homogeneous and that the continuous spectrum found by James Chadwick in 1914 was caused by energy lost after emission. But when C. D. Ellis and W. A. Wooster disproved that view (by measuring the heat evolved by a beta emitter), she and Wilhelm Orthmann confirmed the Ellis-Wooster result by an absolute method. Those measurements gave Wolfgang Pauli the courage to propound the neutrino.

During the first world war she volunteered as an x-ray nurse with the Austrian army but managed to keep her work going through occasional visits to Berlin and was able, with Hahn, to publish the discovery of protoactinium in 1918. She then returned to Berlin, no longer as a "scientific guest" but as head of the radiation-physics department. Lise Meitner was one of the first to use the cloud chamber for the study of alpha-ray straggling and soft beta-ray spectra, to measure the wavelength of a gamma ray by reflection from a crystal, and to find clear deviations (caused by pair production, as was understood later) from the Klein–Nishina formula for the scattering of hard gamma rays.

When Enrico Fermi in 1934 published his results from the bombardment of uranium with neutrons, Meitner persuaded Hahn to take up with her the study of the transuranic elements that appeared to be formed. But work in Hitler's Germany became more and more difficult. In 1938, after the annexation of Austria, she no longer felt safe from racial persecution, having lost the security her Austrian passport had given her. With the help of Dutch colleagues she escaped from Germany and found sanctuary in Sweden, where she continued work until she retired to England in 1960. So she was not in Germany when Hahn and Fritz Strassmann broke the web of error, showing that lighter elements (like barium) were formed when uranium was bombarded by neutrons. But when she heard of this development she worked out and published, with her nephew Otto R. Frisch, the physical details of the process, which was named "nuclear fission" in that paper.

All over the world she made many lifelong friends who appreciated her

PHOTO BY HEKA

LISE MEITNER

open mind and wide interest, and she is fondly remembered by many pupils for her warm-hearted help with personal problems. Among her friends she could be lively and witty, but she never quite lost her shyness, indeed her humility, despite the many honors she received. The culminating honor was the award, given jointly to her and Hahn and Strassmann, of the Enrico Fermi Prize in 1966. She was in excellent health even after her retirement and continued to write, lecture and travel widely until 1965, when her strength began gradually to fail. She died on 27 Oct.

OTTO R. FRISCH
University of Cambridge

Maria Goeppert Mayer: Atoms, molecules and nuclear shells

The mathematical physicist's early work in atomic and molecular physics, and her unfamiliarity with the "fashions" in nuclear physics, gave her the ideal preparation for solving the puzzle of the nuclear "magic numbers."

Karen E. Johnson

PHYSICS TODAY/SEPTEMBER 1986

Maria Geoppert Mayer is frequently mentioned as an example of a women who managed to make significant contributions to science in spite of tremendous obstacles. Robert Sachs, Mayer's first graduate student, has given a personal account of her life and career (PHYSICS TODAY, February 1982, page 247). Joan Dash has written[1] a longer biography of Mayer, focusing on her family life. In this article I look at the distinctive character of Mayer's scientific accomplishments.[2] As a mathematical physicist she worked on a number of apparently unrelated topics during the years 1930–46. I will focus on her work during this period and show how it prepared her for the nuclear physics research for which she received the Nobel Prize in 1963.

Mayer received the prize for her "discoveries concerning nuclear shell structure." Hence I will first outline the essential characteristics of the nuclear shell model. Then I will look back to the period before the shell model and show how the emphasis in Mayer's early work in atomic and molecular physics put her in an ideal position to enter nuclear physics and quickly make a major contribution to the nuclear shell model.

The nuclear shell model

In the Bohr–Pauli model of the atom an electron moves in a potential produced by the atomic nucleus and by the other electrons. The Bohr model, together with the Pauli exclusion principle, explains, among other things, why certain atoms with filled electron shells are particularly stable. The nucleus, by contrast, may be conceived of in two different ways. It may be treated as a homogeneous mass of material, such as a liquid drop, or, in analogy to the Bohr–Pauli atom, it may be treated as a collection of discrete particles. In the latter case, the individual nucleons—neutrons or protons—are considered to be independent particles with independent energy levels, spins and angular momenta, moving in a potential well produced by the action of all the remaining nucleons. This is the basis of the nuclear shell model. In it the interaction energy between individual nucleons is subsidiary to the energy of the central nuclear potential. Again by analogy to the Bohr atom, the primary evidence for the nuclear shell model is that nuclei with certain specific numbers of protons or of neutrons—2, 8, 20, 28, 50, 82 or 126—are particularly stable. Data collected from nuclear binding energies, radioactive-decay energies and isotopic abundances established this pattern. This evidence suggested that nucleons fill nuclear energy levels in a way similar to the way in which electrons fill atomic energy levels. The numbers of nucleons occupying filled shells, however, turned out to be different from those that the usual methods of quantum mechanics predicted for standard central potentials, as the figure on page 46 shows.

Mayer discovered[3] the solution to this puzzle in 1949. Responding to a question by Enrico Fermi, she suggested that there is a strong spin–orbit interaction that splits the energy levels of the nucleons. This interaction occurs for electrons in an atom, but the energy splitting is so small compared with the total electron binding energy due to the central potential that it is generally neglected, except for very heavy atoms. Until 1949, nuclear physicists generally assumed that the same is true of nucleons; Mayer's mathematical argument showing that a strong spin–orbit interaction would account for the magic numbers was a radical departure, and a major contribution to nuclear shell theory.

Mayer once used a simple analogy to explain spin–orbit coupling to her daughter. Think of a room full of couples waltzing. They are moving around the room in circles, each circle enclosed within another. Each circle corresponds to an energy level. In addition to orbiting, though, each couple is also spinning like a top. Now suppose that while orbiting counterclockwise some couples are spinning clockwise, and the rest counterclockwise. Those spinning counterclockwise will find the going easier than those spinning clockwise. As Mayer said,[1] "Everybody who has ever danced the fast waltz knows that it's easier to

Karen E. Johnson is professor in the department of physics at Saint Lawrence University, Canton, New York.

Maria Goeppert Mayer (1906–72) with Max Born. After attending one of Born's lectures on quantum mechanics, Mayer changed her university major from mathematics to physics. (Courtesy AIP Niels Bohr Library.)

dance one way around than the other." Therefore, for a given circle of dancers, the energy necessary to orbit will be different for couples spinning in opposite senses. In the same way, nucleons of a given orbital angular momentum have two possible energies, depending on whether their spin is parallel or anti-parallel to the orbital motion. This splitting of the energy level is called spin-orbit coupling. The figure on page 249 indicates how this effect can explain nuclear shells.

The nuclear shell model thus has two essential features:

▶ The identification of the shell structure itself, based on the evidence for nuclear stability and leading to the basic assumption that the nucleus can be described with a single-particle model.

▶ The assumption of a strong spin–orbit interaction in nucleons, accounting for the splitting of their energy levels.

My thesis is that Mayer's early work led her quite naturally to this particular model.

Study in Göttingen

Maria Goeppert was born 28 June 1906, in Upper Silesia, which is now part of Poland. In 1910 her family moved to the university town of Göttingen. Her father, professor of pediatrics at the university, was the sixth generation of university professors in his family. In later years Mayer was always proud that she was the seventh generation to follow this tradition. Her early schooling was typical for a child of the German middle class. She attended public elementary school until age 15 and then, in 1921, entered the Frauenstudium, a three-year college-preparatory school for girls. She passed the Abitur in 1924 and entered the University of Göttingen.

Goeppert began her university career as a mathematics major, influenced perhaps by the fact that David Hilbert was an old friend of her family. Sometime in 1927, however, she switched to physics after attending one of Max Born's seminars, where she was first exposed to the delights of quantum mechanics. Born became a close friend and advisor to Goeppert for the rest of his life, and he had a pronounced influence on the style of her physics. She always used a strictly mathematical approach to any physical problem, and she showed a definite preference for matrix over wave mechanics.

Goeppert's doctoral dissertation, written in 1929–30, involved a calculation[4] of the probability that an electron will undergo a transition due to the simultaneous emission or absorption of two photons. This problem was suggested in part by some of James Franck's spectroscopic work. Goeppert's calculations indicated that the probability of such transitions was extremely low—so low, in fact, that they were not observed experimentally until the 1960s, after the invention of lasers. Laser experiments confirmed her predictions.

It was while she was at the university that Goeppert met Joseph Mayer, an American chemist who had just completed his doctoral degree with Gilbert Newton Lewis at Berkeley and was in Göttingen in 1929–30 on an International Education Board fellowship. The two were soon engaged to be married and Joe promptly assumed an extremely important role in Maria's scientific career. It was only through his support over the years that she was persuaded to continue in science. Before they were married Joe promised to hire a maid for Maria only if she finished her degree. They were married in January 1930 and she finished her PhD two months later. Joe's importance in encouraging Maria during the years when she found little support from academic science departments should not be allowed to overshadow the fact that he himself was a highly respected chemist and, until Maria received the Nobel Prize, by far the

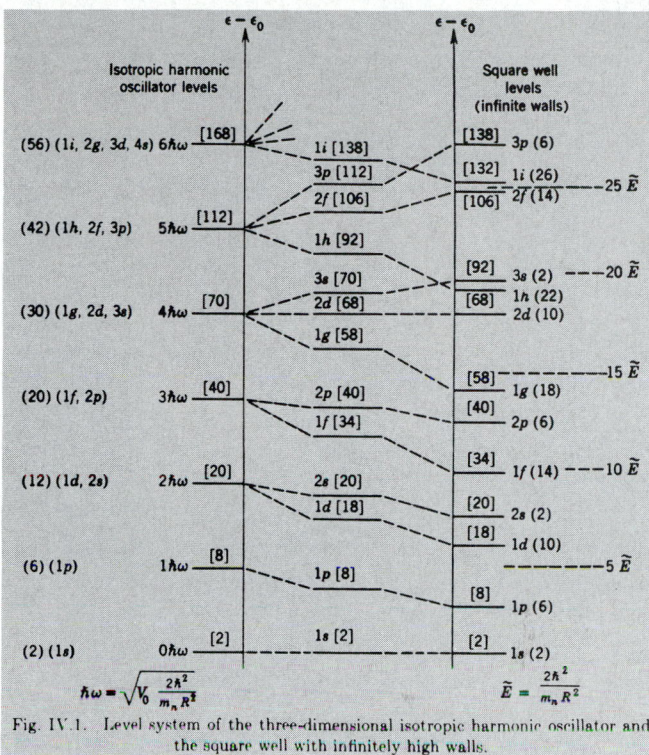

Fig. IV.1. Level system of the three-dimensional isotropic harmonic oscillator and the square well with infinitely high walls.

Shell occupation numbers, predicted for harmonic oscillator and square-well potentials. From Mayer's book with J. Hans D. Jensen, *Elementary Theory of Nuclear Shell Structure* (Wiley, New York, 1955).

better known of the two in the scientific world.

Office space at Johns Hopkins

After Mayer completed her degree, she and her husband came to America where he took up a position in the chemistry department at Johns Hopkins University. Although Maria Mayer was then one of the few people in America who was well trained in quantum physics, Johns Hopkins did not recognize what a resource it had. She received $200 a year, initially to help a member of the physics department with his German correspondence, and later as research associate. She was, however, given office space and the opportunity to undertake, with Robert Williams Wood, the only experimental research of her entire career. The collaboration was not a complete success[5]—Wood distrusted some of Mayer's mathematical techniques, such as reducing data to exponential form—and although Mayer did construct one spectroscope that was still used at Johns Hopkins 15 years later, she never again considered abandoning theory for experiment.

Because there was no other opportunity to do atomic physics at Johns Hopkins, Mayer turned to chemical physics. Most of her work during her nine years at Johns Hopkins was carried out with the two people there who

were willing to work with a woman—her husband and the chemical physicist Karl Herzfeld. Because both worked in chemistry, her research went in this direction as well. Most of it involved the application of quantum mechanics to chemical problems, of which there was no shortage in the 1930s. It was while working with her husband that she learned to take experimental results seriously and to look for patterns concealed in numbers even in the absence of a theoretical interpretation.

Mayer wrote her best-known paper of this period with Alfred Sklar, a student of Herzfeld's. Herzfeld was interested in how chemical structure determines optical properties such as color, and he suggested this as a thesis topic for Sklar. Because the analysis of this problem required complex mathematical techniques, Herzfeld suggested that Mayer assist with the calculations. The complexity of most molecular systems makes it necessary to use various approximations to calculate molecular energy levels. The two most fruitful ones for determining molecular spectra were the Heitler–London–Slater–Pauling approximation and the Hund–Mulliken method. The first involves constructing wavefunctions for the molecule from linear combinations of orbitals of its individual constituent atoms. This assumes a high degree of interaction between the atoms, so that

case in which the spin–orbit interaction becomes significant for electrons.

The same calculations for the eigenfunctions of elements 86, 91 and 93 revealed a similar pattern for the 5f electrons. Mayer's results were therefore consistent with the suggestion that element 93 and the following elements are chemically very similar to the rare earths. When the rest of the transuranic elements were eventually produced, Mayer's predictions concerning their chemical behavior were generally verified.

Mayer's career changed dramatically after America entered the war. Urey asked her to work with his group at Columbia, the so-called Substitute Alloy Materials group, which dealt with problems of isotope separation. Although the Columbia group is best known for the development of the gaseous diffusion process, Urey was not sufficiently optimistic about that method to invest all of his resources in it. One alternative that he considered for separating uranium isotopes was a photochemical method, and this required a thorough understanding of the differences between the spectra of U^{235} and U^{238}. Because by then Mayer was recognized as an expert in spectral analysis, the photochemical separation problem was ideal for her. Urey placed her in charge of the theoretical work. Later she wrote[9] to Born that she had started out as theoretical adviser to a minor project and found herself running an experimental group of 15 people, most of them chemists. She began by collecting and collating all of the available data on the spectra of uranium compounds from the published literature. She found that there were many gaps, and assigned her chemists the task of filling them in. The Atomic Energy Commission published these results after the war. They included a detailed analysis of the fluorescence and absorption spectra of the uranyl compounds, and an appendix with some 75 pages of spectroscopic data, most of which Mayer compiled.

In 1943 the Columbia group abandoned the photochemical-separation project as being too time consuming (it

they cannot be treated as individual units. Sklar used this method in his dissertation, published in 1937.

About a year later Sklar and Mayer published a paper in which they used the second method, the Hund–Mulliken approximation, to calculate[6] the spectrum of the benzene molecule. Their central assumption was that one can determine the Hamiltonian for the molecule by summing the contributions of its individual atoms, with a correction term for their interaction energy. This resembles in many ways the independent-particle approximation for atoms, and Mayer's calculations gave her valuable experience for her later work. Her expertise was recognized; in 1939 Harold Urey listed it as one of the reasons for inviting[7] the Mayers to Columbia University.

At Columbia during the war

Joseph Mayer lost his job with the chemistry department at Johns Hopkins in 1939, in part because of difficulties with administrators. He and Maria decided to accept Urey's invitation to Columbia University. By this time Enrico Fermi was also there, and Urey and Fermi became strong supporters of Maria Mayer and contributed considerably to her career. However, she was offered no position at all at Columbia, not even a nominal one, although Urey asked her to give some lectures in the

chemistry department so that she could have a title, Lecturer in Chemistry, and research opportunities.

Maria Mayer's primary research[8] at Columbia was on a topic that Fermi suggested. In 1940 physicists identified one of the radioactive products of uranium as the transuranic element 93. This element had been shown to behave chemically like the rare earths; however, in other respects it differed somewhat from them, leaving open the possibility that it belonged to a new family of elements. This situation prompted Fermi to suggest that Mayer calculate the eigenfunctions for several rare-earth elements and for several transuranic elements, to see if their predicted chemical behaviors were the same. Fermi had made similar calculations for the rare earths while in Rome.

Using a statistical potential, Mayer first found that the energy and spatial extension of the eigenfunctions change abruptly between the two rare earths lanthanum (atomic weight 57) and neodymium (atomic weight 60). The binding energy of 4f electrons also increases some five times between the two. This behavior can be attributed in part to the decrease in strength of the Coulomb potential for atoms having so many electrons in partially filled shells; the electron–electron interaction energy contributes more to the total binding energy, and this is one

TABLE II. Order of energy levels obtained from those of a square well potential by spin-orbit coupling.

Osc. no.	Square well	Spin term	No. of states	Shells	Total no.
0	1s	$1s_{1/2}$	2	2	2
1	1p	$1p_{3/2}$	4		
		$1p_{1/2}$	2	6	8
2	1d	$1d_{5/2}$	6		
		$1d_{3/2}$	4	12	
	2s	$2s_{1/2}$	2		20
3	1f	$1f_{7/2}$	8	8	28
		$1f_{5/2}$	6		
	2p	$2p_{3/2}$	4	22	
		$2p_{1/2}$	2		
		$1g_{9/2}$	10		50
4	1g	$1g_{7/2}$	8		
	2d	$2d_{5/2}$	6		
		$2d_{3/2}$	4	32	
	3s	$3s_{1/2}$	2		
		$1h_{11/2}$	12		82
5	1h	$1h_{9/2}$	10		
	2	$2f_{7/2}$	8		
		$2f_{5/2}$	6	44	
	3p	$3p_{3/2}$	4		
		$3p_{1/2}$	2		126
		$1i_{13/2}$	14		
6	1i	$1i_{11/2}$			
	2g				
	3d				
	4s				

"Magic numbers" explained. This table from Mayer's 1950 *Physical Review* article (volume 78, page 16) takes spin–orbit coupling into account. Example: The occupation number for the 4th oscillator group is 70. However, the spin–orbit splitting of the 1h level lowers the energy of the $1h_{11/2}$ state enough so that it falls in the 4th oscillator group rather than the 5th, giving an additional 12 nucleons in that group, and shell closure at 82 nucleons.

has been revived in modern times with lasers), and Mayer was reassigned to the gaseous diffusion project. The attention of the entire theoretical group at Columbia was directed toward one set of problems centered around the compound UF_6. Uranium hexafluoride is the only compound of uranium that is a gas at moderate temperatures, and Mayer worked to establish the precise range of temperatures over which it is stable. From spectral measurements she determined the exact chemical structure of UF_6, and from that determined its thermodynamic properties. In part, she worked through analogies to similar compounds such as neptunium hexafluoride. Her research notes show[10] that to calculate the spectrum of NpF_6, she had to take into account a significant spin–orbit coupling for its electrons. She was able to predict the behavior of UF_6 at various temperatures, and the gaseous diffusion project continued.

Shift to nuclear physics

By the end of the war Mayer was extremely familiar with the interpretation of atomic and molecular spectroscopic data in terms of electron shell structures. She was also completely familiar with the individual-orbital approximation, in which one assumes that the interactions between electrons or atoms are small in comparison with the potential due to the atomic or molecular core. She had used this type of approximation in her calculations on molecular spectra with Sklar and in her analysis of the spectra of uranium compounds. Also by the end of the war, because of her calculations on the spectra of the transuranic elements and uranium hexafluoride, she had worked with complex atomic and molecular systems in which the energy of the spin–orbit interaction for electrons is comparable to that of the Coulomb interaction.

It was this theoretical background, coupled with the appreciation of experimental data that she had learned from her husband, that enabled Mayer, with virtually no prior experience in nuclear physics, to formulate one of the most fruitful models of the nucleus since 1937, after only three years in the field. I look briefly now at this transition in her work.

After the war, the University of Chicago invited both of the Mayers to join its new Institute for Nuclear Studies. The physics department appointed Maria Mayer Voluntary Associate Professor of Physics (because of a supposed university antinepotism rule) and made her a member of the Institute. In

July 1946 the Argonne National Laboratory opened with the theoretical division under the direction of Robert Sachs, who had been Mayer's only graduate student at Johns Hopkins in the 1930s. Sachs offered his former professor a paid half-time position in the theoretical group at Argonne. Because both the Institute for Nuclear Studies and Argonne were devoted to nuclear physics, Mayer decided that the time had come to learn something about the subject. As a matter of fact, she learned most of her nuclear physics not from books, but from discussions with colleagues about current problems. This turned out to be most important, because it meant that she was relatively unfamiliar with many of the traditional beliefs in nuclear physics.

The first project that Mayer undertook at Chicago was a joint one with Edward Teller, an attempt to describe a possible mechanism for the origin of the elements. This work required a new list of all available isotopic abundances because more data were available after the war. The task fell to Mayer. In analyzing the data, she found that certain patterns emerged. It appeared that nuclei with 2, 8, 20, 28, 50, 82 or 126 neutrons or protons were unusually stable. She immediately recognized an analogy to electron shell structure, and she began to search for a nuclear potential that would predict the correct nucleon shell numbers. In 1948 Mayer published a paper in which she summarized[11] all of the data supporting the idea that nuclei are composed of shells, but she offered no theoretical interpretation. In this paper she referred to the seven nucleon numbers as "magic numbers"—a term that Eugene Wigner first applied to symbolize his view that the whole idea of nuclear shells was a charming fan-

Joseph Mayer and Maria Goeppert Mayer (center) with Edward Teller and James Franck. Mayer worked with Teller on her first project at the University of Chicago's Institute for Nuclear Studies. (Photograph by Francis Simon, courtesy AIP Niels Bohr Library.)

tasy. Mayer, however, liked the phrase and used it consistently thereafter.

Throughout the 1930s, various physicists attempted to formulate a nuclear shell model, based on just such evidence as that compiled by Mayer. These attempts, however, were largely discounted by nuclear physicists because they violated the fundamental assumption of the liquid-drop model, which was so successful in accounting for nuclear fission and other phenomena. This assumption was that the interaction between individual nucleons is so great that the independent-particle approximation is completely invalid—the nucleus as a whole possesses angular momentum and other collective properties, but the individual nucleons do not, and hence they cannot group themselves into shells.

However, with the publication of Mayer's 1948 paper on the magic numbers, containing so much new evidence for shell behavior, nuclear physicists again turned their attention to the possibility of deriving a model that could explain these patterns. Their attempts all involved assuming some kind of central potential. A harmonic oscillator or a square-well potential, for example, could account for the magic numbers up to but not beyond 20. Eugene Feenberg and Lothar Nordheim, as well as Mayer herself, tried[12] various modifications of these potentials, but any scheme that gave the magic numbers also involved some extremely questionable assumptions.

Finally, one day late in 1948 or early 1949, Mayer was discussing the problem with Fermi in her office. (They were in her office because Fermi would not let her smoke in his office, and she did not like to be without a cigarette for long.) Suddenly, Fermi was called away to answer a phone call, and as he left her asked, "Is there any evidence of spin-orbit coupling?" Mayer immediately recognized that this was precisely the answer to her problem, and by the time Fermi returned she had calculated the splitting involved and the occupation numbers for the various energy levels. They corresponded exactly to the magic numbers. The figure on page 249 shows Mayer's schematic for this solution.

Mayer's interpretation of nuclear shell structure was accepted within a relatively short time. This rapid acceptance was aided by the simultaneous and independent discovery of the same model in Germany by Otto Haxel, J. Hans D. Jensen and Hans Suess. Starting from the same point of departure—an interest in the magic numbers as analogous to chemical data—the German group eventually found[13] the solution of a strong spin–orbit splitting as well. Mayer received the Nobel Prize together with Jensen.

Mayer was particularly well prepared to formulate the shell model of the nucleus. Her experience analyzing spectroscopic data in terms of discrete energy levels for discrete particles and her experience with the strong spin–orbit interaction that occurs in heavy atoms and molecules were important to her contribution. But so was her unfamiliarity with what she later called the "fashions" in nuclear physics. She was, for example, not fully aware of the bias favoring a liquid-drop model over an independent-particle model. She said later that she was not aware that she was violating a long-standing belief when she suggested that spin–orbit coupling for nucleons is a strong effect.

* * *

This article comes out of research that I did for my doctoral dissertation at the University of Minnesota under the supervision of Roger H. Stuewer. I am grateful for his advice. I would also like to thank Robert

Sachs and Jacob Bigeleisen for their help. A doctoral dissertation research grant from the National Science Foundation provided funds for travel to visit archives and conduct interviews.

References

1. J. Dash, *A Life of One's Own: Three Gifted Women and the Men They Married*, Harper and Row, New York (1973).

2. K. E. Johnson, *Maria Goeppert Mayer and the Development of the Nuclear Shell Model*, dissertation, Univ. of Minnesota (1986).

3. M. G. Mayer, Phys. Rev. **75**, 1226 (1949).

4. M. Goeppert-Mayer, Ann. der Phys. **9**, 273 (1931).

5. M. G. Mayer, session III of interviews of James Franck conducted by T. S. Kuhn and Mayer, 11 July 1962; Archives for the History of Quantum Physics. Repositories of the archives are located at the Bohr Institute, Copenhagen; the American Philosophical Society, Philadelphia; the University of California, Berkeley; the Center for the History of Physics, American Institute of Physics, New York; the University of Minnesota; the Accademia dei XL, Rome; the Science Museum, London; and the Deutsches Museum, Munich.

6. M. Goeppert-Mayer, A. L. Sklar, J. Chem. Phys. **6**, 645 (1950).

7. Joseph Mayer Papers, Univ. of Calif., San Diego, library, special collections.

8. M. G. Mayer, Phys. Rev. **60**, 184 (1941).

9. Born *Nachlass*; Staatsbibliothek preussischer Kulturbesitz, Berlin.

10. Maria Goeppert Mayer Papers, Univ. of Calif., San Diego, library, special collections.

11. M. G. Mayer, Phys. Rev. **74**, 235 (1948).

12. E. Feenberg, K. C. Hammack, Phys. Rev. **75**, 1877 (1949). L. Nordheim, Phys. Rev. **75**, 1894 (1949).

13. O. Haxel, J. H. D. Jensen, H. E. Suess, Naturwiss. **35**, 376 (1948). H. E. Suess, O. Haxel, J. H. D. Jensen, Naturwiss. **36**, 153 (1949). J. H. D. Jensen, H. E. Suess, O. Haxel, Naturwiss. **36**, 155 (1949). □

CHANDRASEKHARA VENKATA RAMAN

India's 'great savant' of science made deep contributions to acoustics, physical optics, magnetism, molecular physics and especially to our understanding of the scattering of light by matter.

Aiyasami Jayaraman and Anant Krishna Ramdas

PHYSICS TODAY/AUGUST 1988

The year 1988 marks the centennial of the birth of C. V. Raman, the discoverer of the effect that bears his name. During his extraordinarily creative career, which spanned more than six decades, he had a profound impact on the scientific world.[1] He made significant contributions to acoustics, physical optics, magnetism and molecular physics, including, of course, his extensive studies on elastic and inelastic light scattering. Modern India owes him much for her entry into the world of science, and for the strength of her scientific tradition.

Raman was born in a village near Trichinopoly in the province of Tamil Nadu, India, the second of eight children. His family was of modest circumstances and depended on agriculture for its income. Raman's father, Chandrasekhara Iyer, was the first in the family to receive higher education in the Western educational system introduced by the British, referred to in India as "English" education. India had formally come under the British Crown in 1858; the first three Indian universities based on the Western style—Bombay, Madras and Calcutta—came into existence only in 1857. Iyer obtained a bachelor's degree in the physical sciences and taught physics in a college in Trichinopoly. A scholarly person with intellectual ambitions and strong artistic inclinations, Iyer is reported to have played the violin exceedingly well.

Raman's mother, Parvathi Ammal, the daughter of a Sanskrit scholar, was a gentle, tolerant lady, firmly rooted in the home and family life. In 1892, when Raman was four, Iyer moved his family to the city of Vishakapatnam in the province of Andhra Pradesh, where he had accepted a position as a lecturer in mathematics and physics in Mrs. A. V. N. College, a junior college. Raman spent the next ten years of his life in Vishakapatnam, which lies on the east coast of India, facing the Bay of Bengal with its intensely blue waters.

Raman was clearly special; a precocious child, he finished his primary and secondary education early, matriculating at the tender age of 11. After two years in Mrs. A. V. N. College, he entered Presidency College in Madras and, at the age of 15, passed the examination for his BA at the top of his class, winning gold medals in English and physics. It was evident that Raman's superior intellectual capacities singled him out for a bright career. At the turn of the century this meant studying abroad, and his teachers advised him to go to England. Raman was physically frail, however, and the Civil Surgeon of Madras, who examined him, declared him unfit for the rigors of the English climate. Raman therefore continued his studies in Presidency College, and he received his MA in 1907 at the age of 18, again at the top of his class, with record marks and a gold medal. By this point, his interest in physics had been fully aroused; on his own he studied such works as Lord Rayleigh's scientific papers and treatise on sound, and the works of Hermann von Helmholtz.

While in the MA class Raman published a paper in the November 1906 *Philosophical Magazine* (London) on the "unsymmetrical diffraction bands due to a rectangular aperture, observed when light is reflected very obliquely at the face of a prism." He had observed these bands during a routine class experiment in optics. From his very first research publication Raman demonstrated the ability to discover novel phenomena using simple experiments. He followed this paper with a note in the same journal on a new experimental method for measuring surface tension. These papers were communicated by the author himself and contain no acknowledgment of help from anyone. Presidency College was primarily a teaching institution at that time, and had no tradition at all in research. Whatever Raman did was on his own initiative.

It was abundantly clear that Raman was cut out for a scientific career. However, opportunities in India for such a career were then nonexistent, whereas positions in the civil service were prestigious and financially remunerative. In fact, the civil service was then regarded as the most attractive career for the ambitious, the bright and the young. Because training for the Indian Civil Service involved a trip to England, Raman was ruled out for health reasons. Instead he appeared for the competitive examination for selection into the Financial Civil Service in February 1907. His elder brother C. Subrahmanya Iyer was already a member of the service, and this probably influenced Raman's decision. He secured the first place in the qualifying examination and was selected.

Indian Association for Cultivation of Science

Just before he joined the civil service, Raman married Lokasundari. Accompanied by his wife, he arrived in Calcutta in 1907 to join the finance department as an assistant accountant general. He was only $18\frac{1}{2}$ years old. The Ramans rented a house in Scott's Lane, off Bowbazar Street. In the normal course of events Raman would have had a long career in the finance department of the government of India and, given his abilities, would have moved up the ladder in time, to retire as a respectable accountant general. But Raman was obsessed with physics, and an inner urge drove him to seek an opportunity to fulfill this calling. He then made his first major "discovery"—namely, the presence of the Indian Association for the Cultivation of Science at 210 Bowbazar Street, only a few blocks from his home.

The association had been founded in 1876 by Mahendra Lal Sircar, a leading medical practitioner in Calcutta. Sircar had an abiding interest in science and foresaw the great role it was destined to play in India. He wanted to create an association along the lines of the Royal

Aiyasami Jayaraman is a Distinguished Member of the technical staff at AT&T Bell Laboratories in Murray Hill, New Jersey. **Anant Krishna Ramdas** is a professor of physics at Purdue University.

Institution of London, where young aspirants could pursue scientific interests, and he devoted a good part of his life to collecting funds from Indian princes and affluent citizens to achieve his dream. The association's building had several large halls for laboratories and an excellent lecture theater that could accommodate a large audience. Sircar organized popular-level lecture courses in science for students, which he gave himself or persuaded others to give. However, he was unable to fulfill his dream of a research institution. In those days scientific research in India was nonexistent, and the institution decayed over the following 25 years as rooms grew dusty and laboratories sat idle. In desperation Sircar seems to have declared: "I don't know how to account for the apathy of our people towards the cultivation of science. Younger men must come and step into my place and make this a great institution." Prophetically, that wish was later to be fulfilled by Raman. Sircar died in 1904 a rather disappointed man. His son Amrit Lal Sircar succeeded him as the honorary secretary of the association and guided its affairs.

One day soon after his arrival in Calcutta, Raman noticed the association while riding in a streetcar along Bowbazar Street on his way home from work. He immediately alighted and knocked on the doors of the association, full of excitement. Admitted inside by Ashutosh Dey, who later became his most devoted assistant, Raman met Amrit Lal Sircar and asked if he could conduct research at the association during his spare time. The story goes that Sircar embraced Raman, exclaiming that they had been waiting for a person like him for years, and how happy his father would have been to witness the entry of such a person into the association.

Raman set to work with all enthusiasm and soon started publishing research papers. From 1907 to 1917, except for short absences from Calcutta on temporary transfer to Rangoon and Nagpur, Raman spent all his leisure time at the association, working evenings and very late into the night on his experiments. He had all of the association's facilities completely at his disposal, and the devoted and loyal assistance of Dey as well. Raman soon turned the association into a beehive of activity.

All this meant a different story for Lokasundari Raman, a young bride thrust into the hands of a strange young man with a consuming passion for science. She described the routine of her husband to S. Ramaseshan, the 1978 Raman Memorial Lecturer, as follows:

At 5:30 am Raman goes to the association, returns at 9:45 am, bathes, gulps his food in haste, leaves for his office invariably by taxi to be on time for his work. At 5 pm Raman goes directly to the association on his way back from the office and comes home at 9:30 or 10 pm, after spending the evening hours at the laboratories of the association.

Sundays were spent entirely at the association. Lokasundari ungrudgingly supported Raman in his quest for scientific knowledge throughout his life, endearing herself to all who associated themselves with her husband.

During the period 1907–17, Raman was a full-time civil servant holding a high rank, and was respected and admired as a very capable officer. Despite his full work load he managed to pursue research on problems in acoustics and optics, publishing papers such as "Experimental Investigations on the Maintenance of Vibrations," "Note on the Theory of Subsynchronous Maintenance," "Dynamical Theory of the Motion of Bowed Strings" and "The Experimental Study of Huygens' Secondary Waves" in such journals as *Nature*, *Philosophical Magazine*, *Physical Review*, and the *Bulletin of the Indian Association for the Cultivation of Science*. The publications

convey an image of an extremely energetic scientist with a well laid-out scientific program. Raman drew on his deep insights into vibrations and wave motion to select worthwhile problems that were tractable with the limited resources available to him. In his formulation of problems, his careful experimentation, his physical and mathematical analyses applied with a sure touch, and his transformations of apparently simple, routine issues into deep and unexpected insights, one sees in this early work the rapid evolution of a world-class physicist.

Raman's scientific contributions in acoustics earned him an international reputation. His devotion and dedication began to attract young students, teachers and professors from all over India; they gathered around him to participate in the scientific excitement. The Raman school of physics came into being.

Studies in Light Scattering

Raman's scientific career changed significantly in 1917. He wrote in 1968: "My studies on bowed string instruments represent a phase of my earliest activities as a man of science. They were mostly carried out between the years 1914 and 1918. My call to the professorship at the Calcutta University in July 1917 and the intensification of my interests in optics inevitably called a halt to my further studies of the violin family instruments."

By 1917 Raman had made a profound impression on the leading educators and citizens of Calcutta. Asutosh Mookerjee, the dynamic vice chancellor of Calcutta University at that time, was assiduously engaged in building a first-rate university. He was successful in getting large donations and in attracting eminent scholars and scientists to occupy endowed chairs. He recognized in Raman the ideal person to occupy the Sir Tarakanath Palit Professorship of Physics, which had just been endowed at the University College of Science at Calcutta, and offered it to Raman. However, the appointment required the candidate to have received training abroad. Raman refused to comply with this stipulation, and so the vice chancellor changed the provisions. Finally Raman took the chair and resigned from the civil service, exchanging a lucrative job for one with one-fifth the emolument.

Although he was "under no obligation" to take part in teaching MA and MSc classes according to the "special terms of appointment as the Palit Professor of Physics," Raman took a strong and active interest in modernizing Calcutta's postgraduate curriculum. He fully participated in teaching and inspired his students with his enthusiastic lectures, in which he frequently drew on original papers and classical treatises. M. N. Saha (later famous for the ionization equation named after him) and S. N. Bose (who discovered Bose statistics) were two of the lecturers at the University College in 1917. Raman was led to say, with justifiable pride, "Calcutta University can claim to possess a real school of physics, the like of which certainly does not exist in any other Indian university, and which, even now, will not compare very unfavorably with those existing in the best European and American universities."

In 1921 Raman took his first trip abroad, as a delegate to the Congress of Universities of the British Empire held that year in Oxford. During his brief visit he met the famous physicists of England, including J. J. Thomson, Ernest Rutherford and William H. Bragg. His lecture to the Physical Society on his research in optics and acoustics, which he illustrated with experimental demonstrations, apparently was immensely appreciated by the large body of physicists present.

Many of Raman's scientific investigations were inspired by an intense love of nature and the beauties of nat-

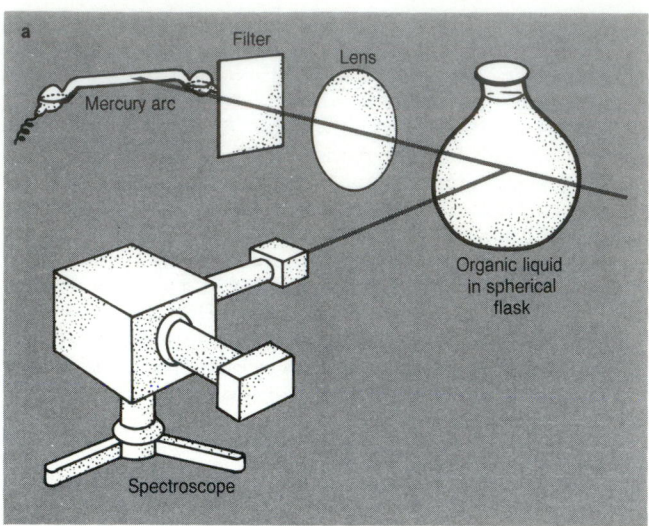
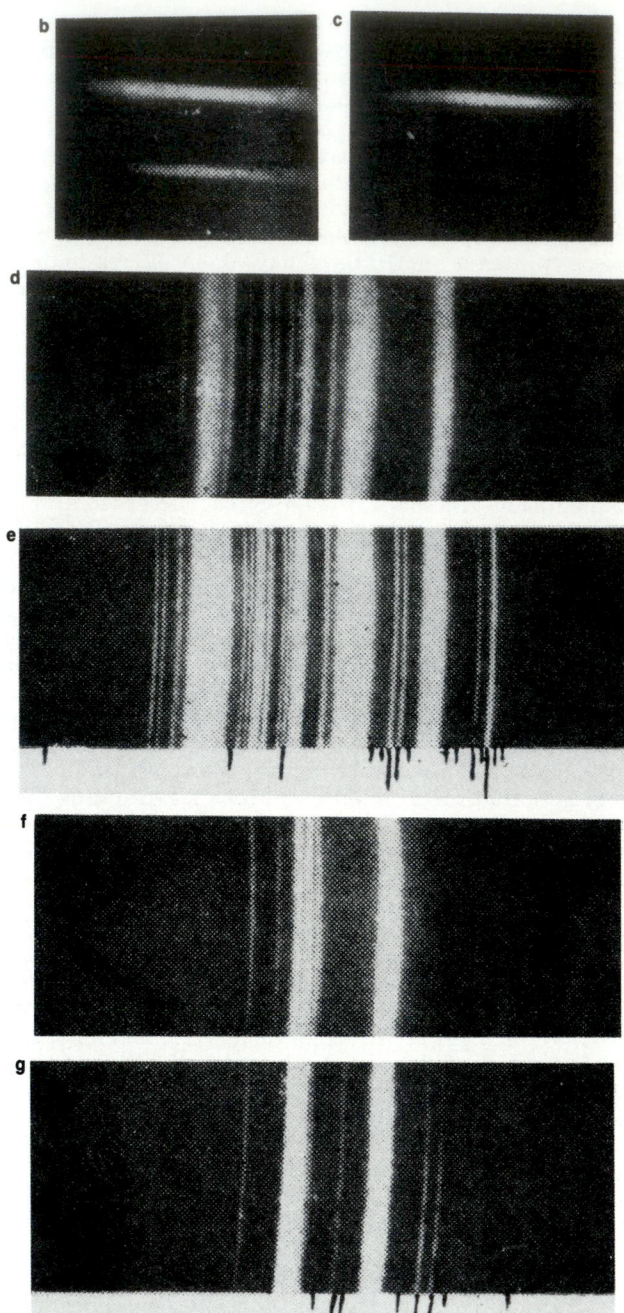

Raman's early spectroscopic studies were carried out with the the apparatus shown schematically in the diagram above (a). Light from the mercury arc passed through a filter and a lens before being partially scattered by the sample. The solid line indicates the path of the light. Such an apparatus led to the first Raman spectra (right, see reference 4). The pair of photographs at the top demonstrate the polarization characteristics of scattering in toluene. In b a beam of sunlight was filtered through a blue-violet glass, with a double-image prism in front of the camera lens; in c the same arrangement was used, except that an additional, complementary green filter was placed in front of the lens. The middle two spectrograms represent the spectrum of the quartz mercury arc lamp in the range 3500–4400 Å (d) and the spectrum of radiation in the same range scattered by liquid benzene (e). The Raman lines are marked by short black lines. For both spectrograms the light was filtered by blue glass. The bottom two spectrograms (f, g) are the same as the middle two, except that the light was filtered by a potassium permanganate solution instead of by blue glass.

ural phenomena. While traveling to and from the Congress of Universities, he was fascinated by the blue of the sky and of the Mediterranean Sea, and his research interests turned toward light scattering. Rayleigh had explained the sky's color as due to the scattering of sunlight by air molecules; he had also conjectured that the blue of the sea was due merely to the reflection of the sky at the water's surface. During his voyage, however, Raman performed a simple experiment, which demonstrated that the color of the sea arises from an intrinsic light scattering by water molecules: He quenched the surface reflection with a Nicol prism at the Brewster angle, and yet the blue of the sea was just as vivid. Convinced by his experiment that the blue of the ocean was a genuine molecular light scattering phenomenon, Raman dashed off several short letters to *Nature* recording his observations. In 1922 he published a comprehensive paper entitled "On the molecular scattering of light in

water and the colour of the sea," in the *Proceedings of the Royal Society*.[2]

Raman then focused his entire scientific program on the scattering of light by matter in all states of aggregation. In 1922 he wrote a memoir entitled *The Molecular Diffraction of Light*, in which he described the measurements he had already carried out with his group, and their plans for the future. The latter included studies of the molecular diffraction of light in systems undergoing phase transitions and in liquid mixtures and solutions and research on the relationship of light scattering to chemical constitution. In his book, Raman carefully marshaled the basic physical concepts and discussed important issues such as the Doppler effect in molecular scattering and molecular diffraction and the quantum theory of light, providing impressive insights. One clearly perceives in this memoir a scientist who has seized upon a first-rate research opportunity, who is fully cognizant of the contemporary status of physics and who, having mobilized his intellectual resources, is ready to proceed with energy and enthusiasm. After 1921, investigations on light scattering became the central theme of Raman's program, which culminated in 1928 with the discovery of the Raman effect.

Raman and his students established careful experimental procedures to observe and characterize the very feeble radiations they typically observed. Their meticulous attention to the purity of the scattering medium, their careful exclusion of all spurious and parasitic radiation, and the ready availability in India of intense sunlight over extended periods led to several fundamental discoveries in rapid succession, including:

▷ They verified the Einstein–Smoluchowski theory of fluctuations under diverse conditions, and made a determination of Avogadro's number.

▷ They made a comprehensive study of the departure from complete polarization of transversely scattered light—so-called depolarization—and its relationship to the anisotropy of molecules, as evidenced, for example, by the contrasting configurations of N_2 and CCl_4.

▷ They correlated the anisotropy of molecules with the optical and magnetic anisotropy of crystals, as well as flow, electric and magnetic birefringence.

▷ They interpreted x-ray diffraction in fluids and amorphous solids in terms of concepts closely akin to the Fourier transform of the pair correlation function.

▷ They observed and explained surface scattering, including the striking case of liquid mercury.

▷ They studied light scattering from colloids (the Tyndall effect) and discovered its intimate relationship to osmotic pressure. This paper formulated the technique for determining the molecular weight of colloidal particles, and predates the classic paper of Peter Debye by 17 years.

A new radiation

In 1923, his student K. R. Ramanathan, under Raman's direction, undertook a detailed study of the scattering of light by water and the possible wavelength dependence of depolarization. When an intense beam of sunlight was passed through an optical filter and focused on carefully purified water, the transversely scattered light exhibited a curious anomaly. Contrary to expectations, on examining the scattered radiation with an optical filter whose transmission characteristics were complementary to those of the filter in the incident beam, Ramanathan found a faint trace of light. Suspecting fluorescence due to some residual impurity, they repeatedly and very slowly distilled the water *in vacuo*, but "in spite of many redistillations, the effect remained practically undiminished."[3] Alcohol also exhibited this "persistent feeble fluorescence." The weak fluorescence was repeatedly encountered in several investigations by Raman and his students in the ensuing years. Attempts in 1924–25 to use spectroscopic techniques to analyze this fluorescence were met with failure.

By the end of 1927 Raman felt that the effect was "some kind of optical analogue to the type of x-ray scattering discovered by Prof. Compton."[4] Spurred by this intuition, he and his student K. S. Krishnan established that the effect was exhibited by a large number of liquids, organic vapors and gaseous CO_2 and N_2O, as well as by crystals such as ice and amorphous solids. "Preliminary observations with sunlight filtered through a combination which passes a narrow range of wavelengths, showed the spectrum of the new radiation to consist mainly of a narrow range of wavelengths clearly separated from the incident spectrum by a dark space."[5] Using monochro-

matic light from a powerful mercury arc, they conducted spectroscopic analyses of the scattered light, initially with a pocket spectroscope and later with a spectrograph, that yielded truly dramatic results. Each incident exciting line of frequency ω_L was accompanied by weaker lines of frequency $\omega_L \pm \omega_i$, $i = 1, 2, 3 \ldots$. Raman attributed the additional lines to an energy exchange between the incident photon and the internal excitations of the scattering medium; the new lines at $\omega_L - \omega_i$ and $\omega_L + \omega_i$, he said, corresponded to a partial exchange of energy in which the medium is excited to or de-excited from an energy level $\hbar\omega_i$ above the ground state. To illustrate, the atoms in a molecule are constantly vibrating about their equilibrium positions with a definite frequency. When monochromatic light hits the molecule, the light can be scattered by a process in which its frequency is decreased or increased by exactly the vibrational frequency of the molecule. Thus a very weak new radiation is produced, which was not originally present in the light. The additional spectral lines occur in pairs displaced equally in frequency to the high- and the low-frequency sides of the exciting line, and are referred to, for historical reasons, as the anti-Stokes and the Stokes shift, respectively.

In the first complete account of the phenomenon, Raman showed that the effect is exhibited by gases, liquids, crystals and amorphous solids, and that it is a "phenomenon whose universal nature has to be recognized."[4] The new lines observed are characteristic of the scattering medium. The shift in frequency from the exciting line is independent of the frequency of the incident radiation. (Interesting exceptions to the rule of frequency independence are Brillouin scattering, Raman scattering associated with sound waves in condensed matter, and some cases in which zone-center optical phonons in crystals free of improper symmetry exhibit linear wavevector dependence.) The inelastic light scattering was soon designated the "Raman effect."

Until 1961 Raman spectroscopy was limited to rotational spectra of gases and vibrational spectra of gases, liquids and solids. When allowed by selection rules, rotational and vibrational transitions occur in the near- and far-infrared spectra. The coupling of the internal excitations with the incident electromagnetic radiation in Raman scattering occurs in a fashion fundamentally different from the direct emission or absorption of radiation caused by the de-excitation or excitation of the system; selection rules for Raman spectra are thus deduced from considerations different from those for infrared spectra. The polarization characteristics in the two phenomena also differ. Thus in many respects Raman spectroscopy is complementary to infrared spectroscopy. Moreover, the ease of conducting spectroscopic studies in the visible range was, and continues to be, an attractive feature of Raman spectroscopy.

With the invention of lasers, Raman spectroscopy reached a new level of sophistication. Raman spectra associated with polaritons, plasmons, magnons, Landau levels and electronic levels of ions and impurities in solids; Raman spectroscopy of molecules adsorbed on surfaces, surface excitations and samples subjected to ultrahigh pressures and magnetic fields; the stimulated Raman effect; and the spin-flip Raman laser—all are examples of the vastly expanded scope that resulted from the introduction of the laser.

Announcement and acclamation

Raman gave a full public account of his discovery in an address before the South Indian Science Association at Bangalore on 16 March 1928. The address, entitled "A

Robert A. Millikan with Raman in Bangalore in 1940. In 1924 Raman spent a few months at Caltech at Millikan's invitation. There the two developed a lifelong friendship. (Photo courtesy of the Archives of Caltech.)

dently by Landsberg and Mandelstam in crystalline quartz and calcite. That inelastic light scattering might be associated with internal excitations of the scattering medium was anticipated by Adolph Smekal; it is implicit in the dispersion theory of Hendrik A. Kramers and Werner Heisenberg.

The announcement of the Raman effect was hailed with great acclamation all over the world. Wood, famous for his important experimental accomplishments in the study of resonance radiation, sent the following cable to *Nature*:

> Professor Raman's brilliant and surprising discovery that transparent substances illuminated by very intense monochromatic light scatter radiations of modified wavelength and that [the] frequency difference between the emitted radiation and the one exciting the medium is identical with the frequency of the infrared absorption bands, opens up a wholly new field of study of molecular structure. I have verified this discovery in every particular.... It appears to me that this very beautiful discovery which resulted from Prof. Raman's long and patient study of the phenomenon of light scattering is one of the most convincing proofs of the quantum theory of light which we have at the present time.

It is of particular interest that the equipment Raman used in making his discovery was very simple, costing 500 rupees at the time—perhaps about $500 in current value.

In the years immediately following the discovery of the Raman effect, there was intense interest in its exploration and exploitation in numerous media as well as curiosity about its theoretical significance, and Raman's group continued to make important contributions. Recognitions and awards now came thick and fast. In 1924 Raman was elected a fellow of the Royal Society of London. He received the Matteucci Medal of the Societa Italiana della Scienza in 1928. The British government in India conferred knighthood on him in 1929. Also in 1929, the Faraday Society held a special symposium on the Raman effect, during which Raman gave an inspiring account of his discovery. The Royal Society awarded him the Hughes Medal in 1930—the same year he received the Nobel Prize in Physics for "his investigations on the scattering of light and the discovery of the effect known after him." He was only 42 years old. Accompanied by his wife, he traveled to Stockholm, where he received the prize on 10 December 1930. In anticipation, he had booked passage in July to ensure that they would have reservations on a ship that could get them to Stockholm by early December.

At this point in his career Raman occupied a very special position in the eyes of the Indian public. During the 1920s the Indian political movement toward complete independence from foreign rule had gathered significant momentum, spearheaded by political leaders like Mahatma Gandhi, Motilal Nehru and Jawaharlal Nehru. There was a great enthusiasm for, and commitment to, efforts to secure and improve national identity. In this context Raman's scientific accomplishments and the international recognition he received were cause for acclaim by his countrymen. He became a national hero, occupying a place in public esteem equal to that of Rabindranath Tagore, who had received the Nobel Prize for Literature in 1913.

Indian Institute of Science

In 1933 Raman was offered the directorship of the Indian Institute of Science, located in Bangalore in south India, in the Karnataka state. Bangalore is one of the most attractive cities in India; having flourished as part of the Princely State of Mysore, it is well laid out, with beautiful

New Radiation," was prepared as an article immediately on his return to Calcutta, and printed overnight. Reprints were posted the next day to scientists all over the world. Although many students participated in the events leading to the discovery of the Raman effect, Krishnan coauthored the short notes preceding the article and several definitive papers following it. In his address, Raman made a special acknowledgment of Krishnan's contribution, stating, "I owe much to the valuable cooperation in the research of Mr. K. S. Krishnan and the assistance of Mr. S. Venkateswaran and other workers in my laboratory."

In the 1920s studies on light scattering were being pursued in several laboratories throughout the world.[6] The younger Lord Rayleigh (England), Robert W. Wood at Johns Hopkins University (US), Jean Cabannes at the School of Science (Montpellier and Paris, France), and Grigorii Landsberg and L. Mandelstam at the Institute of Physics (USSR Academy of Sciences, Moscow) made significant contributions to the field. The possibility of inelastic light scattering was very much "in the air," and the French and Russian workers were close on Raman's heels. Indeed, the Raman effect was rediscovered indepen-

parks and many scientific and educational institutions, today including several "high-tech" industries. The Indian Institute of Science, now a premier scientific institution, was founded in 1909 by J. N. Tata, an industrialist known for his foresight, patriotism and philanthropy.

In 1933 the institute did not have a physics department. After some hesitation, Raman accepted the directorship and proceeded to establish an active center of physical research. He took on a large number of students and introduced them to challenging problems at the frontiers of physics. He had grand visions of transforming Bangalore into a center of excellence by international standards. Around the time he became director, many famous scientists were fleeing Germany, and Raman tried unsuccessfully to secure places for some of them at Bangalore, including Max Born and Erwin Schrödinger.

Despite his efforts to attract distinguished faculty, Raman's strong personality and management style caused serious difficulties with the institute's governing body, and he was forced to resign from the directorship after a few years. He remained at the institute as a professor of physics until 1948, and pursued new avenues. Light scattering continued to figure prominently in his research program.

One of Raman's outstanding contributions during his 15 years at the Indian Institute concerned the diffraction of visible light by high-frequency sound waves generated in fluids by a transducer. Peter Debye and Francis Sears in the US and Lucas and Biquard in France had reported observing beautiful Fraunhofer diffraction when monochromatic light propagated through a liquid cell normal to high-frequency sound waves generated with a quartz transducer. For sound waves of sufficient intensity, numerous orders of diffraction and a wandering of the intensity among the orders were seen in the richly detailed diffraction pattern. Before Raman and Nagendra Nath gave an explanation,[7] all the theories of the phenomenon were off the mark. Raman and Nath realized that the incident plane wave encounters the "phase grating" represented by the fluid in the ultrasonic field and is transformed into a "corrugated wave." All the details observed follow through quantitatively from their theory, which is closely analogous to that developed by Rayleigh in his book *Theory of Sound*. Raman and Nath published several beautiful papers that not only explained the observed effects but also opened up new avenues for development. Today this form of diffraction is of great interest for acousto–optical applications.

Raman and his students made a comprehensive study of diamond, a crystal that fascinated him all his life. His younger brother C. Ramaswamy, reported the first observation of the first-order Raman spectrum of diamond, which consists of a single line at 1332 cm^{-1}. S. Bhagavantam, another of Raman's students, also made extensive studies on the Raman spectrum of diamond, and Nagendra Nath identified the origin of the 1332-cm^{-1} line as a lattice vibration in which the two face-centered Bravais lattices vibrate rigidly against each other. R. S. Krishnan, who exploited the Rasetti technique that involves the 2537-Å resonance radiation of mercury, first observed the second-order Raman spectrum of diamond. Impressed by some of the sharper features of diamond's second-order spectrum, Raman ventured into the dynamics of crystal lattices and immediately involved himself in a bitter controversy with

Born, Debye and others by criticizing their theories concerning lattice dynamics.

According to Raman, the normal modes of a crystal were to be enumerated by considering a supercell eight times the size of the primitive unit cell, yielding $24p - 3$ discrete normal modes, where p is the number of atoms in the unit cell. The vibrational spectrum would then lead to a set of discrete frequencies, and Raman tried to identify these in a few crystals, such as diamond, NaCl and MgO. However, his view that the normal modes consist of only these discrete frequencies conflicted with the widely held Born–von Kármán theory of lattice dynamics, which predicts a continuous frequency spectrum. Indeed, the dispersion relations of lattice vibrations determined since the mid-1950s from inelastic neutron scattering are in excellent agreement with Born–von Kármán theory, and not at all with Raman's set of discrete frequencies. Raman claimed support for his theory from light scattering spectroscopy of crystals. The peaks in the second-order Raman spectrum of diamond that Raman said corroborated his theory are now regarded as consequences of van Hove singularities; they represent phonons at critical points having a large density of states that are typically associated with zone boundary phonons. Raman was incorrect, but he was adamant that he was right in his approach. This attitude made him highly emotional and irrational when it came to lattice dynamics. Further, it also proved counterproductive, as he got sidetracked into an area that was not his forte.

Raman had a lively and lifelong interest in diamond, and he built up an outstanding collection of specimens. He made many studies of its structure, luminescence, lattice dynamics and x-ray diffraction effects. Unfortunately he landed in controversy again when he proposed that diamond exists in two forms, one with tetrahedral and the other with octahedral symmetry, and that certain of its physical properties, such as its infrared and ultraviolet absorption, luminescence and optical birefringence, could be traced to this circumstance. However, in the early 1950s Walter Bond and Wolfgang Kaiser at Bell Labs showed that these properties were due to the presence of nitrogen as a substitutional impurity, and had nothing to do with the fundamental character of diamond. Similarly, the so-called extra spots observed in the Laue pattern of some diamonds, which Raman believed to be due to the excitation of optical phonons accompanying a Bragg reflection, were also precisely those associated with platelets of nitrogen impurities. Raman engaged in a long drawn-out controversy with Kathleen Lonsdale (University of London) over his "quantum x-ray reflections," and many roughly worded exchanges between the two appeared in the literature.

These controversies do not detract from the fine contributions to optics, spectroscopy and crystal physics that Raman made during these years. He and Nedungadi made the first observations and gave the first interpretation of the soft-mode behavior that drives the phase transition in quartz from the α to the β form, almost two decades before William Cochran published his soft-mode theory for phase transitions. Raman and his students made outstanding contributions to the physics of crystals, including, for example, their exquisite observation of conical refraction in naphthalene. During his tenure as professor at the Indian Institute of Science, the physics

Arnold Sommerfeld with C. V. Raman and K. S. Krishnan (left), in Calcutta during Sommerfeld's 1928 visit to Raman's laboratory. (Photo courtesy of Deutsches Museum, Munich, supplied by G. Torkar of Ludwig-Maximilians Universität, Munich.)

department was intensely active in diverse areas of physics, with Raman as its moving spirit.

Raman Research Institute

Raman had visions of establishing an institute where he would continue his scientific research after retiring from the Indian Institute. The Maharaja of Mysore donated a lovely 11-acre piece of land at one of the prime locations in Bangalore for this purpose. When Raman retired in 1948, the building for his new institute was nearing completion. Around this time he had an opportunity to visit the US as a member of the Indian delegation to the World Bank. After discharging his duties, Raman visited a few laboratories, including Bell Laboratories, and bought a lovely collection of minerals and gems that became the nucleus of a crystallographic and mineralogical museum at his new institute. The Raman Research Institute opened in 1949, with Raman as its director and as a professor, assisted by a small staff, including the authors of this article. Jayaraman was a close associate of Raman from 1949 to 1960, and Ramdas was Raman's pupil from 1950 to 1956.

Raman equipped his institute with beautiful museums, a lecture hall, an outstanding library, offices and laboratories, and once more he carried on his scientific work. He took on a few research students, but the institute was established primarily so that he could pursue his interests. Raman utterly enjoyed his investigations on gems and minerals, and between 1950 and 1960 he published a series of papers on the colors of gems and minerals, with some of us as collaborators, but more often by himself. This body of work reflects his taste for aesthetics in physics. He loved to demonstrate his findings to visitors, and to explain in simple language the most esoteric optical phenomena. Maharajas, Prime Ministers, politicians, officials, students and laymen came in throngs to visit the institute and to see Raman. His magnificent collection of gems and minerals and his inspiring tales enthralled them. From 1949 until his death in 1970, Raman enjoyed the research institute he built for himself. He filled the grounds with spectacular flowering trees, shrubs and, especially, roses. The finest roses the Bangalore nurseries could supply were planted under his supervision, and he admired them as a child admires a new toy. Raman knew the scientific names of most of the trees and shrubs on the institute's grounds.

The institute's museum contains a collection of beetles and butterflies, because Raman was fascinated with the colors of these insects. Among the most colorful butterflies are the *Morpho brazilius* and certain Himalayan species, which exhibit a spectacular blue iridescence. Raman studied these and wrote a paper entitled "Iridescent Colors Exhibited by Beetles and Butterflies." Desiring to add some local butterflies to his collection, he went to his country estate outside Bangalore and personally bagged butterflies with a butterfly net bag. Raman was, perhaps, the only Nobel Prize winning physicist to have engaged in chasing butterflies!

In the 1960s Raman became very interested in vision,

and he thoroughly educated himself on the anatomy and physiology of the eye and its performance as a visual apparatus *par excellence*. He talked to visitors at his institute about rod vision, cone vision, color blindness and acuity of vision, and he carried out very simple experiments with color filters using himself and others as guinea pigs. He wrote a series of memoirs on his findings.

Raman bequeathed all his personal wealth to his institute, and he hoped for a bright future for it. He was very much against accepting government grants, for he feared that this practice would destroy the freedom necessary to carry out fundamental research. When India's education minister, M. A. Chagla, offered financial support from the Indian government, Raman said: "Sir, I want this institute to be an oasis in the desert, free from government interference and the application of its rules and regulations. That would destroy my institute. Thank you for your offer." After Raman passed away, his younger son, V. Radhakrishnan, an eminent radioastronomer, succeeded him as the director of the institute. In the past 18 years the institute has grown beyond anything Raman would have imagined, and it is now largely supported by grants from the government of India.

Raman and science in India

Raman created many scientific institutions in India and nurtured them into centers of research. He put new life into the Indian Association for the Cultivation of Science, organized and built the physics departments at the University College of Science and at the Indian Institute of Science, and founded the Raman Research Institute. With his strength of personality, vision and genius, and with discipline and dedicated work, he brought all of them world renown. One of his longest lasting contributions to Indian science is the large number of talented students whom he initiated into the excitement of research. Many of them in turn have created independent schools of research. Raman's influence will thus be felt in India for many generations.

Raman recognized the importance of scientific journals. In *Current Science* in May 1933 he wrote, "While the foundation of the scientific reputation of a country is established by the quality of work produced in its institutions, the superstructure is reared by the national journals which proclaim their best achievements to the rest of the world." Raman's contributions to scientific journalism in India include the *Bulletin of the Indian Association for the Cultivation of Science*, which was later transformed into *Proceedings of the Indian Association for the Cultivation of Science*, and still later into the *Indian Journal of Physics*. When he moved to Bangalore he launched and nurtured the *Proceedings of the Indian Academy of Sciences*. In 1932 he took an active part in starting *Current Science*, a journal styled after *Nature*. Raman strongly believed that the best work done in India should be published in Indian journals, and he set a personal example by publishing all his papers in the journals he started. The pages of the *Proceedings of the Indian Academy of Sciences* are filled with his work and the work of his students.

Recognizing the importance of a scientific body in arousing and maintaining public awareness of science and representing science to those in government, Raman founded the Indian Academy of Sciences in 1934. Outstanding scientists from all over India were elected fellows, and eminent scientists from around the world were invited to join a distinguished body of foreign fellows. Raman was elected founding president and held the office until his death. He also served as editor of the *Proceedings* and established the highest standards for the journal. The *Proceedings* appeared promptly every month in two sections—*A* for physical sciences and *B* for biological sciences—and were dispatched to subscribers in India and abroad. The *Proceedings* are considered among the top scholarly journals of the world; papers in physics currently appear in a new "reincarnation" called *Pramana*.

Raman meticulously organized the academy's annual meetings, with their attendant scientific programs and public lectures. The best work done by the fellows and their associates was presented at these meetings, and Raman's public lectures were usually the high points. The academy met at the invitation of a university serving as the host, so that young students of science were exposed to India's foremost scientists directly. Raman sat through all the sessions, both in the physical and the biological sciences, and added luster and humor to the proceedings. These annual meetings were jocularly referred to as "Raman's circus," and he never failed to be present, except once toward the end of his life. Since his death, the Indian Academy of Sciences has redoubled its activity and multiplied its publication efforts severalfold.

Some reflections on Raman's personality

Raman belonged to the fast-disappearing class of scientists who could properly be called natural philosophers. Far from favoring the extreme specialization that ultimately results in knowing "more and more about less and less," he believed that scientific endeavor should remove, not emphasize, the sharp boundaries between scientific disciplines.

Quoting from Ramaseshan's 1978 memorial lecture: "Raman was, of course, the supreme egotist. But in private conversation he often showed such unbelievable humility as to make one wonder which was his true self". He was a man of emotion and could get greatly excited. But he had an excellent sense of humor and could keep an audience roaring with laughter by describing what might have been a commonplace incident. Above all he was a very simple man, almost childlike. During the memorial meeting for Einstein in Bangalore, Raman choked with emotion before he could begin his talk. He then proceeded to give an eloquent, spontaneous tribute highlighting Einstein's epoch-making achievements.

No single person has done as much for Indian science as Raman. Through his personal example of dedication; through his success as a teacher *cum* leader in training generations of physicists; through the creation of scientific institutions and facilities for research and the founding of scientific academies and journals; and through his gift of eloquence, Raman exercised a tremendous influence on the progress of science in India. In 1954 he was the obvious choice for the first presentation of the highest award given in independent India, the "Bharat Ratna," or Jewel of India. Prime Minister Rajiv Gandhi has declared 28 February "Raman Day," to remember him and to express the gratitude of the nation to this great savant of science.

References

1. S. Bhagavantam, Bio. Mem. R. Soc. London **17**, 565 (1971).
2. C. V. Raman, Proc. R. Soc. London **A101**, 64 (1922).
3. K. R. Ramanathan, Proc. Ind. Assoc. Cult. Sci. **8**, 190 (1923).
4. C. V. Raman, Ind. J. Phys. **2**, 387 (1928).
5. C. V. Raman, Nature **121**, 619 (1928).
6. G. Landsberg, L. Mandelstam, Naturwissenschaften **16**, 557, 772 (1928). A. Smekal, Naturwissenschaften **11**, 873 (1923). H. A. Kramers, W. Heisenberg, Z. Phys. **31**, 681 (1925). I. L. Fabelinskii, Sov. Phys. Usp. **21**, 780 (1978). L. Brillouin, Ann. Phys. (Paris) **17**, 88 (1921). E. Gross, Nature **126**, 201, 400, 603 (1930). L. A. Ramdas, Nature **122**, 57 (1928).
7. C. V. Raman, N. Nath, Proc. Ind. Acad. Sci. **2A**, 406 (1935). ■

EDWIN P. HUBBLE AND THE TRANSFORMATION OF COSMOLOGY

By providing the first widely convincing evidence of the existence of galaxies external to our own, he helped to remake our notions of the origins of the universe.

Robert W. Smith

PHYSICS TODAY/APRIL 1990

In the first three decades of this century the modern science of cosmology was forged from general relativity theory and new observing methods and instruments, particularly large optical telescopes perched thousands of feet above sea level in California and Arizona. These radical changes in the theoretical and observational tools used by astronomers, physicists and mathematicians accompanied revolutionary changes in cosmology itself. The new cosmology of the early 1930s included two key cognitive features absent from the cosmology of the turn of the century: first, the existence of galaxies outside our own stellar system that are visible in Earth-based telescopes, and second, that these galaxies evince the expansion of the universe.

In the literature on the history of cosmology Edwin P. Hubble is the astronomer most closely associated with both of these profound new views of the physical universe. And it is Hubble's status as an observational cosmologist that explains the naming in his memory of the Hubble Space Telescope, whose mission includes further exploration of the distance of external galaxies and the expansion of the universe.

Rather than present a chronological account of Hubble's scientific life, I would like to focus on the two main scientific achievements of his early career and explore their relationship to the rapidly developing observational cosmology of the first three decades of the century (for some recent works see ref. 1). We shall see that the availability of high-powered telescopes, the push to improve the quality of observations, the driving force of aesthetics in describing the universe and the shifting use of language all influenced the choice of problems and solutions in early-20th-century cosmology. We shall also see that Hubble played a central role in addressing new questions about the physical universe and redefining what it meant to pursue cosmology. This, perhaps, is his most enduring legacy.

Language and cosmology

Changes in the language of cosmology in the first three decades of the century both reflected and helped in the framing of new questions about the universe and the interpretation of astronomical observations. Cosmology as we have come to understand it did not exist in the first decade or so of the 20th century, and the term itself was very rarely used. Instead, when astronomers wanted to talk about evolution in the universe, they used "cosmogony." But cosmogony concerned the evolution of the different structures and bodies visible in the heavens— nebulae and stars, for example—whereas the term "cosmology" related to how the universe is currently ordered. In this sense cosmology was, in the words of Agnes A. Clerke, "the elder sister of cosmogony. What *is* must be studied before what *was* can be inferred."[2]

Even "universe" acquired a new meaning during the early part of the 20th century. To astronomers in the first two decades of the century, the term often implied what we would now call our Galaxy or Milky Way system. For instance, when in 1914 the brilliant English astronomer and physicist Arthur Stanley Eddington wrote his famous and widely read *Stellar Movements and the Structure of the Universe*, he was addressing the issue of stellar dynamics in our Galaxy, and by "structure of the universe" he meant structure of the Galaxy. This usage of

Robert W. Smith is a historian at the National Air and Space Museum of the Smithsonian Institution, and also teaches the history of science at The John Hopkins University.

Edwin P. Hubble (far right with (from right to left) his sistor Lucy, brother William and a cousin, in Shelbyville, Kentucky, around 1910. Born 20 November 1889 in Marshfield, Missouri, Hubble later moved with his family to the suburbs of Chicago, where he attended high school and subsequently won a scholarship to the University of Chicago. (Photo courtesy of the estate of Rufus Harrod.)

"universe" was in fact loaded with the generally held view that our Galaxy was the only such aggregation of stars visible in even the largest telescopes. In other words, everything that is visible in the heavens—all of the stars and nebulae—was thought to belong to our Galaxy, a star system then widely suspected to be of the order of ten thousand light years in diameter (one-tenth of our current estimate).

Occasionally astronomers speculated on whether or not other galaxies lay beyond our stellar system, but until the 1910s few astronomers thought there was convincing evidence that external galaxies had actually been sighted. Thus, by the early years of the 20th century it had, according to historian Stanley L. Jaki, "become customary to picture the universe as consisting of two parts: one visible and confined to the Milky Way, and another, truly infinite, which was believed to be forever beyond the reach of visible observations."[3]

The visible part, everyone agreed, contained numerous stars and nebulae. Most of the nebulae were also regarded by astronomers as "spiral nebulae"—nebulae that possess a spiral structure and which are members of the same stellar system as our Sun. The first spiral nebula had been identified in 1845 by Lord Rosse (William Parsons) and his observing colleagues by means of the famous "Leviathan of Parsonstown," a 72-inch reflecting telescope situated at Birr Castle in Ireland. Relatively few were detected in the next several decades, but in 1898 James E. Keeler began photographing bright nebulae and

star clusters with the 36-inch Crossley reflecting telescope at Lick Observatory in California.[4] Although he died in 1900, the photographs Keeler took in his last two years convinced many astronomers that the total number of nebulae visible through the Crossley telescope was well over a hundred thousand, a figure that was an entire order of magnitude higher than that previously accepted. Keeler argued that most of the nebulae possessed a spiral structure, but these "spirals," as they were generally termed, were not interpreted by astronomers as galaxies, and Keeler had not embarked on his program with the aim of examining theories about or properties of galaxies. Rather, spirals were accepted by nearly all astronomers at the turn of the century as protosolar systems in development or perhaps small clusters of stars. Hence the meaning of the term "spiral nebula" and the set of properties astronomers associated with spirals in 1900 were, as we shall see, very different from those of even a decade later.

The shift westward

Not only did the language of cosmology change in the early years of the century, but there was also a major shift in the center of cosmological activities, resulting from the emergence of the United States as a power in observational astrophysics. In the middle of the 19th century, observational cosmology was largely the province of the owners of giant reflecting telescopes. Such instruments were essentially British and Irish in conception, construc-

On a camping trip in southern Indiana around 1914; Hubble is at far right. After receiving his BS in mathematics and astronomy from the University of Chicago and spending three years as a Rhodes Scholar studying jurisprudence and Spanish at Oxford, Hubble returned to the US to teach in New Albany, Indiana. One year later he enrolled as a graduate student at the Yerkes Observatory of the University of Chicago. (Photo courtesy of John Hale.)

tion and use. The builders of the largest reflectors were not industrial concerns, but enthusiastic, wealthy and driven scientific amateurs such as Lord Rosse. These amateurs, historian James A. Bennett has written, "tapped energies characteristic of contemporary British society and had applied them to the current instrumental needs of cosmology."[5]

By the early 20th century the rise of the US as a leading economic power was mirrored by its growing importance in the manufacture and use of giant reflecting telescopes. The high point of this shift was represented by the great 100-inch Hooker telescope on Mount Wilson, completed in 1917. Except for the optics, the Hooker telescope was chiefly the product of large-scale engineering by US companies. Indeed, because of the success of American astronomers in securing research funds (available largely through newly wealthy entrepreneurs), the placing of expensive and powerful telescopes at good sites and the increasingly sophisticated training of American astronomers, the US emerged as the dominant power in observational cosmology.

American astrophysicists and astronomers did not gear up early in the century expressly for an attack on cosmology. Other interests and concerns, chiefly the astrophysics of stars and especially stellar evolution, led the leaders of American astrophysics to build up an infrastructure to support their discipline. Thus when cosmological questions—particularly the study of spiral nebulae—did begin to loom large, the Americans were prepared to tackle them. When Hubble wrote in 1936 that the "conquest of the Realm of the Nebulae is an achievement of great telescopes,"[6] he was surely ascribing too much importance to the instruments and too little to the people using them. He did, nevertheless, grasp a crucial point, for by the late 1910s the observational study of the spiral nebulae had become centered on the small group of astronomers tackling nebular problems at the Lick,

Lowell and Mount Wilson observatories, good sites with powerful telescopes in the Western US. By this time, no matter how gifted, the observational astronomer needed easy access to large and costly telescopes to be at the forefront in studies of spiral nebulae. Such access was available to only a privileged minority, to which Hubble belonged.

Turning-point observations

Science is often viewed as being propelled by theory, but it was the drive to make better observations, along with the desire to prove hypotheses, that fueled early-20th-century cosmology. As astronomers strove to improve the capabilities of telescopes and apparatus, observation sometimes took on a life of its own, apart from theory. Thus it was in some sense serendipitous that two key sets of novel cosmological observations early in this century—Vesto Melvin Slipher's measurements of spectral shifts in spiral nebulae and Hubble's detection of Cepheid variables in nearby spirals—had a profound effect on theoretical cosmology.

To appreciate Slipher's observations, we need to realize that around 1909, when he started his research program on spiral nebulae, Slipher, like many others, still viewed them as protosolar systems. In fact Slipher was hoping that the spiral nebulae would disclose some clues to the origin of the solar system when he first aimed his spectrograph—attached to a fine 24-inch refractor at Lowell observatory—at the Andromeda nebula. By late 1912 he had secured four photographic plates and concluded that shifts of the nebula's spectral lines were discernible. The assumption that the spectral displacements were Doppler shifts permitted him to measure the nebula's speed of approach as 300 kilometers per second, at that date the highest speed recorded for an astronomical body.[7]

By August 1914 Slipher was able to announce to a

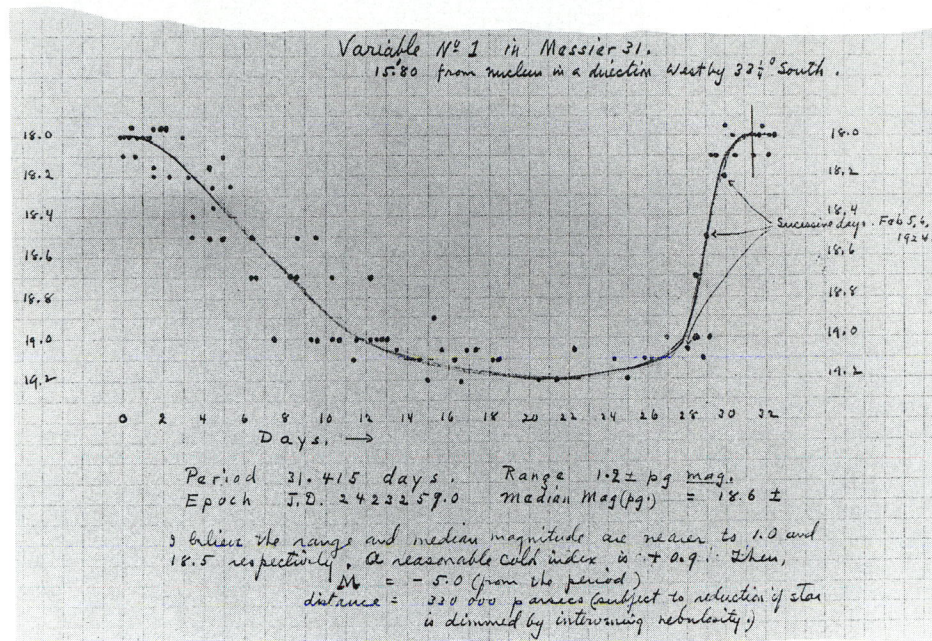

Variable № 1 in Messier 31.
15.80 from nucleus in a direction West by 33½° South.

Period 31.415 days. Range 1.2± pg mag.
Epoch J.D. 2423259.0 Median Mag(pg.) = 18.6 ±

I believe the range and median magnitude are nearer to 1.0 and 18.5 respectively. A reasonable cold index is + 0.9. Then,

M = - 5.0 (from the period)

distance = 330 000 parsecs (subject to reduction if star is dimmed by intervening nebulosity.)

Light curve for the first Cepheid found in the Andromeda nebula. This plot accompanied the letter Hubble sent Harlow Shapley in early 1924. (Courtesy of the Harvard Archives.) Figure 1.

meeting of the American Astronomical Society (among whose number was a 24-year-old graduate student from the Yerkes Observatory of the University of Chicago, Edwin P. Hubble) that he had measured the radial velocities of 15 spiral nebulae, and that most of them were receding. The fastest was traveling—again, based on the interpretation that the redshifts were Doppler shifts—at around 1100 km/sec, several hundred kilometers per second faster than the fastest stars. As one leading astronomer, William Wallace Campbell, told Slipher, the results on the radial velocities of the spirals constituted one of the greatest surprises astronomers had encountered in recent times.[8] But what did the results mean? To those who quickly interpreted the spectral shifts as Doppler shifts, the speeds of the spiral nebulae seemed altogether too great for them to be gravitationally bound to the stellar system. Perhaps, some argued, the spirals were external galaxies after all. Certainly Slipher's results helped to direct increased attention to spiral nebulae and the theory that space is populated by visible galaxies.

If Slipher's measurements of radial velocities spurred new interest in the theory of the spirals as galaxies, Hubble's observations in 1923 and 1924 of Cepheids in the Andromeda nebula effectively ended the debate on the existence or nonexistence of visible galaxies beyond the Milky Way.

From his writings before, during and after 1923, it is clear that ever since he was a graduate student Hubble had been well aware of the ramifications of the theory of the spirals as external galaxies. Although the evidence is somewhat ambiguous, he may even have supported the theory. For example, as a staff member of the Mount Wilson Observatory, Hubble spent a considerable amount of time examining what would now be termed a giant elliptical galaxy, M87, but which in the early 1920s was usually classified as a spiral. Hubble was deeply influenced in the 1920s by the evolutionary scheme of the British mathematician James Jeans,[9] which explained the forms of nebulae in terms of the development of a single basic type, initially formed from a huge cloud of nebulous material, that was gradually converted into stars from the outer regions inward. Hence, when Hubble detected objects he thought indistinguishable from star images in

the outer regions of M87, he was inclined to go a step further and indeed interpret these "condensations" as stars.[10]

Given the brightnesses of these apparent star images and comparing them with bright stars in the Galaxy, one could estimate a distance to M87 that put it far outside our stellar system, although based on such estimates M87 was certainly not as big as the Galaxy. But if Hubble could demonstrate that these objects really were stars (they are now known to be globular clusters), he could measure the distances to the larger spirals relatively easily. He investigated the condensations by taking photographs centered on the outer regions of the spirals, in this way avoiding the off-axis aberrations of his telescopes. As Hubble later recalled in his classic 1936 book *Realm of the Nebulae*, he could indeed detect starlike condensations on the resulting plates. But this did not mean that Hubble was yet fully convinced that he could see individual stars; instead, he judged the images to be indistinguishable from stellar images. Until the images could *definitely* be shown to be stars through the display of some stellar characteristics, the case for the spirals as external galaxies would not be much strengthened.

In 1923 Hubble began, in his invariably thorough fashion, to concentrate on a study of novae in the spiral nebulae. He took numerous plates of the Andromeda nebula, eager to establish its distance accurately. Novae had already been detected in the Andromeda nebula and some other spirals, but their use as distance indicators was problematic because of uncertainties in the measurements of the absolute brightnesses of novae in our Galaxy and in the very wide range of apparent brightnesses of the novae seen in spirals.[11] By observing more novae, Hubble expected to determine with higher precision the mean apparent brightnesses of novae in the nebula. Comparing the apparent brightnesses of novae in the Andromeda nebula with the absolute brightnesses of novae within the Galaxy and employing the principle of the uniformity of nature—according to which novae in the nebula should be, on the average, as bright as those in the Galaxy—he hoped to estimate its distance with more certainty.

Several months after embarking on this program Hubble made a momentous find. As he told the director of the

Harvard College Observatory, Harlow Shapley: "You will be interested to hear that I have found a Cepheid variable in the Andromeda nebula....I have followed the nebula...as closely as the weather permitted and in the last five months have netted nine novae and two variables....The two variables were found last week." The first Cepheid had appeared on a plate taken in October 1923. Hubble's observing program was designed to discover novae, and thus he initially assumed the variable to be a nova. But after checking earlier plates and drawing a light curve—brightness versus time—he realized that the object displayed the characteristics of a Cepheid variable (see figure 1). He quickly calculated an approximate period for the star's light variations.

Armed with the period, he calculated the star's actual brightness by use of the so-called period–luminosity relationship, an empirical relationship between the periods of light variation for Cepheids and their absolute brightnesses. Further assuming that throughout the universe Cepheids with the same periods have the same absolute luminosities—the principle of the uniformity of nature again—he compared the observed Cepheid's apparent and actual brightnesses and secured an estimate of the distance to the Cepheid in the Andromeda nebula—about one million light years. He rapidly discovered other Cepheids in the wake of observing the first. Within a year or so, Hubble had accumulated enough evidence from the Cepheids and the other distance indicators to convince almost all astronomers that the outer regions of the Andromeda nebula consisted of clouds of stars and that the nebula was indeed an external galaxy. Following these finds and similar ones for other, large and so presumably nearby spirals, the debate over the existence of visible galaxies was all but at an end.[12]

Before Hubble's find no one had publicly proposed a program to detect Cepheids in spirals. Indeed, Hubble had initially thought he was dealing with a nova when he found the first Cepheid in the Andromeda nebula; his preliminary misunderstanding was perhaps not surprising, because he had embarked on a research program to detect novae. Yet once they had been found, the Cepheids were enormously influential.

Although the discoveries of both Slipher and Hubble were unexpected, neither was a fluke. Both finds were the products of carefully planned research programs, and in Hubble's case the Cepheids settled the very question he had set out to answer: How far away are the spiral nebulae? Both Slipher and Hubble saw familiar objects in new ways through more powerful or more advanced observational techniques. Slipher made adroit use of an excellent spectrograph, with a camera much faster than those usually used by his contemporaries, to study objects whose faintness had deterred others. Hubble's success was due in part to his skilled deployment of very powerful telescopes—the Mount Wilson 60-inch and 100-inch reflectors—and his focus on the outer regions of the spirals, rather than the inner regions that had attracted the attention of other astronomers. In 1909, when Slipher began his studies of the radial velocities of spirals, he was convinced that the nebulae were familiar objects—single solar systems or small star clusters in the making. Fifteen years later, with Hubble's and others' findings, these objects were transformed into gigantic collections of stars, gas and dust—galaxies far beyond our own.

The expanding universe

Although improving observations was an important motivating factor in cosmology, empirical evidence was not an end in itself, and a sense of aesthetics often drove cosmologists to explain observations in particular ways. Cosmology, by its nature, deals with the faintest and most distant objects, and cosmologists must often work among the noise of their observations. Hence aesthetic criteria—simplicity, beauty, completeness, irreducibility, for example—influenced the selection, rejection and development of theories and observations.

A particularly interesting example of the influence of aesthetics in cosmology is the response to the concept of the expanding universe. It is now usual to trace the idea of an expanding universe, at least in the mathematical sense, to two papers[13] published by the Russian mathematician and meteorologist Alexander Friedmann in 1922 and 1924. Friedmann's starting point was the field equations of general relativity that Einstein had developed in 1917, which, unlike the field equations he had published in 1915, contained the so-called cosmological constant λ. These equations were of the form

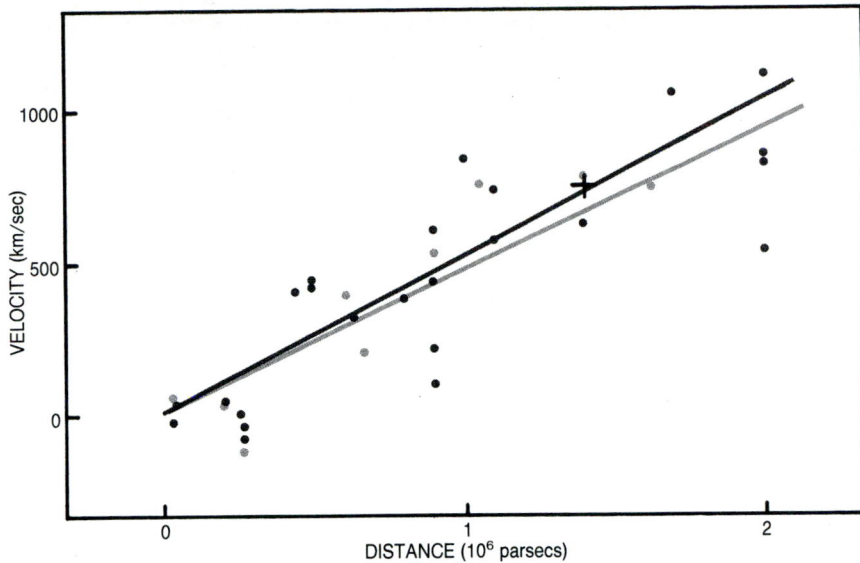

'**Extra-galactic nebulae**' (what we would now call galaxies) had this velocity-distance curve, according to a paper Hubble published in 1929. The heavy dots and line represent the solution to the velocity-distance relation for individual galaxies, and the pale dots and line represent the solution when the galaxies are combined into clusters or groups. The cross represents the mean velocity corresponding to the mean distance of 22 nebulae whose distances could not be estimated individually. (Adapted from ref. 17.)   Figure 2.

$$R_{mn} - \tfrac{1}{2}g_{mn}R - \lambda g_{mn} = -\kappa T_{mn}$$

where T_{mn} is the energy–momentum tensor and R_{mn} the Ricci tensor. Friedmann's simplifying assumptions led him to the metric

$$\mathrm{d}s^2 = \mathrm{d}t^2 - [R^2(t)/c^2]\mathrm{d}\sigma^2$$

where $\mathrm{d}\sigma^2$ is the metric of three-space. Friedmann proceeded to examine cases in which R is either a constant or varies with time, and concluded that nonstationary solutions were possible with both positive and negative spatial curvature. But Friedmann presented his nonstatic solutions to the field equations of general relativity as mathematical solutions to which he did not accord any particular physical significance, nor did he make any concerted attempt to connect them with astronomical observations. Rather, the first person to join theory and observation in a way that would come to be widely seen as physically meaningful within the general framework of the expanding universe was, as Helge Kragh has argued convincingly,[14] a 33-year-old Belgian abbé and professor at the University of Louvain, Georges Lemaître.

In 1927 Lemaître published what would later be recognized as a seminal paper on the expanding universe.[15] But for a brief time, Lemaître's work drew no interest. Even Einstein told Lemaître, at the fifth Solvay conference in 1927, that he did not accept the notion of the expanding universe or the physics underpinning the paper. As Lemaître put it much later, Einstein, who was still wedded to the notion of a static universe, told him, "Your calculations are correct, but your physical insight is abominable."[16]

Certainly the observational evidence at that time for an expanding universe was at best ambiguous. By 1927 there had been a decade's worth of speculation about a redshift–distance relation for external galaxies. A few people had even tried to determine observationally the form (if any) of the relationship. But the evidence presented was not convincing; the published plots of radial velocity versus distance looked like scatter diagrams.

Hubble transformed this situation. At first, using a few radial velocities secured at Mount Wilson by Milton Humason and many other velocities obtained earlier by Slipher, together with his own estimates of the distances of the galaxies, Hubble persuaded his colleagues that there was indeed a readshift–distance relation and that, at least in the first approximation, it was linear (see figure 2). His

initial paper[17] on the redshift–distance relation, published in 1929, was followed by another, much more extensive one, co-authored with Humason, which presented many more redshifts measured by Humason[18] (see figure 3).

Hubble was always careful in print to avoid definitely interpreting the redshifts as Doppler shifts. But the writings of Eddington and others soon meshed the calculations of Lemaître and various theorists with Hubble's observational research on the redshift–distance relation. The notion of the expanding universe was swiftly accepted by many, and the linear relationship between redshift and distance was later widely accepted as Hubble's law.

Even if one accepted the concept of the expanding universe, there was nevertheless the immediately puzzling and fundamental question of what started the expansion. One of those to stress this problem was Eddington, who was probably at that time the world's most influential astrophysicist. For Eddington, the most aesthetically pleasing, or "most attractive," case, to use his term, was that in which the mass of the universe is equal to the mass of the Einstein universe—the mass according to the static solution to the field equations of general relativity discovered by Einstein over a decade earlier. According to Eddington's scheme the world evolved from an Einstein universe and so developed "infinitely slowly from a primitive uniform distribution in unstable equilibrium."[19] Thus infinitesimal perturbations of the static Einstein universe build and start the expansion. But Eddington explicitly rejected the notion of a creation of the universe, as seemed to be implied by a universe with more mass than the Einstein universe, because "it seems to require a sudden and peculiar beginning of things." He contended, "As a scientist I simply do not believe that the present order of things started off with a bang; unscientifically I feel equally unwilling to accept the implied discontinuity in the divine nature."[20]

During the early 1930s several people, including a sometime collaborator of Hubble's, the Caltech mathematical physicist Richard C. Tolman, examined possible physical mechanisms to explain the expansion. Of course an alternative explanation of the expansion was that it really did start with the beginning of the entire universe, and it was Lemaître who introduced this concept into the cosmological practice of the 1930s. In 1931 he suggested the first detailed example of what later became known as

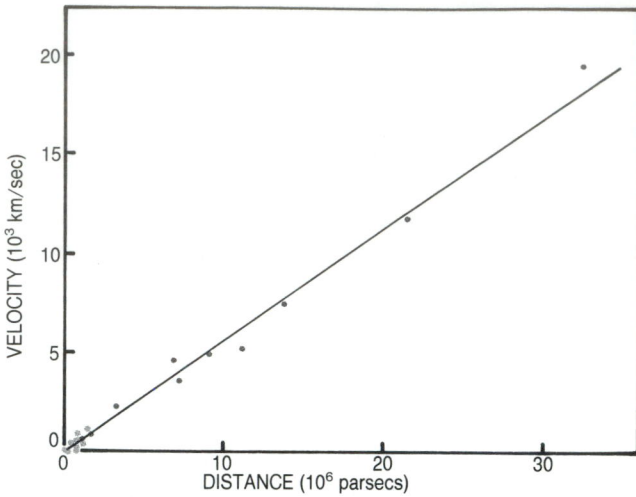

Velocity-distance relation published by Hubble and Milton Humason in 1931. The heavy dots represent mean values for clusters or groups of nebulae. The pale dots near the origin represent individual nebulae, which, together with the lowest two heavy dots, were used in 1929 formulation of the velocity-distance relation. (Adapted from ref. 18.)   Figure 3.

Big Bang cosmology. But unlike the universe of modern Big Bang theories, Lemaître's universe did not evolve from a true singularity but from a material pre-universe, what Lemaître referred to as the "primeval atom."[14] Lemaître understood this primeval atom to be a "unique atom" whose atomic weight was the total mass of the universe. This highly unstable atom would have divided into smaller and smaller atoms by a kind of super-radioactive process. For Lemaître a cosmic singularity was nonphysical notion in which neither time nor space exists, and he insisted that cosmology could and should be understood in physical terms. So, Lemaître wrote in 1931, "The last two thousand million years are slow evolution: They are ashes and smoke of bright but very rapid fireworks."[21] This statement, another example of the changing language of cosmology in this period, introduces the beginning of the universe as a legitimate subject for scientific discourse, even if some—like Eddington—found it distasteful.

During the 1930s theoretical cosmology tended to divide into two streams. One stream, historian John North has written, started from the question of "the stability of the Einstein universe, a question which might, it was thought, throw light on the beginning of the universe—and even, perhaps, on its 'cause.' These questions turned to matters of purely astrophysical interest—condensations in interstellar gas, cosmic radiation and the synthesis of the chemical elements—and to speculation on the so-called primeval atom introduced by Lemaître. Those who followed the other stream of thought argued from geometrical and kinematic premises. They tended (for example, in the appeals to symmetry) to a rigid idealization of the situation. Often this was not accompanied by any but the slightest reference to astronomical practice."[22] Nevertheless, by 1930 there had been produced, at least for those of the first stream, an exchange between theory and data that, although meager by present-day standards, was undreamt of only a few years before, an exchange that owed much to the research of Hubble. Hubble's work helped to breed among astronomers and mathematicians of his day the confidence to discuss and ultimately attempt to explain the entire history of the universe. Indeed, in the first three decades of the century the content of cosmology had been reshaped and extended, and through the efforts of Hubble and others, the practice and the very nature of the enterprise of cosmology had themselves been transformed.

References

1. R. W. Smith, *The Expanding Universe Astronomy's 'Great Debate' 1900–1931*, Cambridge U. P., New York (1982); N. Hetherington, Hist. Stud. Phys. Sci. **13**, 41 (1982); D. R. Osterbrock, R. S. Brashear, J. A. Gwinn, Mercury, January–February, 2 (1990); D. R. Osterbrock, R. S. Brashear, J. A. Gwinn, "Self-Made Cosmologist: The Education of Edwin Hubble," in *The Evolution of the Universe of Galaxies: The Edwin Hubble Centenial Symposium*, R. G. Kron, ed., Astron. Soc. Pacafic, San Francisco (1990); D. E. Osterbrock, "The Observational Approach to Cosmology: US Observatories Pre-World War II," in *Modern Cosmology in Retrospect*, B. Bertotti, R. Balbinot, S. Bergia, A. Messina, eds., Cambridge U.P., New York (1990).

2. A. Clerke, *Modern Cosmogonies*, Adam and Charles Black, London (1905), p. 15.

3. S. L. Jaki, *The Milky Way: An Elusive Road for Science*, David and Charles, Newton Abbott (1973), p. 270.

4. J. E. Keeler, Astrophys. J. **11**, 325 (1900). See also D. E. Osterbrock, *James E. Keeler, Pioneer American Astrophysicist and the Development of American Astrophysics*, Cambridge U. P., New York (1984); D. E. Osterbrock, J. R. Gustafson, W. J. Shiloh Unruh, *Eye on the Sky: Lick Observatory's First Century*, U. California P., Berkeley (1988).

5. J. A. Bennett, in *Human Implications of Scientific Advance*, E. G. Forbes, ed., Edinburgh U. P., Edinburgh (1978), p. 553.

6. E. P. Hubble, *The Realm of Nebulae*, Oxford U. P., Oxford (1936), p. ix.

7. V. M. Slipher, Lowell Observatory Bull. **58** (1913). See also W. G. Hoyt, Biog. Mem. Natl. Acad. Sci. **52**, 411 (1980).

8. W. W. Campbell, letter to V. M. Slipher, 2 November 1914, Lowell Observatory Archives. On Slipher's results from the mid-1910s, see V. M. Slipher, Pop. Astron. **23**, 21 (1915); Proc. Am. Philos. Soc. **56**, 403 (1917).

9. J. H. Jeans, *Problems of Cosmogony and Stellar Dynamics*, Cambridge U. P., Cambridge, England (1919), ch. 9 and 10.

10. E. P. Hubble, Pub. Astron. Soc. Pac. **35**, 261 (1923); letter to H. Shapley, 5 July 1922, Shapley papers, Harvard Univ. Archives.

11. M. A. Hoskin, J. Hist. Astron. **7**, 4753 (1976). See also ref. 12, pp. 42–45.

12. Smith, *The Expanding Universe*, pp. 113–126.

13. A. Friedmann, Z. Phys. **10**, 377 (1922); **21**, 326 (1924).

14. H. Kragh, Centaurus **2**, 114 (1987).

15. G. Lemaître, Ann. Soc. Sci. Bruxelles **47A**, 49 (1927).

16. G. Lemaître, Rev. Questions Sci. **129**, 129 (1958).

17. E. Hubble, Proc. Natl. Acad. Sci. **15**, 168 (1929).

18. E. Hubble, M. Humason, Astrophys. J. **74**, 43 (1931).

19. A. S. Eddington, Month. Not. R. Astron. Soc., **90**, 672 (1930).

20. A. S. Eddington, *The Nature of the Physical World*, Cambridge U. P., Cambridge, England (1932), p. 85.

21. G. Lemaître, Nature **128**, 700 (1931).

22. J. North, *The Measure of the Universe*, Clarendon, Oxford (1965), p. 126.

PHYSICS TODAY/FEBRUARY 1984

A history of the synchrotron

The events surrounding the origin of the synchrotron—the machine that made high-energy physics possible—narrated by a discoverer of the phase-stability principle that made the synchrotron possible.

Edwin M. McMillan

Speaking not as a historian but from a personal point of view, I would like to tell the story of the origin of the synchrotron as I saw it. The beginning, for me, was in the spring of 1945, when I was on the staff at Los Alamos, the wartime atomic-bomb laboratory. The Trinity test was in preparation, and I was already thinking about what to do on my return to Berkeley—from which I was on leave—after the war ended. I had spent a great deal of time and effort before the war on the design and operation of cyclotrons, I had a reasonably good understanding of the limits on the particle energies attainable by cyclotrons, and it seemed like a worthy goal to find ways to exceed these limits. The cyclotron, as you know, is a resonance accelerator; it pushes particles to high energies by the repeated applica-

Edwin N. McMillan was professor emeritus of physics at the University of California, Berkeley. He was awarded the Nobel Prize in chemistry in 1951.

tion of a moderate voltage, which must be applied at the proper instant each time the particle comes around in its circular orbit.

In the simple case of a particle of fixed mass in a uniform magnetic field, the frequency of rotation is constant and easily matched to a fixed accelerating frequency. But things are always more complicated in the real world. The mass of the accelerated particle is not fixed; it increases by the mass equivalent of the added energy. The magnetic field cannot be uniform or the particle orbits will not be stable. Bethe and Rose had pointed out these things in 1937, but at that time the economic limits on the size of machines were more important than limitations in principle. By 1945 this situation was reversing. One way to avoid the timing problem was to use an induction accelerator or betatron, in which the acceleration is independent of timing. So it happened that in May 1945 I started trying to design an air-core betatron. The reason for the air-core was that the

absence of an iron core allowed the use of a high magnetic field and reduced the size of the machine for a given energy.

Discovery of phase stability

This design never got very far. One night as I was lying in bed thinking about the problem of getting high-energy particles, my mind returned to the concept of resonance acceleration. If there were only some way to keep the motion of the particles in step with the alternating electric field that was pushing them along! I was tracing out in my imagination the motion as it unfolded in time when I suddenly realized that it had a natural tendency to lock into step with the accelerating field, if certain simple conditions were satisfied. I felt like the inventor in a cartoon with a lightbulb flashing on over his head. I did not record the date of that night, but it must have been close to the first of July. The next day, I started to tell my colleagues at Los Alamos about my idea. I remember vividly the reaction

Berkeley synchrotron after completion in 1948. Walter Gibbins, who supervised the construction, is standing in front. The machine reached its design energy of 300 MeV in January 1949. Figure 1

Lower yoke of the Berkeley synchrotron magnet and the coils that carry pulses of current from the condenser bank to excite the rectangular-design magnet.   Figure 2

Capacitor bank for Berkeley synchrotron was assembled from surplus units at other installations.   Figure 3

Vacuum chamber of fused quartz doughnut type was successfully used in Berkeley synchrotron.   Figure 4

of Don Kerst, who said: "I am kicking myself that I didn't think of it." Soon I had a name for the locking-in phenomenon, which I called "phase stability" because the word "phase" is used to describe the timing relation, and a name for the accelerator that would use that principle, which I called the "synchrotron."

On 4 July I communicated my thoughts to Ernest Lawrence in Berkeley by a letter which concluded, referring at first to the air core betatron,

> In any case, it is pretty much of a "brute force" machine, and it is not the sort of thing that one would want to build if a neater way could be found to do the job. I believe that I have a much neater way of accelerating electrons. A brief description of its principle is enclosed. I will send further details.

The "neater way" was the synchrotron, already called that in the enclosed brief description, which starts:

> This is a device for the acceleration of particles to high energies. It is essentially a cyclotron in which either the magnetic field or the frequency is varied during the acceleration, and in which the phase of the particles with respect to the high energy electric field automatically adjusts itself to the proper value for acceleration.

Today, the possibility of varying both field and frequency together would be specifically mentioned under the name "proton synchrotron," and the version with frequency variation alone would be called a synchrocyclotron. Lawrence and I had further discussion when he came to New Mexico to witness the Trinity test on 16 July, and he agreed that the construction of a synchrotron in Berkeley should be seriously considered. There were still some theoretical worries about the loss of energy by radiation (what is now called "synchrotron radiation"), and when the answer to this problem came—in the form of a calculation by Julian Schwinger, brought to me by I. I. Rabi—I went ahead with the publication of a Letter to the Editor of the Physical Review entitled "The synchrotron—a proposed high energy particle accelerator"; this was submitted for publication on 2 September 1945. (Rabi tells me that he persuaded Schwinger to make the calculation because of his concern over my problem.)

Later in September I returned to Berkeley. The war was over, but the Manhattan Engineer District was still providing funds for the Radiation Laboratory. General Groves was supportive of Lawrence's plans for conversion back to peacetime research activities, including the construction of a synchrotron, and design work was started at once, along with searches for surplus materials that might be usable. The actual directive authorizing construction was issued by the Manhattan District Office in Oak Ridge, Tennessee, on 29 August 1946. This authorized a total cost of $500 000, of which $225 000 was in the form of actual expenditures, while the rest represented the value of capacitors that existed as surplus at other installations, and that would be needed for storing energy to power the magnet. It did not include the building, for which $61 052 had already been authorized under another directive. All of this went on before the formation of the Atomic Energy Commission; the synchrotron was authorized and its basic funding was arranged while the Army was still in charge.

International efforts

Some time late in October of 1945 I got a telephone call from Charlotte Serber, who was then the librarian at Los Alamos. She reported that a Russian journal that had come into the library had in it an article, in English, describing an idea for an accelerator that was much like the synchrotron. I wrote to her on 30 October and requested a copy of that article, and thus I learned that Vladimir I. Veksler of the Soviet Union had developed the idea of phase stability in much the same way I had. A few months later, there appeared in the *Physical Review* a letter by Veksler complaining of my failure to give reference to his previous publications. In reply to this I sent a personal letter to Veksler and a letter to the editor of the *Physical Review*, in which I said: "It seems to be another case of the independent occurrence of an idea in several parts of the world, when the time is ripe for the idea." Veksler sent me a very friendly reply, dated 27 June 1946, in which he said:

> I fear that the English translation of my letter was somewhat more gruff than the Russian original. You are quite justified in saying that the history of science affords many examples of the simultaneous appearance of similar ideas in several parts of the world, as in our own case.

When Veksler used the word "simultaneous" he was being generous, as he had made three publications on the subject, his first being over a year ahead of mine, but when communications are almost non-existent, the concept of simultaneity is modified. I must admit that communications did not get much better for some time, and although it seemed likely to me that Veksler was building a synchrotron in Moscow, I had very few details about it.

I had even less information about the proposal that Mark Oliphant made in 1945 for the construction of a machine

Radiofrequency oscillator that supplied accelerating potential for electrons in Berkeley synchrotron is the brass structure at left with tube extending into center of machine.   Figure 5

Target inside bore of vacuum doughnut consisted of platinum strip (at left) which produced x-rays when struck by electron beam; scintillating crystal on same mounting enabled measurements of the beam intensity. Figure 6

in Birmingham, England. There were some rumors among the British contingent at Los Alamos about such a proposal, but no one seemed to know much about it. Oliphant had talked about it with Lawrence during visits to Berkeley, but apparently in very general terms, so that Lawrence's knowledge of what Oliphant was planning was neither clear nor specific. During the design period of the Berkeley synchrotron, there was no interaction with the Birmingham group; it was only later that I found out that the original unpublished proposal, which contained little in the way of design detail or theoretical analysis, was for what would now be called an air core proton synchrotron. This was modified to an iron core design before construction was started in Birmingham.

The first electron synchrotron to operate was that of F. G. Goward and J. E. Barnes, who modified an existing 4-MeV betatron to give 8 MeV as a synchrotron at the Woolwich Arsenal in England in 1946. Incidentally, Goward told me later that they got the idea from my publication, which they saw before they saw Veksler's. The second synchrotron was that of Herbert C. Pollack and his colleagues at the General Electric Laboratory in Schenectady, which was made from parts originally intended for a betatron, and which gave 70-MeV electrons. It was with this machine that the phenomenon now known as "synchrotron radiation" was first observed in 1947. Even before these two pioneer synchrotrons, however, the principle of phase stability was shown to be valid by experiments conducted by J. Reginald Richardson and collaborators at Berkeley, using the old 37-inch cyclotron with the addition of a rotating variable condenser to modulate the frequency. The success of these experiments led to the redesign of the 184-inch cyclotron (its construction had been halted by the war) as a synchrocyclotron, using the synchrotron principle with frequency modulation, and it was brought into operation late in 1946.

Berkeley synchrotron

Now let me return to the construction of the synchrotron at Berkeley. The design energy had been set at 300 MeV in the published letter, but no design details had been established; therefore we had much to do, and many people became involved (far too many to list here). For the magnet core, a rather conventional rectangular design was used (see figures 1 and 2). It was to be excited by the energy stored in a large capacitor bank (see figure 3) and discharged through the magnet by a set of ignitrons, giving pulsed operation, with a batch of electrons accelerated at each pulse. The original vacuum-chamber design, however, was far from conventional. It depended on the magnet pole tips and the plastic walls supporting the pole tips being made vacuum tight; this proved to be impossible, as the plastic used was too porous, and this design had to be abandoned. We went to a more conventional design that used a fused-quartz doughnut-type of vacuum chamber (see figure 4), and

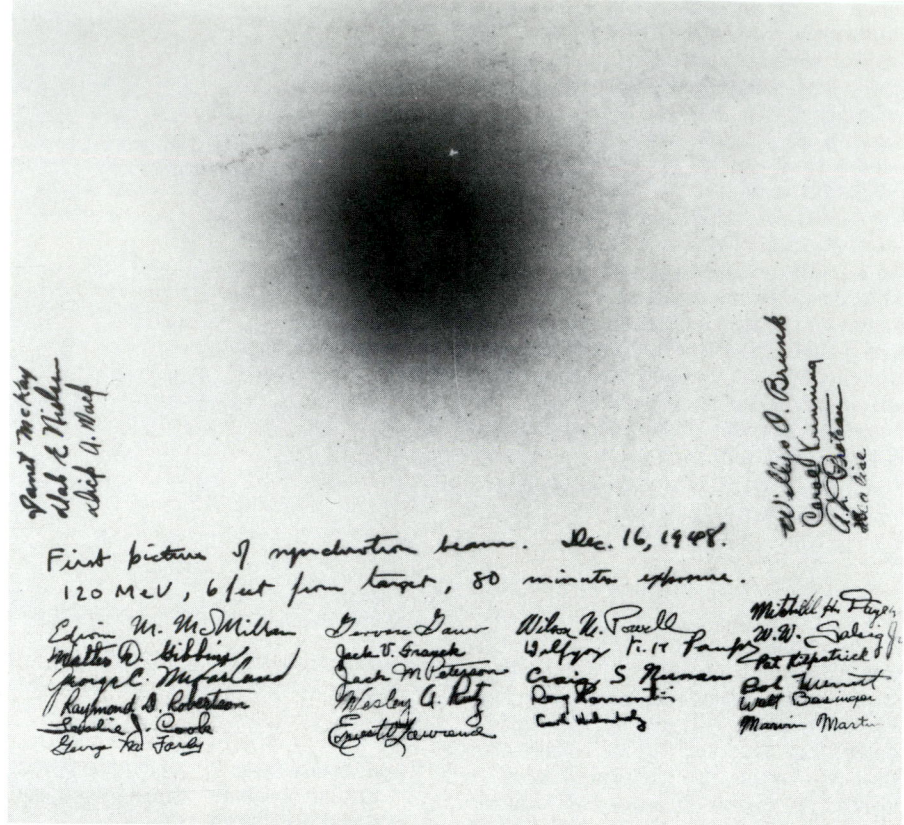

First acceleration of beam in Berkeley synchrotron was documented by film exposed to narrow cone of x rays produced by electron beam.   Figure 8

this arrangement worked out fine.

Another serious problem was caused by irregularities in the shape of the magnetic field, due to remanence in the laminated iron pole tips. This was particularly bad at the instant when electrons were injected into their orbits, when the field was weak and the errors due to remanence were relatively large. Other groups who had started to build 300-MeV synchrotrons at about the same time—Robert R. Wilson at Cornell, Ivan A. Getting at MIT and Robert C. Haxby at Purdue—had the same problem, and a great deal of gloomy correspondence went on between the groups. At Berkeley, Wilson M. Powell, our expert on magnet design, set out to correct these field errors in detail with hundreds of little wires cemented onto the pole tips. This massive effort turned out to be unnecessary, however, and all of Powell's wires were finally removed. The shape of the orbit is determined primarily by the low harmonics of the azimuthal field distribution; in the system that was finally used, the field was corrected octant by octant with individual controls brought into the control room so that one could adjust the field shape during operation.

With these adjustments it would be possible to optimize a beam of electrons once it was found; the problem was to find the beam the first time, when we did not know where to set the adjustments. We were trying various things when, on 20 November 1948, a telephone call came in from Wilson at

Operator in Berkeley synchrotron control room adjusting controls while observing the signal from the "divining rod" scintillating crystal (see figure 6).   Figure 7

Cornell; he told me that he had found a beam by operating the magnet at very low voltage. Three days later we found a beam at Berkeley, using the same procedure. Then the magnet voltage was raised bit by bit, optimizing the adjustments at each stage, and the full design energy was reached on 17 January 1949.

Figure 5 shows the oscillator that supplied the accelerating potential, and figure 6 shows the target that the electron beam was supposed to strike to make x rays. In this view, the actual target is the platinum strip at the left, which is inside the bore of the doughnut when the assembly is in place. Next to the target is a scintillating crystal that makes a flash of light when the beam hits it. This light would travel down a transparent lucite rod to the photocell in the box at the right. The signal from the photocell was displayed in the control room. We called this device the "divining rod" because it served to detect and measure the presence of a beam in the machine. I believe that this represents the first use of what is now called a "light pipe" in connection with particle detection; it was proposed by Emilio Segrè and built by Clyde Wiegand, and without it I don't know how we would have gotten the synchrotron into operation.

Figure 7 shows a scene in the control room, with the operator (myself) watching the signal from the "divining rod" while making adjustments with his two hands. At the extreme right of the picture are the sixteen knobs (eight for the upper magnetic pole and eight for the lower pole) that controlled the

magnetic-field corrections I mentioned earlier. As soon as a high-energy beam was found and allowed to strike the target, we could look for the x rays produced by the impact. The x rays would be expected to emerge in a narrow cone and to make a dark spot when they struck a photographic film. On 16 December 1948, when a sufficiently high energy was reached, we put a film in the path of the x rays and exposed for 80 minutes; the result is shown in figure 8. This film was signed by all present at the occasion.

Figure 9, taken ten years later, shows the "business end" of the synchrotron as it appeared during most of its life as a research instrument. The x-ray beam from the platinum target, which was inside the donut, emerged toward the viewer through a hole in a lead collimator, a little to the right of center. Two years later, in 1960, the Berkeley electron synchrotron was retired. It is now in the Smithsonian Institution, as part of a very fine exhibit of nuclear research equipment. The last figure shows me with Vladimir Veksler, taken at a meeting in Berkeley in 1959, and illustrates the fact that we did not allow our initial lack of communication to persist forever.

Proton synchrotron

Finally, let me consider the development of the proton synchrotron, again as seen from Berkeley. As I noted earlier, Oliphant proposed a machine of this type in 1945, but in Berkeley we had no clear notion at that time what was going on in Birmingham. William M. Brobeck, the chief engineer at the

Radiation Laboratory, quite independently had the idea of designing a proton accelerator of the synchrotron type with a time-varying magnet field, but with the addition of a time-varying frequency to keep the orbit radius constant or nearly constant. This was some time in 1946, but because Brobeck apparently kept no records of the inception of the idea, the exact date cannot be fixed. I recall that Robert Serber and I were both consulting with Brobeck on the design, but we did not keep records either.

The earliest tangible record is a drawing by Brobeck dated 12 November 1946, labeled "10 billion volt proton accelerator." This drawing shows many features that were embodied in the Bevatron, such as the use of four straight sections in the orbit, allowing space for injection, acceleration, and ejection of the beam. There were also features that were changed, including the energy. Lawrence thought that the cost would be too high and insisted that the size, and therefore the energy, of the machine should be reduced. I recall that sometimes during this stage of the design both Panofsky and I independently insisted with Lawrence that the energy should not be reduced below the threshold for making antiprotons, which is about 6 GeV. A drawing made in October 1947 and labeled "Study No. 2 of 50 foot bevatron" shows the next stage of development. The energy was to be 3 or 6.5 BeV, depending on the magnet gap and aperture used. The orbit radius, which was 80 feet in the original design, had been reduced to 50 feet, and that

Business end of Berkeley synchrotron—x-ray beam emerges toward viewer through a hole in lead collimator, to right of center.

Figure 9

Vladimir Veksler to the left of the author at a meeting in Berkeley in 1959.   Figure 10

became the radius used in the final design for the Bevatron.

The design work that I am describing was well known in other laboratories. I remember one occasion when Rabi was visiting Berkeley and was shown Brobeck's first drawing, with which he was greatly impressed, and was given a copy to take home. Thus it came about that when the time came to make serious proposals for construction to the Atomic Energy Commission, now in charge of funding for the laboratories, both Berkeley and Brookhaven were in contention. In November 1947 and February 1948 the General Advisory Committee discussed the matter at length, debating how many machines should be built, what size and where. The final decision of the Commission was to build two machines, one at Brookhaven to give 3 BeV and one at Berkeley to give a little more than 6 BeV. The formal authorization was sent to Berkeley on 20 May 1948. Note that by this date the electron synchrotron at Berkeley was still not yet operating, but the 184-inch synchrocyclotron had been running with great success for over a year, so there was no doubt that the principle was sound. And long histories of success also were achieved with the Cosmotron and Bevatron (as the machines at Brookhaven and Berkeley were called, owing to a lack of agreement at the time on a generic name) and with the still more powerful accelerators made possible by the later invention of strong focusing.

* * *

Based on the Morris Loeb Lecture presented at Harvard University, 13 April 1982. □

Von Kármán: Fluid dynamics and other things

He was one of the pioneers of aerodynamics, contributing the concepts of vortex streets and explaining the sand ripples in the desert, and he practically invented consulting.

William R. Sears

PHYSICS TODAY/JANUARY 1986

It is a pleasure to write about Theodore von Kármán. I am somewhat concerned, however, as I consider the audience for this article, many of whom are also fluid dynamicists: I can imagine quite a few old friends who knew von Kármán—perhaps as well as I—saying, "Oh, that isn't the way it really was." Well, I can only say that what I am going to write about is the way I remember it. (There must be some kind of "statute of limitations" that says that when you get to a certain age—and I suppose I have—you're entitled to tell it the way you remember it, and let the chips fall where they may.)

It is very fitting that we claim von Kármán as a part of our fluid-dynamics heritage—and a very colorful part indeed. "Colorful" is the right adjective for him all right. *Time* magazine used it once when they described him, and I think it pleased him. He asked me, "Am I colorful?" I told him, "Yes, I think that might be right!"

However, such a colorful person attracts stories, so there is a pitfall in talking about a man like von Kármán: There is a danger that when you quote his clever remarks and describe his personality you will end up making him look like an eccentric or a clown. His memory has already started at-tracting all the best and most famous absent-minded-professor stories, even those that have been told for a hundred years—with equal justice about equally colorful people, no doubt—and bear no resemblance to him.

He was an engineer, physicist, applied mathematician and solid-mechanicist, an administrator and consultant, adviser to governments and military services, author, editor, organizer of international science, *bon vivant* and much else, but reference to the five volumes of his *Collected Works* will confirm that his greatest published output was in fluid mechanics.

He is memorialized by buildings, laboratories, wind tunnels, lectureships, medals, fellowships, even an international research college, all named after him. Seventeen years after his death, meetings and symposia were held in his memory in Washington, Aachen, Pasadena, Long Beach (at the meeting of the American Institute of Aeronautics and Astronautics) and probably elsewhere. That was at the time of his 100th anniversary. (APS apparently preferred to celebrate his 103rd birthday. I'm not sure whether 103 has a special significance, but it is, for one thing, a prime number.)

It is appropriate to ask, especially on behalf of our younger colleagues, who may wonder if they too may be so remembered: Who was von Kármán? What did he do? What, in fact, is the legacy that he left?

Martians

First of all, he was a Hungarian! He was born and raised in Buda, which is the part of Budapest on one side of the Danube, with Pest on the other. There is a theory, proposed by the distinguished physicist, my friend Philip Morrison, that Buda was infiltrated by a colony of Martians! It is the only logical explanation of the unearthly language of the people there, the unearthly beauty of their women (the Gabor sisters), their music (Bartók), and their superhuman intelligence—the quite unreasonable impact the Hungarians have had in science and the arts!

As evidence, Morrison offered the following: Besides von Kármán, John von Neumann, Edward Teller, Leo Szilard and Eugene Wigner all came from Buda, and all, I believe, went to the same high school! The list, no doubt, includes others of comparable fame whom I've forgotten; it also includes my colleagues in fluid dynamics Nicholas Rott and Nicholas Hoff, and probably Les Kovásznay. That high school had a prize, given to the senior who showed the greatest mathematical talent, and I think all those I've listed won that prize.

I had the incredible pleasure of bringing Morrison and von Kármán together one day in Ithaca, and hearing Morrison tell von Kármán of his discovery—that he had blown the Martians' cover. They had never met before, but obviously charmed each other at first sight, there in the Statler Club. I said, "Dr. von Kármán, Phil has made a discovery about you Hungarians that I want you to hear." So Morrison proceeded to present the evidence and the conclusion—

W. R. Sears is professor of aerospace and mechanical engineering at the University of Arizona, Tucson. He was a student of von Kármán's at Caltech (1934–38), his collaborator in research (1937–41), and his close friend, coworker and confidant until von Kármán's death in 1963.

Theodore von Kármán in 1961, at the International Academy of Astronautics in Paris.

which he had arrived at while working with Teller at Los Alamos. As Morrison talked, von Kármán walked back and forth in the room chuckling and saying, "Funniest thing I've heard!" Then he turned to me and said, "Mind you, I do not deny!"

From a brilliant start as a student in Hungary, von Kármán went to Göttingen as a graduate student under the great Ludwig Prandtl. That was in the greatest days of the Prandtl school. Prandtl had just "discovered" inviscid fluid mechanics. Not literally, of course, but he had revolutionized fluid mechanics by discovering that classical, mathematical fluid mechanics and practical, engineering fluid mechanics could be brought together by the radical concepts of the boundary layer and the inclusion in the flow of the "inviscid" fluid of such features as circulation and vortices, which dominate the behavior of the fluid and whose existence depends on viscosity.

Prandtl didn't *discover* the boundary layer either—you can already find it in William Froude's work, for example—but it was Prandtl who tamed it, who recognized it as a singular perturbation and thus made the whole subject of inviscid irrotational flow meaningful. We now say, in fancy language, that he invented matched asymptotic expansions, but Prandtl was not a mathematician (nor was von Kármán, as I'll point out later), and what he did was not the formal mathematics of matched asymptotic expansions. He left that to his successors.

What a school Prandtl built at Göttingen! His students included Albert Betz, Jakob Ackeret, Max M. Munk, Adolph Busemann and Theodore von Kármán. All of practical fluid mechanics seemed to be yielding to Prandtl's wonderful ideas at the time von Kármán went there: the aerodynamics of monoplanes and multiplanes, induced drag and the slope of the lift curve, the ground effect, wind-tunnel boundary interference, propellers, windmills and much more besides.

Von Kármán's first big success in fluid mechanics, in 1910, was in the same spirit, but so novel that it could not be called a mere application of Prandtl's ideas; in fact Prandtl was so surprised and impressed by it that he told his young student: "You've got something there! Write it up and I'll present it to the Academy."

Vortex streets

What had happened was this: Prandtl was intrigued and bothered (as many workers in fluid dynamics still are) by the drag of bluff bodies—the

---

Theodore von Kármán (1881–1963)

Theodore von Kármán, who always identified himself as an engineer, became renowned not only in that profession but also as an applied mathematician and physicist. He was world-famous as a teacher, research worker, author, consultant and raconteur.

Born in Budapest, he received his education there before proceeding to Göttingen for graduate studies under the great Ludwig Prandtl. He became director of the Aerodynamics Institute at Aachen, then emigrated in 1928 to Pasadena, where he was made director of Caltech's Guggenheim Aeronautics Laboratory (GALCIT). He became an American citizen in 1936.

Von Kármán's earliest scientific work was in solid mechanics: He developed a theory for the effects of plastic deformation in the case of a buckling column. In later life, he returned to problems of stability and buckling of structures, including buckling of shells. He collaborated with Max Born on a famous paper on the molecular structure of solids and specific heat. His publication list is also peppered with purely mathematical papers and book chapters. But he is probably best remembered for his many contributions to fluid mechanics.

He was a pioneer in the statistical treatment of turbulence and turbulent flows, especially the turbulent boundary layer. He was also a pioneer in the theories of transonic and supersonic flows—flows involving speeds near and exceeding the speed of sound, respectively.

Von Kármán was a great teacher. His influence upon the character and quality of engineering education in America was profound. His books, including *Aerodynamics: Selected Topics in the Light of Their Historical Development* (Cornell U. P., Ithaca, N. Y., 1954) and *Mathematical Methods in Engineering* (McGraw–Hill, New York, 1940), are considered as classics. There is a five-volume collection of his published works.

In his later years, he was devoted to the subjects of international science and cooperation. He was founder of NATO's Advisory Group for Aeronautical Research & Development. The list of the honors he received is long and impressive: He was elected to academies and received awards and honorary degrees in many nations; he was also selected in 1963 to be the first recipient of the National Medal of Science of the United States.

---

phenomenon of oscillatory separation and vortex production. He asked Karl Hiemenz, a doctoral candidate, to build a water channel and a precisely circular cylinder so they could study the basic, symmetrical flow. Hiemenz worked with great precision, but when von Kármán asked him, every morning, "How goes it?" the answer was, sadly, "It always oscillates." Von Kármán thought about it and had the really startling idea that maybe only the asymmetrical, oscillating flow configuration was stable. Again it was modeled as an inviscid irrotational flow, but with an infinite train of ideal vortex filaments—later called the "Kármán vortex street."

Actually, von Kármán's earliest scientific triumph was not in fluid mechanics but in structures: the theory of inelastic buckling, that is, the behavior of columns whose buckling is accompanied by plastic deformation. This work originated a whole branch of structures research that is still going on (as is the subject of bluff-body drag, because the details are so complex).

Later, he and G. I. Taylor more or less founded the whole subject of statistical turbulence. They defined the averaging, identified the important statistical quantities, employed tensor techniques and drew many conclusions from what they found. They made just

about all the enticing mistakes—wrong guesses—about the array of correlations that crop up, saving their successors a lot of trouble. Their cooperation, as well as their work with Sydney Goldstein and others, is in delightful contrast with the well-publicized story of the search for the DNA structure.

But surely von Kármán's greatest legacy to us is not inelastic buckling, the vortex street or the statistical theory of turbulence. It is the legacy of a lifetime, the accumulation of a hundred timely, ingenious attacks on a hundred real, intransigent problems, in each of which he provided remarkable, original insight. Some of the problems in fluid dynamics he wrote on are:
▶ skin friction, both laminar and turbulent
▶ theory of helicopters
▶ vortex theory of propellers
▶ resistance of supersonic projectiles
▶ the rolling of metals such as steel
▶ sand ripples in the desert
▶ stalling and the maximum lift coefficient
▶ open-channel water flow and the gas-dynamic analogy
▶ detonation and deflagration waves.

I think von Kármán believed that any problem in engineering (and perhaps in a much broader category) could profitably be attacked mathematically.

Family portrait. From left to right: von Kármán, his mother Helene, his sister Josephine and an unidentified visitor at the von Kármán home in Pasadena, around 1934.

He asked: What are the dominant phenomena? How can they be modeled most simply? (Often this meant linearly, but if not, how? At best, it would be with a *differential* model!) His ability to use the principle of similitude—that is, the principles of sophisticated dimensional analysis—was incredible. Some of this magic was obviously learned by his students Clark Millikan, Francis Clauser and H. W. Liepmann, and passed along by them to Don Coles and to their students.

One of my fellow students at Caltech's Guggenheim Aeronautics Laboratory (generally referred to as GALCIT) told me: "The old man is hooked on approximations. I think he'd rather have an approximation than an exact solution." I resented the remark at the time, but now I think it might have been true. The beautiful, simple formula, embodying only the really important effects, was certainly what fascinated him. He was willing to let others carry out the details. How many PhD dissertations at GALCIT were the results of competent carrying out of the beautiful ideas jotted by von Kármán, often on the back of a menu or concert program, and left on his student's desk!

Von Kármán didn't believe you had even qualitative understanding of an observed phenomenon until you could write at least the equations of its first

approximation! (I guess an intelligent layman won't understand that sentence; he'll say "qualitative" and "equations" are contradictory.)

This, I believe, is his legacy: the conviction that a mathematical description—the simpler the better—is *necessary* in good engineering. He used to quote (I know not from whom), "There is nothing more practical than a good theory."

This was radical in the 1930s and 1940s. It was not how engineering was taught or practiced, either in the US or in the United Kingdom. A great transformation was accomplished between the 1930s and the 1960s, and I think von Kármán led it.

How he did it

How did he do it? Well, first of all, he had the credentials—the papers and books written, the honors received. Second, he was a great teacher, not only in the classroom, but also in the offices of generals, admirals and industrialists and in the meetings of engineering societies. Some of his papers and books are pedagogical masterpieces: *Mathematical Methods in Engineering*, which he wrote with Maurice Biot; *Aerodynamic Theory* (Volume II of the Durand series), "The engineer grapples with nonlinear problems" (*Bulletin of the American Mathemat-*

ical Society **46**, 615, 1940); and a real classic: *Aerodynamics: Selected Topics in the Light of Their Historical Development*. (An indication of the force of von Kármán's personality is the fact that *Mathematical Methods in Engineering* is always referred to as "*Karman–Biot*," pronounced BEEoh, because that is how any Hungarian pronounces the distinguished coauthor's distinguished Belgian name. The same happens with "*Karman–Tsien*," where the coauthor's distinguished Chinese name also became Hungarian. Von Kármán once went to a ticket window to buy a ticket to Schenectady, but returned to tell me, "Bill, I asked the fellow for one-way *Schen*ectady, and he gives me a one-way ticket to Kansas City.")

I urge our young colleagues who haven't read these marvelous books and papers to do so. They might also enjoy von Kármán's "Atomic engineering?" (*Mechanical Engineering* **67**, 672, 1945), a paper not really devoted to that subject but to engineering education in general; he deplores the fact that engineering students are not given more intellectual challenge. It includes the remark, "At the institution where I have been teaching in the last fifteen years . . . [the atmosphere for engineering students is one] in which the beam on three supports is considered a most

Last-minute calculations, on 23 August 1941, just before a test of the rocket-powered plane on whose wing von Kármán is writing. The flight was one of a series that demonstrated the potential of jet-assisted takeoff. From left to right: Clark Millikan, Martin Summerfield, von Kármán, Frank J. Molina and Homer A. Boushey, the pilot. (From T. von Kármán, L. Edson, *The Wind and Beyond*, Little, Brown, Boston, 1967; reproduced with permission of the Caltech Archives.)

difficult problem."

There are two papers in von Kármán's bibliography that cast important light on his attitude toward mathematics: "Some remarks on mathematics from the engineer's viewpoint" and "Tooling up mathematics for engineering." The latter was the first paper of the *Quarterly of Applied Mathematics*, and has the form (learned from Galileo) of a dialog between two speakers, of whom one does and the other doesn't believe in applied mathematics. It's a technique used by writers who are not, themselves, sure which side they are on. Actually, von Kármán was not very well trained in mathematics; his mathematical skill was mostly intuitive. I know, from writing papers with him, that he didn't approve(!) of analytic continuation, and thought it a huge joke when the solutions for mixed transonic flow in the hodograph plane came out in terms of what he called "confluent hypergeometric functions á la Whittaker–Watson!"

Third, he had infinite energy. He always taught classes, carried on original research, masterfully directed GAL-CIT, consulted with industry, served on governmental boards and committees, attended scientific meetings and lived an active social life. One of the unfounded legends now told about von Kármán is that he was an absent-minded administrator. On the contrary, he was a very able administrator. Many times I stood with Mabel Rhodes, the departmental secretary, in her little office on the second floor of GALCIT, and overheard conversations in which Clark Millikan told von Kármán that some distinguished person—Carl-Gustaf Rossby, Leslie Howarth or whoever—proposed to come to GALCIT for a period. When von Kármán agreed that it was desirable, Millikan would ask what we would use for funds. After a moment's thought, the director would always know how much was available and where it was coming from. Sometime each spring he would ask us younger staff members what we planned to do in the coming summer; if we said we'd like to continue our

research (and salary) right there, he'd say OK. No proposal, no contract—those were the good old days!

In his work and in his social life he was at all times "assisted" by his beloved sister Josephine. (He said "assisted," but to some of us it seemed the word should be "harassed.") She is in part responsible for what is likely to be a puzzle for many generations of librarians and bibliographers: One lone paper among von Kármán's is by "Theodore de Kármán and Leslie Howarth." Josephine didn't like Germany or anything German, and was unhappy about Theodore's use of the Germanic "von"; she was always "Josephine de Kármán." Both "von" and "de," of course, are translations of some Hungarian honorific that reflected the honor bestowed upon their father. When her brother and Leslie Howarth were writing up their fine paper on isotropic turbulence in 1937, she seems to have badgered him into changing his professional name from the German to the French form. Obviously the change didn't last beyond that one paper—it wasn't a very practical idea. In the *Collected Works*, which were assembled in 1956, the Kármán–Howarth paper is presented with the "de" changed back to "von."

Finally, he had the advantage of a charming, winning, colorful, Hungarian personality. Let me use the rest of my space on the personal side of this man; it was an important part of his impact in his field.

He loved people. Wherever he went, he struck up conversations with taxi drivers, waiters and waitresses, and chambermaids. He found out where their accents or surnames came from and tried to converse in their native languages or discuss their major fields of college study—because he was sin-

cerely interested. He also loved parties, drinks, girls, jokes, the *bon mot*. All his life he played the part of the dangerous Hungarian bachelor. He succeeded in shocking some of the young wives (and their husbands), but charmed many more. He told us: "I have decided how I want to die. At the age of 85 I want to be shot by a jealous husband."

He had a story for every occasion, an apropos joke to take the edge off any unpleasant episode. Whenever he met an Air Force general whose name he didn't know, he would greet him as "General Anderson," whereupon the general would explain: "Oh, I'm not General Anderson, Doctor, I'm General —." Von Kármán's reply was, "Oh, of course. You see, I know three General Andersons, so when I meet an Air Force general whose name I've forgotten, I call him General Anderson to maximize the probability."

He was a great admirer of the legendary Sam Goldwyn and loved to emphasize points in meetings by quoting such Goldwynisms as "OK, but include me out," "Wait a minute, I've got a wonderful idea, but I don't think much of it" and "I can give you my opinion in two simple words: Im Possible!"

Biting wit

He could, however, be bitterly sarcastic, especially when he encountered sham or scientific fakery. I remember when a fellow "generalized" the famous Kármán–Tsien linear approximation to the adiabatic gas law by using an exponent whose value depended on which family of NACA airfoil profiles was under consideration. Von Kármán's comment at an Institute of Aeronautical Sciences national meeting was: "This speaker reminds me of my boyhood in Hungary; we always

Von Kármán in 1928, shortly after moving to the United States.

AIP NIELS BOHR LIBRARY

had gypsy magicians who did tricks. The difference is that the gypsies only *pretended* to violate the laws of physics!"

A Czech scholar annoyed von Kármán by saying, constantly, that the Czechs had been oppressed by the Austro-Hungarian Empire. Von Kármán's father was an important official in the Empire's educational system— he was knighted for his work—and von Kármán always retained a sentimental attachment to it. He pinned the scholar down at a cocktail party: "Now, Dr —, tell us how you were oppressed?" All that the gentleman could come up with was, "For ten years I was assistant professor, and never promoted." Von Kármán turned to the other guests, saying, with great sarcasm, "Oh, that was terrible oppression indeed, but hardly worth fighting a World War over!"

He often encountered engineers who said, "Well, professor, I don't follow all of your fancy mathematics. I am a practical engineer." To which von Kármán's reply was: "Yes. You know what a practical engineer is? He's one who perpetuates the mistakes of his predecessors!"

Unfortunately, his clever remarks and analogies were not always understood by his hearers. I came across the memoirs of Stanislaw Ulam, the mathematical physicist, and looked in the index to see if von Kármán was mentioned. (Actually, of course, I looked first to see if I was mentioned. I wasn't. Von Kármán had already taught me that one always looks up oneself first and that the greatest number of references is invariably to the author himself.) I found three mentions of von Kármán. One is a delightful story: Ulam saw von Kármán at a party and asked John von Neumann who that little old guy was. "What, you don't know Theodore von Kármán?" said von Neumann. "Why, he invented consulting!"

But Ulam didn't quit there. He went on to say that von Kármán was one of the earliest European airplane pilots and held one of the oldest pilot's licenses. (He never flew an airplane.) And then quoted von Kármán as defining an *engineer* as one who perpetuates the mistakes of his predecessors. Surely that is blasphemy!

I hope that by now you have had a glimpse of von Kármán's personality. Let me close by telling you something about his character. He was absolutely committed to personal and intellectual honesty and to the highest ethical standards in both the academic and business worlds. I was with him once when an ethical matter arose in connection with the consulting activities of one of my contemporaries. It involved reneging on a consulting commitment because the activity had led the consultant to some very promising results, which he now wished to keep for himself, and, after all, he hadn't yet been paid any fee for the work. Von Kármán was shocked, and let my friend know it in no uncertain terms. But he was not nearly as angry as he had been once at Caltech when a graduate student suggested that he should have some special favor "because we Central Europeans should stick together."

There was only one category in which he was not, I'm afraid, strictly reliable: the recommendations he wrote for us, his students, were much too kind. I think they must have originated in Hungarian, where *nem* (no) often comes out sounding like *igen* (yes). Duncan Rannie traveled with von Kármán through Western Europe in 1946. He says that wherever they stopped, von Kármán was given the keys to the city and offered a directorship in the local university. Duncan says, "As far as I could make out, he accepted them all."

I don't know whether I should describe von Kármán as modest. He came from a 19th century European tradition that valued scholarship, science and college professors. He surely knew that in those terms he was something special, but he didn't expect to receive special privileges as a result. Within the academic world, on the other hand, he believed in special privileges (and responsibilities) for professors, but that meant all professors and not special favors for himself.

I remember when he was visited in Pasadena by the son of one of his former students, who said: "I am going to study engineering, Professor. If I can be half as great an engineer as you, I will be happy." Von Kármán told me this, in private, and added, with a twinkle in his eye: "Now, Bill, what do you think? Fifty percent is a modest wish?"

Theodore von Kármán was too big, too complex, too many-sided to describe in 700 lines! I apologize for all that I've left out.

* * *

This article is adapted from an invited lecture delivered at the annual meeting of the APS division of fluid dynamics on 17 November 1984 at Brown University. □

Chapter 4
____Physicists Primarily as People _____

Physics is done by so many kinds of people, with such a wide variety of personal characteristics, that it may seem that all they have in common is their devotion to science and the drive to work at it. In the history of scientific development the personal aspects of the process are usually omitted or played down to emphasize that the thing discovered is independent of the discoverer and that the result can be checked. But, as Einstein has pointed out, scientific concepts are "created in the minds of men," and in some way the nonprofessional aspects of life and mind are inevitably related to the professional. In this chapter scientific achievements are by no means slighted, but the emphasis is on the human side.

Not one of the people featured in this chapter is still alive; this is hardly accidental, for people are not complete until they have died. Death provides the frame, so to speak, for any picture of life. Not that these articles supply full portraits: Fritz Stern's piece on Einstein and Germany, for example, is confined to a single subject, but it is a complicated and profound one. Other pieces are personal reminiscences, typically warm and appreciative. The brief remarks of Paul Ewald on several physicists he had known may seem insignificant at first glance, but it does not take careful reading to see they are not trivial. Taken together they also make evident Ewald's own character—few physicists

have been more respected and beloved than he.

Four articles on Landau complement each other in imparting some sense of one of the most amazing scientists of this century. His career, like Fermi's, was cut off in his early 50's, not by death in this instance but by an automobile accident that left him brain damaged. Already the diversity and abundance of his ideas seemed astonishing. As Ginzburg and others have pointed out, the physicist he most resembled was Richard Feynman. Most *Physics Today* articles on Feynman himself are being reserved for other publication, but we include one piece which most nearly reveals his personality. Like Mozart in music, these men possessed a kind of genius in physics that is beyond ordinary understanding!

Those featured in other retrospective pieces were all notable physicists, each in some special way or in a variety of different ways. We have other gifted theorists, talented and inventive experimentalists, and those whose most creative achievements were in physics education. For some (few compared to the number who merit the distinction) *Physics Today* has provided obituaries that convey in summary some sense of a life and work. All in all, the contents of this chapter, along with many of the articles in earlier chapters, give compelling evidence that scientists are never the inhuman machines often envisaged by non-scientists.

Einstein and Germany

The native German physicist, unlike many of his colleagues, had an early antipathy to German nationalism, so that for him, Hitlerism was a confirmation of an earlier intuition.

Fritz Stern

PHYSICS TODAY/FEBRUARY 1986

There was nothing simple about Albert Einstein, ever. His apparent simplicity concealed an impenetrable complexity. Even the links to his native Germany were prematurely ambiguous. At a time when most Germans thought their country a hospitable home, a perfect training ground for their talents, Einstein was repelled: In 1894, as a 15-year-old, he left Germany and became a Swiss citizen. Twenty years later, a few weeks before the outbreak of the Great War, he returned to Germany and remained for 18 years of troubled renown, years in which he appreciated what was congenial and opposed what was antipathetic in Germany. Long before Hitler's rise, he felt unease.

Einstein's fame, his capacity for homelessness, and the degradation of his country made him a citizen of the world, seemingly detached from Germany. But I believe that his early encounters with Germany and his consequent hostility to its official culture shaped his public stance. The German experience haunted Einstein to the very end, as it haunted so many of his generation later. It was the text of his political–moral education, the background against which he came to mold his unorthodox views and play his controversial public role.

In Einstein's time, Germany was the promise and later the nemesis of the world, a country that had decisive bearing on world politics and where, for a moment that seemed a lifetime, the moral drama of our era was enacted. At certain critical moments, Einstein's responses differed radically even from those of his closest colleagues. Documenting this diversity will complicate our understanding of Germany, and this is desirable because Germany's past has often been treated with didactic simplicity. Einstein and Germany: They illuminate each other.

A rebel from the start

Einstein grew up in southern Germany. We know little of his early life. He was no child prodigy; rather, his reticence in speaking for the first three years, his difficulty with learning foreign languages and his mistakes in computation have been a source of endless comfort to the similarly afflicted and their parents—though affinity in failure may not suffice for later success. He went through a brief but intense religious phase, the end of which, he said, left him suspicious of all authorities. His parents, secularized Jews, had little to do with his intellectual development; an uncle fed his mathematical curiosity. His father was an amiable failure, mildly inept at all the businesses he started. In 1894 his parents went to Italy to start yet another business, leaving the 15-year-old Albert behind in a well-known Munich *Gymnasium*. The authoritarian atmosphere and the mindless teaching appalled him. There was more than a hint of arrogance about the young Einstein, and hence it does not strain one's credulity that a teacher exclaimed, "Your mere presence spoils the respect of the class for me." He was a rebel from the start.

Encouraged by his teachers' hostility, he decided to quit school and leave Germany. His unsuccessful career facilitated his later fame in Germany: Erik Erikson has rightly referred to "the German habit of gilding school failure with the suspicion of hidden genius." It is often said that Einstein left school because he objected to its militarism. I find this unpersuasive: Bavarian militarism? I would suppose

Addressing a meeting. Einstein is at the podium delivering the principal address at a meeting sponsored by the Academic Assistance Council, a group formed to aid refugees. Seated at the right are Ernest Rutherford, Austin Chamberlain and the Bishop of Exeter. The October 1933 meeting was held in Royal Albert Hall, London.

that there might have been a stifling Catholicism; an insolent, thoughtless authoritarianism; a repulsive tone—any of which would have sufficed to discourage a youth like Einstein.

I suspect Einstein left Germany so precipitously to escape serving in the German army; by obtaining Swiss citizenship in time, he could do so without incurring the charge of desertion. His first adult decision, then, was to escape the clutches of compulsion—the image of Einstein as a recruit in a field-gray uniform does boggle the mind. He left Germany without regrets. His encounters with that country had not been happy.

There followed the obscure and difficult years in Switzerland, the failures, the marginal existence, the Zurich Polytechnic and, finally, the security of the patent office in Berne. From there in 1905 emerged the four papers destined to revolutionize modern physics and cosmology. They were published in the *Annalen der Physik*, and Max

Planck was the first man to recognize the genius of the unknown author. The international scientific community took note as well, and Einstein finally received his first academic appointments. In 1913, while he was a professor at the Zurich Polytechnic, two German scientists appeared, Walter Nernst and Planck, to offer him an unprecedented position: salaried membership in the Prussian Academy of Sciences and a professorship at the university, but without the obligation to teach. When Nernst and Planck left, Einstein turned to his assistant, Otto Stern, and said, "The two of them were like men looking for a rare postage stamp." The remark was perhaps an early instance of that self-depreciatory humor, that modesty of genius, that was to characterize Einstein.

Einstein began his new German life in April 1914. Berlin was the world's preeminent center of the natural sciences, and Planck, Fritz Haber and a dazzling array of talent rejoiced at

having this young genius at the head of their circle. Three months later the war shattered the idyllic community. Einstein had returned to Germany in time to see the country seized by the exaltation of August 1914, when almost all Germans were caught up in an orgy of nationalism, gripped by a joyful feeling that a common danger had at last united and ennobled the people.

The intoxication passed; the business of killing was too grim to sustain the

Fritz Stern is Seth Low Professor of History at Columbia University, where from 1980 to 1983 he served as provost. He is the author of *Gold and Iron: Bismarck, Bleichröder, and the Building of the German Empire* (Knopf, New York, 1977) and other books on European history. He is on the editorial committee of the Collected Papers of Einstein and is currently working on a book on Einstein and the German public.

unbridled enthusiasm of August 1914. The elite rallied to the nation, as it did elsewhere too. In the fall of 1914, 93 of Germany's best-known scientists and artists, including Planck, Haber and Richard Willstätter, signed a manifesto that was meant to repudiate Allied charges of German atrocities, but that by tone and perhaps unconscious intent argued Germany's complete innocence and blamed all misfortunes and wrongdoing on Germany's enemies. The manifesto of the 93 has often been seen as a warrant for aggression, a declaration of unrestrained chauvinism. I suspect it was as well the outcry of people to whom the outside world mattered and who intuitively sensed that the Allies would come to cast Germans as pariahs again. Some of the 93 probably hoped for continued respect across the trenches—and signed a document that had the opposite effect. It was not the last time that Germans confirmed the sentiments they set out to deny.

With but few exceptions, intellectuals everywhere joined in this chorus of hatred and in the cry for blood. So did the guardians of morality and the servants of God, the priests who sanctified the killing as an act of mythical purification. In time, some of the 93 turned moderate, or perhaps remained the patriots they had been, but others passed them on the right in the nation's wild leap to pan-German madness.

Politicized by the war

Einstein was alone and disbelieving. The war that was to politicize everyone as the cause of universal grief politicized him as well. Before 1914 Einstein had never concerned himself with politics; his very departure from Germany had been a youthful withdrawal from the claims of the state. Now, for the first time, he ventured forth from his study, convinced of the insanity of the war, shocked by the ease with which people had broken ties of international friendship and mutual respect. A pacifist asked him to sign a countermanifesto addressed to Europeans, demanding an immediate, just peace, a peace without annexations. It was the very first appeal he ever signed. It was published only in 1917 and then only

abroad, for want of sufficient signatures. Somewhat later he joined a tiny group of like-minded democrats and pacifists. In November 1915 the Berlin Goethebund asked for his opinion about the war, and he sent a message with this rather special ending: "But why many words when I can say everything in one sentence, and moreover in a sentence that is particularly fitting for me as a Jew: Honor your Master Jesus Christ not in words and hymns, but above all through your deeds."

Intermittently Einstein forsook his work—his central passion—to bear witness in an unpopular cause he took to be right. He had been a pacifist and a European of the first hour, never touched by the frenzy that ravaged nearly all. Convinced of Germany's special responsibility for the outbreak and the continuation of the war, he hoped for the nation's defeat.

To understand Einstein's isolation, one must look at the responses of his friends and colleagues to the war. Haber, for example, became the very antithesis of Einstein. Einstein's senior by 11 years, Haber was a chemist of genius, a born organizer and, in wartime, an ardent patriot. Without Haber's process for fixing nitrogen from the air, discovered just before the war, Germany would have run out of explosives and fertilizers in the first six months of the war. During the war he came to direct Germany's scientific effort; in 1915 he experimented with poison gas and supervised the introduction of the new weapon at the western front. To enable him to operate within a military machine that had no understanding of the need for a scientist, Haber received the effective rank of captain. He relished his new role; marshaling all one's talents and energies in a cause one believes in and under the shadow of danger is a heady experience. Einstein, the lonely pacifist who had come to feel his solidarity with Jews, and Haber, the restless organizer of wartime science and a convert from Judaism—the contrast is obvious. For all their antithetical responses, Haber and Einstein remained exceptionally close and, on Haber's side, loving friends. Haber's

life was a kind of foil to Einstein's, and it encompassed the triumphs and the tragedy of German Jewry. I shall return to him because his relations with Einstein were so important—and because he happened to have been my godfather and paternal friend of my parents.

Einstein had been horrified at the beginning of the war, but I doubt that even he could have imagined the full measure of disaster: the senseless killing and maiming of millions, the starving of children, the mortgaging of Europe's future, the rupture of a civilization that appeared ever more fragile. For what? Why? Einstein blamed it on an epidemic of madness and greed that had suddenly overwhelmed Europe—and Germany most especially. The old German dream of greatness had turned into a nightmare of blind and brutal greed. During the later phases of the war, Einstein was again totally absorbed in his work, but whiffs of hysteria would reach him—and always from the German side. I doubt that he knew of the excesses on the other side.

Einstein had been right about the war. At its end many felt as he had at the beginning. The war was a great radicalizing experience, pushing most people to the left and some to a new, frantic right. If there had been no war, Bolshevism and fascism would not have afflicted Europe. The war discredited the old order and the old rulers; antagonism to capitalism, imperialism and militarism appeared everywhere. Lenin's Bolsheviks offered themselves as the receivers of a bankrupt system; Bolshevism was a speculation in Europe's downfall. Liberal Europeans pinned their hopes on Woodrow Wilson, but those hopes faded in the vengeful spirit of Versailles. The logic of events had brought many Europeans to share Einstein's radical–liberal, faintly socialist, thoroughly internationalist views.

A public figure

For a short time Einstein had hopes for Germany. Defeat had brought the collapse of the old and the rise of a new, democratic regime, as he had expected. In November 1918, at the height of the

Fritz Haber and Einstein. The two men first met in 1911 and remained friends until Haber's death in 1934.

Über
Relativitätsprinzip, Äther, Gravitation

von

P. Lenard

in Heidelberg

Neue, vermehrte Ausgabe

Ladenpreis
für Deutschland und Deutsch-Österreich 5 Mk.;
für das Ausland nur nach der Valutaordnung
des deutschen Buchhandels

Verlag von S. Hirzel in Leipzig 1920

Anti-relativity pamphlet. This is the title page of Philipp Lenard's 1920 pamphlet "On the relativity principle, ether, gravitation," in which he argues that relativity theory is false. This eminent physics professor attacked Einstein not only as a publicity-seeking theorist but also as a Jew.

German Revolution, he cautioned radical students who had just deposed the university rector: "All true democrats must stand guard lest the old class tyranny of the right be replaced by a new class tyranny of the left." He warned[1] against force, which "breeds only bitterness, hatred and reaction," and he condemned the dictatorship of the proletariat in the first of his occasional bitter denunciations of the Soviet Union as the enemy of freedom. However, at other times and in different contexts, he would sign appeals of what we have come to call "front organizations."

We now come to a fateful coincidence in the rise of the public Einstein. In March 1919 a British expedition headed by Arthur Stanley Eddington had observed the solar eclipse. In November it was announced that the results confirmed the predictions of the general theory of relativity. In London the president of the Royal Society, Nobel laureate J. J. Thomson, hailed Einstein's work, now confirmed, as "one of the greatest—perhaps *the* greatest of achievements in the history of human thought." Almost overnight Einstein became a celebrated hero—the scientific genius, untainted by war, of dubious nationality, who had revolutionized man's conception of the universe, redefined the fundamentals of time and space, and done so in a fashion so recondite that only a handful of scientists could grasp the new, mysterious truth.

The new hero appeared, as if by divine design, at the very moment when the old heroes had been buried in the rubble of the war. Soldiers, monarchs, statesmen, priests, captains of industry—all had failed. The old superior class had been found inferior; *Disenchantment* was, appropriately, the title of one of the finest books written about the war. "Before 1914," Noel Annan has asserted,[2] "intellectuals counted for little." After the war, and in a sense in the wake of Einstein, they counted for more. Einstein now became a force, or at least a celebrity, in the world.

After 1919 Einstein appeared more and more often as a public figure. His views were continually solicited, and

Einstein and Zionism

During World War I, Einstein became a champion of Zionism, the effort to found a Jewish homeland in Palestine as a secular haven for the persecuted and as a means of moral regeneration. By the early 1920s he had become a public advocate of Zionism—to the surprise and likely dismay of many of his colleagues. Assimilated Jews must have found this reminder of Jewish apartness painful; internationalists must have boggled at the implied argument for a new national community. But Einstein had come to feel a sense of solidarity with other Jews, especially with Jewish victims of discrimination, and he appeared to believe in the existence of an ineradicable antagonism between gentiles and Jews, especially between Germans and Jews—with the fault by no means all on one side. Hence his view that Jews needed a spiritual home and a possible haven. He specifically cited the discrimination that talented Jews from Eastern Europe and from Germany suffered at German universities.

In 1921 Chaim Weizmann persuaded Einstein to join him on a trip to the United States to raise money for the projected Hebrew University in Jerusalem (see the photograph at right). For Weizmann, Einstein's support was critical; for Einstein, the visit to Jerusalem in 1923 was a deeply moving experience. Still, there were conflicts. Einstein railed against the mediocrity of the American head of the university; he saw him as a creature of the crass American–Jewish plutocrats for whom Einstein had contempt even as he helped to lighten their financial burden. He quarreled publicly with Weizmann over the policies of the Hebrew University and repeatedly threatened to withdraw his sponsorship. He urged a Jewish presence in Palestine that would promote, not injure, Arab interests. In 1929, at a time of major attacks on Jewish settlements, he again pleaded with Weizmann for Jewish–Arab cooperation and warned against a "nationalism à la prussienne," by which he meant a policy of toughness and a reliance on force:

If we do not find the path to honest cooperation and honest negotiations with the Arabs, then we have learned nothing from our 2000 years of suffering and we deserve the fate that will befall us. Above all, we should be careful not to rely too heavily on the English. For if we don't get to a real cooperation with the leading Arabs,

Chaim Weizmann (bearded) and Einstein at a reception at City Hall, New York, April 1921.

BROWN BROTHERS

then the English will drop us, if not officially, then *de facto*. And they will lament our debacle with traditional, pious glances toward heaven, with assurances of their innocence, and without lifting a finger for us. [Letter to Chaim Weizmann, 29 November 1929, in the Weizmann Archives, Yad Chaim Weizmann, Rehovot, Israel]

Weizmann replied instantly, at the height of the violence in Palestine, with a four-page handwritten letter. He expounded his views, which were somewhere between those of Zionist extremists and those of the irenic Einstein—who, in the meantime, had criticized the Jewish stance publicly. Weizmann pointed to the recalcitrance of the Arab leaders, their fanaticism, their inability to understand anything but firmness. He pleaded with Einstein to cease his injurious attacks on the Zionists. Of course they would negotiate in time, Weizmann insisted, but "we do not want to negotiate with the murderers at the open grave of the Hebron and Safed victims." Einstein remained skeptical. Weizmann, desperate to retain his support, had written to Felix Warburg a year earlier, "There is really no length to which I would not go to bring back to our work the wonderful and lovable personality—perhaps the greatest genius the Jews have produced in recent

centuries and withal so fine and noble a character."

At the time of the greatest need for a Jewish home in Palestine, immediately after Hitler's seizure of power, Einstein formally broke with the Hebrew University and with Weizmann. The correspondence between the two men suggests all the intractable issues about Jewish–Arab relations, all the differences between the safe outsider and the practical statesman. In April 1938 Einstein resigned his position on the Governing Body of the Hebrew University and again warned against a "narrow nationalism." Once again Weizmann explained that at the moment when five million Jews faced, as he put it, "a war of extermination," they needed the support of the intellectual elite of Jewry, and not, by implication, public criticism.

Einstein was not an easy ally. To some he must have appeared as a man of conscience and of unshakable principle; to others, as an uncompromising fanatic impervious to practical exigencies. As Robert Oppenheimer put it in his memorial lecture on Einstein (*New York Review of Books*, 17 March 1966, page 4): "He was almost wholly without sophistication and wholly without worldliness.... There was always with him a wonderful purity at once childlike and profoundly stubborn."

---

he obliged with his ideas about life, education, politics and culture. He had a special kinship with other dissenters from the Great War. Like Bertrand Russell, Romain Rolland and John Dewey, he became what the French call *un homme de bonne volonté*. His views—rational, progressive, liberal, in favor of international cooperation, condemnatory of the evils of militarism, nationalism, tyranny and exploitation—described as well a cast of mind

characteristic of the Weimar intelligentsia.

The intellectuals of Weimar—and this needs to be said at a time when Weimar is often portrayed as some sort of Paradise Lost—were a shallow lot in their moralizing politics. Their views often were utopian and simplistic, pious and fiercely polemical by turns. They were cynical because, as Herbert Marcuse once said to me about himself, they knew how beautiful the world

could be. They lived in a world peopled by George Grosz caricatures and three-penny indictments of bourgeois falsehood. It is perhaps too simple to say that they lived off the bankruptcy of the old order, but they did rather revel in the crudity of their opponents. It is not good for the mind to have dumb, discredited enemies. The real strength of Weimar lay in clusters of talent: in Heidelberg around Max and later Alfred Weber; at Göttingen in mathemat-

Einstein in a motorcade on the occasion of his arrival in New York City, 1921.

ics; the Bauhaus and the Berlin circles.

Einstein stood above these progressive intellectuals, in consonance with them, but usually more complicated and less predictable and always more independent than they. But he too was a theorist without a touch of practical experience. Einstein offered his prescriptions the more readily because he had been so overwhelmingly right when the multitudes had been wrong. By 1919 he had not only overthrown the scientific canons of centuries; he had also defied conventional wisdom and mass hysteria in wartime. His views were often deceptively simple; they were not so naive as has often been alleged nor quite so profound as admirers think. There is no reason to think that a scientific genius will have special insights into other realms. He had reflected on some issues and felt strongly about others; as for the rest, his views showed clearly that genius is divisible and can be compartmentalized.

Einstein's views and prescriptions were unassailably, conventionally well intended, but they often lacked a certain *gravitas*, a certain reality—in part, I think, because he approached the problems of the world distantly, unhistorically, not overly impressed by the nature or intractability of the obstacles to ideal solutions. He was not a political thinker; he was a philosopher, moralist, prophet, and the travails of the world would prompt him to propose or support social remedies. Sometimes those remedies would be blueprints of utopia addressed to people who had lost their footing in a swamp and were sinking fast.

Much later, in fact at a moment when Einstein had attacked the Nazi government, Max von Laue questioned whether the scientist should deal with political issues. Einstein rejected such considerations:

... you see especially in the circumstances of Germany where such self-restraint leads. It means leaving leadership to the blind and the irresponsible without resistance. Where would we be if Giordano Bruno, Spinoza, Voltaire and Humboldt had thought and acted this way?

Laue pointed out in a letter to Einstein that Einstein's examples were not exact natural scientists and that physics was so remote as not to prepare its practitioners for politics in the same way that law or history did. On that letter Einstein simply scribbled, "Don't answer."

Like so many thinkers of the 1920s Einstein underestimated the force of the irrational, of what the Germans call the demonic, in public affairs. That is what left them so ill prepared for an understanding of fascism. In their innocence they thought that men were bribed to be fascists, that fascism was but frightened capitalism; in its essence it was something much more sinister and elemental.

A democratic rebuke to authority

What gave Einstein's views exceptional resonance was the magic of his person and his incomparable achievement. He was taken by many as a sage and a saint. In fact, as I have said before, he was an unfathomably complex person. In the complexity of nature he found simplicity; in the complexity of his own nature, the principle of simplicity ranked high. Indeed, it was his simplicity, his otherworldliness, that impressed people. His clothes were simple, his tastes were simple, his appearance was meticulously simple. His modesty was celebrated—and genuine, as was his unselfishness. He was a lonely man, indifferent to honors, homeless by his own admission, solicitous of humanity and diffident about his relations with those closest to him. At times he appeared like a latter-day St. Francis of Assisi: a solitary saint, innocently sailing, those melancholy eyes gazing distractedly into the distance. At other times he played with the press, finding himself in the company of the famous and the powerful despite himself.

In some ways, I believe, he came to invest in his own fame, perhaps unconsciously to groom himself for his new public role. He lectured in distant lands, "a traveler in relativity." In 1921, after his first visit to the United States, he said:[3]

The cult of individuals is always, in my view, unjustified.... It strikes me as unfair, and even in bad taste, to select a few [individuals] for boundless admiration, attributing superhuman powers of mind and character to them. This has been my fate, and the contrast between the popular estimate of my powers and achievements and the reality is simply grotesque.

This admiration would have been unbearable except that "it is a welcome symptom in an age which is commonly denounced as materialistic that it makes heroes of men whose goals lie wholly in the intellectual and moral sphere.... My experience teaches me that this idealistic outlook is particularly prevalent in America." He knew that he had become a hero—and was endlessly surprised by it. In 1929 he described himself as a "saint of the Jews." He played many roles by turns, each, I think, completely genuinely; he was a simple man of complex roles.

In the simplicity and goodness that were his, I detect, perhaps wrongly, a distant echo of his encounters with German life. Could one imagine a greater contrast between the Germans surrounding him—people so formal in their bearing, so attentive to appearance, so solicitous of titles, honors and externals—and himself? Did the insolence of office, the arrogance of the uniform, push him into ever greater idiosyncratic informality? Was not his appearance a democratic rebuke to

League of Nations commission. Einstein is seated fourth from the right at this 1927 meeting of the International Commission on Intellectual Co-operation. Hendrik A. Lorentz is at the far right.

authority?

In the immediate postwar era, Einstein was friendly to the governments of Weimar and appalled by the vindictiveness of the Allies, who seemed to have caught what he had thought was a German disease. In all his public stands he had what Gerald Holton has called a "vulnerability to pity," and in the early 1920s he had a fleeting moment of pity for Germany. He refused to leave it in its time of trial. For years he was an uncertain member of the International Commission on Intellectual Co-operation of the League of Nations, intermittently resigning whenever he thought the commission too pro-French, too *Allied*. He hoped to restore an international community, Germans included. In the end he asked Haber to take his place. Successive German governments regarded him as a national asset, perhaps their sole asset in a morally and materially empty treasury. They saw in his travels and in his fame the promise of some reflected glory. But his own hopes gradually faded. He had warned Walter Rathenau against assuming the foreign ministry; Jews should not play so prominent a role, he felt. When right-wing assassins killed Rathenau and were widely hailed in Germany as true patriots, Einstein had reason to fear for his own life. The inborn servility of the Germans, he thought, had survived the successive shocks of 1918.

Immediately after the war and at the beginning of his popular fame, Einstein embraced several causes. Having embraced them, he would often embarrass and repudiate them as well. He was the antithesis of an organization man. Unstintingly he would help individuals and chosen causes, but I doubt that he

listened to them. He remained a detached theorist who thought the rational order of the world wantonly violated, but at times his commandments contained visionary practicality. A pacifist during the war, he now became Germany's most prominent champion of organized pacifism. He hated militarism—as blindly as its defenders loved it. He condemned[4] "the worst outgrowth of herd life, the military system. . . . I feel only contempt for those who take pleasure marching in rank and file to the strains of a band. . . . Heroism on command, senseless violence and all the loathsome nonsense that goes by the name of patriotism—how passionately I despise them!" This, surely, is exemplary of the spirit of the 1920s, formed by the experience of the first war and soaked in the we–they antithesis that precludes understanding. It precluded the understanding that had led William James to plead for a moral equivalent of war, for something practical that would make peaceful use of the old martial virtues. Einstein insisted that "the advance of modern science has made the delivery of mankind from the menace of war . . . a matter of life and death for civilization as we know it." But Einstein did not grapple with the psychological issues, with people's desire for danger and comradeship. In his exchange with Freud about the nature of war he acknowledged[5] that "the normal objective of my thought affords no insight into the dark places of human feeling and will." For Einstein war was a disease, a disorder planted by men of greed, to be abolished by men of good will through the creation of international sovereignty or through a revolutionary pacifism, that is, through the refusal of men to bear arms in peace or

war. He called for resistance to war, but in 1933, almost immediately after Hitler's assumption of power, he renounced pacifism altogether—to the fury of his doctrinaire followers. In fact he urged the Western powers to prepare themselves against another German onslaught.

Humane collegiality

Unlike many academics, Einstein took education with the utmost seriousness—and academics with magnificent irreverence. He had great faith in the possibilities of primary and secondary education; at one point he said that if the League of Nations improved primary education, it would have fulfilled its mission. His ironic contemplation of universities found expression in private letters. He once complimented[6] his close friend Max Wertheimer, the Gestalt psychologist: "I really believe there are very few who have been so little harmed by learning as yourself." In 1924 he wrote:[7] "In truth, the university is generally a machine of poor efficacy and still irreplaceable and not in any essential way improvable. Here the community must take the point of view that the biblical God took toward Sodom and Gomorrah. For the sake of very few, the great effort must be made—and it is worth it!"

Einstein's success—the enormous acclaim, especially abroad at a time when most German scientists were still banished from international meetings—caused much ill will at home. His opinions enraged the superpatriots. Some physicists condemned the fanfare surrounding the dubious theory of relativity; one fellow Nobel laureate, Philipp Lenard, attacked it as "a Jewish fraud." For anti-Semites, Einstein became a favorite and obvious target.

Meeting at Harnack House, Berlin, 1931. Left to right: Max Planck, British Prime Minister Ramsay MacDonald, Reichsminister Gottfried Treviranus (from behind), Einstein, Privy Counselor Hermann Schmitz (?) of I. G. Farben, Vice-Chancellor Hermann Dietrich and (partially obscured) Foreign Minister Julius Curtius (?).

The waves of hatred spilled from the streets into the lecture halls, and Einstein's occasional and sometimes ill-considered deprecations made things worse.

It would be hard to imagine three causes less pleasing to the bulk of the German professoriat than liberal internationalism, pacifism and Zoinism (see the box on page 285).

Germany frightened Einstein again. His hopes for the Weimar Republic had dimmed. As early as 1922 his life was threatened. He traveled even more than before, but still he refused handsome offers from Leiden and Zurich, the universities with which he had the closest ties. He stayed in Germany despite his misgivings; he stayed because Berlin in the 1920s was the golden center of physics; he stayed because proximity to Planck, Laue, Haber and others was a unique professional gift, because, as he wrote Laue in 1928, "I see at every occasion how fortunate I can call myself for having you and Planck as my colleagues." In 1934 he wrote Laue that "the small circle of men that earlier was bound together harmoniously was really unique and in its human decency something I scarcely encountered again." In 1947 he wrote[8] Planck's widow that his time with Planck "will remain among the happiest memories for the rest of my life."

The unpublished correspondence among these men suggests even more than a professional tie. The letters bespeak a degree of humane collegiality, a shared pleasure in work, as well as a delicacy of sentiment, a candid avowal of affection, that in turn allowed for confessions of anguish and self-doubt, of melancholy as well as high spirits. They spoke of joys and torment, in close or distant friendship, in an enviable style. The letters also breathe a kind of innocence, as if science was their insulated realm, nature the great, enticing mystery and one's labors of understanding exclusively an intellectual pursuit, remote from social consequences. Such clusters of collaboration and of friendship have always existed, I suppose, and they have made life better and infinitely richer. Germany may have had a special knack for breeding them.

Einstein's Germany included gentiles and Jews working together in extraordinary harmony. Still, one can state categorically that none of the Jewish scientists escaped the ambiguity, the intermittent hostility, that being a Jew entailed in imperial and Weimar Germany. Neither fame nor achievement, neither the Nobel Prize nor baptism, offered immunity. Passions were fiercer in Weimar, that cauldron of resentments. Official barriers against Jews had been lowered, but new fears and hatreds came to supplement old prejudices.

Three incidents may illustrate the uncertain temper of the time. In 1921 Haber begged Einstein not to go to America with Chaim Weizmann on the ground that Germans would take amiss Einstein's traveling in Allied countries with Allied nationals at the very time when the Allies were once again tightening the screws on Germany. To persuade Einstein, Haber warned that German anti-Semites would capitalize on Einstein's seeming desertion and that innocent Jewish students would be made to suffer; anti-Semitism, rampant as it was, did not need to be goaded. Einstein's warning to Rathenau originated in a similar apprehension. Or take another incident. In 1920 a well-known physicist opposed the university appointment of the later Nobel laureate Otto Stern: "I have high regard for Stern, but he has such a corrosive Jewish intellect."

Or consider this last example. In 1915 the king of Bavaria, confirming the Nobel laureate Willstätter's appointment to a professorship at Munich, admonished his minister, "This is the last time I will let you have a Jew." Ten years later, discussing with his colleagues a new academic appointment, Willstätter proposed a candidate. A murmur arose: "another Jew." Willstätter walked out, resigned his post and never entered the university again, the unanimous pleas of his students notwithstanding. For the next 14 years he had daily, hour-long telephone calls with his assistant so that she could conduct the experiments in a laboratory that he would no longer enter. A man of conscience and of courage, someone who did not blink at the reality of anti-Semitism. But his stand in 1924 was his undoing a decade later. A devoted German, but no longer a civil servant, he assumed that the Nazis would leave untouched a private scholar. He believed that some Jews had contributed to this new storm. He could not comprehend the radical newness of the phenomenon. In February 1938 he wrote my mother urging her not to leave Germany without the most careful reflection. He himself refused exile until the aftermath of the *Kristallnacht* forced him into it.

I cite Willstätter's example in particular precisely because of its contradictory nature: Awareness of anti-Semitism could cloud one's perception of Nazism. If anti-Semitism had always existed, then perhaps Nazism was but an intensification of it. It is not uncommon these days to hear summary judgments about German Jewry, about their putative self-surrender, their cravenness or their opportunism. These judgments often have a polemical edge and they are likely to do violence to the past and to the future: The myth of yesterday's self-surrender could feed the delusion of tomorrow's intransigence. If our aim is to understand a past culture, we must note that German–Jewish scientists thought Germany their only and their best home, despite the anti-Semitism that crawled all around them. They may have loved not wisely but too well, and yet their sentiments are perhaps not so much an indictment of themselves as a tribute to the appeals of Germany. We owe that past no less than what we owe any past: a sense of its integrity.

The denouement

Let me hasten to the denouement. In 1932 Einstein left Germany provisionally, with the intention of returning to Berlin for one semester each year. Hitler's accession to power the next year changed all that. Einstein immediately denounced the new regime, and in response the Prussian Academy expelled him. His books were burned, his property seized. The first Nazi decrees on the purification of the universities would have allowed some Jews to maintain their positions. Einstein's non-Aryan friends spurned such sufferance and resigned. German physics was decimated, and a few remaining masters battled to defend some shreds of decency, some measure of autonomy. Laue once wrote Einstein that in teaching the theory of relativity he had sarcastically added that it had of course been translated from the

Hebrew. Even such jokes—to say nothing of Laue's eulogies of Jewish colleagues—aroused Nazi wrath. The Nazis proscribed the very mention of Einstein, even in scientific discussions. They wished him to be a nonperson.

For most, exile was hard; the habits of a lifetime are not easily shaken. For others, as the physicist Max Born put it,[9] "a disaster turned out to be a blessing. For there is nothing more wholesome and refreshing for a man than to be uprooted and replanted in completely different surroundings." Resiliency was a function of age and temperament. For Haber exile was a crushing blow and led to a final irony in his relations with Einstein. In mid-1933 he wrote to Einstein that as soon as his health would allow it, he would go to Palestine, but in the meantime he begged Einstein to patch up the public quarrel Einstein had had with Chaim Weizmann. Einstein replied[10] at length: "pleased...that your former love for the blond beast has cooled off a bit. Who would have thought that my dear Haber would appear before me as defender of the Jewish, yes even the Palestinian, cause. The old fox [Weizmann] did not pick a bad defender." He then lashed out against Weizmann and concluded:

I hope you won't return to Germany. It's no bargain to work for an intellectual group that consists of men who lie on their bellies in front of common criminals and even sympathize to a degree with these criminals. They could not disappoint me, for I never had any respect or sympathy for them—aside from a few fine personalities (Planck 60% noble, and Laue 100%). I want nothing so much for you as a truly humane atmosphere in which you could regain your happy spirits (France or England). For me the most beautiful thing is to be in contact with a few fine Jews—a few millennia of civilized past do mean something after all.

The German patriot Haber died a few months later in Basel, en route to Palestine. And Einstein found a refuge at the Institute for Advanced Study at Princeton under conditions not dissimilar from what the Prussian Academy had offered him 20 years earlier. For as Erwin Panofsky has said[11] of the institute, it "owes its reputation to the fact that its members do their research work openly and their teaching surreptitiously, whereas the opposite is true of so many other institutions of learning."

Einstein's public life continued to be

Henry A. Wallace, Einstein, journalist Frank Kingdom and singer Paul Robeson during the 1948 Presidential campaign. Wallace was the Progressive party candidate.

BETTMANN ARCHIVE

Rudolf W. Ladenburg and Einstein, on the occasion of Ladenburg's retirement from Princeton University.

Y. Elkana, eds., Princeton U.P., Princeton, New Jersey (1982).

I benefited from conversations with Marshall Clagett, Felix Gilbert, Gerald Holton, Martin Klein, I. I. Rabi and Malvin Ruderman. It was in long and frequent talks with Otto Stern that I first sensed how extraordinary those early days in Zurich must have been.

References

● In writing this essay I found *Einstein on Peace* and *Ideas and Opinions*, references 1 and 3 below, particularly pertinent. I was also fortunate enough to be allowed to use the Albert Einstein Archives when they were still at the Institute for Advanced Study in Princeton, New Jersey, a treasure made still more valuable by the always helpful advice and recollections of Helen Dukas, who was in charge of them. I also read the unpublished correspondence of James Franck and Albert Einstein, deposited at the University of Chicago Library. In addition to the books cited in the preceding notes, I found the following particularly useful: A. D. Beyerchen, *Scientists under Hitler: Politics and the Physics Community in the Third Reich*, Yale U.P., New Haven (1981); G. Holton, *The Scientific Imagination: Case Studies*, Cambridge U.P., New York (1978); R. Willstätter, *Aus Meinem Leben*, Verlag Chemie, Weinhein (1949); and H. Zuckerman, *Scientific Elites: Nobel Laureates in the United States*, Free Press, New York (1977).

1. O. Nathan, H. Norden, eds., *Einstein on Peace*, Schocken, New York (1960), p. 25.

2. N. Annan, Daedalus, Fall 1978, p. 83.

3. A. Einstein, *Ideas and Opinions*, new trans. and rev. by S. Bargmann, Crown, New York (1954), p. 4.

4. O. Nathan, H. Norden, eds., *Einstein on Peace*, Schocken, New York (1960), p. 111.

5. A. Einstein, S. Freud, *Why War?*, International Institute of Intellectual Cooperation, League of Nations, Paris (1933), p. 12.

6. A. Einstein, letter to Max Wertheimer in the Einstein Archives, Boston Univ., Boston, Mass.

7. A. Einstein, letter to Julius Schwalbe, 18 July 1924, in the Einstein Archives.

8. Letters in the Einstein Archives.

9. M. Born, *My Life and My Views*, Scribner's, New York (1968), p. 38.

10. Letter in the Einstein Archives.

11. E. Panofsky, *Meaning in the Visual Arts*, U. of Chicago P., Chicago (1983), p. 322.

12. Copy in the Einstein Archives.   □

dominated by his fear of Germany. He warned the West against a new German onslaught. He abandoned the pacifism he had so fervently espoused and in 1939 signed the famous letter to President Franklin D. Roosevelt urging the Administration to prepare the United States because Germany might develop nuclear fission for military purposes. In the winter of 1945, when Germany was desolate in defeat and when the Morgenthau spirit, if not the plan, had a considerable grip on American thinking, a fellow laureate and old friend, James Franck, asked Einstein to sign a manifesto of exiles that would appeal to the United States not to starve the German people. Einstein vowed that he would publicly attack such a plea. Franck pleaded with him that to give up all hope for a moral position in politics would be tantamount to a Nazi victory after all. But Einstein, who had signed so many appeals that he himself once said he was not a hero in no-saying, scathingly rejected Franck's plea. For him, genocide was Germany at its most demonic; after Auschwitz he could muster no magnanimity. Even the righteous could not redeem the "country of mass murderers," as he called Germany. He rebuffed Laue's plea to help a young German physicist. He knew that Planck, who lost one son in the first war, had now lost another, whom the Nazis murdered because of his participation in the plot against Hitler. The serene Einstein, always the champion of the rights of the individual against the collectivity, now proclaimed the principle of collective guilt. At that moment, of course, the world shared Einstein's horror at German inhumanity. But in him the violence of senti-

ment, the total absence of that vulnerability to pity, puzzles, for it shows how desperately deep and all-consuming his antipathy to Germany had become.

Even Einstein's postwar laments about America, his horror at McCarthyism, were shaped by his image of Germany. America, he believed, was somehow following the path of Germany. The world of politics he saw through German eyes—always.

But his deepest feelings also retained something of a German cast, echoed some very German themes. When Rudolf Ladenburg, a physicist and fellow exile, died in 1952, Einstein spoke[12] at the graveside:

Brief is this existence, as a fleeting visit in a strange house. The path to be pursued is poorly lit by a flickering consciousness, the center of which is the limiting and separating "I."

The limitation to the I is for the likes of our nature unthinkable considering both our naked existence and our deeper feeling for life. The I leads us to the Thou and to the We—a step which alone makes us what we are. And yet the bridge which leads from the I to the Thou is subtle and uncertain, as is life's entire adventure.

When a group of individuals becomes a We, a harmonious whole, then the highest is reached that humans as creatures can reach.

* * *

This essay is a revised version of a lecture I gave at the Einstein Centennial Symposium in Jerusalem in 1979. Papers from the symposium appear in Albert Einstein, Historical and Cultural Perspectives, *G. Holton,*

Born and educated in Italy, Enrico Fermi came to this country in 1938 instead of returning to Italy from Stockholm, where he had just received the Nobel Prize. Shortly before his death he was named by the AEC to receive a special $25,000 award for his outstanding contributions to the atomic energy program.

ENRICO FERMI
1901 - 1954

The untimely death of Enrico Fermi on November 28th, 1954, deprived the world of one of its most brilliant and productive physicists. The following remarks by three of Fermi's friends and colleagues were made on the occasion of a memorial service held on December 3rd in the University of Chicago's Rockefeller Memorial Chapel. Samuel K. Allison, professor of physics and director of the University's Institute for Nuclear Studies, presided at the ceremony.

PHYSICS TODAY/JANUARY 1955

A Tribute to Enrico Fermi by S. K. Allison

WE are here to honor the memory of Enrico Fermi, Charles H. Swift Distinguished Service Professor of Physics at this University during the last decade. I shall try to express the sentiments of his associates here in the Institute for Nuclear Studies. Actually, the Institute is *his* Institute, for he was its outstanding source of intellectual stimulation. It was Enrico who attended every seminar and with incredible brilliance critically assayed every new idea or discovery. It was Enrico who arrived first in the morning and left last at night, filling each day with his outpouring of mental and physical energy. It was Enrico's presence and calm judgment, and the enormous respect we had for him, which made it impossible to magnify, or even mention, any small differences among us, such as can arise in any closely associated group. It was at Enrico's personal and urgent request that I took on the chore of directing the Institute in its routine affairs.

It is a completely objective statement, not at all prompted by the emotion of this occasion, to remark that every one who had more than a trivial acquaintance with Mr. Fermi recognized at once that here was a man who possessed a most extraordinary endowment of the highest human capabilities. We may have seen his physical energy before, or his basic balance, simplicity, and sincerity in life before, or even possibly his mental brilliance, but who in his lifetime has

ever seen such qualities combined in one individual?

In my attempts to understand him, with his completely successful adjustment to the life of today, and his leadership in it, I conclude that one reason such men are so rare is that it is so improbable that such a combination should be formed.

I would like to recount one incident showing Enrico in action. During the war, Professor A. H. Compton, Enrico Fermi, and I were travelling together to visit the Hanford Plutonium Plant in the State of Washington. Mr. Compton and Mr. Fermi were so valuable that they were not allowed to travel by air; I was expendible, and could have flown, but was on the train with them for company. The hours seemed to drag crossing the mountains, and Enrico, who always disliked travelling, was restless and bored. After some long silences, Mr. Compton said:

"Enrico, when I was in the Andes mountains on my cosmic-ray trips, I noticed that at very high altitudes my watch didn't keep good time. I thought about this considerably and finally came to an explanation which satisfied me. Let's hear you discourse on this subject."

Enrico's eyes flashed. A problem! A challenge! Something to work on! Having been in several such situations before, I relaxed and prepared to enjoy the fireworks that would surely follow. He found a scrap of paper and took from his pocket the small slide rule he always carried. During the next five minutes he wrote down the mathematical equations for the entrainment of air in the balance wheel of the watch, the effect on the period of the wheel, and the change in this effect at the low pressures of high altitudes. He came out with a figure which checked accurately with Mr. Compton's memory of the deficiencies of his timepiece in the Andes. Mr. Compton acknowledged the correctness of the calculation, and I shall not forget the expression of wonder on his face.

It is with such a man that we in the Institute could consult daily, and it is such a man that we have lost.

Let us pause a moment and ask ourselves why a man of this calibre abandoned a comfortable professorship and great honors in his own country to join us and become a citizen of the United States. Many other intellectuals of the highest type have done likewise. There is really only one reason, namely, that the limitations placed on the range and freedom of activity of the mind had become intolerable in the countries which they left. They could not tolerate politicians proclaiming and acting upon pseudo-racial doctrines that could not for a moment stand the light of rational analysis. They could not tolerate a climate in which responsible

and vigorous criticism of political actions was rewarded with defamation of character and possibly with imprisonment and death. Let us be sure that our freedoms here in these respects remain unimpaired. As long as men like Enrico Fermi turn to us and join us, though hosts be against us, we shall prevail.

THE speakers on our program have been chosen because of long and intimate association with Mr. Fermi. Professor Emilio Segrè, who will speak first, was the recipient of the first Doctor of Philosophy degree awarded under Fermi's sponsorship at the University of Rome. He was one of the group who associated there with Mr. Fermi in his classic researches on the properties of slow neutrons. Professor Segrè comes from the Department of Physics of the University of California for this occasion.

The final speaker will be Professor H. L. Anderson, who is a member of the staff of our Institute for Nuclear Studies. He was a student of Mr. Fermi's at Columbia University, and worked continuously and closely with him during the great effort of the war years, which led to the controlled release of nuclear energy from uranium, twelve years ago yesterday, here on our campus.

A Tribute to Enrico Fermi by Emilio Segrè

WE are here to commemorate and honor one of the greatest scientists of the century and it is appropriate that the highlights of his achievements, some of which seem likely to become of transcendental importance for mankind, be properly mentioned. But the choice of speakers, not from his peers in science, but from pupils and friends, seems to me to indicate a desire to have his human traits remembered also.

However, for Enrico Fermi physics was almost synonymous with life, and the man and the scientist are one. Any effort to separate them would be futile and irreverent.

He was born in Rome on September 29, 1901, and hence his much too brief life spanned only 53 years. He studied at Rome, and at Pisa at the Scuola Normale, an institution stemming from Napoleonic times which gave many illustrious scientists to Italy.

He obtained his Doctor's degree in 1922 with a thesis on X-rays. However, he was essentially self-taught, or better, his real spiritual teachers were a strange assortment of books ranging from a natural philosophy of the Jesuit Father Caraffa, written in 1840, the *Mécanique Rationelle* of Poisson, to Kelvin and Tait, Richardson's *Theory of Electrons,* and, above all, Sommerfeld's *Atombau* for the more modern subjects. These he read between the end of childhood and the end of adolescence.

His first published works are concerned with relativity, mechanics, and electrodynamics. We see him trying his forces on several interesting subjects, but soon he moves to deep reflections on thermodynamics and statistical mechanics. Thus, he was all prepared to

discover in 1926, immediately after the formulation of the exclusion principle by Pauli, the statistical laws followed by the antisymmetrical particles now called fermions.

This work brought him at once to a pre-eminent place among theoretical physicists, and it was promptly followed by numerous other studies in atomic physics. In all of his work of the time we find his personal scientific style already fully developed. Really brilliant ideas are developed with such apparent simplicity of theoretical means that the results seem to flow without effort. The theory of the Raman effect, of the hyperfine structure, of the intensity of the alkali doublets, of the pressure shift of spectral lines, of the latitude effect in cosmic rays, the concept of the virtual quanta accompanying a moving charge, the statistical atom and many more, bear testimony to the universality of his interests and to the power of his genius. He initiated many a line of thought which was to be pursued by a whole generation—and the mine is not yet fully exploited.

I first met Fermi at this time and I remember the experience shared later by others of beginning a conversation with him, which ended by his taking a piece of chalk and improvising on a blackboard a theory that needed only to be written up and published. The last time that I saw this was, alas, on February 11th of this year when I was telling him about some nuclear experiments in which I was involved.

In 1927 the school of Rome was also founded and I beg to be excused if I am too personal in my remem-

Laura and Enrico Fermi in the latter's study at the Institute for Nuclear Studies at the University of Chicago. A warmly human panorama of Fermi's life is to be found in *Atoms in the Family*, a biography written by Mrs. Fermi.

brances. Fermi's exceptional ability had been recognized, not without some struggle, by a professorship at Rome, a coveted position and quite exceptional for a man of only 26 years. However, he decided then that he needed some help, and some co-workers, and in very characteristic fashion proceeded to create them. He selected a small nucleus of young men, by his own criteria, and trained them in his own unorthodox way. I do not think he ever spoke of scientific ideals or that he used any moralizing words, but by force of example inspired in everybody such a burning devotion to science that I venture to say that for this group of young men between ages 20 and 25, with a leader of 27 or 28, science was the greatest passion, none excluded. And the Fermi influence of their scientific outlook was indelibly impressed and persisted even after they lost Fermi's mannerisms of speech and deep voice which they had unwittingly acquired in their daily common life.

In the early 30's more theoretical work followed. A reformulation of Dirac's theory of radiation led Fermi in the abstract paths of second quantization from which his rather practical mind at first recoiled. But his feeling changed after he had developed in 1933 what he considered a "practical" application, namely the theory of beta decay, one of the milestones of theoretical nuclear physics. With this he began his career as a nuclear physicist.

However, 1934 was to be the wonder year in which, without abandoning theory, he entered professionally into the experimental field. Indeed he had always, even from his childhood, dabbled a little in experiment and some of his work with Rasetti is quite first class, but the plan he had nurtured for some time of attacking experimentally some important problem concerning the nucleus materialized when news came of Curie-Joliot's discovery of artificial radioactivity. Fermi realized at once that neutrons would be more powerful projectiles than charged particles and tried them immediately; it is characteristic of the man that he tried in order all available elements beginning with hydrogen, and did not give up when the first eight were unsuccessful. The ninth, fluorine, finally gave a positive result. It was also characteristic that he summoned his young pupils and friends, mostly busy with their own problems, to come, help, work hard, and share the conquests with him.

A series of startling discoveries followed. The letters to the *Ricerca Scientifica,* sent to many nuclear scientists as what we would call today "preprints", elicited great interest, and Rome became, for a short period, the capital of the nuclear world. Lord Rutherford in person congratulated the young experimentalist for his debut. If I remember correctly, "Not bad for a beginner", were his own words in a congratulatory letter that he wrote at the time to Fermi.

In rapid succession all the elements, including uranium, were bombarded, but God, for his own inscrutable ends, made everybody blind to the phenomenon of fission. Chance confronted us with the strange phenomena undergone by neutrons in passing through hydrogenous substances; Fermi's mind grasped what was going on in a couple of hours. Thus, slow neutrons were discovered and these first steps, by a logical development, led him to study the diffusion of neutrons.

It was at this time more than at any other that I saw the full application of one of Fermi's outstanding human, or I would almost be tempted to say superhuman, characteristics, namely his unbelievable physical and mental strength. We were working quite methodically from 8 in the morning to 1, followed by lunch, siesta, and then again from 3 to 8 in the evening; but the intensity of the work was such that this practically represented the limit of our forces—and we were not weaklings.

However, every morning at 8, Fermi came back with some piece of theory concerning the neutron, ready to test it experimentally and to change it according to the results of the work of the day. This performance puzzled us a little, even knowing with whom we were dealing, but we soon discovered that the miracle occurred between 4 A.M. and 8 A.M. because he had insomnia and had decided to lengthen his day's work. I wanted to mention this because this strength and indomitable vitality was one of his fundamental characteristics.

By this time, intolerant persecution was rampant in Germany. We had had as visitors, guests, and friends, many brilliant young colleagues from central Europe, attracted to Rome by Fermi. Bethe, Bloch, Placzek, Peierls, London, and several others stayed with us for a few months, an ominous warning of impending catastrophe, and when in 1938, Italy also was submerged, Fermi departed as had many others, for the New World.

The aging Sommerfeld from Germany commented in a moving letter, "Sic transit gloria mundi veteris" (Thus passes the glory of the old world), and added, "To the greater glory of the New World."

In 1939, the Power which had initiated this tremendous chain of events opened the eyes of man to fission and Fermi, who had just arrived at Columbia, started a new group of young people, and, using his mastership of the neutron, embarked on that trip which was to land him, 12 years ago almost to the day, in that new world so properly indicated in the historic message announcing the criticality of the pile. One of his companions in this trip will tell you about it.

I cannot terminate this brief tribute without mentioning some things which I find I have omitted, because in writing with Fermi in mind they just did not occur to me: he had had all the honors that a scientist can have, none excluded. He was part of great councils, and for a large group of scientists his word was final. I have not mentioned these facts because for him they were really unimportant. Nothing altered his simplicity, which did not arise from false modesty—indeed he knew quite well how much he was intellectually above other men—but from charity. Nothing altered his unceasing interest in Science and his will to work humbly and indefatigably on the study of nature. If he had foreseen the cruel destiny that was to deprive us of him so unexpectedly early, he could not have husbanded his time to give more than he gave.

A Tribute to Enrico Fermi by H. L. Anderson

TODAY we are gathered here to pay our respects and to honor Enrico Fermi. He was our friend, our colleague, and our teacher, and he was a great man.

When he came to America 16 years ago, Enrico Fermi was already a celebrated scientist. He had a long list of scientific achievements. He had discovered new and fundamental laws of nature. The Fermi-Dirac statistics, his theory of the beta rays and his statistical atomic model stand out among many other great accomplishments. For his mastery of the neutron he had been awarded the Nobel prize and he stood at the pinnacle of his profession.

But now the Fascist mold had begun to infect the free and fruitful development of science in Italy. Fermi, with characteristic courage and decision, turned his back on his native land and set out to America to establish the American branch of his family. In America, Fermi could expect to find fertile ground for his ideas and a receptive climate for his genius. For the sanctuary which we gave him then, Fermi repaid us a thousandfold. We can be forever grateful that, when he came to us, our gates were open.

His needs were few. Chalk, a blackboard, and an eager student or two were enough for a start. Teaching was an essential part of his method. Through teaching he would sharpen his wits, clarify his thoughts, develop his ideas. Students and colleagues soon learned that no one could touch him when it came to clarity and brilliance of lectures. It was usually "Standing Room Only" when Fermi spoke—but he would lecture with equal brilliance to a lone student. And he would make a deal —if you would correct his English and teach him Americanisms—he would teach you physics.

The eternal scholar, Fermi was always eager to learn. He was always grateful when he found out something new. What he learned he felt he should enrich. Having enriched what he learned he felt he should teach it to others. Thus, he prepared the fertile ground out of which arose the new solutions and new ideas which kept his subject bright, fresh, and exciting.

At Columbia he had hardly settled his family when news of the discovery of the fission of uranium arrived. "Let me explain this business of the fission of uranium," he said. "The neutron enters and causes an instability in the uranium nucleus and it's split apart. A great deal of energy is released, as Otto Frisch has shown. But the circumstances are those in which, in all probability, neutrons will be emitted as well, and this is at the root of the matter. For if the neutrons are emitted in greater number than they are absorbed, a chain reaction will be possible and the way to a new source of energy will have been found. Come and help me find these neutrons. Let us measure their absorption and their emission with some care so that we can understand these processes in detail and know how to proceed."

To explore the mysteries of nature with Enrico Fermi was always a great adventure and a thrilling experience. He had a sure way of starting off in the right direction, of setting aside the irrelevancies, of seizing all the essentials and proceeding to the core of the matter. The whole process of wresting from nature her secrets was for Fermi an exciting sport which he entered into with supreme confidence and great zest.

No task was too menial if it sped him towards his goal. He thoroughly enjoyed the whole of the enterprise: the piling of the graphite bricks, the running with the short-lived activated rhodium foils, and the merry clicking of the Geiger counter which effected the measurement. All was done with great energy and obvious pleasure, but by the end of the day, in accordance with his plan, the results were neatly compiled, their significance assessed, and the progress measured, so that early in the morning on the following day, the next step could begin.

It was a feature of the Fermi approach never to waste time, to keep things as simple as possible, never to construct more elaborately or to measure with more care than was required by the task at hand. In such matters his judgment was unerring. In this way, step by step, the work sped forward until in less than four short years Fermi had reached his goal. A huge pile of graphite and uranium had arisen in the West Stands of the University of Chicago Campus. When, on December 2, 1942, 12 years ago just yesterday, Enrico Fermi stood before that silent monster he was its acknowledged master. Whatever he commanded it obeyed. When he called for it to come alive and pour forth its neutrons it responded with remarkable alacrity; and when at his command it quieted down again, it had be-

come clear to all who watched that Fermi had indeed unlocked the door to the Atomic Age.

By now the Manhattan Project had grown enormously in size. The exigencies of war required it to produce atomic bombs and it assumed a character not at all in keeping with the Fermi style. Administrative and organizational responsibilities usually charged with controversy he avoided. But there were those who came to recognize that in scientific and technical matters the words of Fermi were golden. Such advice he gave generously and freely and so in unobtrusive ways he helped guide the whole enterprise to its successful conclusion.

There followed a period at Los Alamos where Fermi had been asked to go to participate more directly in the atomic bomb work. Here the work had been already well advanced by a group of distinguished scientists among whom were many former colleagues. They too, like himself, had found sanctuary in America from the oppressions of Europe. Here, in a remarkable cooperative effort, the atomic bomb was designed, made, and tested.

One unerasable picture of Fermi had to do with the fateful morning set aside to test the first atomic bomb. It showed Fermi standing in the blinding glare of that explosion, methodically dropping small bits of paper to the ground. Some of these were carried forward by the arrival of the blast. Impatient to know the strength of the atomic explosion, Fermi had devised his own simple means for measuring it.

What Fermi missed at Los Alamos was the University. This he regained at Chicago, whose faculty he joined at the end of the war. Here, in the Institute for Nuclear Studies, established essentially according to his own design, he was again free to explore nature according to his fancy. Students flocked to his classes while physics itself, flushed with its success on the field of battle, surged on in new directions.

New particles had been discovered and huge electronuclear machines had been constructed to produce them. Here, at Chicago, under his guidance, we built a huge synchrocyclotron. This became Enrico's newest plaything. This machine could produce the mesons which had come to occupy the center of the stage. These were the particles which were responsible for the nuclear force. A new great challenge for Enrico. What were the facts? What was their meaning?

Counters, liquid hydrogen, magnets, all the paraphernalia of modern physics were brought to bear with Enrico in the thick of it. It was in the midst of this work that from an unexpected quarter, Enrico Fermi was suddenly and unaccountably struck down.

We all know what a pleasure it was to have Enrico around; what a privilege it was to work with him. We all know how considerate and thoughtful he was; how helpful he could be. He was the center of our Institute around whom all revolved and for whom we all tried to do something good enough to win his praise. Well, he isn't going to be around any more. We're going to miss him awfully but we can all try to keep the spirit that he had.

WOLFGANG PAULI 1900–1958

PHYSICS TODAY/JULY 1959

in commemoration of

Wolfgang Pauli

*Tributes by **Marcus E. Fierz** and **Victor F. Weisskopf***

A memorial service for W. Pauli, whose death on 15 December 1958 saddened the world of physics, was held at Fraumünster in Zürich, Switzerland, on Saturday, December 20, 1958. Among those who came to pay their last respects were Prof. Fierz, whose address is printed in a translation below, and Prof. Weisskopf, whose words of tribute follow. A former assistant to Pauli (1936–40), Marcus Fierz was professor of theoretical physics at the University of Basle, and editor of Helvetica Physica Acta. Victor Weisskopf, Pauli's collaborator in Zürich from 1933 to 1936, is professor of physics at the Massachusetts Institute of Technology. He flew to Switzerland to attend the memorial service as a representative of the National Academy of Sciences.

A photograph from the files of the Brookhaven National Laboratory, where Pauli was a visitor in 1958. The snapshot, last page, was taken in 1949 during a gathering of physicists in Holland.

WOLFGANG Pauli's death came unexpectedly. Indeed, we knew that something distressed and tormented him for some time, but what it was we did not know. Now we know that with a severe and latent physical affliction the shadow of death encompassed him. In Pauli we not only lost a great theoretical physicist, but also a most impressive and remarkable man. For all of us the world grew poorer.

Wolfgang Pauli was born in Vienna on the 25th of April, 1900. His father was a professor of physical chemistry there. Pauli studied physics in Munich under Arnold Sommerfeld. In his Nobel Prize acceptance speech, Pauli spoke of his great teacher as having been guided by a feeling of internal harmony, as was Johannes Kepler, in his search to explain the laws of spectra as a relationship between integers. In comparing the thoughts of his teacher with those of Kepler, he acknowledged himself not only as a pupil of a physicist, but also of a natural philosopher. As a student of Sommerfeld, hardly twenty years of age, he wrote the well-known encyclopedia article on the theory of relativity where, in 200 pages, a vast and widespread field is lucidly and beautifully presented. Even today this work has lost neither its merit nor its power to convince. Here Pauli showed for the first time his strength in critical and concise formulation which was to prove itself again in his two *Handbuch* articles on quantum theory.

He wrote the second of these articles in 1932 while in Zürich, having been called there from Hamburg in 1928. The general principles of wave mechanics which had emerged after a stormy development are presented in that article. No other similar presentation has revealed the physical meaning of this theory manifesting itself in its mathematical structure. This classical work, there-

fore, was reprinted almost unchanged in the new *Handbuch* which appeared recently.

The new quantum theory is of course not Pauli's work alone. We owe basic ideas to Louis de Broglie, Werner Heisenberg, and Erwin Schrödinger, physical interpretations to Max Born and above all to Niels Bohr. However, Pauli not only took an active part in the whole development, but he exerted an immeasurable and decisive influence in its evolution through his scientific publications, his extensive personal correspondence, and his discussions. Never willing to be satisfied with plausible but basically unclear ideas, his sharp criticism forced many people to think neatly and clearly. His criticism was always accepted, for it was motivated by a true philosophical search for truth. Even though his colleagues feared his criticism, they still presented their ideas to him, for everyone hoped that in a conversation with him their true meaning and value would manifest themselves. Moreover, Pauli was capable beyond all

A photograph from V. F. Weisskopf's album. Otto Stern is at left.

Informal speech by Pauli during 1956 meeting of Nobel Prize winners in Lindau at Lake Constance.   (*Ullstein photo*)

Passport photograph taken in July 1940, after the fall of France, in preparation for trip to United States. Mrs. Pauli writes of this picture of her husband, "To my opinion, this is the best existing photo of W. Pauli."

others of synthesizing the rich results obtained by a whole generation of ingenious theoreticians and in placing their ideas into proper perspective in the framework of physics.

This is not the time and place to look into his own numerous contributions to theoretical physics, in which he played an important part in the advancement of many fields. These contributions were often only short works which were then followed by a series of investigations by other authors. His remarks in a discussion, like the hypothesis of the neutrino, were capable of the greatest consequences. However, we do want to mention that, together with Jordan and Heisenberg, Pauli developed a quantum theory of fields. This subject, of fundamental importance, is even today full of unsolved problems. Here Pauli followed each advance with alert and critical interest, and always participated anew in helping to resolve the unresolved.

One of Pauli's discoveries must be stressed at this point because it will always be associated with his name: the Pauli Exclusion Principle, which he formulated in 1925. In its original formulation this principle says that two electrons in an atom can never exist in the same state. Pauli discovered the law in connection with his studies of the anomalous Zeeman effect. One of the difficulties which had to be overcome was that of finding a sufficiently precise and yet completely general definition of the concept "state of an electron", which was accomplished only when Pauli realized that the quantum numbers characterizing a state which were used at that time were not enough. Electrons possess another property which leads to doubling the number of states known in those days. This new property has turned out to be the spin of the electron, but at that time the fundamental relation between spin and the exclusion principle was not yet known. In 1940, however, Pauli was able to show under very general assumptions that all particles with half-integer spin have to obey the exclusion principle. The proof was based on the quantum field theory and is, in Pauli's own words, one of the most important deductions which can be made by applying the relativistic theory to quantum mechanics. The Pauli principle is much more far-reaching than simply its application to atomic spectra. It is a deciding factor in the structure of matter. Above all, it extends beyond the framework of quantum mechanics as understood by the correspondence principle and points to relations still obscure to us.

Great scientific honors were bestowed on Pauli for his contributions, among them the Lorentz, Nobel, Franklin,

Mateucci, and Planck Medals. He was also a member of numerous academies, such as the Accademia dei Lincei, of which Galileo was a member, and the Royal Society of London. The many letters of condolence which arrived from all over the world reflected the impact of his death. On behalf of the Max Planck-Gesellschaft, Max von Laue said, "Pauli was one of the greatest physicists of the Twentieth Century; his influence on theoretical physics will never be forgotten." Nobel Laureate I. I. Rabi writes, "With the whole world of physics we mourn the death of Wolfgang Pauli, my teacher, colleague, and friend. The Pauli principle, which covers the beautiful simplicity of atomic structure, places Pauli with the highest in the history of science. Kind, forthright, and of uncompromising principles, he was the conscience of physics." All this applies not only to a successful physicist, but also to a philosopher, critic, and teacher.

If we ask ourselves what above all was Pauli's calling, the answer would be: he was a natural philosopher in the classical sense of the phrase—as it applies to Kepler, Galileo, and Newton. Like hardly another of today's physicists, Pauli was imbued with a deep feeling that the scientist's efforts in reading Nature's book would lead man towards recognition of his own image in Nature itself. This intuitive feeling was not based on more or less plausible intellectual considerations; it came from deep experiences. For Pauli, the basic questions of physics were in their deepest meaning also the questions of human life. He was convinced that the harmony manifested in Nature should have a corresponding counterpart, a harmony of the world within us.

(Marcus Fierz)

Relaxed portraits were taken in Pauli's office at the Physikalisches Institut der ETH in Zürich on the occasion of his becoming a Foreign Member of the Royal Society in April 1953.

DEAR Friends: I am here today to express the most profound sorrow and condolence in behalf of the National Academy of Sciences of the United States and the American Physical Society. I am sure too that I could include all the academies and scientific institutions around the world, for wherever scientists are at work, the name Pauli is a great one.

Just what Pauli means to us and to the physicists of the whole world cannot yet be appreciated so shortly after the tragic blow. He was for such a long time an integral part of our world, we cannot yet understand what it means to be without him. Only when we plunge once again into our problems and our work, will we begin to realize what we all have lost.

This is not the time to speak about Pauli's achievements and contributions to the world of science, as great and significant as they have been. I would like to speak about what Pauli has meant to us, to the physicists of Europe, of America, of the East, of the whole world.

Pauli had a very special way of working in science. He introduced an original style of thought that has profoundly influenced and guided all the physicists of the world. His style brought into focus the fundamental ideas and the symmetries in the laws of nature as conceived in mathematical formulas, without many words or much talk. His manner, his thinking, and his way of life were unusual in every respect. He was regarded as the conscience of physics. How often did we ask ourselves, when considering our work, "what would Pauli say to this"? How often did we fear that, "Pauli would not accept it"?

But it is more than the work alone: it is the profound and human clarity and directness that radiated from Pauli and which determined his relations with colleagues and fellow men. We are all familiar with his acid criticism, his irony and the vigor with which he fought false ideas, the wit and the contempt with which he met what seemed to him unsound or half-truths. All of this was the expression of his constant struggle for final clarity and purity in science and in human relations.

To a very great extent it is due to Pauli that in the world community of physics there is still something simple, pure, and healthy to be found, in spite of all newspaper publicity, politics, and public to-do that surrounds science nowadays and which Pauli always deeply abhorred. May we succeed in preserving something of his manner without him!

Pauli was not only a great physicist, he was a great man. In human matters as well as in science, he was endowed with a clearer understanding and a more profound insight than most of us. He looked more deeply into human and philosophical questions than many of his friends would have thought. The depths of the human soul were not unfamiliar to him. He never made life easy for himself when trying to live in harmony with his ideals. And this he succeeded in doing. In his mature years Pauli led a happy and a profoundly rewarding life from which he was so suddenly and so unfortunately torn. He lived his life before us as a man and a scientist in quiet contemplation and simplicity. May he serve all of us as an example in these disturbed and difficult times, of how to uphold purity of mind and human understanding in all aspects of learning and living.

(Victor Weisskopf)

Personal memories of Pauli

Looking back half a century, a disciple of Wolfgang Pauli recalls how Pauli's extreme honesty and directness expedited work on fundamental problems in quantum mechanics and made for unusual human relations.

Victor F. Weisskopf

PHYSICS TODAY/DECEMBER 1985

We older disciples of the great Wolfgang Pauli remember with pleasure and nostalgia the prewar years when we had the privilege to work with him. It was one of the most interesting, stimulating and productive periods in physics. At the same time, however, Europe saw events that were among the most terrible and depraved actions of man against man. The coincidence in the history of humankind of the greatest achievements and the worst evils has always impressed and depressed me deeply. As Dickens said: "It was the best of times, it was the worst of times."

Let me begin my tale a little earlier. In 1932 I received a Rockefeller grant for one year, to study at places of my choice. I wanted to divide my time between Copenhagen and Cambridge, England, to learn from the two great men Niels Bohr and Paul A. M. Dirac. In Copenhagen, I not only profited greatly from Bohr's overwhelming personality—everyone who spent some time with him was deeply influenced by his way of thinking and living—but I also met my wife there, so the division of my Rockefeller time was somewhat biased toward Copenhagen.

My stay in Cambridge was also very important to me, but not so much

because of Dirac. It was not easy to learn from him; he worked for himself and did not have much contact with other physicists or would-be physicists. It was in Cambridge that I met Rudolf Peierls, who also had a Rockefeller grant. He was a few years older than I, and I learned much from him. Two or three years are not much, but they make a big difference when one is young and, sometimes, at the end of one's life. He introduced me to relativistic field theory—how to make calculations with the Dirac equation, a skill that was referred to as "alpha gymnastics."

Peierls's stipend was $200 a month, whereas, to my slight annoyance, mine was only $150. I hasten to add that either amount represented at that time undreamed-of riches for an average European. The reason for the difference was that he was married, whereas I was only on the way to be. The Rockefeller Foundation asked us to send in reports on our achievements during the time of support. As proof of some of his activities, Peierls sent the foundation an announcement of the birth of his first child; of course, I was unable to match this. The officials in New York didn't appreciate Peierls's sense of humor, we heard.

How to deal with Pauli

At that time—it must have been May or June of 1933—came a letter from Pauli asking me to be his assistant in

Zurich, replacing Hendrik Casimir, whom Paul Ehrenfest, shortly before his suicide, had called back to Leiden. What could be better for a young physicist than to work with Pauli? It was the fulfillment of a dream. Why did he take me and not other, more experienced people such as Hans Bethe? I found out later.

Now I had something else to learn from Peierls, who had been Pauli's assistant before. Of course, I asked him about his experiences. He said, "It is a great thing, but you must be prepared." Hence, Peierls taught me not only quantum electrodynamics, but also "how to deal with Pauli." He gave me much good advice; here is one example: "Be very careful when you give a talk at the Zurich colloquium. Pauli likes to interrupt a speaker when he thinks he is wrong or inconsistent. The best method to counteract this is: The day of your talk go to Pauli in the morning and tell him what you are planning to say. If he does not like it he will tell you in the strongest terms how silly it is, that it is all wrong or that it is trivial and known to every child, etc., etc. Then in the afternoon at the colloquium," continued Peierls, "you say exactly what you intended to say in the first place. You don't need to change anything, except if you really have been convinced by him. Pauli will sit in the first row, and when you come to the critical points, he will almost inaudibly

Victor F. Weisskopf is institute professor emeritus at Massachusetts Institute of Technology, in Cambridge. He was director general of CERN from 1961 through 1965.

Arnold Sommerfeld (left) with Wolfgang Pauli at an October 1934 metals conference in Geneva. (CERN photograph, courtesy AIP Niels Bohr Library.)

mumble to himself: 'I've told him already, I've told him anyway!' So it won't be so bad at all."

I found out why[1] Pauli took me instead of Bethe when I came to Zurich to begin my duties in the fall of 1933. I knocked several times at the door of Pauli's office until I heard a faint "Come in." I saw Pauli at his desk at the far end of the room and he said, "Wait, wait, I have to finish this calculation" (*Erst muss ich fertig ixen*). So I waited several minutes. Then he said, "Who are you?" "I am Weisskopf, you asked me to be your assistant." "Yes," he said, "first I wanted to take Bethe, but he works on solid-state theory, which I don't like, although I started it." This, then, was the reason.

I made a contract with him. I said, "Of course, I am more than delighted to work for you but, please, that new stuff you are working on, the Klein–Kaluza approach to general relativity, that I am unable to understand. I don't want to deal with it, but I am ready to work on everything else." He accepted the conditions because he was already somewhat bored by it. (Today it seems to be the *dernier cri* of the most sophisticated particle theorists.) Pauli then gave me some problem to study—I have forgotten what it was—and after a week he came and asked me what I did. I showed him and he said, "I should have taken Bethe after all." I was well

Niels Bohr and Franca Pauli in 1936, at a conference at the Bohr Institute in Copenhagen. (Photograph by P. Ehrenfest Jr, AIP Niels Bohr Library, Weisskopf Collection.)

prepared by Peierls for events like this and I took it for what it was: a challenge to get a deeper understanding of physics.

The numerous Pauli anecdotes circulating among physicists give a distorted impression of Pauli's personality. He is seen as a mean character who wanted to hurt his weaker colleagues. Nothing is further from the truth. Pauli's occasional and highly publicized roughness was an expression of his dislike of half-truths and sloppy thinking, but it was never meant to be directed against any person. Pauli was an excessively honest man; he was of an almost childlike honesty. What he said were always his true thoughts, directly expressed. Nothing is more reassuring than to live and work with somebody who says everything that is on his mind—but you must get accustomed to it. Pauli did not want to hurt anybody, although he sometimes did, without intention. He disliked half-truths or ideas that were not thought through, and he did not tolerate talking around a half-baked idea. He was, as many people said, the conscience of physics. He wanted people to understand things thoroughly and express them correctly. He never tired of answering questions and explaining problems to anybody who came asking. He was not a good lecturer before an audience because he did not have the ability to judge how much the crowd could take in, and his listeners did not often dare to interrupt him with questions. Once a student did so and said, "You told us that conclusion is trivial, but I am unable to understand it." Then Pauli did what he did frequently when he had to think things over during a lecture: He left the room. After a few minutes he came back and said: "It is trivial!"

When you came to him and said, "Please explain this to me, I don't understand it," he would explain it with great patience and pleasure. We often said, "For Pauli every question is stupid, so don't hesitate to ask him whatever you want." He loved simple and illustrative explanations, but they had to be correct, and not misleading. Once his colleague in experimental physics, Paul Scherrer—an excellent lecturer and a lover of simple conclusions—came to him and said: "Look, Pauli, I would like to show you how I explained that effect in my course. You see, here the spin is up, and there it is down and then they interact,... isn't that simple?" Whereupon Pauli answered: "Simple it is, but it is also wrong!"

Pauli loved people and showed great loyalty to his students and collaborators. All of Pauli's disciples developed a deep personal attachment to him, not only because of the many insights he gave us, but because of his fundamentally endearing human qualities. It is true that sometimes he was a little hard to take, but all of us felt that he helped us to see our weaknesses. Ehrenfest expressed it well after J. Robert Oppenheimer came to him as a young postdoc in the late 1920s. Ehrenfest was unhappy because Oppenheimer always gave quickly an answer to any question, and Ehrenfest felt that the answer was not always correct but was unable to reply fast enough. So he wrote to Pauli: "I have here a remarkable and intelligent American but I cannot handle him. He is too clever for me. Couldn't you take him over and spank him morally and intellectually into shape?" (*Zurecht prügeln*). We all were spanked into shape by dear Pauli and we loved it.

There was one person to whom Pauli acted quite differently. When Arnold Sommerfeld, his former teacher, came to Zurich for a visit, it was all, "Yes, Herr Geheimrat, yes, this is most interesting, but perhaps I would prefer a slightly different formulation, may I formulate it this way...." It was much fun for us victims of his aggression to see him well behaved, polite and subservient; a completely different Pauli.

Calculations and publications

There were not many regular duties for Pauli's assistants to perform. Pauli himself made up the homework problems for his course, and we assistants only had to grade them. Our main duty was to be ready for discussion of his work and of new developments. He took this very seriously and it was not easy to get permission to leave Zurich during the term. Once I asked him with great trepidation, "May I go to Copenhagen for a week?" "Why?," he asked impatiently. I answered, "I intend to marry and come back with my wife." To my great relief he replied, "I approve of that, I am going to get married also!" (*Ich heirate nämlich auch*).

Pauli's assistants had another pleasant duty. Pauli's wife asked him to change his eating habits because of his proverbial bulk. But Pauli loved sweets and cakes, and many afternoons he wanted to continue our discussions in a nearby *Konditorei* where one could get delicious pastries. One of my duties was to promise never to mention these secret outings to his wife.

Of course there were also more serious duties. A lively correspondence was taking place between Werner Heisenberg and Pauli about the problems of quantum electrodynamics. Some of these problems, such as the unavoidable infinities, were quite serious and were solved only much later. Many of them, however, could be straightened out at the time. Whenever a letter came from Heisenberg, Pauli discussed it with me and frequently asked me to draft an answer: "You write it, I will correct it and then we will send it to

Paul A. M. Dirac and Rudolf Peierls (right) with Pauli (center). (AIP Niels Bohr Library.)

him." A letter once came from Heisenberg and Pauli was terribly dissatisfied with its content: "Such silly statements; it is all stupid and wrong. You must tell this to him in your letter!" What could I do? Well, I started out explaining our disagreements as well as I could, and then I quoted Leporello in Don Giovanni: "My master wants to tell you, myself I would not dare to." Then I was able to repeat literally all of Pauli's curses.

It was the time of an unhappy episode in my career as a physicist. Pauli asked me to calculate the self-energy of the electron on the basis of the positron theory to see if this energy is less divergent in that theory. I found that it diverges equally badly and I published this result. A few weeks after the publication I received a letter from Wendell Furry, who worked with Oppenheimer at the time, informing me that I had made a simple mistake of a sign in my calculation. If it is done correctly, the divergence is only logarithmic. The positron theory improved things considerably, in contradiction to my paper. I was down and depressed to have made and published a silly mistake in such a fundamental problem! I went to Pauli and said that I wanted to give up physics, that I would never survive this blemish. Pauli tried to console me: He said, "Don't take it too seriously, many people published wrong papers; I never did!"

What followed shows how decent the relations between physicists were at that time. I asked Furry by letter to publish his result under his name, or, at least, to coauthor a paper correcting the mistake. But Furry was a gentleman. He answered, no, I should publish a correction in my name only and mention him as the person who drew my attention to the error. Since then, the logarithmic divergence of the self-energy of the electron goes with my name and not with Furry's. Yes, times have changed and so have the attitudes of physicists toward publication by others. I remember having shown Pauli a newly published paper on a subject of his interest. He said, "Yes, I thought of that too, but I am glad he worked it out, so that I don't need to do it myself."

Let me now say a few words about the origin of the paper that Pauli and I wrote[2] about quantum electrodynamics with Bose particles. In 1934, I was playing around with the so-called Klein–Gordon equation, which is the relativistic wave equation of particles with zero spin. I was struck by the fact that the wave intensity $|\psi|^2$ is not conserved in the presence of electromagnetic fields, whereas the expression for the charge density is different from $|\psi|^2$ and fulfills the charge conservation laws. I felt there might be something like a lack of conservation of particle number in that equation, and that this might lead to pair creation or annihilation. In spite of what I learned from Peierls in Cambridge I was not able to deal with this problem. It required what was called "second quantization," or the quantization of the wave field, something with which I was not very familiar at that time.

I went to Pauli for help. It was just after his separation from his first wife, and he was in a very bad mood. I tried to explain to him my difficulties and my tentative conclusions, but he was very impatient and repeated over and over how silly my remarks were. Finally I quoted to him a verse from Wagner's opera *Die Meistersinger* that said approximately, "Oh master why so much excitement and so little repose; I believe your judgment would be more mature if you would better listen." He looked up to me and asked, "What is that?" I said it is from *Meistersinger*, whereupon he replied, "Wagner, I

don't like at all!" So ended the discussion.

The next day he was in a better mood and I repeated my story. He said: "That's interesting. Why didn't you tell it to me yesterday?" This began a wonderful time for me, when I learned from Pauli in detail how to deal with second quantization. We found that the quantum electrodynamics of spinless particles indeed leads to antiparticles, pair creation and annihilation—and all of it came out without the necessity of filling up negative-energy states with particles, as Dirac did to get these phenomena from his equation. Pauli never liked that trick and frequently referred to our paper as the anti-Dirac paper. Of course, later on it was shown that the trick of filling up negative-energy states was not necessary even with the Dirac equation.

Pauli asked me to calculate the pair creation and annihilation cross sections of spinless particles according to our theory. The calculation was not too different from the one for ordinary electrons and positrons that Bethe and his collaborators had carried out shortly before. I met Bethe at that time at a conference in Copenhagen and asked him to show me how to do the calculations. I wondered how long it would take to get to the final result and he told me, "It would take me a few days, it will take you a few weeks." It did. Moreover, I made a mistake of a factor of four. Again a proof that Pauli should have taken Bethe.

The exclusion principle. Our paper was little more than an interesting formal exercise, because at that time no particles with zero spin were known to exist. However, some years after I left Pauli, he used some of the ideas for his famous proof of the connection between spin and statistics: Particles with half-integer spin must fulfill the Pauli exclusion principle; those with integer spin

Eating. This photo from the 1934 Bohr Institute conference shows Pauli (center) with Gerhard Dieke (right) and an unidentified conferee who wanted to tell his ideas to Pauli. He got the reply, "First I must eat!" (Photograph by P. Ehrenfest Jr, AIP Niels Bohr Library, Weisskopf Collection.)

must have Bose statistics. This achievement shows Pauli in his greatness. He discovered the exclusion principle in 1925 from a careful analysis of atomic spectra and was not satisfied until, fifteen years later, he was able to show how it followed by necessity from quantum field theory.

One day in the course of my work as assistant I came across an interesting note that Pauli had made a couple of years before his discovery of the exclusion principle. It was one of my lighter duties to keep Pauli's collection of reprints in reasonable order. In doing so, I browsed through older papers and found a copy of Bohr's famous paper about the *Aufbauprinzip*, in which he explained the periodic system of elements as an effect of electronic shell structure. Everybody who teaches this wonderful triumph of quantum theory uses the Pauli principle to explain the way in which the electron shells are filled when going from one atom to the next one with an additional electron. However, when Bohr published his paper on the periodic system, the exclusion principle was not yet known! It was a testimony to Bohr's unfailing intuition that he nevertheless got the right results. Browsing through Pauli's copy of that paper, I looked at the page on which Bohr says, "Going

from Neon to Sodium, we must expect that the eleventh electron goes into the third shell ..." and my attention was caught by a remark that Pauli had written in the margin in big letters: "How do you know this? You only get it from the very spectra you want to explain!" Three heavy exclamation marks followed. (What I have just said differs slightly from my earlier account, but is the correct version. See PHYSICS TODAY, August 1970, page 17.) It took Pauli two more years to tell us why.

Moral integrity

Pauli was in Princeton during the Second World War. To stay in Switzerland would have been too dangerous because he was not yet a Swiss citizen; he carried an Austrian passport and was considered a German national after the Nazis took over Austria. The Swiss consented to give him citizenship only after his Nobel prize in 1946. There was some discussion in Los Alamos about whether or not to ask him to join the atomic bomb project. As far as I could find out, he was never asked. I am sure he would have refused to join for many reasons. He had a strong, fundamental aversion to any work connected with weapons, and he would not have felt at ease in a large

team. He was a pure character who instinctively would have stayed away from such work. Moreover, nuclear physics never interested him very much, even though he discovered nuclear spin.

When the war was over, Pauli was of great help to all of us. He was asked to give a series of lectures about the latest ideas in particle theory. He brought us back to fundamental physics. Shortly afterward he returned to Zurich. He wanted to keep in touch as closely as possible with American physics, and he asked me and others to write to him about the newest developments in theory and experiment. I remember a letter of mine reporting Chien-Shiung Wu's preparations to test the conservation of parity in weak interactions. Pauli wrote back that, in his opinion, this was a waste of time; he would bet any amount of money that parity is conserved in any process. When the letter arrived, I had just heard the news that parity was strongly violated. My better self won, and I did not send a telegram saying, "Bet for $1000 accepted," but reported to him the surprising result of Wu's experiment. Overseas telephone calls were not yet used for physics. Pauli was completely flabbergasted. He wrote back expressing his astonishment that "God is a weak left-

Hans Albrecht Bethe. Because of Pauli's lack of interest in solid-state theory, Weisskopf got an assistantship that Pauli had considered giving to the more-experienced Bethe, who had done work in that field. (AIP Niels Bohr Library.)

Hendrik Casimir (left) with Weisskopf, at the 1934 Bohr Institute conference in Copenhagen. (Photograph by P. Ehrenfest Jr, AIP Niels Bohr Library, Weisskopf Collection.)

hander" and added: "I am glad that I did not conclude our bet. I can afford to lose some of my reputation but not some of my capital."

Pauli mellowed much in his later years. This was mainly due to his second wife Franca, who was able to make his life bearable and even pleasant. This was not an easy task. Pauli had a very difficult character, was easily depressed and often felt thoroughly unhappy. Franca succeeded in creating a comfortable and protected home for him, in which he could feel at ease and pursue his many interests that reached far beyond physics.

He became very interested in various forms of mysticism, mainly through his connection with the Swiss psychologist Carl Gustav Jung. Later he developed a deep friendship with Gershon Scholem, the great scholar and world authority on Jewish mysticism, the Kabalah. (The Kabalah ascribes a number to each word of the Hebrew language, a number that has a deep symbolic significance. The number corresponding to the word Kabalah happens to be 137.) Pauli and Scholem saw each other frequently and exchanged their views in letters. With a few notable exceptions, Pauli rarely spoke about this side of his interests to his physicist friends. He did not speak much about

it with me, except that he urged me to visit Scholem when I went to Jerusalem. It was a unique experience to meet a great man and to be introduced to ideas that are so alien to those of our science.

Pauli's interest in these different avenues of human experience was in many respects a natural expansion of his involvement in modern physics. He was a disciple of Bohr—perhaps Bohr's closest disciple. Bohr often applied his concept of complementarity to human concerns beyond natural science. The rational scientific approach is only one way of dealing with the world around us. There are other, seemingly contradictory approaches—as contradictory as the particle and wave pictures within physics—that deal with aspects of our thoughts and emotions. A given approach seems fragile and senseless when analyzed within the framework of another, but is forceful and convincing within its own frame. Pauli was very much attracted by this generalization of complementarity.

Pauli created a style of theoretical thinking and research that influenced physics all over the world. It is a style that emphasizes the essential roots and the symmetries of the laws of nature in their mathematical form without much talk or handwaving. His clean way of

thinking and working appears to all of us as an ideal to be emulated. We often ask ourselves, "What would Pauli say to this?" We often come to the conclusion, "Pauli would not accept that."

However, Pauli set his example through more than the character of his work. He personified the striving for utmost clarity and purity in science and human relations. We owe it, in part, to Pauli that in the community of physicists there is still a certain amount of healthy simplicity, honesty and directness, in spite of all the politics, publicity and ambitions, attitudes that were so foreign to Pauli. He was not only a physicist, he was also a great personality, able to see deeper than others into scientific and human problems. The dark riddles of the human psyche were not unknown to him. He is an example to all of us of how to live a quiet and contemplative life of intellectual and moral integrity in these unruly times.

References

1. I have told the story that answers this question, before; however, an author is allowed to republish his own results. See V. Weisskopf, *Physics in the XX Century*, MIT Press, Cambridge, Mass. (1972), and "The Joy of Insight," Basic Books, New York (1991).
2. W. Pauli, V. F. Weisskopf, Helv. Phys. Acta **7**, 709 (1934). □

GEORGE UHLENBECK AND THE DISCOVERY OF ELECTRON SPIN

How two young Dutchmen, one with only a master's degree, the other a graduate student, made a most important finding in theoretical atomic physics.

Abraham Pais

The owl depicted on the signet ring George Uhlenbeck used to wear—"Uhlenbeck" in German means "owl's brook"—derives from his family's coat of arms. The shield reads, in the language of heraldry: Azure, on a tree trunk proper rising from water argent, an owl contourné, head affronty. In plain language, it depicts an owl with its head turned toward you, sitting on a tree trunk in natural color, which rises up out of a silvery brook. (I owe the transcription of the Dutch blazon into English heraldry to Michael Maclagan, the Richmond Herald in the College of Arms, in London.)

The Uhlenbeck ancestry can be traced to German roots. Records for the years 1634 and 1656 kept at the Staatsarchiv in Düsseldorf, at one time the capital of the duchy of Berg, mention that at those times a Jan in der Ulenbeck was the proprietor of the estate Üllenbeck, situated near the township of Velbert in the district of Angermund. The next four generations of George's ancestors were born and raised on that same estate. A great-great-grandson of Jan in der Ulenbeck, Johannes Wilhelmus Uhlenbeck, went into military service under King Frederick II—"the Great"—of Prussia. On account of a duel he had to flee the country. In 1768 he entered the military service of the Dutch East India Company on the island of Ceylon, a Dutch colony from 1658 until 1796. He is the first of the Dutch branch of the Uhlenbeck family. (I am deeply grateful to Else Uhlenbeck for information on her husband's family background.)

Eugenius Marius Uhlenbeck, a great-great-grandson of Johannes Wilhelmus, born in 1863 on the island of Java in the Dutch East Indies (now Indonesia), served with the Dutch East Indian Army, eventually as lieutenant colonel. During the Bali wars (1847–49), two of his uncles, also army officers, perished, one by his own hands to avoid capture by

Abraham Pais is Detlev Bronk Professor Emeritus at Rockefeller University, in New York. He based this article on his presentation at APS's Uhlenbeck Memorial Symposium, held in Baltimore on 3 May 1989.

George Uhlenbeck (left) with Hendrik Kramers (middle) and Samuel Goudsmit in Ann Arbor, Michigan. This photograph was taken around 1928, three years after Uhlenbeck and Goudsmit proposed the idea that each electron rotates with an angular momentum $\hbar/2$ and carries a magnetic moment of one Bohr magneton, $e\hbar/2mc$.

brutal tribes. In 1893 he married Anne Beegers, who was born in Sumatra in 1874, the daughter of a Dutch major general.

George Eugene Uhlenbeck, born in Batavia (now Jakarta) on 6 December 1900, was one of the six children of that marriage. Two of the other children died very young in the Indies, of malaria. Military duties caused the family to move about a good deal. Thus it happened that George received his first schooling at a kindergarten in Padangpandjang, on Sumatra.

Early interest in physics

In 1907 the family moved permanently to Holland and settled in the Hague, where Uhlenbeck's regular education began, first at an elementary school, then at a higher burgher school (what is now called an atheneum). A three-year course in physics first drew him to the subject that was to be his life's devotion. Eager to learn more, he would bicycle to the Royal Library in the Hague to seek further information. There he absorbed Hendrik Lorentz's *Lessen over de Natuurkunde* (*Lectures on Physics*), an undergraduate university text. Uhlenbeck's particular interest in kinetic gas theory dates from those early days. His knowledge of physics, uncommonly deep for a high school student, brought him to the attention of his physics teacher, A. H. Borgesius, who discussed science with him and gave him books from which to study differential and integral calculus.

In July 1918 Uhlenbeck passed his final high school examination. He could not enter a Dutch university,

however, because his school had not provided training in Greek and Latin, at that time a prerequisite by law for university study in any discipline. (Johannes van der Waals and Jacobus van't Hoff, in similar positions at earlier times, had been able to enter a university only by special governmental dispensation.) In September 1918 he therefore entered the Technische Hogeschool (Institute of Technology) in Delft, intent on studying chemical engineering. However, almost immediately thereafter a new law was enacted dispensing with the Greek and Latin requirements for university training in the sciences. In January 1919 Uhlenbeck left Delft and enrolled in the University of Leiden to study physics and mathematics.

At that time, the professors at Leiden were Paul Ehrenfest, Heike Kamerlingh Onnes and J. P. Kuenen. Every Monday, Lorentz, Ehrenfest's predecessor, would come from Haarlem to give a physics lecture.

Uhlenbeck received his undergraduate education in physics, both theoretical and experimental, from Kuenen. Every physics undergraduate had to take a laboratory course. Uhlenbeck's laboratory reports had a strongly theoretical bent. "Even for the simplest electromagnetic experiments I started from the Maxwell equations." (Throughout this article, quotations without attribution are from private conversations with Uhlenbeck.) Kuenen's laboratory assistant, who could not quite follow those reports, showed them to Kuenen, who was impressed. Largely through Kuenen's advocacy, Uhlenbeck obtained a scholarship, a quite welcome development because his father did not find it easy to raise four children on a military pension. (From September 1921 until June 1922 Uhlenbeck partially supported himself by teaching ten hours a week at a high school in Leiden; the flirtatious young girls there contributed greatly to his difficulties in keeping order in his classes.)

In addition to his physics studies, Uhlenbeck took courses in mathematics by J. Droste, Jan Cornelis Kluyver and W. van der Woude, in astronomy by J. Woltjer, in crystal structure by K. Martin, in physical chemistry by F. Schreinemaker and in inorganic chemistry by Willem Jorissen. Among his fellow students who later would make names for themselves in physics were Dirk Coster, Gerard Dieke and Samuel Goudsmit.

All through his student years, Uhlenbeck commuted by train between Leiden and his family's home on the Lübeckstraat in the Hague. His mother would pack his lunch and give him a *kwartje* (25 cents) for coffee. He saved the money until one day he spent it on a secondhand copy of Boltzmann's *Vorlesungen über Gastheorie* (*Lectures on Gas Theory*), lecture notes that he found hard to grasp. Not long thereafter his brother-in-law, a chemist, introduced him to Paul and Tatiana Ehrenfest's encyclopedia article on statistical mechanics.[1] "That was a revelation. I began to see what Boltzmann was up to."

Ehrenfest

After graduating in December 1920, Uhlenbeck attended courses by Ehrenfest and Lorentz and also the celebrated Wednesday evening "Ehrenfest colloquium," which one could attend by invitation only, but to which one had to go once admitted. Ehrenfest even took attendance.

Ehrenfest was by far the most important scientific figure in Uhlenbeck's life. In all the years I knew Uhlenbeck, in Utrecht, in Ann Arbor and in New York, a single picture always stood on his office desk: a small photograph of a warmly smiling Ehrenfest. In 1956, upon receiving the Oersted Medal from the American Association of Physics Teachers, Uhlenbeck publicly expressed[2] his veneration for his respected and beloved teacher, whose life had long since come to a tragic end. In his acceptance lecture he recalled some characteristic Ehrenfest sayings:

> *Was ist der Witz* . . . ? [Do you say that to make a point, or only because it happens to be true?] *Weshalb habe ich solche gute Studenten? Weil ich so dumm, bin.* [Why do I have such good students? Because I am so stupid.]

Uhlenbeck also described some typical traits of Ehrenfest's style of lecturing and conducting seminars:

> First the assertion, then the proof. . . . His famous clarity, not to be confused with rigor. . . . He never gave or made problems; he did not believe in them; in his opinion the only problems worth considering were those you proposed yourself. . . . He worked essentially with only one student at a time, and that practically every afternoon during the week. . . . In the beginning, at the end of the afternoon one was dead tired.

Uhlenbeck also added a personal touch:

> One of the compliments I treasure most is when some of [my own students] told me of the identical experience they had working with me, especially the fact of the extreme exhaustion in the beginning.

It is one of the good fortunes of my own life to have been in a position to pay Uhlenbeck this compliment myself, with feeling.

Let us return to the early 1920s. Ehrenfest's graduate lectures consisted of a two-year course: Maxwell theory, ending with the theory of electrons and some relativity, one year; and statistical mechanics, ending with atomic structure and quantum theory, the other. Uhlenbeck attended these lectures and took additional instruction in mathematics. One day toward the end of his second graduate year, Ehrenfest asked in class whether anyone might be interested in a teaching position in Rome. Uhlenbeck raised his hand. So it came to pass that from September 1922 until June 1925 he became the private tutor in mathematics, physics, chemistry, Dutch, German and Dutch history of the younger son of the Dutch ambassador J. H. van Royen. The summers were spent in Holland, however, and in September 1923 Uhlenbeck obtained the degree of "doctorandus," the equivalent of a master's degree.

Fermi

For about a year right after his arrival in Rome, Uhlenbeck took Italian language lessons at the Berlitz school. Thereafter he continued his Italian studies by taking private tuition two hours a week, eventually reading Dante's *Divina Commedia* with his teacher. (He reread Dante in later years and was fond of occasionally reciting passages of this work.) By the fall of 1923 he had mastered the language sufficiently to attend mathematics courses taught at the University of Rome by Federigo Enriques, Tullio Levi-Civita and Vito Volterra. He also made contact with Roman physicists. When Uhlenbeck was in Holland during the summer of 1923, Ehrenfest told him of a young Italian physicist by the name of Enrico Fermi who had written a paper on the ergodic theorem. Ehrenfest had not understood Fermi's reasoning and asked Uhlenbeck to carry a letter to Rome with questions for Fermi. Thus it came about that Uhlenbeck and Fermi, who was nearly one year younger, met for the first time in the autumn of 1923. Their acquaintance grew into a friendship that lasted throughout Fermi's life. Together with a few other young Italian physicists they organized a small colloquium. "Fermi was the born leader and did most of the talking."

Fermi wrote his paper on the ergodic theorem in 1923 during his stay in Göttingen, Germany, a visit that adversely affected his self-confidence. The learned Göttingen style did not agree with him. At Uhlenbeck's urgings, Fermi went to Leiden for three months in 1924; he even published a paper in Dutch. One of Uhlenbeck's contributions to physics lies in initiating the personal contact between Ehrenfest and Fermi, which helped greatly to restore Fermi's self-confidence.

Thus Uhlenbeck stayed in touch with the sciences during his Italian period. Yet they receded from the center of his attention. He became deeply involved in history, especially cultural history. He became a regular visitor of the Nederlandsch Historisch Instituut te Rome; befriended a Dutch contemporary, Johan Quiryn van Regteren Altena (who later became a professor of art history in Amsterdam); and studied the works of Johan Huizinga, a professor in Leiden, and other cultural historians. The first article Uhlenbeck ever published is historical and is written in Dutch.[3] It deals with the Dutchman Johannes Heckius, one of the four cofounders of the Acedemia dei Lincei in Rome, in 1603. That it was the Dutchman Uhlenbeck who introduced Fermi, born and raised a Roman, to Michelangelo's Moses in the church of San Pietro in Vincoli says something about the personalities of the two physicists.

When Uhlenbeck left Rome for good to return to Holland, in mid-June 1925, he was seriously considering giving up physics to become a historian. He called on Huizinga in Leiden, who gave him a friendly reception; and he discussed the matter with his uncle, the distin-

Paul Ehrenfest (front, center) in Leiden, 1924. In January 1919 Uhlenbeck enrolled at the University of Leiden, where Ehrenfest was a professor. Ehrenfest became by far the most important scientific figure in Uhlenbeck's life. In the photograph with Ehrenfest are, from left to right, Gerard Dieke, Goudsmit, Jan Tinbergen, Ralph Kronig and Enrico Fermi. (Brookhaven National Laboratory photograph, courtesy AIP Niels Bohr Library.)

guished linguist Christianus Cornelis Uhlenbeck, an expert on American Indian languages and a professor at Leiden. His uncle was sympathetic to the idea, but suggested that it might be best to obtain first a PhD in physics, because he had already progressed quite far. Ehrenfest also responded benevolently to the new project but suggested that Uhlenbeck first find out what was currently happening in physics. He proposed that Uhlenbeck work with him for a while, and also that he learn from Goudsmit what was going on in *Spektralzoologie*, as Pauli used to call the study of spectra. Uhlenbeck accepted both suggestions, at the same time arranging for lessons in Latin from a friend in the Hague. His work with Ehrenfest on wave equations in multidimensional spaces (with special emphasis on the differences between odd and even numbers of spatial dimensions) led to a mathematical paper,[4] followed by a joint paper with Ehrenfest.[5] Uhlenbeck enjoyed this collaboration. So did Ehrenfest, who in the fall of 1925 appointed Uhlenbeck to succeed the mathematician Dirk Struik as his assistant.

Throughout the summer of 1925 Goudsmit came to the Lübeckstraat to educate Uhlenbeck in spectra. In his later years Uhlenbeck would refer to this period as the "Goudsmit summer."

Then, in mid-September 1925, doctorandus Uhlenbeck and graduate student Goudsmit discovered spin.

Gone were Uhlenbeck's aspirations of becoming a historian.

Work with Goudsmit

Samuel Abraham Goudsmit—"Sem" to his friends—was born in 1902 in the Hague, the son of a prosperous wholesale dealer in bathroom fixtures. His mother owned a fashionable hat shop. He got his first taste of physics at the age of 11 when browsing through an elementary physics text; he was particularly struck by a passage explaining how spectroscopy shows that stars are composed of the same elements as the Earth. As Goudsmit recalled, "Hydrogen in the sun and iron in the Big Dipper made Heaven seem cozy and attainable."[6] After finishing high school in one year less than the usual time, he became a physics student in Leiden, where Ehrenfest turned his interest into devotion. It soon became evident that he had a bent for intuitive, rather than analytical, thinking, starting from empirical hunches. Uhlenbeck later said of Goudsmit: "Sem was never a conspicuously reflective man, but he had an amazing talent for taking random data and giving them direction. He's a wizard at cryptograms." I. I. Rabi said: "He thinks like a detective. He is a detective."[6] In fact, Goudsmit once took an eight-month course in detective work, in which he learned to identify fingerprints, forgeries and bloodstains. A two-year uni-

At the Kamerlingh Onnes Laboratory in Leiden, 1926. Uhlenbeck is at the far left. Goudsmit is at the far right, next to Kramers. Ehrenfest is at the right, rear, next to his wife. Paul Dirac is the one in the dark coat at the left.

versity course taught him to decipher hieroglyphics. In physics the decoding of spectra became his passion. At age 18 he completed his first paper, on alkali doublets.[7] Uhlenbeck called it "a most presumptuous display of self-confidence but . . . highly creditable."[6]

In August 1925 the two men started their regular meetings in the Hague. George was the more analytic one, better versed in theoretical physics, a greenhorn in physics research and an aspiring historian with a paper on Heckius to his credit. Sem was the detective, thoroughly at home with spectra (on which he had already published several papers), known in the physics community and a part-time assistant to Pieter Zeeman in Amsterdam. In almost no time Sem's tutelage of George turned into joint research and publication, and their relationship into a close and lasting friendship. I know, more from my own later personal friendships with both than from their writings,[8,9,10] how each remained forever beholden to the other for his share of the work during those months. (See Goudsmit's article in PHYSICS TODAY, June 1976, page 40, and Uhlenbeck's article in PHYSICS TODAY, June 1976, page 43). There was no politesse, but deep appreciation.

Among the topics that Sem taught George that summer was Alfred Landé's theory of the anomalous Zeeman effect, those splittings of spectral lines that do not follow the patterns predicted much earlier by Lorentz on the basis of classical theory. In 1921 Landé had found it possible to explain those anomalies by the new and quite daring assumption that angular momentum quantum numbers can take on half-integer values. Sem went on to tell the story: How Werner Heisenberg, in his first published paper, had gone further by proposing that in alkalis the valence electron and the residual atomic

Rumpf, the core, each have angular momentum $\frac{1}{2}$ (in units of $h/2\pi$). How then Landé deduced from this that g, the gyromagnetic ratio, should have the value 2 for the core instead of 1, the classical prediction. How next Pauli had shown that the core had to have zero angular momentum. How he, Sem, had written that Landé's $g = 2$ is "completely incomprehensible" but that, using this assumption, one nevertheless "masters completely the extensive and complicated material of the anomalous Zeeman effect."[11] How Pauli thereupon—we are now in January 1925—had proposed to assign a new, a fourth, half-integer-valued quantum number, not to the core but to the electron itself. And how Pauli was thereby led to the discovery of the exclusion principle.

Another subject Sem taught George was Arnold Sommerfeld's formula for the fine structure of the hydrogen spectrum: how it worked very well, how there was no problem with the Zeeman effect that experimentally appeared to be (but of course was not) normal at that time.

George was unhappy. "He knew nothing; he asked all those questions which I never asked," Goudsmit would later recall.[8] Why two distinct models if the alkalis and hydrogen were so much alike? Why not try the half-integer quantum numbers on hydrogen as well? In August 1925 this led to their first joint paper, a little-known but quite good piece of work, written in Dutch, in which they modified the quantum number assignments Sommerfeld had given to the atomic levels and reported an improved treatment of He^+ fine structure.[12]

Goudsmit wrote about what happened next: "Our luck was that the idea [of spin] arose just at the moment when we were saturated with a thorough knowledge of the

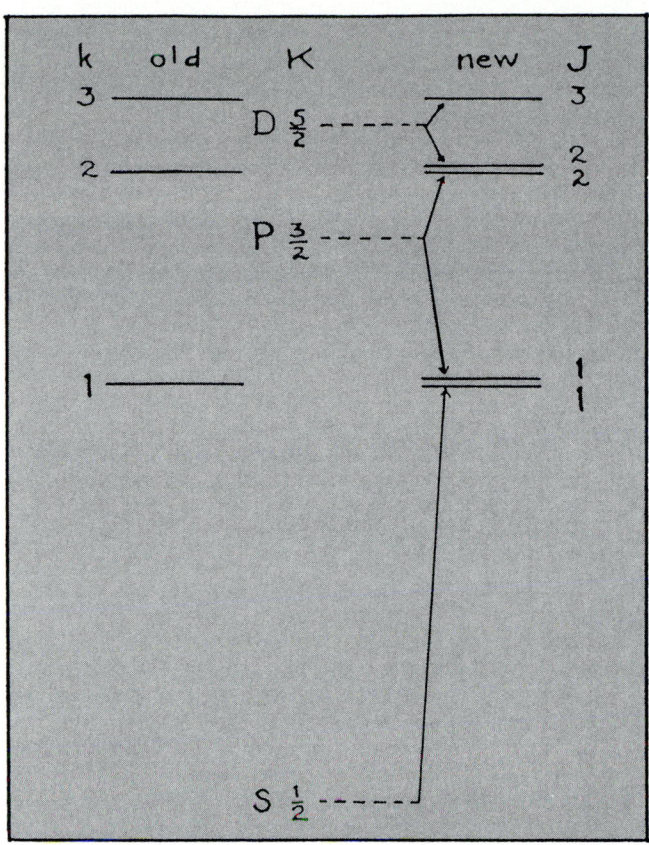

Diagram from L. H. Thomas's February 1926 *Nature* paper explaining the extra factor of 2 in the formula for fine-structure splitting in hydrogen-like spectra as derived from a semiclassical treatment of spin precession. Thomas's relativistically correct calculation reduced the angular velocity of the electron (as seen by the nucleus) by the needed factor of 2. The dotted lines represent the levels calculated without spin. (From *Nature* **117**, 514, 1926.)

structure of atomic spectra, had grasped the meaning of relativistic doublets, and just after we had arrived at the correct interpretation of the hydrogen atom."[9] Uhlenbeck recalled: "It was then that it occurred to me that, since (as I had learned) each quantum number corresponds to a degree of freedom of the electron, the fourth quantum number must mean that the electron had an additional degree of freedom—in other words the electron must be rotating!"[10]

Everything fell into place. The electron had spin $\frac{1}{2}$. Landé's $g = 2$ does not apply to the core but to the electron itself!

Sem asked whether this g value could be given a physical meaning.[10] Following a hint by Ehrenfest, George found in an old article by Max Abraham[13] that an electron considered as a rigid sphere with only surface charge does have $g = 2$. All this was written up in a short note[14] that includes the Abraham model, but with a caveat: If that model were the explanation of $g = 2$, then the peripheral rotational velocity should be much larger than the velocity of light, assuming the electron to be an extended object with "classical radius" e^2/mc^2.

That last comment is quite important. It makes clear that the discovery of spin, made after Heisenberg had already published the first paper on quantum mechanics, is an advance in the spirit of the old quantum theory, that wonderfully bizarre mixture of classical reasoning supplemented by *ad hoc* quantum rules.

The discovery note was published with Uhlenbeck as first author and Goudsmit second because (George told me) Ehrenfest suggested that this order would avoid the impression that George was only Sem's student, while Sem himself preferred to come second because it was George who had first thought of spin.

The discovery note is dated 17 October 1925. One day earlier Ehrenfest had written to Lorentz asking him for an opportunity to have "his judgment and advice on a *very*

witty idea of Uhlenbeck about spectra."[15] Lorentz listened attentively when George went out to see him soon thereafter, and then raised an objection. The spinning electron should have a magnetic energy on the order of μ^2/r^3, where μ is its magnetic moment and r its radius. Equate this energy to mc^2. Then r would be on the order of 10^{-12} cm, too big to make sense. (The weak point in this argument was to be revealed years later by the positron theory.) George, upset, went to Ehrenfest to suggest that the paper be withdrawn. Ehrenfest replied that he had already sent off their note, and he added that its authors were young enough to be able to afford a stupidity.[10] Some time later Lorentz handed Uhlenbeck a sheaf of papers with calculations of spinning electrons orbiting a nucleus. This work was to become the last paper[16] by the grand master of the classical electron theory. It was presented to the Como conference in September 1927.

No sooner had George and Sem's note appeared when Goudsmit received a letter from Heisenberg congratulating him on his "*mutige Note* [brave note]" and inquiring "wie Sie den Faktor 2 losgeworden sind [how you have got rid of the factor 2]" in the formula for the fine-structure splitting in hydrogen as derived from a semiclassical treatment of spin precession.[17] The young Leideners had not even thought of calculating this splitting. After some struggle they found that Heisenberg was right: The fine structure came out too large by a factor of 2. That puzzle was still unresolved when, in December 1925, Niels Bohr arrived in Leiden to attend the festivities for the golden jubilee of Lorentz's doctorate. Late one evening in 1946, Bohr told me in his home in Gamle Carlsberg what happened to him on that trip.

Bohr's train trip

Bohr's train to Leiden made a stop in Hamburg, where he was met by Wolfgang Pauli and Otto Stern, who had come to the station to ask him what he thought about spin. Bohr must have said that it was very interesting (his favorite way of expressing his belief that something was wrong) but he could not see how an electron moving in the electric field of the nucleus could experience the magnetic field necessary for producing fine structure. (As Uhlenbeck admitted later, "I must say in retrospect that Sem and I in our euphoria had not really appreciated [this] basic difficulty."[10]) On his arrival in Leiden, Bohr was met at the train by Ehrenfest and Albert Einstein, who asked him what he thought about spin. Bohr must have said that it was very, very interesting but what about the magnetic field? Ehrenfest replied that Einstein had resolved that. The electron in its rest frame sees a rotating electric field; hence by elementary relativity it also sees a magnetic field. The net result is an effective spin–orbit coupling. Bohr was at once convinced. When told of the factor of 2 he expressed confidence that this problem would find a natural resolution. He urged Sem and George to write a more detailed note on their work. They did; Bohr added an approving comment.[18]

After Leiden, Bohr traveled to Göttingen. There he was met at the station by Heisenberg and Pascual Jordan, who asked what he thought about spin. Bohr replied that it was a great advance and explained about the spin–orbit coupling. Heisenberg remarked that he had heard this remark before but that he could not remember who made it and when. (I will return to this point shortly.) On Bohr's way home the train stopped at Berlin, where he was met at the station by Pauli, who had made the trip from Hamburg for the sole purpose of asking Bohr what he now thought about spin. Bohr said it was a great advance, to which Pauli replied, "Eine neue Kopenhagener Irrlehre" (a new Copenhagen heresy). After his return home Bohr wrote to Ehrenfest that he had become "a prophet of the electron magnet gospel."[19]

Two additional comments:

▷ The mysterious factor of 2 was supplied in February 1926 by L. H. Thomas and has since been known as the Thomas factor.[20] Thomas noted that earlier calculations of the precession of the electron's spin had been performed in the rest frame of the electron, without taking into account the precession of the electron's orbit around its normal. Inclusion of this relativistic effect reduced the angular velocity of the electron (as seen by the nucleus) by the needed factor of 2.

▷ In March 1926 Hendrik Kramers received a letter from America written by Ralph Kronig,[21] a young Columbia University PhD who had spent two years studying in Europe, including a stay in Copenhagen from January to November 1925. Kronig reminded Kramers that prior to Goudsmit and Uhlenbeck he, Kronig, had already had the idea of spin, though he too had an extra factor of 2 in the fine structure, and that he and Kramers had discussed those matters in Copenhagen. Heisenberg's hazy recollection, mentioned a few lines earlier, of having heard part of the spin story before must refer to a discussion with Kronig. In his letter, Kronig told Kramers that he had not published because "Pauli ridiculed the idea, saying 'that is indeed very clever but of course has nothing to do with reality,' " and added, "In the future I shall trust my own judgment more and that of others less."

After Kramers told this story to Bohr, the latter wrote to Kronig, expressing his "consternation and deep regret."[22] Kronig replied, "I should not have mentioned the matter at all [to Kramers] if it were not to take a fling at the physicists of the preaching variety who are always so damned sure of, and inflated with, the correctness of their own opinion."[23] He asked Bohr to refrain from public reference to the affair since "Goudsmit and Uhlenbeck would hardly be very happy about it." Kronig is an eminent physicist and a gentleman. So was Uhlenbeck, who has written, "There is no doubt that Ralph Kronig anticipated what certainly was the main part of our ideas."[10]

I should like to conclude with a few remarks of a personal nature. In my undergraduate years in my native Amsterdam I began taking courses in physics, chemistry and mathematics, in a rather unfocused way. Then in the winter of 1938 Uhlenbeck, at that time a professor in Utrecht, came for a visit and gave two lectures on beta decay. I did not understand much. I had not yet heard about neutrinos. Nevertheless, listening to those talks given in a calm yet ever so compelling way I knew, I just knew: That is what I want to do. After graduation I moved to Utrecht, where I did my warm-up research exercises with Uhlenbeck and took his courses. Since then I have met many other physicists of distinction but never a better lecturer than George. We became personal friends in later years. We published two joint papers.[24] He had a great and lasting influence on me for which I shall remain forever grateful.

References

1. P. Ehrenfest, T. Ehrenfest, in Enzyklopädie der Mathematischin Wissenschaften vol. 4, part 2, Teubner, Leipzig (1911), section 28; English translation by M. J. Moravcsik, in *The Conceptual Foundations of the Statistical Approach in Mechanics*, Cornell U. P., Ithaca, New York (1959).
2. G. E. Uhlenbeck, Am. J. Phys. **24**, 431 (1956).
3. G. E. Uhlenbeck, Commun. Dutch Hist. Inst. Rome **4**, 217 (1924).
4. G. E. Uhlenbeck, Physica (Utrecht) **5**, 266 (1925).
5. P. Ehrenfest, G. E. Uhlenbeck, Proc. R. Acad. Amsterdam **29**, 1280 (1926).
6. D. Lang, The New Yorker, 7 November 1953, p. 47; 14 November 1953, p. 45.
7. S. Goudsmit, Arch. Neerl. Sci. Exactes Nat. **6**, 116 (1922).
8. S. Goudsmit, Ned. Tijdschr Natuurk. **37**, 386 (1971); English translation in Delta, Summer 1972, p. 77.
9. S. Goudsmit, Physica B1 **21**, 445 (1946).
10. G. E. Uhlenbeck, PHYSICS TODAY, June 1976, p. 43.
11. S. Goudsmit, Physica (Utrecht) **5**, 281 (1925).
12. S. Goudsmit, G. E. Uhlenbeck, Physica (Utrecht) **5**, 266 (1925).
13. M. Abraham, Ann. Phys. (Leipzig) **10**, 105 (1903), section 11.
14. G. E. Uhlenbeck, S. Goudsmit, Naturwissenschaften **13**, 953 (1925).
15. P. Ehrenfest, letter to H. A. Lorentz, 16 October 1925, Lorentz Archives, University of Leiden. Microfilmed as part of the Archives for the History of Quantum Physics project and available at the AIP Niels Bohr Library in New York and at the 16 other AHQP libraries of deposit around the world.
16. H. A. Lorentz, *Collected Works*, vol. 7, Martinus Nijhoff, The Hague (1936), p. 179.
17. W. Heisenberg, letter to S. Goudsmit, 21 November 1925; reproduced in ref. 8 and available at AHQP libraries of deposit.
18. S. Goudsmit, G. E. Uhlenbeck, Nature **117**, 264 (1926).
19. N. Bohr, letter to P. Ehrenfest, 22 December 1925; available at AHQP libraries of deposit.
20. For more details on the spin story see A. Pais, *Inward Bound*, Oxford U. P., New York (1985), chap. 13, sections c and d.
21. R. de L. Kronig, letter to H. A. Kramers, 6 March 1926; available at AHQP libraries of deposit.
22. N. Bohr, letter to R. de L. Kronig, 26 March 1926; available at AHQP libraries of deposit.
23. R. de L. Kronig, letter to N. Bohr, 8 April 1926; available at AHQP libraries of deposit.
24. A. Pais, G. E. Uhlenbeck, Phys. Rev. **79**, 145 (1950); **116**, 250 (1959). ■

PHYSICS TODAY/APRIL 1979

Samuel A. Goudsmit
1902–1978

Samuel A. Goudsmit died on 4 December on the campus of the University of Nevada, Reno, where he had been a Distinguished Visiting Professor since 1974. Sam—as he was called by his friends—was a man of many parts and possessed an interesting and complex personality.

He was born in The Hague, Netherlands on 11 July 1902. He studied physics at the University of Leiden and received his PhD in 1927. As a beginning student, only nineteen years old, Goudsmit published his first paper (Naturwiss. **9**, 995, 1921); he pointed out that the relativistic Sommerfeld formula for x-ray doublets was also valid for the alkali doublets. As George Uhlenbeck wrote to me: "It showed his remarkable instinct for finding empirical regularities. At that time because of the 'Rumpf-Modell,' Sam's paper was a kind of heresy, so that he never got the proper credit for it. Still, I think, it foreshadows the spin!"

After his first examination, he became a kind of "house theoretician" of Zeeman at the University of Amsterdam. While both Goudsmit and Uhlenbeck were still students they realized that the fourth quantum number introduced by Pauli in formulating his exclusion principle could be interpreted as a new degree of freedom of the electron. They found that a spin $\frac{1}{2}$ and a magnetic moment of one Bohr magneton could explain the spectroscopic results (Naturwiss. **13**, 953, 1925). Goudsmit and Uhlenbeck charmingly and instructively told of their outstanding discovery at the American Physical Society Annual Meeting in January 1976 (PHYSICS TODAY, June 1976, page 40). After some resistance—as any important new idea is likely to meet—and especially after Llewellyn H. Thomas had explained a missing factor of two in the doublet splitting as a relativistic effect, the electron spin was generally accepted and its role in atomic and solid-state physics recognized as basic. The generalization to protons and neutrons ensures its role also in nuclear and elementary particle physics. It is now generally believed that all matter contains spin-$\frac{1}{2}$ constituents, leptons and quarks (fermions), which communicate with each other via particles of integer spin (bosons). For electrons, Dirac's relativistic theory (1928) "predicts" the spin $\frac{1}{2}$.

Among Goudsmit's many other spectroscopic contributions was the determination with Ernst Back of the first nuclear spin and its Zeeman effect in Bi^{209} from

GOUDSMIT

an analysis of its hyperfine structure (1927–28). Sam said that he was more thrilled by this discovery than by the electron spin.

In 1927 Goudsmit and Uhlenbeck joined the physics department of the University of Michigan at Ann Arbor. This was the first place in the United States to have a Physics Summer School, an idea inspired by Harrison M. Randall which has since spread all over the world. Goudsmit helped greatly in the running of the summer school. While at Ann Arbor, he had a number of PhD students, including Robert F. Bacher, R. A. Fischer, David R. Inglis, and Ta-Yau Wu. With Bacher he introduced fractional parentage coefficients for the treatment of many-electron problems, an idea which Racah and his students later applied fruitfully to nuclear physics. Bacher and Goudsmit published a book, *Atomic Energy States* (McGraw-Hill, New York 1932), an important source book for many years. Sam also published a lucid introduction to atomic physics with Linus Pauling (*The Structure of Line Spectra*, McGraw-Hill, 1930) for which Goudsmit's Leiden PhD thesis supplied important background. Besides spectroscopy his main scientific interest was in statistical problems.

After World War II broke out, Sam joined the MIT Radiation Laboratory to work on radar and was detailed to the Army on a scientific intelligence mission for which he was uniquely qualified: To find out where the German scientists stood in the race for the atomic bomb. He found that they were far behind the American–British effort—not to the

surprise of every nuclear physicist but to the relief of all. He reported on his mission by its code name in a widely-read (and now sold-out) book, *Alsos,* (H. Shuman, New York, 1947). In the course of this mission he had the devastating experience of seeing the ruins of his childhood home in The Hague and of learning that his parents had been deported by the Germans to an extermination camp. These tragic events depressed Sam and his ebullience and enthusiasm for research suffered.

In 1946 he joined Northwestern University, Evanston, Illinois. In 1948 he came to Brookhaven National Laboratory, where he remained until his retirement in 1970. From 1952 to 1960 he was the Chairman of the Brookhaven Physics Department and in this capacity attracted many outstanding physicists who played a vital role in the development of the Laboratory. Here also he carried out his only experimental work, the building and use of a magnetic time-of-flight mass spectrometer of his design (with E. E. Hays and Paul I. Richards, Phys. Rev. **84**, 824, 1951).

Sam was Editor-in-Chief of the American Physical Society from 1951 to 1974. He founded *Physical Review Letters* in 1958, a much imitated letter journal and still the most widely known. His sense of humor helped him deal with the many idiosyncrasies of authors: without mentioning names he enjoyed telling some of his experiences such as the story of a manuscript which was followed by a telegram saying simply, "I am worried about equation two." He also liked to tell the joke, provoked by the explosion of physics articles, that simple extrapolation shows that by the year 2000 the speed of growth of the *Physical Review* on a shelf would exceed the velocity of light; but this would not contradict the special theory of relativity since the information transfer would by then go to zero!

While at Brookhaven he also found time to teach courses at Rockefeller University. He especially enjoyed teaching physics to "humanists" and continued to do this at the University of Nevada. His warm interest in young people, his enthusiasm, his scholarship and his wit made his lectures a great success.

Among his honors, many of which he shared with his friend George Uhlenbeck, are the Max Planck Medal of the German Physical Society (1965), the Karl T. Compton Award for Distinguished Statesmanship in Science, American Institute of Physics (1974), Commander of the Order of Orange-Nassau (1977) (one of the highest Dutch distinctions), and the

National Medal of Science (1977).

Sam had a passionate interest in Egyptology which began when he was a graduate student, and he did original research on papyri. On Sam's first visit to Copenhagen in 1926, Niels Bohr took him to the collection of Egyptian sculpture at the Glyptotek. As Bohr started to translate the Danish labels, Sam quietly told him that this was not necessary; he could read the inscribed hieroglyphs. He had a talent for solving puzzles, be they scientific, historical, artistic or crossword.

Sam will be remembered as a man who stood up for his convictions. In his will, in his usual self-effacing way, he asked that no memorial session be held for him, a request not easy for his friends to honor.

MAURICE GOLDHABER
Brookhaven National Laboratory

Physicists I have known

Paul P. Ewald

PHYSICS TODAY/SEPTEMBER 1974

Paul E. Ewald, pioneer solid-state physicist and crystallographer, was born in Berlin in 1888. After obtaining his PhD in Munich in 1912, Ewald taught and did research at Munich, Stuttgart, Cambridge, and The Queen's (Belfast) Universities and at the Polytechnic Institute of Brooklyn, where he headed the physics department. Ewald gave a talk at the New York State Section of the APS at Rensselaer Polytechnic Institute in 1974 on which, along with notes he generously supplied to PHYSICS TODAY, this article is based. After retirement Ewald and his wife Ella lived in Ithaca, New York, where they were close to their daughter and son-in-law, Rose and Hans Bethe. The picture of Ewald below shows him as a *Feldröntgenmechaniker* (field x-ray technician)—see also the WWI-vintage mobile x-ray unit, next page.

My first encounter with science came when I was eleven. My mother and I were living in Berlin, but often spent summer vacations with friends in Cambridge, England. In their house we met **Siegfried Ruhemann,** holder of the first chair in organic chemistry in England, at Cambridge's Gonville and Caius College.

Ruhemann, who was then doing delicate research on snake-venom alkaloids, took me to his little laboratory. It was in a one-story shed separated from the other college buildings by a cobblestoned yard. I was amazed to see hundreds of neatly-labelled bottles on the shelves, and benches filled with faucets, glassware and other implements.

The professor took a piece of glass tubing and heated it in the flame of a Bunsen burner, turning it in his hands. I could not understand how he was able to hold the tube while the ends were glowing red. Closing one end with his finger, he brought the other end up to his mouth and, in several stages, blew the midpart to a magnificent bubble five or six inches in diameter. More marvels were to come. After cooling the bulb in the luminous flame, making it (to my dismay) temporarily quite sooty, Ruhemann heated the lower part of the tube in a noisy pointed flame, and quickly severed it. After the now perfectly rounded bulb had cooled thoroughly, Ruhemann poured clear liquids from two different bottles into the bulb and shook it thoroughly. Lo and behold, the inside became the brightest silver! In my enthusiasm I resolved then and there to become a chemist.

I treasured the beautiful silver bulb beyond all my other possessions. But on the midnight journey home, the drowsy boy, carrying his treasure like a priceless relic, hit the railing of the steamer's gangplank. Only a silver star of thin glass was left. Back home in Berlin, however, I started my own laboratory.

Although the boy had just completed high school, Sommerfeld
allowed him to enroll in his course. "May I also come to the seminar?"
asked Pauli. Reminiscences of the heroic age of physics.

On another visit to Cambridge, our
friends introduced me to **J. J. Thomson,** a
lively man and splendid conversational-
ist who seemed to be acquainted with
every novel that was mentioned. I
think he said that he read a novel a
night. He was very proud of his son,
now **Sir George Thomson** of electron diffraction fame. George, with the help of
his father, constructed scale models of
warships—one of every type in the British navy.

When I returned to Cambridge in
1905 as an undergraduate at Gonville
and Caius, I attended a very impressive
evening lecture by J. J. Thomson. It
was entitled "The Structure of Atoms."
In his model of the atom, negative electrons swam freely in a uniformly filled
sphere of positive electricity. They
were thus attracted to the center and
repelled by one another—all by coulomb forces. The idea was illustrated
by J. J.'s famous assistant-mechanic in
a demonstration of intriguing simplicity. Little cork boats, each carrying a
compass needle, floated in a round dish
of water. One pole of a bar magnet suspended above the center of the dish
provided the central force. Watching
the image projected by an overhead
mirror, I saw a single boat (for hydrogen) place itself at the center. Two
boats, representing the helium atom,
came to rest on opposite sides of the
center. Three formed an equilateral
triangle and four a square. With the
addition of a few more boats, however,
these symmetrical arrangements clearly
became unstable; in addition to the
symmetrical inner array, an incomplete
outer circle began to form. To Thomson, the relevance of this phenomenon
to the periodic system was obvious.
What a pity the model was a wrong
guess!

The next day I repeated Thomson's
demonstration in my "digs"—I am sure
I was not the only one. I was very
proud when the experiment in my wash

BOHR LIBRARY

J. J. THOMSON

SOMMERFELD . . .

. . . AND FELLOW HIGH-SCHOOL GRADUATES

basin, using magnetized sewing needles, succeeded.

In the spring of 1901, when I was 13, my mother took me to the village of Glion on Lake Geneva to recover from bronchitis. The mountain meadows were full of crocuses and narcissi and the air was balmy. As my mother sketched the scenery, a rather haggard young lady walked up to chat about the painting with her. She was in her twenties, with a beautiful pale face dominated by startling blue-grey eyes. **Edith Stoney** was a lecturer in physics at a women's college in London, and was staying in a big hotel higher up the road to recover from overwork and a tuberculosis attack. I instantly fell in love with her and could hardly wait to meet her again. Our stay came to an end too soon, but she invited us to visit her on our next trip to England.

Next year we went to London and were invited to tea with Miss Stoney and her parents. I was deeply disappointed when Miss Stoney opened the door—the angelic invalid with the translucent features had turned into a red-cheeked, somewhat buxom young woman. She introduced us to her lively mother and her retired-physicist father, **Johnstone Stoney.** Little did I know at the time that this was the man to whom we owe the name electron for J. J. Thomson's "corpuscles." The bearded old gentleman took me to his attic room as the sun set over a sea of roofs and chimneypots, and showed me the large spectrograph with which he studied light absorption in the atmosphere.

In 1906 I enrolled in the chemistry course at Göttingen. In the interests of the medical students, the course was given at seven in the morning. Chemistry was taught as a jumble of facts with as little theoretical connection as the recipes in a cookbook. When I found myself dozing off for the third time, I gave up attendance.

My interest changed to mathematics, for which Göttingen was the Eldorado. **David Hilbert** gave me my first paid appointment—to write out, for 100 marks per semester (about $24 at that time), his lecture course on differential and integral calculus. Going over these notes carefully with Hilbert's assistant **Hellinger,** a devoted teacher, taught me practically all the analysis I ever needed. Through Hellinger I met a group of advanced students, including **Herman Weyl** and his classmate **Wink,** who had composed a rhymed play for the mathematician's excursion.

On this yearly occasion, some 100 mathematics students and their teachers, from the great **Felix Klein** down to the lowliest assistant, wound through fields and meadows to Mariaspring, a

HILBERT . . .

. . . AS DRAWN BY HOLM

DEBYE

linden-shaded rural inn about four miles from Göttingen. There we had supper, easy approach to our teachers and entertainment. The play showed a timid young student seeking advice on the various branches of mathematics from a professor. In the role of the professor I unwittingly presented a caricature of the formidable Felix, but no harm came of it.

Hilbert was not a smooth lecturer; he made mistakes. He would walk about the dais and think until you became so nervous that you wanted to shake him to find out what was going to happen. What finally emerged, however, was so impressive you did not easily forget it.

In a student's mess where I took my middle meal (the main meal of the day) I met **Ragnar Holm**, who was later to be-

come the chief authority on electrical contacts and the father of plasma physics. Holm, who had a keen sense of humor, once gave a talk at the Göttingen Physical Society, a rather dignified, formal group. He spoke in a German strongly laced with a musical Swedish accent; so he ended by saying, "I have heard many a bad talk in this society, but today I have taken my vengeance." It was Holm who, during a lecture, drew the caricature of Hilbert shown on this page—he even put an integral sign on the enormous forehead!

After two years at Göttingen I transferred to Munich to study the new rigorous approach to mathematics originated by **Karl Weierstrass**. My professor was **Alfred Pringsheim**, a dapper little man with a bald crown flanked by bristling rows of gray hair and the convivial red face of a *bon vivant*. His lectures were works of perfection. It was a pleasure to see Pringsheim's neat writing extending over the blackboard, and to listen to his occasional sarcastic remarks on other trends of mathematics. Yet, much as I enjoyed these lectures, I later found that they had really not

penetrated under my skin—as Hilbert's certainly had.

When a friend dragged me into a lecture course by **Arnold Sommerfeld** I was quite unwilling to be distracted from pure mathematics into the field of hydrodynamics. Besides, I had not taken the course in mechanics which generally precedes it. In a few lectures, however, Sommerfeld introduced the notions of vector algebra and vector analysis, illustrating them with the behavior of liquids. The immediacy of the correspondence between concept and phenomenon was a delight. My resistance gone, I joined the Sommerfeld school, renouncing the pale shade of pure mathematics.

Sommerfeld's school of theoretical physics was unique until about 1924,

BORN ...   ... 60 YEARS LATER   KARMAN

when **Max Born** in Göttingen and **Niels Bohr** in Copenhagen formed comparable schools. Aside from the obvious requirement of a scientific leader of great experience, fertility and integrity, environment and the snowball effect also played a role in the formation of such schools, which attracted so many devoted, highly qualified students.

Before his world-wide lecture tours acquainted him with life in other countries (particularly the US), Sommerfeld's understanding of the world was rather restricted. This cannot be said of his relations with his students, the development of whose faculties he loved to watch. He stimulated beginners by giving them small tasks, such as drawing diagrams, calculating tables or discussing special cases for his own research papers, or giving seminar talks.

He gave his students the freedom to find their own way. There were no formal examinations except the final ones: the doctorate or the qualifying exam for high school teachers.

Sommerfeld made what he called his greatest discovery among the engineering students at the Polytechnic School in Aachen, where he was professor of technical mechanics after leaving Göttingen. This was **Peter Debye**, a keen, bright-eyed young man who commuted from the nearby Dutch town of Maastricht. Sommerfeld made him his assistant and, when he was offered the chair of theoretical physics in Munich, accepted it on the condition that he could bring Debye with him.

When I joined Sommerfeld's group in 1909, Debye was still working on his voluminous thesis on the theory of the rainbow. In it he replaced the old Cartesian theory of refraction and dispersion of the sun's rays in raindrops with a theory based on the diffraction of light by small drops of water. He promised the manuscript to Teubner for publication as a book, but more urgent problems intervened and the book never appeared.

Max Planck's quantum theory of

blackbody radiation had introduced the enigmatic new universal constant h. Was h a property of the atom or of the radiation? Were energy quanta produced in emission, absorption or both? **Albert Einstein** had already applied the quantum theory to the vibrations of atoms in a solid, thereby obtaining a relation for the temperature dependence of the specific heat. Unfortunately, Einstein's model, in which all vibrations were assumed to be of the same frequency, was unsuccessful in rendering the accurate measurements of **Walther Nernst** and his school. Debye, applying Planck's distribution of energy quanta to the entire spectrum of frequencies in a solid, obtained a different formula for specific heat. This turned out to be well in accord with the measurements. It was a daring innova-

PAULI

LORENTZ

EWALD

tion to consider quantum effects to be spread throughout the solid body, instead of being localized at a single atom.

Born and **Theodore von Kármán** used a similar approach in Göttingen at about the same time. Using a crystal lattice as a model of a solid, they determined its vibrational spectrum. This was the beginning of crystal dynamics as we know it today.

In 1919, with the development of atomic and quantum physics in full swing, a slim, young-looking boy appeared in Munich. The boy, **Wolfgang Pauli**, had a letter from his father, a chemistry professor at the University of Vienna, asking Sommerfeld to place his son where he saw fit. Pauli has just completed high school, but had also studied physics on his own. Sommerfeld told the young man that he could

enroll in his current lecture course, but doubted whether he could understand it. "Certainly," replied Pauli, and added, "May I also come to the seminar?" Sommerfeld could see no sense in his coming to the seminar, which was for advanced graduate students, but nevertheless gave Pauli permission to sit in. It soon turned out that Pauli had the quickest grasp, the most profound understanding and the greatest ability of the participants.

My wife and I had Pauli, who could not go home to Vienna, with us for Christmas 1919. He told us how he used to wear himself out during his high school vacations by reading mathematics until two in the morning. He needed the school term, when his father insisted on his going to bed at eleven, for recovery.

I was deeply impressed by the great Dutch physicist **H. A. Lorentz**, the originator of the electron theory of metals, which formed the foundation of my thesis and later work. It was many years after the first World War before German physicists were welcome in the Allied countries. Through **A. D. Fokker** I was invited to lecture in Holland in 1923, and on that occasion I was re-

ceived by Lorentz in the Teyler Stichting in Haarlem. To my surprise, Lorentz the physicist was most proud of his engineering achievements as the head of the state commission for the reclamation of land from the Zuider Zee. Goethe's Faust, who obtained the greatest satisfaction from the fact that he had reclaimed land from the sea, came to my mind.

When I attended a lecture of his Amsterdam University course, I was struck by the personal, rather fatherly way Lorentz spoke to his class, a contrast to the more aloof lecture style of most German university courses.

The discovery of x-ray diffraction and the development of the "new crystallography" brought friendship with many more memorable personalities than I can include here.   □

Paul P. Ewald 1888–1985

Paul P. Ewald, a key figure in the evolution of modern physics, died at his home in Ithaca, New York, on 22 August 1985, at the age of 97.

Ewald was born in Berlin, Germany, in 1888. His father, who died shortly before Ewald was born, was a historian at the University of Berlin. His mother was an internationally successful portrait painter. He learned to speak English and French before starting school, and his classical education at the *Gymnasium* gave him a lifelong love of literature and languages, especially classical Greek; he readily quoted Homer throughout his life. Soon attracted to the sciences, he tried chemistry at Cambridge, then mathematics at Göttingen and Munich, where he was finally drawn to physics. He thought of Arnold Sommerfeld, David Hilbert and Alfred Pringsheim as his most important teachers. From among the thesis topics proposed to him he chose what Sommerfeld considered the "least promising" one, namely "Dispersion and double refraction of electron lattices (crystals)." He had been fascinated since boyhood by light and its interaction with solid matter, and he was to be preoccupied by this topic for the rest of his life.

Shortly before submitting the thesis, he sought an interview with Max von Laue to clarify some details. In the course of this discussion Laue conceived the idea of x-ray diffraction in crystals. Experiments quickly established x rays as waves and confirmed the existence of crystal lattices. This concept was basic to Ewald's thesis, which also contained most of the mathematical formalism of the dynamical theory of x-ray diffraction in perfect crystals. The theory was worked out fully shortly thereafter.

During his professorship in theoretical physics at the Technische Hochschule, Stuttgart (1921–37), Ewald's department became an international center for x-ray diffraction and solid-state physics. In 1932 he became *Rektor* (equivalent to university president), but resigned this post soon after the Nazis came to power. He continued in his position as professor until 1937, when he was pensioned after walking out of a faculty meeting in protest over a speaker's statement: "Objectivity is no longer a valid or acceptable concept in science." Soon thereafter he left Germany for Cambridge, England, where he had been offered a small research grant. Subsequently he held academic positions at Queen's University, Belfast, Northern Ireland (1939–49), and the Polytechnic Institute of Brooklyn (now of New York, 1949–59), where as department head for seven years, he created a new center for research. Retirement in 1959 did not stop his research nor his many other endeavors relating to crystallography.

Ewald's celebrated theory of x-ray diffraction (1917) remains a masterpiece of a self-consistent theory of normal modes including many-body interactions, and its treatment of optical boundary conditions at the microscopic level (the extinction theorem) is unsurpassed.

Forty years passed before the semiconductor industry produced crystals of sufficient perfection to permit detailed confirmation of the theoretical predictions. Today, this same theory enables industry to verify the vitally important perfection of its crystals, and in such applications as x-ray interferometry, leads to precise values for many solid-state parameters. The theory's influence, however, extends beyond x rays: The original theory of electron diffraction in crystals (by Hans Bethe) drew on Ewald's concepts, and its subsequent development as a quantitative tool was essentially dynamical. It may be surprising to learn that even the analysis of many of the contrast details in the electron microscope, which are generated by strain fields of dislocations, relies heavily on Ewald's conceptual description of two-beam dynamical interactions, including anomalous absorption (the Borrmann effect). An offshoot, the Ewald sum procedure, was originally invented to calculate the electrostatic energy of an ionic crystal and is widely used, for instance, in modern band-structure calculations.

For 60 years Ewald was a prime mover in x-ray crystallography. His book *Kristalle und Röntgenstrahlen* (1923) gave the first comprehensive treatment of the subject, while *Fifty Years of X-Ray Diffraction* (1962) surveyed the mature field. Together with C. Hermann he founded *Strukturbericht* (first published in 1931), a collection of results on crystal structures, which, with its successor volumes *Structure Reports*, is the standard structure-data repository for industry and science. The *International Tables*, also conceived by him in the 1930s, set the uniform nomenclature, units and standards for the specification of these data. After World War II, Ewald initiated *Acta Crystallographica* and acted as one of its chief editors from 1948 to 1959. He was president of the American Crystallographic Association for 1951–52.

Ewald was very much concerned with the international character of his science. After World War II he was instrumental in reestablishing the International Union of Pure and Applied Physics, serving as its first secretary-general and later as vice-president. He was instrumental in founding a separate International Union of Crystallography, of which he served as president for some time.

The success of these diverse and wide-ranging science-policy initiatives rested largely on Ewald's personal qualities as a scientist of high vision and standards and as a diplomatic and convincing negotiator, but ultimately on his disarming honesty and modesty. To his students and colleagues he was not only a window into the larger world of science and the intellect, but also a model physicist in his scrupulous search for the physically correct and lucid formulation of ideas. He was equally demanding in his insistence, honed by long years as an editor, on the precise language in which these ideas were to be expressed. At the same time, he always enjoyed his science and was delighted when progress was made, either by himself or by others. He remained receptive to new ways of thinking and was ready to get them a fair hearing when normal processes for doing so broke down. Finally, there is no better testimony to his personal qualities—harmoniously complement-

ed by those of his wife, Ella—than their enormous international circle of friends, who made the Ewald residence, wherever it was, an eagerly sought-out stopover point for travelers from far and wide.

In 1979 Ewald received the first Gregori Aminoff Medal of the Royal Swedish Academy, in honor of his life-long accomplishments. In 1985 IUC established the Ewald Prize, which was awarded for the first time in 1987.

In his late eighties, Ewald told one of us that he would like to "finish his doctor's thesis" by finding a way to deduce the structure of a crystal direct-ly from the intensities of the x-ray diffraction spots. He would have been

delighted that the 1985 Nobel Prize in chemistry was awarded to Jerome Karle and Herbert Hauptman for find-ing a solution to this problem.

H. J. JURETSCHKE
Polytechnic Institute of New York
Brooklyn, New York
Royal Melbourne Institute of Technology
Melbourne, Australia
A. F. MOODIE
CSIRO
Clayton, Victoria, Australia
H. K. WAGENFELD
Royal Melbourne Institute of Technology
Melbourne, Australia
H. A. BETHE
Cornell University
Ithaca, New York

PHYSICS TODAY/MARCH 1961

LEV DAVYDOVICH LANDAU

WINNER OF THE SECOND FRITZ LONDON AWARD

An address presented at the 7th International Conference on Low Temperature Physics (Toronto, Aug. 29 to Sept. 3, 1960) on the occasion of the 2nd Fritz London Award ceremony. Dr. Landau was unable to attend.

By J. R. Pellam

Photo by L. Alvarez

I HAVE been asked by the Committee for the Second Fritz London Award to give an account of the life and work of this eminent recipient of the Award, Lev Davydovich Landau. I was very honored that I had been asked to undertake this task but felt rather overwhelmed by the responsibility it entailed. Because Landau has contributed to so many fields of physics, an award could have been made to him at any one of several conferences in any one of several fields. The main problem, I found, was to limit myself primarily to Landau's work in the field of low-temperature physics for which this Award is made. My own work in this field has been so strongly influenced by these significant contributions that I, like so many of us similarly influenced, feel that I do know him, although I have never met him personally.

A considerable wealth of material is available describing Landau's work in the many fields of physics to which he has contributed. The following outline of Landau's career is drawn from two articles [1], [2] published in Soviet scientific journals commemorating his fiftieth birthday, which he kindly arranged to have fall two years before winning the Fritz London Award.

Lev Davydovich Landau was born on January 22, 1908, in Baku, the capital of Azerbaijan on the Caspian Sea. His father was an engineer; his mother a doctor. His mathematical talents were apparent at a very early age and he can scarcely remember not being able to differentiate and integrate. At the age of fourteen he entered Baku University, from which he transferred two years later to the University of Leningrad, where he completed his studies in 1927 at the age of nineteen. Scientific writing did not await the completion of his studies, however, for he published twice during each of his last two school years. He developed an active interest in the new science of quantum mechanics, and at the age of nineteen introduced the concept of the density matrix for energy which is now so widely used in

John R. Pellam was professor of physics at the California Institute of Technology, Pasadena, California.

[1] Soviet Physics—JETP **7**, 1 (1958).
[2] Uspekhi Fizicheskikh Nauk **64**, 616 (1958).

quantum mechanics. His active scientific research career began in the Leningrad Physicotechnical Institute where he stayed from 1927 to 1929, working on the theory of the magnetic electron and on quantum electrodynamics. In 1929 he was sent abroad and spent a year and a half as guest of the Danes, the Germans, the Swiss, the Dutch, and the English. Of particular importance to Landau's development was his work at the Institute in Copenhagen during this period, and he considers himself a student of Niels Bohr. (At Bohr's invitation, Landau was in Copenhagen again in 1933 and in 1934, participating in theoretical conferences.)

SOME measure of his personality can be gained by the following quotations from letters which I have received from two physicists associated with Landau during this period. The first is from Professor Niels Bohr, his teacher:

> It is a great pleasure indeed to learn that the Fritz London Award will be presented to Landau. Of course we all here share in the appreciation of Landau's great work and have vivid remembrances from the time about thirty years ago when he joined our group in Copenhagen. From the very beginning we got a deep impression of his power to penetrate into the root of physical problems and his strong views on all aspects of human life, which gave rise to many discussions. In the booklet which was published at my seventieth birthday, Rosenfeld has given a vivid picture of the stir at the Institute caused by the paper of Landau and Peierls on the measurability of field quantities, which eventually gave rise to a long treatise by Rosenfeld and myself. Also from our visits to Russia before the war my wife and I have many treasured remembrances of Landau's personal attachment and his striving for promoting mathematical physical research in Russia, in which he since has had so great success. In the years after the war we have constantly hoped to see Landau here again, but so far he has not been able to come. However, my son Aage and several of the other members of the Institute have, on visits to Russia, met and spoken with Landau and not only learned about the admiration in which he naturally is held by his colleagues, but in him found the same warm and enthusiastic personality, which we all here hold in so deep affection.

The other letter is from Professor Edward Teller, a contemporary of Landau:

> I met Landau in Leipzig in 1930 and later I spent some time with him in Copenhagen in 1934. My most vivid visual memory of him is the red coat he wore in Copenhagen. Mrs. Bohr teased him that he was wearing precisely the correct outfit for a postman. You will understand the somewhat strange circumstances that I would have forgotten about the red coats of the Copenhagen postmen except for this incident. I liked Landau very much and learned from him a great deal of physics. He enjoyed making statements calculated to shock members of the bourgeois society. While we were both in Copenhagen I married. He approved of my choice (and played tennis with my wife). He also asked both of us how long we intended to stay married. When we told him that our plans were definitely for a rather

long duration and, in fact, we had given no thought to terminating the marriage, he expressed most strong disapproval and argued that only a capitalistic society could induce its members to spoil a basically good thing by exaggerating it to this extent. In Copenhagen Landau had many arguments with James Franck about religion. He considered his religious belief incredibly outmoded for a scientist and expressed himself in immoderate terms both in the presence and absence of Franck. Franck always laughed at him. It was very nice that when Landau left Copenhagen he made a very special point to say good-bye to Franck. It was quite clear that if he meant what he said about Franck, he did mean it in rather a peculiar way and, in fact, he meant perhaps the opposite of what he said.

> I continue to have a great deal of affection for Landau and I am glad that he is getting the Fritz London Award; he fully deserves it.

One should remember that Landau was very young at this time; he may have mellowed some since.

During this period abroad there occurred the first step which represented a transition of his interests and was destined to confront him with the major problems of low-temperature physics. The interesting pattern which had dominated his previous work provided the ammunition for tackling new problems. This became a cumulative process. At the age of 22 he developed the theory of "Landau diamagnetism" of metals, showing that a degenerate ideal electron gas possessed a diamagnetic susceptibility equal to $\frac{1}{3}$ the paramagnetic susceptibility. Some years later (1937–38) this led to the explanation of the de Haas-van Alphen effect. In this very case of diamagnetism, the proficiency which in his early years Landau had developed in manipulating Fermi systems has been basic to his latest theory predicting "zero sound" in liquid helium-3, involving distortions of the Fermi surface. Landau's ease of handling this situation is quite understandable considering the mastery of Fermi systems which he gained thirty years earlier.

HIS return to Leningrad was of short duration, for at the age of 24 he went to Kharkov to head the theoretical section of the Physicotechnical Institute (1932–37), where versatility both in achievement and outlook began to appear. His publications during the first year at Kharkov range from a paper "On the Theory of Stars" to a paper "On the Theory of Energy Transfer in Collisions". The latter characterizes a Landau specialty: the solution of difficult theoretical problems by brilliant mathematical flank attacks. The same methods have held him in good stead—his mastery of collision problems reached a peak in 1949 when he considered roton-roton and roton-phonon collisions (with Khalatnikov) to predict (correctly) the attenuation of second-sound waves.

Landau's convictions that independent creative work in any field of theoretical physics must begin with a sufficiently deep mastery of all its branches took root at Kharkov, where he developed the special program widely known among his physics students as the "theoretical minimum". Here also he began to accumulate a following among students, of whom the best known in low-temperature physics include Lifshitz and Pomeranchuk. His versatility is illustrated by quoting the titles of the papers which he wrote during his last two years at Kharkov:

> Theory of Photo-emf in Semiconductors, Theory of Monomolecular Reactions, Theory of Sound Dispersion (with E. Teller), Kinetic Equation of the Coulomb Effect, Properties of Metals at Very Low Temperatures, Scattering of Light by Light, Theory of Phase Transitions.

All these were published in 1935. In 1936 he published:

> The Kinetic Equation for the Case of Coulomb Interaction, Absorption of Sound in Solids, Theory of Phase Transitions, Theory of Superconductivity, Statistical Model of Nuclei, Scattering of X-Rays by Crystals Near the Curie Point, Scattering of X-Rays by Crystals with Variable Structure, Origin of Stellar Energy.

Of deeper consequence to the field of low-temperature physics, however, was a direction of interest which he developed at Kharkov and continued after moving to Moscow, during the organization of the P. L. Kapitza Institute for Physical Problems. Landau's attention to diamagnetism proved transitional between quantum mechanics and the theory of metals. Besides explaining the de Haas-van Alphen effect, Landau's applications of thermodynamics to electronic systems at low temperatures included the following:

1. He introduced the concept of antiferromagnetic ordering as a new thermodynamic phase;
2. He developed the thermodynamic theory of magnetic domains (with Lifshitz), providing a foundation for theories of magnetic permeability and resonance of ferromagnetics;
3. He studied phase transitions and determined the profound relation between transitions of the second order and variation of symmetry of the system. He gave a detailed thermodynamic theory of the behavior of systems near the transition point;
4. He studied the intermediate state of superconductors and proposed a theory of laminar structure of superconductors.

Also during this Kharkov period, Landau started the series of now well-known monographs on theoretical physics.

IT was only natural upon his arrival at Moscow in 1937, where he was appointed head of the theoretical section of the Institute for Physical Problems, that his interests turned to the subject of superfluidity which was then being investigated experimentally by Kapitza himself. This marks an all-out assault by Landau on pure low-temperature physics, and under his attack the major problem of the nature of the helium II phase of liquid helium-4 soon withered. This work was close to the well-known interests of Fritz London, who solved the problem using another approach. The crux of Landau's cracking the helium problem (published in 1941) was his ability to deduce semiempirically the energy spectrum [3] of the Bose excitations in this liquid. The shape of the now well-known curve of energy versus momentum for such quasi-particles included a valley occurring at an energy height (equivalent kT) of $8 - 10$ °K. Such a spectrum permitted these quasi-particles to exist in equilibrium at this level, and these, following a suggestion by I. Tamm, Landau named "rotons". The energy gap, Δ, inherent to these rotons, permits the existence of superfluidity.

As a consequence of Landau's interpretation of superfluidity, he was able to predict independently the existence of the "second-sound" mode of wave propagation in liquid helium II (independently, because Tisza somewhat earlier had predicted second sound on the basis of Fritz London's approach).

Two aspects of Landau's manner of handling the second-sound problem are particularly noteworthy, in that they may also bear on his most recent predictions of "zero sound" in liquid helium-3:

1. Landau's presentation shows certain detachment from the problems of experimental generation and detection of second sound. Early efforts by Shalnikov and Sokolov before the war were unrewarding because they attempted to detect second sound using standard acoustical methods. In fact, the problem was clarified by a subsequent publication by Lifshitz, who pointed out the essential thermal nature of second sound. On the basis of this prescription, Peshkov observed second sound experimentally in 1944.

2. In the same 1941 paper, Landau correctly predicted the magnitude of the velocity of second sound in the vicinity of absolute zero as $c_1/\sqrt{3}$, where c_1 is velocity of ordinary sound. He produced this result only after complicated mathematical acrobatics, and one wonders how much faith could possibly be placed in such a conclusion. Landau's own faith in his result was eloquently expressed, in a 1949 Letter to the Editor of *The Physical Review* defending his theory:

> ... I have no doubt whatever that at temperatures of $1.0 - 1.1$ °K the second-sound velocity will have a minimum and will increase with the further decrease in temperature. This follows from the thermodynamic quantities in helium II calculated by me.

Who could be so certain? This clearly demonstrates Landau's extraordinary physical intuition. Despite the intricate mathematics he recognized the situation at absolute zero, not as an extrapolation, but as an end position for buttressing the results. Thermodynamic complications dissolved as $T \rightarrow 0$ °K. With only phonons of first sound present, the root-mean-square velocity component along any particular propagation direction of any more subtle propagation could occur only $1/\sqrt{3}$ as fast. This was perhaps Landau's ace-in-the-hole and private little joke besides. We will later

---

[3] A purely quantum-mechanical derivation of this spectrum has been achieved recently by Feynman.

recall these two facets in connection with the theory of "zero sound" in liquid helium-3, and how they may bear on this subject.

IT is quite out of the question to consider all aspects of Landau's accomplishments. Typical of his versatility is a series of five papers published in 1945 concerning shock waves at large distances from their place of origin, and related subjects. (This work was carried out under the Engineering Committee of the Red Army.) Then, in 1946, papers appeared on oscillations of plasmas, which, it is stated, "received specially large notice recently in connection with the study of the properties of plasmas". A large amount of work in this field has been carried out recently by a group under A. E. Akhasier in Kharkov.

During the late 1940's Landau devoted his efforts to a whole gamut of activities. Efforts in the field of low-temperature physics consisted primarily of further applications of his spectrum of excitations in liquid helium to examining various kinetic processes. This included viscosity, thermal conductivity, and attenuation of second-sound waves (with Khalatnikov). In recent years his efforts have included a series of papers (with A. A. Abrikosov and M. I. Khalatnikov) on quantum electrodynamics. During the period when nonconservation of parity in weak interactions had been proposed by Lee and Yang, but before experimental verification, Landau proposed the hypothesis of the *conservation of combined parity*. He transferred his attention to the fact that nonconservation of parity does not, without fail, require violation of the properties of symmetry of space, if it is assumed that also ⟨charge conjugation⟩ is not conserved simultaneously but the product of these quantities, named by him "combined parity", is conserved. This puts definite restrictions on the general hypothesis of conservation of parity. He predicted the polarization of the neutrino, as did Lee and Yang, who did not however connect it with the principle of combined parity. He also discussed the polarization of β particles.

The theory of "zero sound" in liquid helium-3 may quite possibly develop into Landau's greatest contribution to low-temperature physics. This combines Landau's talents in the fields of diamagnetism and of the properties of quantum liquids. Essentially it is a treatment of oscillations of the surface of the Fermi sea, and Landau is quite at home navigating waves on the Fermi sea.

As in his successful approach to the helium-4 problem, Landau considers not the individual particle motion, but instead the collective motion of particles, i.e., the "elementary excitations" or quasi-particles. Also, as in the case of his second-sound predictions, the precise nature of "zero sound" in the sense of the experimental techniques for generation or detection is not discussed; at least, this is the case for the experimentalist who is speaking! The ubiquitous $\sqrt{3}$ shows itself again, and, as before, I feel sure that it carries more physical significance than the limiting form of a complicated for-

mula. But here the velocity of "zero sound" equals the velocity (c_1) *times* $\sqrt{3}$, rather than (c_1) *divided by* $\sqrt{3}$. Probably this is the key to the reason Landau has named this mode of propagation "zero sound" rather than "third sound", for example. It evidently represents [4] a turning back of the crank to arrive at an even more elementary excitation than first sound!

The scientific accomplishments of Lev Davydovich Landau have received due recognition within his own country. In 1946 he was elected an active member of the Academy of Sciences of the USSR. He has been awarded the Stalin prize three times (once in 1941 for his theory of liquid helium and work on phase transitions). Outside his own country, Landau has been elected to membership in the Danish and the Dutch Academies of Sciences; he has recently been elected a foreign member of the Royal Society of London and of the US National Academy of Sciences. He has published well over a hundred papers in more than a dozen scientific journals, and is the author or coauthor of a total of ten books. I will conclude with two excerpts from the *JETP* article [1] written on the occasion of his fiftieth birthday, which to me appear particularly appropriate:

It is not without significance that at the weekly seminar which Lev Davydovich conducts at the Institute for Physical Problems, reports are presented not only on theoretical researches but also on the results of experimental work on the most varied problems in physics. Participants in the seminar are repeatedly amazed to see Lev Davydovich show equal enthusiasm and thorough knowledge in discussing, for example, the energy spectrum of the electrons in silicon, directly after dealing with the properties of the so-called "strange" particles. . . .

The breadth of Lev Davydovich's grasp of contemporary physics is even more convincingly shown by the course of theoretical physics which he has written together with E. M. Lifshitz.

Taken together, these books are a fundamental treatise on theoretical physics. In originality of exposition and broad grasp of the material they are unprecedented in the whole world-wide literature of physics, and so have attained wide popularity not only in this country but also abroad.

The contribution for which theoretical physics is indebted to Lev Davydovich is not exhausted by his own scientific writings. We have already spoken of another side of his activity—his founding of a broad school of Soviet theorists. His inextinguishable enthusiasm for science, his acute criticism, his talent and clarity of thought attract many young people to Lev Davydovich. The number of those, both young and mature scientists, who turn to Dau (as his pupils and associates have come to call him) is very large. Lev Davydovich's criticism is hot and merciless, but behind this outer sharpness is hidden devotion to high scientific principles and a great human heart and human kindness. Equally sincere is his wish to aid the success of others with his criticism, and equally warm is his expression of approval.

---

[4] Zero sound appears distinguished from first sound primarily as a distortion, rather than a displacement, of the Fermi surface.

REMINISCENCES OF LANDAU

Landau envisioned theoretical physics as one indivisible science. By imparting this philosophy to his students, he set the tone of modern Soviet theoretical physics.

I. M. Khalatnikov

PHYSICS TODAY/MAY 1989

In 1932 Lev Davidovich Landau moved from Leningrad to Kharkov. The lectures he started giving at Kharkov University immediately drew the attention of the students. Those who knew Landau can readily imagine the charisma he emanated. Besides, those years were the golden age of theoretical physics. Quantum mechanics had already been created, but the wide field of its applications was still to be explored. In particular, the quantum theory of the solid state began to sprout around that time. Landau's sociable disposition and his readiness to discuss physics immediately stimulated the formation of a group of young physicists and students who were keen on working with him.

Not all of them, however, had had a sufficiently good grounding in theoretical physics for advanced study under Landau. Landau knew that. Already at that time he envisioned theoretical physics as one indivisible science, with its own logic based on certain general principles. Later, he converted this vision into the course on theoretical physics he developed with Evgenii M. Lifshitz. The plan of the course became the "theoretical minimum" for the students; it also involved a number of mathematical problems that called for what Landau regarded as indispensable knowledge from every theorist. Now young people eager to work with Landau had to pass an examination based on that course. Later and in jest, Peter L. Kapitsa of the Institute for Physical Problems called Landau's course the "technical minimum."

The Landau school

Although much has been written about the theoretical minimum, I would like to touch on it again because it was in fact the seed from which emerged what we now call the Landau school. Virtually all his students and associates were tested on the theoretical minimum. The first exam Landau gave anyone eager to become his student was in mathematics. The exam required the applicant to be able to calculate any indefinite integral that could be expressed in terms of elementary functions, to be able to solve any ordinary differential equation, and to have knowledge of vector analysis, tensor algebra and the principles of functions of a complex variable. Landau believed that tensor analysis and group theory should be studied together with the fields of theoretical physics in which they find application. Only after passing this exam could an applicant move successively on to the study of the seven sections of the theoretical minimum. This study demanded basic knowledge of all fields of theoretical physics. Landau thought all theorists should master this basic knowledge, regardless of their eventual specialty fields.

At first Landau gave the exams himself. Later, when the number of students who wanted to take the exams became too large, he entrusted the task to his closest associates. But Landau always reserved for himself the first meeting with each new applicant.

Of course, not everyone had the ability or the persistence to complete the study of the theoretical minimum. All in all, 43 physicists have passed the exam. The effectiveness of the selection is evident from the fact that 10 of these people have already become members of the Academy of Sciences of the USSR and 20 have obtained the Doctor of Sciences degree.

The Landau school did not spring into being spontaneously: It was conceived—programmed, as we might say now—and the theoretical minimum became a mechanism that enabled talent to be collected selectively. The school became the birthplace of many Soviet theoretical physicists, some of whom have founded schools of their own, imparting to them specific characters. Gradually, with the advance of theoretical physics, Landau's school also evolved. I would like to tell you about Landau's working style and that of his students in the postwar years, when I had the lucky opportunity to study and work with him.

I met Landau for the first time in the fall of 1940 when I came to see him at the Institute for Physical Problems in

I. M. Khalatnikov is director of the L. D. Landau Institute for Theoretical Physics in Moscow.

The physicist as a young man: Lev Davidovich Landau.

Moscow with a letter of recommendation from my first physics teacher, Boris N. Finkel'stein of Dnepropetrovsk University, and expressed my wish to take the theoretical minimum exam. I passed the exam in two stages, in the fall of 1940 and in the spring of 1941. I had known about the theoretical minimum as a student at Dnepropetrovsk University. The way theoretical physics was taught in Dnepropetrovsk was based on the lectures Landau gave in Kharkov, and some who had graduated before me had gone on to Kharkov and taken the theoretical minimum. One might say that at that time Landau's fame had already spread far and wide.

After I passed the last test, Landau recommended me for the postgraduate course. But the war interfered with my studies. Only in the fall of 1945 did I become a postgraduate student at the Institute for Physical Problems; I cooperated closely with Landau until the tragic car accident in which he was involved in January 1962.

The Landau seminar

The Landau seminar was held each Thursday in the conference hall of the Institute for Physical Problems.

Attendance at the seminar was governed by one of the unwritten laws that everybody strictly observed—although, naturally, there was no way to control it: Attendance was compulsory for Landau's students working either in the institute's theoretical division or in other institutes where they already headed theoretical groups. The seminar always started at 11 o'clock sharp. But usually everybody came earlier, and at 2 or 3 minutes to 11, when most of the participants, usually about 10 to 12 in all, were seated at a rectangular table, Landau would say as a joke, "There is still one more minute. Let us wait, maybe Migdal will come." As a rule, at that very moment the door would open and Arkadii B. Migdal would turn up. This joke became part of the distinctive seminar ritual.

Reports on original papers were sometimes delivered at the seminar, but more often articles from authoritative scientific journals were surveyed. Landau called the seminar participants in alphabetical order, asking each for the current issue of a journal (very often someone would bring *Physical Review*). Lev Davidovich looked through the journal and marked the articles he thought of particular interest. Because his scientific interests were

Family portrait. The young Landau is pictured with his parents and sister in a photograph from the family album.

not confined to any specific field of physics, the articles he chose ranged over all fields, from solid-state physics to general relativity theory. Sometimes the selected papers covered very narrow, specific problems of solid-state physics. About such papers Landau usually said something like, "Well, this is about alum!" Yet the papers about alum were analyzed at the seminar as thoroughly as the papers on the fundamentals of quantum field theory. Landau admired all that is physics.

Presenting a report at the seminar was not a simple task; preparation was very time consuming and required an extensive background. One was to summarize the contents of the chosen paper based on a complete understanding of the subject. No one could plead unfamiliarity with a subject to justify a failure to survey this or that paper. This is where the training ensured by the theoretical minimum manifested itself: Landau was grounded in all fields of theoretical physics, and he required the same of his students and colleagues.

As long as Landau or any other participant in the seminar had questions, the speaker had no right to abandon the "stage." After the presentation Landau would give an evaluation of the results obtained in the reviewed article. If the results were outstanding, they were inscribed into the "Golden Book." If in the course of the discussion there arose problems requiring further investigation, these were introduced into the "Book of Problems." (This book, from which young physicists regularly gleaned subjects for serious research, was kept regularly until 1962.) Some articles Landau denounced as "pathological," which implied that principles of scientific analysis were violated either in the solution of the problem or in its formulation. Landau himself did not read scientific journals, and thus the seminar was converted

into a "creative laboratory" where Landau's students, while feeding him scientific data, were taught his deep critical analysis and understanding of physics.

As the years passed, the circle of participants at the seminar was gradually enlarged as more physicists passed the theoretical minimum exam. The members of the seminar became too numerous to be seated at a table on the stage; now they filled the hall of the Institute of Physical Problems to the brim.

Each physicist who passed the theoretical minimum exam acquired both rights and duties. He acquired the right to be backed by Landau, but at the same time he made a commitment to give reports at the seminars. If a speaker failed to give intelligible answers to the questions pertaining to the reviewed material or could not clearly expound his thoughts, his situation was not enviable. Sometimes the unlucky fellow's name was excluded from the list of speakers, and he was deprived of the right to review articles from scientific journals (this happened rather seldom). In Landau's circle this measure was regarded as capital punishment: Landau despised such a theorist and immediately denied him backing.

Not all seminars were devoted to reviewing articles. Landau's students and physicists from other institutes and cities also made reports on original work. As a rule Landau would acquaint himself with original papers before the seminar; if he found a paper interesting, it would be presented. Landau personally spoke on all his own works at the seminar.

It was difficult, but a great honor, to deliver a talk at the seminar. The speaker was subjected to severe interrogation. The audience had the right to interrupt; the presentation was more a dialogue between the speaker and the audience (led by Landau) than a report. Often in the course of the dialogue, errors, gaps of logic, discrepancies and points of disagreement on basic assertions of the paper were brought to light. Landau was a man of great critical intellect: His criticism always helped find the truth. If an author was a success at the seminar, he could be sure that his work was not logically inconsistent and that he had new results. This is why theorists were anxious to report their work at Landau's seminar: They knew they would always get an impartial, unprejudiced assessment of their work, and from the highest possible authority.

Critical analysis of research is important in any field of science, and particularly so in theoretical physics. Investigation in theoretical physics is a chain of logical constructions that can sometimes be ruptured. In beginning his work an author may make assumptions whose validity is not always confirmed at the end; often these assumptions are not explicit. At Landau's seminar it would sometimes happen that after exhausting all his arguments, an author would unsuccessfully resort to his trump card: showing that his results coincided with the observed experimental data. This argument invariably provoked only laughter in the audience, since no coincidence of theory with experiment can justify logical gaps in the theorist's work.

Being a man of great critical intellect, Landau was also self-critical. It is well known that he had a weakness for classifying everything around, including physicists, but in his ranking of physicists the place he assigned to himself is more modest than he deserved. When I once told Landau how I admired his critical intellect, he replied: "You have not met Pauli! That was a great intellect!"

Landau's interactions with his students

The basis for any relationship with Landau lay in his interest in physics. His working day always started with a visit to the experimental laboratories on the ground floor of the Institute for Physical Problems. There he rushed through the laboratories, found out the latest news and lingered on in case anyone wanted his immediate theoretical assistance. Landau believed the problems experimenters were currently solving had priority over the problems of theorists. He was always willing to cut short any activity whenever an experimenter asked him even for some minor calculation. His cooperation with experimenters gave rise to many of Landau's outstanding works. Indeed, his main work of art—the formulation of the theory of superfluidity—was the fruit of his close day-to-day cooperation with experimenter Kapitsa, who had discovered and studied this phenomenon in helium-4.

Working relationships with experimenters were also standard for Landau's closest colleagues. Immediately after I became a postgraduate student, I came into contact with the liquid helium group at the institute, where Vasili P. Peshkov and E. L. Andronikashvili had gotten some interesting results from their work with superfluids. They needed an explanation for the effect of viscosity they observed in supposedly "viscosity-free" superfluid. My first, tentative calculations based on Landau's superfluidity theory provided a qualitative explanation. It took some time, however, to convince Landau that the calculations were correct. In the final analysis, the temperature dependence of the kinetic coefficients in quantum liquid proved to be different from what one obtained from the well-known kinetic theory of gases.

For the sake of "economy of thought" Landau would often employ fundamental general principles, rejecting anything that could not be confined within these principles. But any new nontrivial result plunged him deep into thought. In such a case, Landau would apply his methods to the problem, either confirming or rejecting the result. It was in just this fashion that Landau became interested in the kinetic equation for elementary excitations in a quantum liquid; soon we found its exact solution. Our joint work on the theory of the viscosity of superfluid helium emerged from this research.

This collaboration was rather typical of Landau's relationships with his students. A student would find a problem to solve and would carry out the preliminary calculations. Later, often at the most difficult stage, he would seek help from Landau and his mighty technique. Sometimes the advice would be minor, but Landau often would perform serious, thorough calculations.

A contribution such as this did not necessarily mean that Landau would give permission to include his name on the paper's list of authors. He was generous and often gave away his results. Only when his result was valuable and he judged his contribution to be important would he consent to becoming a coauthor.

Landau never did for his pupils what he believed they should do themselves. Sometimes, after many unsuccessful attempts to solve a problem, a student would ask Landau for help and hear the reply: "This is your problem. Why should I do it for you?" After Landau's flat refusal it would become clear that no outside help would arrive; if one was lucky, enlightenment would dawn and the problem would soon be solved. Neither did Landau formulate problems for his students, or supply thesis topics to his postgraduates: They were responsible for these tasks themselves. He thus trained them to be independent, educating them as future leaders of science.

I shall touch upon another example that is characteristic of collaboration with Landau. In the early 1950s a huge leap in quantum electrodynamics came with the invention of Feynman diagrams and the elimination of infinities. Landau was not familiar with this new technique in theoretical physics. At that time I was working in close contact with Alexei A. Abrikosov, with whom I eventually published many joint papers. There were not many theorists in physics at that time; maybe because of that, or because of our habit of reading journals, we were the first in Moscow to master the relativistic theory of perturbations and to study the Feynman works. Perhaps it was because we were young that we made a bold, reckless attempt to find the exact solution of the quantum electrodynamic equations. We also had in mind to make use of the theory's gauge invariance for this purpose. We started calculations, which we regularly discussed with Landau. But when we finished deriving the final equations for the mass and charge of the electron, it became clear that our idea would not work out, because of a very fine effect. This was where Landau rushed into action. He suggested selecting and summing up the most important diagrams (or terms of the perturbative series). Then it was simply a matter of applying the technique Abrikosov and I had mastered. As a result there appeared a series of works by the three of us on the asymptotic behavior of Green's function in quantum electrodynamics.[1] The methods developed in these papers later became widely adopted in statistical physics and other fields of physics.

In 1956 I witnessed Landau's creation of the quantum Fermi-liquid theory. By then much experimental data had accumulated on the liquid phase of helium-3. This evidence did not agree with the behavior expected for an ideal gas of fermions. One day Landau showed up in my

Дау сказал . . . .

Caricature by A. Yusefovich. The caption translates as "Dau said. . . ."

Landau posing on a pedestal while on holiday in Palanga on the Baltic Sea, 1961.

office at the Institute for Physical Problems and started rapidly writing down conservation laws on the blackboard, starting with the kinetic equation. From his calculations it appeared that the momentum conservation law was not automatically fulfilled. By the next day he had the solution to the problem. The ideal-gas picture did not hold; it was necessary from the very beginning to take into account the interactions between the fermions. This is how one of Landau's most elegant theories was created. Since we, Landau's students, actually witnessed its creation and took part in discussions about it, we felt somewhat like Landau's accomplices. Abrikosov and I soon applied the Landau theory to studying concrete properties of the Fermi liquid.

There are grounds to believe that thorough homework lay behind Landau's creation of this theory, although at the time we thought it had been created before our very eyes. Indeed, Landau often did obtain his results through artful improvisation. He usually gave away such improvised calculations to those who formulated the problem.

Clear-cut logic and simplicity were characteristic of Landau's works. He thoroughly thought over his lectures and articles. As is well known, he did not write his articles himself: His associates—most often Lifshitz—were entrusted with this respected task. I was lucky to have a chance to write with Landau two of his famous papers, one on two-component neutrinos and the other on combined parity conservation. Landau would think about and discuss each phrase as I wrote it down; only the best,

clearest version was allowed in the final paper. By working this way he both perfected his literary style and spotted points requiring additional explanation. This approach was typical of the way Landau operated. In any case his relationships with his students were not those of a maestro who generates ideas to pupils who snatch them and develop them further.

Institute for Theoretical Physics

When in 1962 it became evident after the car accident that Landau would never be able to resume the study of theoretical physics, his closest colleagues and associates faced the serious problem of preserving the Landau school and its traditions. Although among Landau's pupils there were already mature physicists and noted scientists, no one dared to think he could take Landau's place as leader. The most important and most difficult problems were to preserve the high scientific standards inherent in the school and to maintain the team of scientists who ensured these high standards. Finally we arrived at the natural conclusion that only our combined collective intellect could to some extent substitute for the powerful critical intellect of our teacher. Our idea received backing from the Academy of Sciences of the USSR, and in the fall of 1964 the Institute for Theoretical Physics came into being.

Our institute became part of the Noginsk Scientific Center of the Academy of Sciences at the time the Institute of Solid State Physics was being set up there. It was natural that our institute would confine itself to solid-state theory, despite the universal range of interest that was characteristic of Landau and his school. Little by little, however, research into nuclear physics, relativistic astrophysics, quantum field theory, plasma physics and other fields began to grow within the institute, and the divisions of mathematics and mathematical physics were established.

Because we worked on such a wide range of problems at the institute, one of our main tasks involved ensuring mutual understanding among experts in different fields of science. We had to admit that the age of such broadly educated people as Landau was over. Physics had become such an extensive, all-embracing science that universality seemed possible only within a select group of people who spoke a common scientific language. The evolution of theoretical physics during the previous decades had shown how significantly different fields of physics could affect one another. To take one well-known example, the methods developed in quantum field theory played a determining role in solid-state theory, particularly in solving the phase transition problem.

A common scientific language could be reached only within a small collective of carefully selected experts. One example in particular serves to confirm our success in achieving this common language. In recent years, a joint effort by theorists and mathematicians from our institute resulted in significant advances in quantum field theory and in the theory of the recently discovered phenomenon of superfluidity of helium-3. Both cases involved applying methods from topology. We owe these achievements to a new generation of theorists trained at the Institute for Theoretical Physics. This new generation—the pupils of Landau's pupils—is a guarantee that Landau's cause will live on for some years to come.

Mathematics

In mathematics Landau always set greater store by methods that enable one to solve concrete physical problems than he did by existence theorems. As an example of "real" mathematics he always cited the Hopf-Wiener method for solving semi-infinite integral equa-

Walking in Kiev during a 1955 conference on low-temperature physics are (left to right) Evgenii M. Lifshitz, I. M. Khalatnikov, Landau, H. Akhiezer and A. Akhiezer.

tions. In the mid-1950s Gerald Reuter and Ernest Sondheimer applied this method, based on the theory of functions of a complex variable, to the anomalous skin effect (in which the mean-free path of the electron is long compared with the penetration depth of an electromagnetic field into a metal). As a result the names of Eberhard Hopf and Norbert Wiener enjoyed great popularity in the 1950s among physicists involved in the quantum theory of metals; Landau especially admired the elegance and efficiency of the method.

Not long before the car accident, Landau met Wiener at a lunch at Kapitsa's house in Moscow. At that time Wiener was preoccupied with information theory, and the conversation during the meal did not impress Landau. After the lunch he ran into my office at the Institute for Physical Problems and said: "I have never met a more narrow-minded man. It is quite clear that he could not have thought up the Hopf–Wiener method. It is Hopf who did it!"

Landau had underestimated abstract mathematics, which was not yet widely adopted in physics. Sometimes in jest he would tell me, "We know that the mathematics of the 20th century is nothing but theoretical physics." At the time I shared his point of view; however, 20 years later, modern mathematical methods—topology, algebraic geometry, manifold theory—have become useful in modern physics. I do not know what Landau would say about this development, but I do not doubt that if he were alive today, he would master these methods and appreciate their importance.

Landau was a highly skilled mathematician. He had a very good command of the methods of complex variable function theory, group theory and probability theory. He also made a major contribution to ensuring the stability of numerical integration methods for the equations of hydrodynamics and thermal conductivity (simultaneously with, but independently of, John von Neumann).

But certain new methods of theoretical physics remained out of Landau's reach. For instance, when Abrikosov and I were trying to use the Feynman diagram methods in quantum electrodynamics to find out the asymptotic behavior of Green's functions at high energies, Landau immediately grasped the main point of the matter and gave us the idea of summing up the most important diagrams, yet he made no calculations of his own. And when the job was done, Landau, a rightful coauthor of our work, said to our mutual friend Naum Meiman, "This is

the first work where I could not carry out the calculations myself." This was said by a man who had the right to be regarded as a master of modern theoretical physics techniques of his time. When he called himself a champion of technique, he meant that he could solve a theoretical physics problem faster than anyone else, when it was properly formulated. In retrospect, it should be added that the problem had to be solvable by the methods with which he was familiar.

At that time Landau believed that Lars Onsager's work in which he calculated the thermodynamic characteristics of the two-dimensional Ising model was the acme of theoretical physics. That work included the exact solution of the phase transition problem; Landau admitted that he never could have solved it.

Landau's opinions of himself should not be accepted unquestionably: They have their own limitations.

Art

Landau read voraciously; he was fond of painting and was a fan of the movies. A rational man, he accepted only realistic art. In the USSR in the 1950s the books of the German writer Erich Maria Remarque were extremely popular. I remember how excited Landau was about Remarque's *The Spark of Life*; Remarque made a very strong impression on him. Enraptured, he often repeated, "This is the book." Landau liked poetry and was fond of reciting poems. In the spring of 1962, after the car accident, he was moved to the Institute of Neurosurgery in Moscow. I became hopeful: His condition was improving rapidly. I remember how, sitting in a wheelchair at the institute, he would recite poems by Nicholas Gumilev to me. But he would often repeat himself, reciting a poem from beginning to end and then starting over again.

In the mid-1950s our young poets became very popular, particularly Yevgeny Yevtushenko. Once Yevtushenko came to the Institute for Physical Problems to recite his poetry. He was extremely good at reciting, and the social content of the poems had a uniquely strong effect. The audience received Yevtushenko warmly and did not want to let him go; they asked him to read more and more. When Yevtushenko ceased reading, however, and invited the audience to put questions to him, no one did. Yevtushenko misinterpreted the reaction of the audience (and of Landau) to his poems and was disappointed. As it was, we felt we understood the world we were living in well enough: We wanted Yevtushenko's poetry, not

Lifshitz and Landau in 1948.

his explanations. In spite of this, Landau was very enthusiastic about the poems he had heard and said to me, "We must all take off our hats before Yevtushenko!" Could there be higher praise for a poet?

I have already mentioned that Landau acknowledged only realistic art; this should not be taken to imply that he never accepted the Impressionists. He particularly admired Claude Monet. But he thought Henri Matisse a rather minor artist. He often repeated his opinion that "Matisse is nothing but a house painter. The only thing he can do is to paint fences."

In our close circle we always discussed new films; at one point we were focusing on Italian neorealism. We all were delighted by *Near the Walls of Malapaga*—a film some people did not understand. Igor Evgeievich Tamm formulated a sort of intelligence test based on one's attitude toward this movie: Only those who liked it were acknowledged to be intellectuals. Landau also had a high opinion of a film by Grigori Chukhrai called *Ballad of a Soldier*.

Landau was fond of traveling. He usually spent vacations with Lifshitz, traveling in Lifshitz's car. He sometimes went to the mountains: In the winter he would ski in the Vorobiev hills. (His friends would say in jest that he stood on his skis eyeing pretty girls more often than he skied.) In the summer he would play tennis on the courts at the Institute of Physical Problems. He went in for sports for fun rather than for achievement. Landau did not play chess, although he knew the rules; he thought the game a pure waste of time. In this he disagreed with Kapitsa, who, to the end of his long life, looked upon chess seriously and passionately. Kapitsa played chess as a way of asserting himself.

Landau had been friendly with the noted Soviet physicist George Gamow, the author of the alpha-decay theory. But when Landau saw Gamow a while after he had emigrated, Landau was disappointed with how he had

changed. In the West, Gamow apparently had acquired a passion for what we sometimes call "thingism:" the desire to acquire things. He took little interest in science and paid too much heed to conveniences and the comforts of life. Landau spoke about the "new" Gamow with regret and even some loathing.

Kapitsa

The names of Landau and Kapitsa were closely linked in science and in life. When the Institute for Physical Problems was being established, Kapitsa offered Max Born the chance to head the institute's theoretical division; after emigrating from fascist Germany Born was looking for a job. In the long run Born got a chair in Edinburgh, and Kapitsa suggested that Landau head the division. In 1937 Landau moved to Moscow and accepted that position, which he held until the end of his life. It was at the institute that Kapitsa discovered superfluidity, and it was there that he and Landau created the theory of this fundamental phenomenon. For this work Landau was awarded the Nobel Prize in 1962, after the accident. Their work on superfluidity forever linked the names of Landau and Kapitsa, although they were never on very close terms. Landau always regarded Kapitsa the way a younger man regards his elder. But it is known that it was only after Kapitsa's insistent appeals to the government that Landau was set free from prison in 1939, where he had been incarcerated in 1938 on false charges.

Kapitsa was not a particularly delicate man, and he sometimes would make mean jokes—not about Landau but about theorists in general. During the sessions of the Scientific Council he used to say, "Ask a theorist and do the opposite." I believe Landau would not ordinarily have permitted such jokes in his presence, but he did not react to Kapitsa. Instead he would say, "He [Kapitsa] saved my life, so there was no offense."

Landau thought highly of Kapitsa as an organizer of science. But there was another—Artemii Isaakovich Alikhanian—whom Landau thought to be a great organizer of science, and with whom he had long and friendly relations. "Artusha" was his confidant and was familiar with all of Landau's personal problems.

Merciless?

Much has been written about the theoretical minimum and its role in the creation of the Landau school. Landau kept track of all who passed the theoretical minimum exams, putting down only the date when the exam was passed, without any marks. In some cases he put exclamation and question marks next to a student's name; if the student acquired three question marks, he was rejected as not suited to the study of theoretical physics under Landau. Then would come a most unpleasant moment: telling the student. Landau always did this himself; he entrusted this mission to no one else. One can imagine what it meant for a beginner to hear such a bitter pronouncement from Landau.

Once I said to Landau that he was merciless—I thought that for a kind man such a task would have been absolutely unbearable. This made Landau indignant. He ran away from me and told everyone he saw in the corridors of the Institute for Physical Problems, "You know, Khalat thinks I am merciless." I once asked him how he would behave if he fell out of love with a woman. He replied that he would straightforwardly declare it. I again said that he was merciless.

During scientific discussions Landau never took pity on well-respected physicists; his criticism was always sharp, but well-deserved and just. For instance, he did not always hold a very high opinion of John Bardeen, and

Landau (center) and colleagues at the Ukrainian Physicotechnical Institute in Kharkov, 1934, during a visit by Niels Bohr (pictured at Landau's right).

often explicitly expressed his attitude at seminars. Not until 1957, after Bardeen created the theory of superconductivity with J. Robert Schrieffer and Leon Cooper (for which Bardeen was awarded his second Nobel Prize), did Landau change his mind and admit the high scientific level of this physicist.

On the other hand, Landau was an extremely delicate and civil person in everyday life; he would stop in the street to give detailed directions to a stranger.

'Deception of the working people'

My last scientific discussion with Landau took place in my little office at the Institute for Physical Problems on Friday, 5 January 1962. We spoke about singularities in cosmology: He was happy with the results Lifshitz and I had obtained. On 7 January the tragedy took place; after the accident Landau could not do science again. There are different opinions about Landau's mental abilities following the accident. A few months afterward, a psychiatrist came to examine Landau. When the psychiatrist started asking him questions usually put to mentally retarded children, Landau demanded that "this idiot" immediately be taken away. But in the years after the accident, Landau avoided scientific discussions and the people who might offer them, pleading a pain in his leg. He made an exception for me: Usually, as I was about to leave, he would ask me to come back again. Our conversations mostly involved standard jokes and cliches.

There was one episode when Landau might have shown some scientific interest, but the evidence is not quite clear. In 1967 a friend of mine defended his doctorate at the Institute for Physical Problems. The thesis subject was approximate numerical calculations of electron spectra in metals. It should be pointed out that at the time such calculations were undervalued in theoretical physics: We thought more highly of analytical formulas. Landau had been attending sessions of the Scientific Council because Kapitsa thought it useful for his rehabilitation. I do not know whether Kapitsa was right or not. In any case, for each session of the Scientific Council, Landau came to the institute accompanied by a nurse called "Twin" (she was one of twin sisters—an often

repeated subject for Landau's jokes). He would sit there, hardly moving: This was not very pleasant to see. At these sessions, each had his own place. Landau always was seated in the third chair of the first row, next to me. In the middle of this particular report, when the speaker was telling us about his numerical calculations, Landau whispered in my ear, "Deception of the working people." This was an absolutely reasonable appraisal of the work, one that might have come from a healthy Landau. "Deception of the working people" was a favorite phrase of his; he used it when he felt there had been some kind of fraud. It is quite possible that if he were alive now he would have changed his mind about this work: Such calculations have now become rather conventional, and this particular work is often referred to.

On 22 January 1968 Landau turned 60. At the time I was away on business in India. Kapitsa decided to wait for me to come back to celebrate—he wanted me to organize the jubilee. On 5 March 1968 all of Landau's friends got together at the Institute for Physical Problems to mark this anniversary and to congratulate him. The general feeling was not very cheerful: All of us felt we were bidding farewell to him. Landau was gone in less than a month.

The last time I saw Landau was on 31 March, after he had had an operation. His health had greatly deteriorated. Lifshitz and I were summoned by the hospital; we were informed that there was practically no chance that he could be saved. When I entered his ward, Landau was lying on his side, his face turned to the wall. He heard my steps, turned his head and said, "Khalat, please, save me." These were the last words I heard from Landau; he died that night.

*　　*　　*

This article is adapted from a talk the author gave at the Landau Birthday Symposium at NORDITA, Copenhagen, 13–17 June 1988. The talk is also published in the book Landau: The Physicist and the Man, *edited by I. M. Khalatnikov (Pergamon, Oxford, 1989)*

Reference

1. A. A. Abrikosov, I. M. Khalatnikov, L. D. Landau, Dokl. Akad. Nauk SSSR **195**, 497, 773, 1117 (1954); **96**, 261 (1954). ∎

LANDAU'S ATTITUDE TOWARD PHYSICS AND PHYSICISTS

Lev Davidovich Landau was a unique physicist and teacher of physicists.

Vitaly L. Ginzburg

PHYSICS TODAY/MAY 1989

On 22 January 1988 in Moscow, we celebrated Lev Davidovich Landau's 80th birthday in the same hall at the Institute of Physical Problems where Landau held his seminars and where I used to see him. Perhaps that is why the thought haunted me at that memorable meeting that Landau might have been among us that day, sitting there in the front row as he used to, and I expressed it in my opening remarks. But, alas, more than 27 years have already passed since Landau conducted his last seminar: On 7 January 1962, Landau met with a car accident and was disabled for the rest of his life. He died on 1 April 1968.

Today only those older than 45 or so retain any personal impressions of Landau as physicist. Younger physicists and, more generally, those who had no personal contact with Landau know him mostly from his papers, many of which have not lost their freshness or become out of date with the passage of time. The large number of references to Landau's publications that one encounters in the current literature confirms this. For this reason and because of the remarkable graduate-level course in theoretical physics that Landau developed with Evgenii M. Lifshitz (a new edition is now being prepared by Lev P. Pitaevsky) Landau's contributions to physics continue to be of interest. But Landau's legacy cannot be summed up merely by recording his publications. Understandably, there is considerable interest in both Landau as a teacher and person and in his peculiarities as a physicist. One can get acquainted with these aspects of the "Landau phenomenon" from a wonderful paper by E. M. Lifshitz[1,2,3] and also from the volume of memoirs that at last is expected to

appear.[3] That volume contains my reminiscences as well. In addition I wrote a paper in connection with Landau's 60th-birthday celebration, but it appeared instead as an obituary because of Landau's death.[4] In this article I would like to concentrate on some examples that illustrate Landau's attitude towards physics. But I make no claim whatsoever to have developed detailed analyses and generalizations.

I should note that several times in the past when I wrote reminiscences of outstanding physicists, I was criticized for focusing the discussion partly on myself. Such criticism is quite understandable and, in principle, absolutely right. The reader or hearer of reminiscences of, say, Landau is interested in Landau himself, not in the author of those reminiscences. Unfortunately, this clear understanding of the matter does not help much. Hard as I may try, I cannot put myself aside altogether and in my reminiscences get rid of the "I." The situation is made somewhat comprehensible, however, by noting that one remembers better those things that are important to oneself. In addition, I have a peculiar memory, sometimes very poor or with a high threshold, which exercises a kind of selectivity that I cannot help. So, I ask you, dear reader, to please regard any shortcomings you find in this article as not arising from any desire to use this opportunity to say something about myself.

Theory of superconductivity

I will start with an example that is interesting in many respects. In the only paper I wrote with Landau[5] we constructed a phenomenological, or macroscopic, theory of superconductivity. (Perhaps it would be better to refer to this theory as quasimacroscopic, for I think Landau preferred that term.) The crucial point in our paper is the equation for a certain order parameter Ψ, "the effective wavefunction of superconducting electrons." The term of

Vitaly L. Ginzburg is in the theoretical department at the P. N. Lebedev Physical Institute of the USSR Academy of Sciences, in Moscow.

Lev Davidovich Landau. Photo taken in 1960.

this equation that depends on the vector potential **A** takes the form

$$\frac{1}{2m^*}\left(-\,\mathrm{i}\hbar\nabla - \frac{e^*}{c}\mathbf{A}\right)^2\Psi$$

which is obviously quite similar to the corresponding term in the Schrödinger equation for a particle of charge e^* and mass m^*. The mass m^* can be chosen arbitrarily because the quantity $|\Psi|^2$ cannot be measured directly. So we can put—this is supported by the idea of pairs—that $m^* = 2m$, where m is the free-electron mass. But what is the meaning of the charge e^*? Since Landau and I were dealing with a phenomenological theory, it seemed to me from the very beginning that e^* was some effective charge that might well be different from the free-electron charge e. But Landau rejected that idea, and in our 1950 paper there is a statement, typical of Landau, that "there is no reason to believe that $[e^*]$ differs from the electron charge." At the time we wrote that paper I had no concrete reasons for insisting on the introduction of an effective charge e^*, but several years later I came to the conclusion that the introduction of an effective charge $e^* = (2–3)e$ made the agreement between theory and experiment much closer. The possibility of determining e^* experimentally arises because our theory involves a dimensionless parameter

$$\kappa^2 = \frac{2e^{*2}}{\hbar^2 c^2}H_\mathrm{c}^{\,2}\delta_0^{\,4}$$

where H_c is a critical magnetic field and δ_0 the penetration depth of the external magnetic field into the superconductor. The quantities H_c and δ_0 can be measured directly (in reference 5 we were concerned with what are now called type I superconductors, for which $\kappa < 1/\sqrt{2}$). The parameter κ can also be determined directly from the data on the surface energy between the superconducting and the normal phases and from the limiting field for "supercooling." Thus, knowing H_c, δ_0 and κ, one can determine the value of e^*.

Naturally, I informed Landau of my result, but he once again objected to the introduction of the effective charge e^*. He argued—he probably had the argument formulated even at the time we wrote our paper—that the effective charge, like the effective mass, might depend on pressure, temperature, metal composition and so on. This meant that e^* could depend also on the coordinates (by virtue, say, of sample inhomogeneity or the dependence of temperature on the coordinates). But such a behavior of e^* would break gauge invariance. I tried to somehow avoid this difficulty but did not succeed, and it was clear that Landau's thoughts were fairly set. So I wrote a paper[6] in which I pointed out the possibility of substantially improving agreement between theory and experiment by introducing the charge $e^* = (2–3)e$, and I also mentioned in it, with Landau's permission and of course with a reference to him, Landau's arguments against an effective charge different from the free-electron charge. As is well known, the microscopic theory of superconductivity of John Bardeen, Leon Cooper, and J. Robert Schrieffer, in which the charge $e^* = 2e$ arises due to pair formation, appeared soon thereafter. Even now I feel sorry, and to some extent ashamed, to think that I did not consider the possibility of pair formation. Landau's objection to introducing an effective charge no longer arises if the effective charge is universal and equals $2e$ irrespective of temperature, composition and so on. But neither Landau nor anyone else who knew of my paper[6] thought of the pos-

Evgenii M. Lifshitz (left) and Landau on holiday, Borzhomi, 1960.

sibility of introducing a universal charge of $e^* = 2e$. There are reasons why the idea of pairing, which seems trivial now, was not at all obvious at that time, and I will come back to them later.

Publishing papers under Landau

It has often been said—I have heard it myself—that Landau's sharp criticism of the work of others prevented them from creating something great or at least from obtaining or publishing exceedingly important results. Indeed, Landau did criticize others, hotly and not always politely, but that was his style, and those who knew him were convinced that even his sharpest expressions usually did not imply any personal hostility. As for Landau's preventing others from carrying on their work or from publishing papers, it was out of the question, at least in the cases I know of. The story of the effective charge is not atypical: Though Landau was resolutely opposed to the introduction of the effective charge, not only did he not put obstacles in the way of my publishing my paper, but as I have already mentioned, he allowed me to present our respective arguments.

Important scientific achievements and discoveries do not appear from nothing. More than one person, and sometimes even many people, may think of the same thing—at times, they may even attain the same objective. On the other hand, if someone leaves something out or does not understand a problem to the bottom, what is to blame for that? First, chance plays an exceedingly large role. Of course, I do not mean that the theory of relativity or, more generally, great and profound ideas arise merely by random occurrence. But when we speak of some particular effect, phenomenon or theorem, there may be an infinity of reasons why one or another physicist did not think it through to the bottom, did not "hit" it and appreciate and publish the results. Second, an author's own underestimation of results he obtained at the time he obtained them is the best

evidence that he made the discovery accidentally or that he didn't really make it at all but only claimed later that he had. (When this happens, as it does sometimes, it is in some cases not a deliberate deception but rather a not uncommon psychological effect.) Let me relate a story in this connection: Physicist A mentioned to physicist B that he had derived the Schrödinger equation before Schrödinger but had not published this result because he had not thought it to be of sufficient importance. To this B replied, "I advise you not to tell anybody else about this, for it is no shame to fail to derive the Schrödinger equation, but to obtain such a wonderful result and fail to appreciate its importance really is a shame." I think I heard this story from Landau, and I believe it was he who had played the role of physicist B; at any rate, Landau was of precisely the same opinion as B. In short, Landau might have failed to comprehend or support some obscure idea brought to him for his opinion, or he might have criticized it, but it would simply be foolish to make him responsible for the fact that afterwards, in someone else's hands, that idea proved very fruitful.

On the whole Landau was very tolerant. (Some exceptions cannot change this conclusion because most people have some things about which they are anomalously sensitive.) I will give some examples below, but for the moment I will note only that Landau was rather liberal regarding publications. He objected to publication only of those papers that contained no new results but presented, say, another way of obtaining known results. Generally, Landau spoke with contempt and irritation about "giving new grounds" (from the German word *Neubegrundung*, which Landau liked to use in this connection). I am not sure, however, that he interfered actively with publication even of such papers. Of course his approval was not required for publication of most papers as long as the paper did not contain blatant errors. However, not to interfere and to approve are two different things. I don't remember whether or not Landau approved the publication of my paper[6] suggesting the value of $(2-3)e$ for e^*, but in general he used to reproach me, sometimes even with irritation, for "publishing everything [you] know." He didn't like one's being too casual in publishing papers, and he, obviously, didn't publish or rush into print everything he did.

It is my opinion that the question of whether to be casual or restrained about what one allows oneself to publish cannot be decided unanimously, for the answer depends on the author's taste and style. I think that Landau's restraint was in some measure determined by, among other things, the fact that he didn't like to write; it is common knowledge that even papers of which he was given as the sole author were usually written by someone else. It also seems to me that Landau was affected by the thought that a physicist of his rank should not publish trifles. I, for one, write rather easily, and moreover, until I write a paper I don't feel that the work is done even in the rough. And once a paper or a note is written and seems to be of some interest, why not try to publish it? This is what I have customarily done and still do, although I long ago came to realize that publishing small notes in no way raises my renown but gives rise to ill-disposed criticism. But I repeat that in my opinion this is a matter of taste and that to me not to publish something only out of fear of the "clamor of Boeotians" does not look like valor at all.

Landau's scientific publications

The two volumes in Russian of Landau's "collected works" contain 98 papers. (The English edition is somewhat different, but not in any essential way.) Seventeen papers and notes are not reprinted in the collected works, but are listed in it. Of these 17 publications, some are short reports on the papers included in the collection. Those papers that Landau himself regarded as erroneous are not included in the collection (see reference 1). Among these is a paper that attempts to explain superconductivity on the basis of the spontaneous current hypothesis.[7] That paper may seem erroneous, but in fact it contains a very interesting idea.

Landau considered in that paper a phase transition into a state with a nonzero spontaneous current density $\mathbf{j}$.

This breaks the gauge invariance, which must be maintained, but we must remember that the paper appeared before Landau developed the general theory of phase transitions in 1937. Instead of j, however, one can take for the order parameter the density of the toroidal dipole moment $\mathbf{T}$, which possesses the same transformation properties as the current density. This order parameter leads us to a new type of magnetics—toroidal magnetics.[8] I turned to the paper[7] in 1978, almost 45 years after it was published. I was glad I knew it existed.

The fate of this paper of Landau's seems to me to be in some respects typical of his activities. Like any other man, Landau was in the wrong sometimes. But the number of mistakes he made, and especially their proportion relative to his correct statements, was very small. And even Landau's mistakes often were very instructive about physics.

The number of papers Landau published could have been much larger. As I already mentioned, Landau did not publish everything he did; he obviously did not in the least try to publish papers just to lengthen his publication list. Also many results reported in papers by other authors in fact belong partly to Landau. I am not saying that these were instances of plagiarism, but rather that without Landau's inestimable advice and criticism, some of the papers his students and colleagues wrote would not have appeared at all or would have been of much less value. Sometimes Landau simply renounced co-authorship, refusing to be included even when he had contributed substantially to the work reported.

In 1943, for example, I was working on the problem of the field acting in a rarefied plasma. Some authors at that time believed that the effective field $\mathbf{E}_{\text{eff}}$ was equal to the mean macroscopic electric field $\mathbf{E}$.[9] Others supposed that $\mathbf{E}_{\text{eff}}$ did not equal $\mathbf{E}$ but, rather, for instance, that

$$\mathbf{E}_{\text{eff}} = \mathbf{E} + \frac{4\pi}{3}\mathbf{P} = \frac{\epsilon + 2}{3}\mathbf{E}$$

where

$$\mathbf{P} = \frac{\epsilon - 1}{4\pi}\mathbf{E}$$

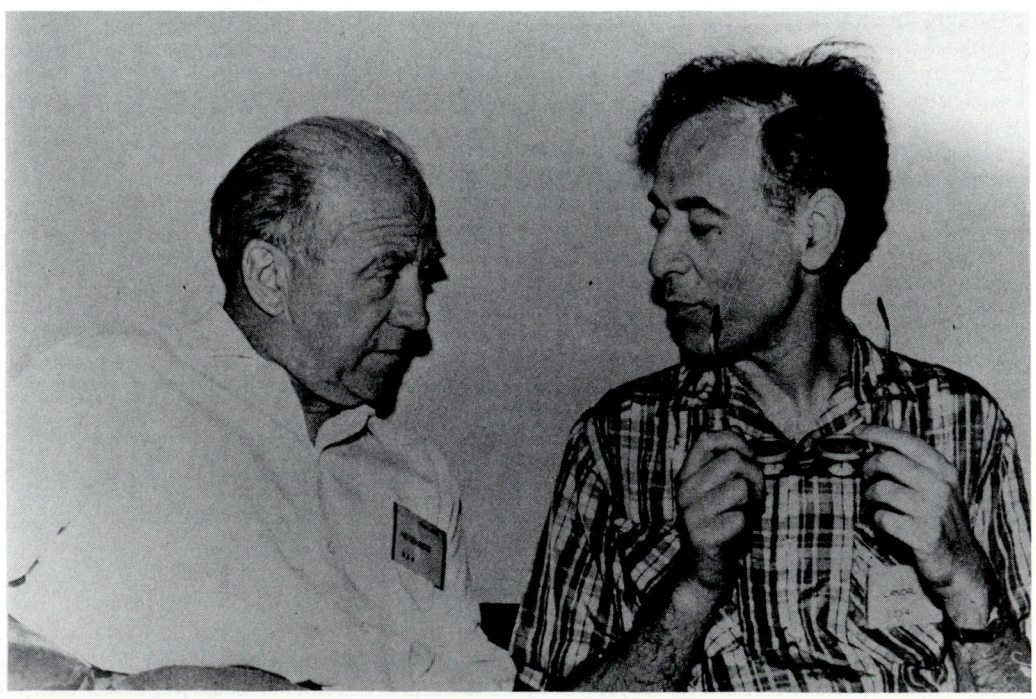

Werner Heisenberg (left) and Landau at Kiev, 1960.

is the polarization and ϵ is the dielectric permittivity. To sort out the right answer was not an easy task, given the state of plasma theory at that time. I got confused, didn't know how to achieve consistency in my analysis of the problem and asked Landau for advice. He was convinced from the very start that $\mathbf{E}_{eff} = \mathbf{E}$, but he did not think that this was obvious and could go without proof. With Landau's help I proved that $\mathbf{E}_{eff} = \mathbf{E}$. I then wrote a paper reporting the proof and brought it to Landau, whom I had included as my coauthor. But he did not accept the offer, and the paper appeared under my sole authorship.[9] In the paper I acknowledged Landau for "a detailed discussion of the problem" and for instructive advice on how to take close collisions into account. (In reference 3 I give another example of Landau's renouncing coauthorship. But I would not like to dwell here on that case, which is unpleasant for me, although it is less trivial than the one narrated above.)

I do not know exactly why Landau renounced authorship when he did; I suppose that in these cases Landau simply did not consider the results interesting or valuable enough. By the way, some who sought Landau's advice or help could not appreciate it (which was indeed not always easy to do), and for this reason published the results under their name only. I know of results, some of them popular in the literature, that in all fairness should bear Landau's name and not the names of his colleagues only.

Superfluidity

I now return to the theory of superconductivity and superfluidity. It was no accident—or so it seems to me—that Landau did not surmise the existence of pair formation following my suggestion that the effective charge in our phenomenological theory might be twice the free-electron charge. Landau had long been of the opinion that Bose statistics and Bose–Einstein condensation had nothing in common with the superfluidity of He II (the superfluid phase of He^4). He based his assertion on the fact that an ideal Bose gas is not superfluid.[10] Besides, it seemed to Landau that one did not need to invoke the Bose statistics of He^4 atoms to explain the superfluidity of helium II. But it is essential for the superfluidity of He II that atoms of He^4 obey the Bose statistics . As far as I know, this was first clearly shown by Richard P. Feynman.[11] Unfortunately, I don't remember how Landau reacted to the fact that liquid He^3, first obtained in 1948, is not superfluid down to very low temperatures $T > 0.1$ K. (In the early 1970s it was discovered that He^3 is superfluid at temperature $T \ll 0.1$ K.) In any case, before the formulation in 1957 of the BCS theory of superconductivity, the idea of electron pairing was alien to Landau, as well as to very many other people. I have written this in more detail elsewhere.[12] Here I wish merely to note that if reproaches are admissible in such cases, then I, perhaps more than Landau, am the one who deserves them.

Fermi-liquid theory

For some time Landau believed that plasmons in solids cannot be "good" quasiparticles because their damping time must be small, that is, of the same order of magnitude as the period of oscillations. (See reference 3.) This view reflected one of Landau's favorite theses, namely that electrons cannot form an almost ideal Fermi gas in a normal (nonsuperconducting) metal "because nobody has yet repealed Coulomb's law." It was, of course, none other than Landau who explained all this later in the theory of the Fermi liquid.[13] As for plasmons, they do exist in simple metals—their damping is relatively small.[14]

General relativity

Landau was a great admirer of general relativity theory, which he called "the most beautiful of the existing physical theories." But he categorically denied the possibility of introducing the Λ term into Einstein's equation, as well as of somehow changing or generalizing general relativity theory, even when such changes did not violate the good agreement between the theory and the known observational data. Though I fully share his admiration, I could not understand why it should rule out the existence of the Λ term. I don't remember that Landau ever put forward any physical arguments against the Λ term, but at that time there undoubtedly were no realistic arguments for its existence, and Einstein himself believed, I think, that his introducing this term had been a mistake. I haven't been able to find in Einstein's papers the place where he expressed this sentiment, but in the appendix to his book *The Meaning of Relativity* we read: "The introduction of this additional term made the theory more complicated and thus did much harm to logical simplicity."[15] Wolfgang Pauli, too, expressed a negative attitude toward the use of the Λ term.[16] It is now common knowledge that introducing the term Λ is equivalent to using the equation of state $P = -E$, and the term is now the subject of much discussion in the theory of the early universe.

What do I wish to illustrate by this discussion of the Λ term in general relativity? Landau, like his great older contemporaries Einstein and Pauli, attached much importance to logical simplicity and beauty in a fundamental theory. He realized that emphasizing logic and beauty is inevitable and necessary when considering problems for which there is not enough experimental data, and for which the theoretical possibilities are numerous. This attitude was above all a manifestation of Landau's pragmatism: In those days research in general relativity and especially in relativistic cosmology was not yet in a position to justify the introduction of the Λ term or to extend the general theory of relativity.

First physics encounter with Landau

For a short time starting toward the end of 1939, two groups of theorists—one headed by Landau, at IFP, the Institute of Physical Problems of the USSR Academy of Sciences, the other headed by Igor Tamm at FIAN, the P. N. Lebedev Physical Institute of the academy—met for joint seminars, held alternately at the two institutes. Both groups were very small; of Landau's coworkers at the time I now remember only Evgenii Lifshitz. I recall in particular two joint meetings. At one of those, held at IFP, Landau reported on the theory of superfluidity and Tamm proposed the term "roton," as mentioned in reference 10. At the other meeting, at FIAN, Tamm started discussing one of my first papers, devoted to the quantum theory of Vavilov–Čerenkov radiation. I had shown that the condi-

Landau as a young man.

tions for emission of this radiation follow when one applies the laws of conservation of energy and momentum to a particle radiating a photon of energy $\hbar\omega$ and momentum $\hbar\omega n/c$ in a medium of refractive index $n(\omega)$. I had also calculated the intensity of the VC radiation, but we did not come up to the intensity in that seminar.

Landau immediately expressed skepticism about my results, saying that they were of no interest since the effect was classical and there was no point in treating it quantum mechanically. He was right in some sense, because quantum corrections to VC radiation are of the order of $\hbar\omega/mc^2$ (m is the particle mass) and therefore are small in the optical region. But very often a new interpretation, approach or conclusion turns out to be fruitful. This was the case in my work on the VC radiation: The quantum approach and the use of the conservation laws appeared to yield new results in, for example, studies of the Doppler effect in a dispersive medium. All this is discussed in reference 17 and in the literature cited there, so I will leave out the details.

I have dwelt on this experience with Landau, first, to demonstrate once again his pragmatism and his dislike for *Neubegrundung*. Second, this example is a striking illustration of the role of tastes and feeling in science. I literally love problems dealing with the VC radiation and, generally, with radiation from uniformly moving sources.[17] Landau was quite indifferent to this range of questions and did not consider the VC effect beautiful. The above incident is not the only illustration

Portrait by Mogilevsky.

of this predilection: I remember telling Landau about a paper—I think it was the paper by David Bohm and David Pines—in which the plasma (longitudinal) wave damping discovered by Landau[18] was interpreted as the inverse VC effect. But Landau remained quite indifferent to that interpretation.

I have already discussed the "accusations" that Landau's bitter criticism might have been an obstacle in someone's way. The claims that Landau was a conservative and that he "considered himself the most clever man" are groundless as well. I am not inclined to idealize Landau and to pass in silence over his weak points. Landau himself was not beyond criticizing even those he regarded as great, and he was critical of himself. The latter trait showed up in many ways. For example, Landau considered himself inferior "in class" to a number of other physicists and his contemporaries.[1,3,4] This was particularly true of his attitude toward Feynman, who was 10 years younger than Landau. In 1962 I met Feynman at a conference in Poland. Feynman was concerned about Landau's health and inquired after him. (The two had never met.) I mentioned how highly Landau thought of Feynman's results, and that he ranked them higher than his own. Feynman was somewhat embarrassed by this and resolutely declared that Landau was wrong in so judging him. Such comparisons are not my main point, however. Even Landau mentioned his "classification" less often as the years passed, and he started to take a more sober view of classifying people. By the way, of all the physicists I have known personally, nobody resembled Landau more than Feynman. The resemblance extends to many things, scientific style, aspects of their personalities

and behavior, and interest in pedagogical ideas. Even among famous physicists there is a diversity of talents. Niels Bohr and Landau, for instance, were on opposite extremes. Landau and Feynman I see as having had very similar talents. The similarity seems to me to be genetic. The differences, which one could attribute to their different surroundings and different upbringing, are certainly also very large. What a pity it is that these two remarkable physicists never met. It really pains me to think of this "product" of our past.

Landau's talent was so bright, and he commanded the techniques of theoretical physics so skillfully, that one is inclined to think he could have done a lot more and solved still more difficult problems had he lived longer. Once, in a conversation with him I expressed my feeling that he could achieve more than he had. He replied immediately and very definitely, as if he had already thought about it, "No, this is wrong; I have done everything I could." I believe he was right. He generally worked very hard and tried to solve very difficult problems. For example, he spent much effort trying to formulate a theory of second-order phase transitions beyond the scope of the self-consistent field approximation. He once told me that no other problem had taken so much of his strength as this one, with which he still had not made much progress.

Landau often also asserted that he was not an inventor, that he had invented nothing. As far as the invention of devices is concerned, it is true that he had no talents as a designer. The sober mind of a highly educated physics theorist and analyst works in a way that is somehow orthogonal to the search-in-the-dark, trial-and-error methods of the inventor. But Landau was fairly

Landau at the low-temperature physics conference, Kiev, 1955, with (from left to right) unidentified, S. Kapitza, E. M. Lifshitz, A. Akhiezer and I. Akhiezer.

Landau, playing tennis, 1934.

inventive when it came to solving complicated problems and searching for new methods.

Landau's highly critical attitude stemmed from his sober mind and his comprehension and profound knowledge of physics. Besides, Landau didn't care a cent about presenting his remarks and opinions disingenuously. That often made him look opinionated and unwilling to adopt new ideas. But that would have been a misimpression, because Landau very often agreed, maybe not immediately, with disputable hypotheses and, in general, with new trends. So I do not support the opinion about Landau's conservatism that I sometimes hear. It is certainly difficult to weigh the degree of conservatism on a precision balance. It is also difficult to decide where the boundary lies between the true conservatism and the "healthy" conservatism—by the latter I mean, for example, the recognition that the old order should not be broken without good reason.

Landau's last published paper "On the Fundamental Problems," is to me vivid proof that he was not a conservative. It appeared in 1960, in a collection in commemoration of Pauli.[19] In this paper Landau expressed the opinion that the "Hamiltonian method for strong interactions is dead and must be buried." So Landau was ready to face the breakdown of fundamental theory. (The Hamiltonian method was later on found to be far from exhausted and now forms the basis of quantum chromodynamics.)

In this article, I realize, I could promote the understanding of Landau's style and his scientific image only to a very small degree. But I find consolation in the thought that to provide real insight into the nature of this remarkable physicist is extremely difficult. I would, however, like to make one final remark. Few are those to whom I have returned in my memory as often as I have to Landau, even though he left us many years ago. I cannot explain this fact only by my friendly feelings toward him and by his tragic and bitter demise. Another fact might provide a clue to this phenomenon: Landau was a unique physicist and a teacher of physicists. So my attitude toward him is inseparably linked with our attitude toward physics, which is so dear to all of us.

* * *

This article is adapted from a talk the author gave at the Landau Birthday Symposium at NORDITA, Copenhagen, 13–17 June 1988. The talk is also published in the book Landau: The Physicist and the Man, *edited by I. M. Khalatnikov (Pergamon, Oxford, 1989)*

References

1. E. M. Lifshitz, "Lev Davidovich Landau (1908–1968)," in L. D. Landau, *Collected Papers*, vol. 2, Nauka, Moscow (1969).

2. L. D. Landau, *Collected Papers* (two volumes), Nauka, Moscow (1969). L. D. Landau, *Collected Papers*, Pergamon, London (1965).

3. *Memoirs About L. D. Landau*, Nauka, Moscow (1988); English ed. to published by Pergamon.

4. V. L. Ginzburg, Usp. Fiz. Nauk **94**, 181 (1968) [Sov. Phys. Usp. **11**, 135 (1968)].

5. V. L. Ginzburg, L. D. Landau, Zh. Eksp. Teor. Fiz. **50**, 1064 (1950); reprinted in L. D. Landau, *Collected Papers* [2], vol. 2, p. 126; [Eng. ed., p. 546].

6. V. L. Ginzburg, Zh. Eksp. Teor. Fiz. **29**, 748 (1956) [Sov. Phys. JETP **2**, 589 (1956)].

7. L. D. Landau, Phys. Zh. Soviet Union **4**, 43 (1933).

8. V. L. Ginzburg, A. A. Gorbatsevich, Yu. V. Kopaev, B. A. Volkov, Solid State Commun. **50**, 339 (1984).

9. V. L. Ginzburg, Izv. Akad. Nauk SSSR, Ser. Fiz. **8**, 76 (1944). The content of this paper is presented also in V. L. Ginzburg, *The Theory of Radio Wave Propagation in Ionosphere*, Gostchkhizdat, Moscow (1949), section 6.

10. L. D. Landau, Zh. Eksp. Teor. Fiz. **11**, 592 (1941); J. Phys. USSR **5**, 71 (1941).

11. R. P. Feynman, Phys. Rev. **91**, 1301 (1953). R.P. Feynman, *Statistical Mechanics*, Benjamin, Reading, Mass. (1972).

12. V. L. Ginzburg, Prog. Low Temp. Phys. **12**, in press.

13. L. D. Landau, Zh. Eksp. Teor. Fiz. **30**, 1058 (1956).

14. D. Pines, *Elementary Excitations in Solids*, Benjamin, New York (1963) [Russian ed.: D. Pines, *Elementarnye vozbyzhdeniya v tverdykh telakh*, Mir, Moscow (1965)].

15. A. Einstein, *The Meaning of Relativity*, 5th ed., Princeton, U. P., Princeton, N. J. (1956).

16. W. Pauli, *Theory of Relativity*, Pergamon, London (1958), note 19. [Russian ed.: W. Pauli, *Teoriya Otnositelnosti*, Nauka, Moscow (1983), note 19.]

17. V. L. Ginzburg, *The Lesson of Quantum Theory*, Proc. Niels Bohr Centenary Symposium, 3–7 October 1985, North-Holland, New York (1986), p. 113. This is an abridged revision of the talk I gave at the Bohr centenary symposium. For the complete text of that talk, see *Proc. Lebedev Physics Institute*, Nova Science, New York (1988). [Trudy Fian **176**, 3 (1986).]

18. L. D. Landau, Zh. Eksp. Teor. Fiz. **16**, 574 (1946); J. Phys. USSR **10**, 25 (1946).

19. L. D. Landau, in *Theoretical Physics in the Twentieth Century*, M. Fierz, V. F. Weisskopf, eds., Interscience, New York (1960), p. 245. (See also L. D. Landau, *Collected Papers*, vol. 2, Nauka, Moscow (1956), p. 421; L. D. Landau, *Collected Papers*, Pergamon, London (1965), p. 800.

My years with Landau

The discoverer of "type-II" superconductivity
lets us in on the excitement of an
important time for low-temperature physics

A. A. Abrikosov

PHYSICS TODAY/JANUARY 1973

During a decade or so of my life I had the opportunity to talk with Lev Davidovich Landau almost every day, and to profit from his advice. This may be why the period from about 1950 to 1960 was so successful for me, and I would like to share my memories of it with you.

In 1950 Vitali L. Ginzburg and Landau wrote their well known paper[1] on superconductivity. Without the microscopic theory, developed later by John Bardeen, Leon Cooper and J. Robert Schrieffer,[2] the meaning of several quantities entering the Ginzburg–Landau work remained unclear, above all the meaning of the "superconducting electron wave function" itself. Nevertheless this theory was the first to explain such phenomena as the surface energy of electrons at the superconducting–normal phase boundary and the dependence of the critical field and current in thin films on temperature and thickness.

A new type of superconductor

Experimental verification of the Ginzburg–Landau predictions concerning the critical fields of thin films

Alexei Alexeivitch Abrikosov, the winner of the eighth Fritz London Award for low-temperature physics, is at the Landau Institute for Theoretical Physics in Moscow.

was undertaken by my friend N. V. Zavaritski, who was then a young research student of A. I. Shalnikov. I often discussed the matter with Zavaritski. Generally his results fitted the theoretical predictions well, and he even managed to observe the change in the order of the phase transition with decreasing effective thickness (ratio of thickness to penetration depth at a given temperature). For that experiment he used the hysteresis of the dependence of the resistance $\rho(H)$ on the field. One day, Zavaritski slightly altered his technique of sample preparation. Usually he evaporated a metal drop on a glass plate and put the resulting mirror into a Dewar flask. This time instead, he began to do the evaporation inside the Dewar flask, with the glass plate at helium temperature.

We know now that in this case atoms reaching the plate are trapped at the place where they hit and are unable to move to form a regular structure. Therefore an amorphous substance is produced, which at every effective thickness will be a "type-II" superconductor. But at that time this was, of course, not known.

The dependence of critical field on thickness, as measured by Zavaritski,[3] did not follow the Ginzburg–Landau formulas. This seemed to be a para-

dox: Apart from its beauty, the theory really explained many things, and we were surprised to see that suddenly it had failed.

When Zavaritski and I discussed the possible origin of this discrepancy, we came to the idea that the approximation $\kappa \ll 1$ based on the surface-tension data (where κ is the adjustable Ginzburg–Landau parameter), could be incorrect for such new objects as the low-temperature films. In particular one could suppose that $\kappa > 1/2^{1/2}$. According to Ginzburg and Landau, the surface energy should be negative under these conditions. Intuitively, they felt that in this case the phase transition in the magnetic field would always be of second order, and this was in fact what Zavaritski observed.

When I calculated the dependence of critical field on effective thickness with $\kappa > 1/2^{1/2}$, the theory appeared to correspond to the experimental data. This gave me the courage, in a 1952 paper[4] containing this calculation, to state that apart from ordinary superconductors whose properties were familiar to everybody working on the subject, there exist in nature superconducting substances of another type, which I propose to call "superconductors of the second group" (now called type-II superconductors). The division between the first and the second group

The theoretical group at the Institute for Physical Problems in 1955: (seated, left to right) L. Prozhorova, A. A. Abrikosov, I. M. Khalatnikov, L. D. Landau, E. M. Lifshitz, (standing, left to right) S. Gerstein, L. P. Pitaevski, L. Veinstein, R. Arkhipov, I. Dzialoshinski.

was defined by the relation between the quantity κ and its critical value $1/2^{1/2}$.

After that I tried to investigate the magnetic behavior of bulk type-II superconductors. The solution of the Ginzburg–Landau equation in the form of an intinitesimal superconducting layer in a normal sea of electrons was already contained in their paper. Starting from this solution I found that below the limiting critical field, which is the stability limit of every superconducting nucleation, a new and very peculiar phase arose, with a periodic distribution of the ψ function, magnetic field and current. I called it the "mixed state."

Landau showed a notable interest in this work and wanted me to publish my results for the vicinity of the upper critical field, which I named H_{c2}. But I wanted to understand how the new mixed state looked in the total range of fields.

Landau has his doubts

At this time I became ill and had to stay in bed for almost three months. One day Landau visited me. The conversation, as in most cases, concerned everything but physics, and Landau sipped with great pleasure from a glass of *glühwein*, which was not at all like him. And then suddenly

I destroyed all this paradise by telling him what I had wanted for the mixed state, namely the elementary vortices. As Landau's eyes fell on the London equation with a delta function on the right-hand side, he became furious. But then, remembering that a sick person should not be bothered, he took possession of himself and said: "When you recover we shall discuss it more thoroughly." He hastily bade me farewell and disapoeared.

He did not come to see me any more. When I felt better and appeared at the Institute and tried to tell him again about the vortices, he swore rather ingeniously. At that time I was still very young and did not know the temper of my teacher well enough. He had seen many kinds of pseudoscience in his life, and this made him suspicious toward unusual statements. However, by making some effort and disregarding the noise he made, one could always "drag through" him any reasonable idea. But at that time I sadly put my calculations in my table drawer until better times.

But in fact the idea was not so bad. Analyzing the solution that I got close to H_{c2}, I saw that in the plane perpendicular to the field there are points where ψ becomes zero. The phase of the ψ-function changes by 2π along a path around such a point. I thought

about why such singularities should appear, and saw that it could not be otherwise. Indeed the Ginzburg–Landau equation contained not the magnetic field but the vector potential. If the magnetic field does not vary in sign over the whole sample, then the vector potential must increase with the coordinate. But the physical state in a uniform field (as is the situation close to H_{c2}) must be either uniform or, at most, vary periodically in space. So the increase of the vector potential must be compensated for by a change in the phase of the ψ-function.

The figure on page 58 helps us to see this. Let the field be along the z-axis and let us choose the vector potential **A** such that $A_y = H_x$. Imagine the xy plane. The gray points are those I have mentioned, where ψ becomes zero. If we want to have a unique determination of the phase we must draw cuts in the plane. We draw them through the gray points parallel to the y-axis. From the figure it is evident that when we go around the points the phase increases by $(\Delta\psi)_1 = \pi y/a$ if we move along the lower path, and by $(\Delta\psi)_2 = -\pi y/a$ if we move along the upper one. That means that at every cut the gradient of the phase $\partial\psi/\partial y$ undergoes a jump $2\pi/a$. If we use ordinary units (at that time I used the dimensionless Ginzburg–Landau units)

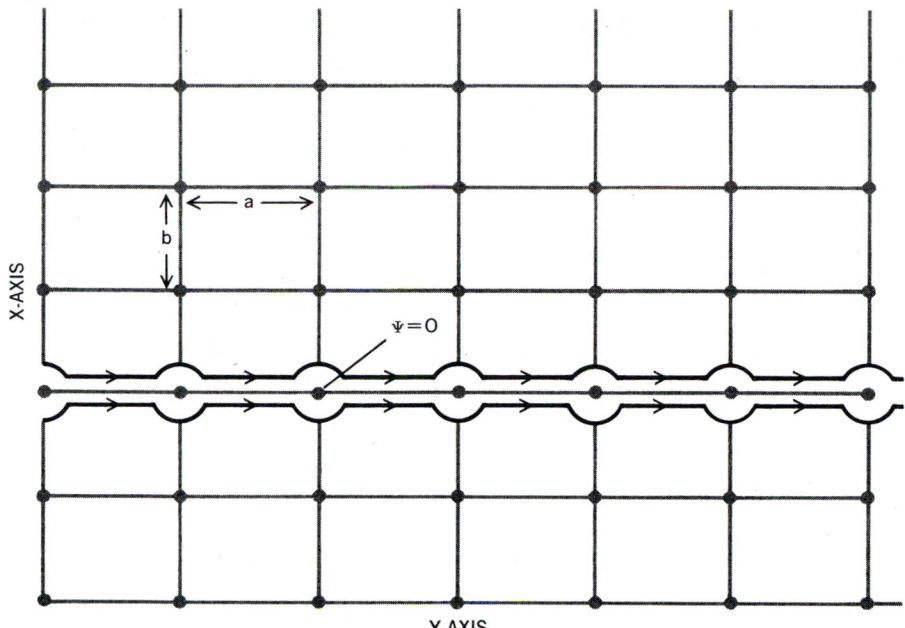

Vortex lattice of a type-II superconductor is seen in a schematic diagram with magnetic field along the z-axis and unit cell dimensions a and b. To determine the phase of the ψ-function uniquely, draw a cut through the gray points parallel to the y-axis. Around each point, the phase changes by π if we follow the lower (clockwise) path and by $-\pi$ along the upper path, so that, with y/a the number of points, $(\Delta\psi)_{\text{lower}}$ is $\pi y/a$ and $(\Delta\psi)_{\text{upper}}$ is $-\pi y/a$. At each cut then, $\partial\psi/\partial y$ jumps by $2\pi y/a$. In the limit, with decreasing magnetic field, one vortex exists, so that ψ must equal zero along the z-axis.

we see that compensation for the increase in A_y demands

$$\frac{2e}{c}\, Hb = \frac{2\pi\hbar}{a}$$

or

$$Hab = \frac{\pi\hbar c}{e} = \Phi_0$$

which is the flux quantum. Because I used dimensionless quantities I did not mention the flux quantum on the right, but I understood that with a decreasing magnetic field the cell dimensions "ab" must increase, and as a limit one vortex must be considered, in which case the phase of ψ changes by 2π in going around it. On the z-axis ψ must equal zero. Otherwise the ψ-function is not uniquely defined. Such a picture allowed me to find the lower critical field H_{c1} and the magnetization curve $M(H)$.

A worthwhile diversion

But, as I have noted (and this was in 1953), after the fuss my professor made I dropped the matter. There was also another reason. Interesting news appeared in a completely different field— quantum electrodynamics.

After the wonderful work of Julian Schwinger, Richard Feynman and Freeman Dyson, many people were interested in knowing whether it would be possible to sum up all the higher-order corrections and to find formulas for the Green's functions and physical phenomena without developing in pow-

ers of the fine-structure constant $e^2/\hbar c$. My friend Isaac Khalatnikov as well as myself had an old interest in the problem. At this time a paper[5] by S. F. Edwards was published, in which an attempt was made to sum up a ladder sequence of Feynman graphs for the electron-photon vertex part. We studied this paper and finally came to the conclusion that Edwards had done what he could but had not done what was really necessary, since he had no reason to choose this particular sequence. We tried to do something better and finally wrote some relation between the electron Green's function and the vertex part, which, as it soon became clear, was completely wrong. However when we substituted it into Dyson's equations we began to obtain various interesting consequences, as for example expressions for the electron mass and for the renormalized interaction.

Landau became extremely interested, but being busy with other problems he had no time to study the new technique of the quantum field theory. So he asked Khalat and me to teach him. I must confess that at that moment we were able to do calculations but had no true understanding of the fundamentals of the theory. Landau swore heavily, but after a month he said that he understood everything. He explained to us that we needed to find the main sequence of graphs having the highest power of the big logarithm with a given power of the inter-

action constant. This simple idea put everything in its place. We were indeed successful in calculating the asymptotic expressions for the Green's functions and various physical phenomena at high energies.[6] Moreover the principle of summing the main Feynman graphs proved afterwards to be extremely useful in various statistical problems.

Vortices from Feynman

Being occupied with such interesting things I of course did not turn my mind back to the work on type-II superconductors. But Landau had a long-standing interest in the state of He II in a rotating vessel. On the one hand, the helium should not be dragged along by the wall, but on the other hand, this was energetically favorable. In 1955 Landau and E. M. Lifshitz published a paper[7] in which they proposed a layer-type structure with velocity jumps on the layer boundaries. After a year they discovered Feynman's paper[8] in which it was shown that elementary vortices appear in rotating helium. Landau immediately said that Feynman was right, and that he and Lifshitz were wrong. Of course it was true. In terms of our superconducting notation, the He II could be considered as an extreme case of a type-II superconductor, with a correlation length of the order of interatomic distances and with an infinite penetration depth. But at that time this was not so evident.

When Landau began to praise Feynman's work I asked him: "Dau, why are you ready to accept the vortices from Feynman and flatly rejected the same idea from me." Landau answered: "You had something different." "Well then look, please," I said, and produced my calculations from the drawer. This time no objections followed. We discussed the subject very thoroughly and Landau's remarks were very useful.

When everything was put in order I remembered that I had already seen very similar magnetization curves with two critical fields, namely those of superconducting alloys. Digging for the corresponding experimental data I found the old work of 1937 by L. V. Shubnikov, W. I. Khotkevich, J. D. Shepeliov and J. N. Riabinin[9] on the magnetization curves of Pb-Tl alloys. They had prepared their samples very carefully, annealing them for a long time close to the melting temperature. So their samples were probably sufficiently uniform, and this was also confirmed by a rather small hysteresis. But at that time, and during the subsequent 25 years, everybody explained this form of the magnetization curve in terms of the formation of a "Mendelssohn sponge," that is, a nonuniform structure with a distribution of critical parameters.[10] It is worth mentioning that even many very good experimentalists finally believed in the mixed state only after they saw the powder figures of a vortex lattice obtained in 1966 by Uwe Essmann and Hermann Träuble.[11]

So my work was published in 1957 in *JETP*.[12] In the same year I reported it at a low temperature conference in Moscow in which some physicists from Oxford and Cambridge also took part. Nobody understood a single word. This could be explained however, by the fact that I had a terrible cold with high temperature and had hardly any idea myself of what I talked about. The translation of the paper was then published in the *Journal of Physics and Chemistry of Solids,*[13] but with more than one hundred errors in the formulas and text, and this of course did not improve the situation.

As you know, in the same year—1957—the famous paper by John Bardeen, Leon Cooper and J. Robert Schrieffer appeared. Everybody became enthusiastic about its ideas—we ourselves among the others. Therefore my paper did not attract attention. Of course it would be unjust for me to complain, because by using BCS theory we managed to get a lot of interesting results and were able to develop and improve greatly the methods of statistical calculations.

Then, in 1961 John Kunzler and his colleagues discovered that the alloy Nb_3Sn possesses a critical field of about 100 kOe.[14] Shortly after that, alloys with high critical fields began to be used for constructing superconducting coils. This drew attention to the theory of superconducting alloys.

Type-II rediscovered

In my work of 1957 I noted the connection between the quantity κ and the free path length; as I have said, nobody knew about this paper. However there did exist a series of papers by Brian Pippard[15] in which he qualitatively established the connection between the sign of the surface tension on the one hand and the ratio of the correlation length to the penetration depth on the other. He also mentioned the decrease of the correlation length with the free path length. In 1961, on the basis of Pippard's ideas, Bruce Goodman rediscovered that alloys with high critical fields have a negative surface tension.[16] Goodman calculated the magnetic properties of such alloys supposing a simple layer model for the distri-

bution of the normal and superconducting phases. The results were in qualitative agreement with experiment.

I don't know how it happened, but probably somebody told Goodman about my work. What followed was completely incredible. In 1962 Goodman published another paper[17] in which he gave a short presentation of my theory and analyzed the experimental data for type-II superconductors, comparing them with the predictions of both theories, mine and his own. He came to the conclusion that the vortex model fits experiment much better than the laminar one. So the aim of Goodman's paper was to prove that his theory was worse than mine! I have never in my life seen another example of this kind, and took the first opportunity to express my admiration to Goodman.

After this paper by Goodman, physicists working on superconductivity finally developed an interest in my work, and I suppose that this was to a considerable extent the cause for the favor done to me by the London Award Committee. Therefore I would like to express once more my gratitude to Bruce Goodman.

Of course the fact that the award

Alexei Alexeivitch Abrikosov is seen here in a moment of relaxation. This photograph was taken in 1967 during a visit to the Kamchatka peninsula in the northeastern USSR.

A more formal portrait of Abrikosov is seen in this photograph, taken in 1970.

given me has the name of Fritz London is particularly pleasant for me, because he, together with Heinz London, invented the first phenomenological equations of superconductivity. As became apparent much later, they had described the electrodynamics of just the type-II superconductors. Also it was Fritz London who introduced the notion of the magnetic flux quantum, which has a direct relation to the subject.

Finally I would like to mention some of my other activity in the low-temperature field. With one exception it had a much quicker reception. My work with L. P. Gor'kov on the microscopic theory of superconductivity (high frequency behavior,[18] Knight shift[19] and the influence of impurities, particularly magnetic impurities,[20] on the properties of superconductors) was accepted by readers rather soon after publication, and has been developed further.

It appeared, by the way, that the gapless superconductivity that we predicted for magnetic alloys could also exist under rather different conditions. My work on the Kondo effect[21] and the studies with Khalatnikov[22] of the properties of liquid He³ had a lucky fate too.

The exception I have mentioned is the theory of semimetals of the bismuth type that L. A. Falkovski and I constructed ten years ago.[23] This theory explained such peculiarities as the strange crystal structure, the small number of free carriers and the large dielectric constant and gave formulas for the energy spectrum of bismuth that fitted experiment very well.

Recently I got some new results based on this theory and these have been published in the *Journal of Low Temperature Physics*.[24] I hope that they will be interesting to those who work on semimetals and on the metal-insulator transition. In conclusion I would like to express my gratitude to the Fritz London Award Committee for the honor it has given to me.

* * *

The author is the recipient of the 8th Fritz London Award for Distinguished Contributions in Low Temperature Physics. This article is adapted from his acceptance speech, delivered in his absence by Pierre Hohenberg at the 13th International Conference on Low Temperature Physics in Boulder, Colorado in August 1972.

References

1. V. L. Ginzburg, L. D. Landau, Zh. Eksp. Teor. Fiz. **20**, 1064 (1950).
2. J. Bardeen, L. N. Cooper, J. R. Schrieffer, Phys. Rev. **106**, 162 (1957); **108**, 1175 (1957).
3. N. V. Zavaritski, Dokl. Akad. Nauk. SSSR **86**, 501 (1952).
4. A. A. Abrikosov, Dokl. Akad. Nauk, SSSR **86**, 489 (1952).
5. S. F. Edwards, Phys. Rev. **90**, 284 (1953).
6. L. D. Landau, A. A. Abrikosov, I. M. Khalatnikov, Dokl. Akad. Nauk. SSSR **95**, 773 (1954).
7. L. D. Landau, E. M. Lifshitz, Dokl. Akad. Nauk. SSSR **100**, 669 (1955).
8. R. P. Feynman, *Progress in Low Temperature Physics*, Volume 1 (C. J. Gorter, ed.), Interscience, New York (1955), page 34.
9. L. V. Shubnikov, W. I. Khotkévich, J. D. Shepelióv, J. N. Riabínin, Zh. Eksp. Teor. Fiz. **7**, 221 (1937).
10. K. Mendelssohn, Proc. Roy. Soc. **A152**, 34 (1935).
11. H. Träuble, U. Essmann, J. Appl. Phys. **39**, 4052 (1968).
12. A. A. Abrikosov, Zh. Eksp. Teor. Fiz. **32**, 1442 (1957) [Sov. Phys.—JETP **5**, 1174 (1957)].
13. A. A. Abrikosov, J. Phys. Chem. Solids **2**, 199 (1957).
14. J. E. Kunzler, E. Buehler, F. S. L. Hsu, J. H. Wernick, Phys. Rev. Lett. **6**, 89 (1961).
15. A. B. Pippard, Proc. Roy. Soc. **A216**, 547 (1953).
16. B. B. Goodman, Phys. Rev. Lett. **6**, 597 (1961).
17. B. B. Goodman, IBM J. Res. Devel. **6**, 63 (1962).
18. A. A. Abrikosov, L. P. Gor'kov, I. M. Khalatnikov, Zh. Eksp. Teor. Fiz. **35**, 265 (1958) [Sov. Phys.—JETP **8**, 182 (1959)].
19. A. A. Abrikosov, L. P. Gor'kov, Zh. Eksp. Teor. Fiz. **42**, 1088 (1962) [Sov. Phys.—JETP **15**, 752 (1962)].
20. A. A. Abrikosov, L. P. Gor'kov, Zh. Eksp. Teor. Fiz. **39**, 1781 (1960) [Sov. Phys.—JETP **12**, 1243 (1961)].
21. A. A. Abrikosov, Physics **2**, 5 (1965).
22. A. A. Abrikosov, I. M. Khalatnikov, Rept. Progr. Phys. **22**, 329 (1959).
23. A. A. Abrikosov, L. A. Falkovski, Zh. Eksp. Teor. Fiz. **43**, 1089 (1962) [Sov. Phys.—JETP **16**, 769 (1963)].
24. A. A. Abrikosov, J. Low Temp. Phys. **8**, 315 (1972). □

DICK FEYNMAN—THE GUY IN THE OFFICE DOWN THE HALL

A brilliant, vital and amusing neighbor, Feynman was
a stimulating (if sometimes exasperating) partner in discussions
of profound issues. His sum-over-paths method may turn out
to be not just useful, but fundamental.

Murray Gell-Mann

PHYSICS TODAY/FEBRUARY 1989

*I hope someday to write a lengthy piece about Richard
Feynman as I knew him (for nearly 40 years, 33 of them as
his colleague at Caltech), about our conversations on the
fundamental laws of physics, and about the significance of
the part of his work that bears on those laws. In this brief
note, I restrict myself to a few remarks and I hardly touch
on the content of our conversations.*

When I think of Richard, I often recall a chilly afternoon
in Altadena shortly after his marriage to the charming
Gweneth. My late wife, Margaret, and I had returned in
September 1960 from a year in Paris, London and East
Africa; Richard had greeted me with the news that he was
"catching up with me"—he too was to have an English
wife and a small brown dog. The wedding soon took place,
and it was a delightful occasion. We also met the dog
(called Venus, I believe) and found that Richard was going
overboard teaching her tricks (leading his mother, Lucille,
with her dry wit, to wonder aloud what would become of a
child if one came along). The Feynmans and we both
bought houses in Altadena, and on the afternoon in
question Margaret and I were visiting their place.

Richard started to make a fire, crumpling up pages of
a newspaper and tossing them into the fireplace for
kindling. Anyone else would have done the same, but the
way he made a game out of it and the enthusiasm that he
poured into that game were special and magical. Meanwhile, he had the dog racing around the house, up and
down the stairs, and he was calling happily to Gweneth.

Murray Gell-Mann is Robert A. Millikan Professor of
Theoretical Physics at the California Institute of Technology.

He was a picture of energy, vitality and playfulness. That
was Richard at his best.

He often worked on theoretical physics in the same
way, with zest and humor. When we were together
discussing physics, we would exchange ideas and silly
jokes in between bouts of mathematical calculation—we
struck sparks off each other, and it was exhilarating.

What I always liked about Richard's style was the lack
of pomposity in his presentation. I was tired of theorists
who dressed up their work in fancy mathematical
language or invented pretentious frameworks for their
sometimes rather modest contributions. Richard's ideas,
often powerful, ingenious and original, were presented in
a straightforward manner that I found refreshing.

I was less impressed with another well-known aspect
of Richard's style. He surrounded himself with a cloud of
myth, and he spent a great deal of time and energy
generating anecdotes about himself.

Sometimes it did not require a great deal of effort. For
example, during my first decade at Caltech there was a
rule at our faculty club, the Athenaeum, that men had to
wear jackets and ties at lunch. Richard usually came to
work quite conventionally dressed (for those days) and
hung his jacket and tie in his office. He rarely ate lunch at
the Athenaeum, but when he did, he would often make a
point of walking over in his shirt sleeves, tieless, and then
putting on one of the ragged sport coats and one of the loud
ties that the Athenaeum provided in the cloakroom for
men who arrived unsuitably attired.

Many of the anecdotes arose, of course, through the
stories Richard told, of which he was generally the hero,
and in which he had to come out, if possible, looking
smarter than anyone else. I must confess that as the years
went by I became uncomfortable with the feeling of being a

Portrait of Richard Feynman by Jirayr Zorthian, one of Feynman's art teachers. The painting hangs in Jadwin Hall at Princeton University.

rival whom he wanted to surpass; and I found working with him less congenial because he seemed to be thinking more in terms of "you" and "me" than "us." Probably it was difficult for him to get used to collaborating with someone who was not just a foil for his own ideas (especially someone like me, since I thought of Richard as a splendid person to bounce my ideas off!).

At first, none of that was much of a problem, and we had many fine discussions in those days. In the course of those talks not only did we "twist the tail of the cosmos," but we also exchanged a good many lively reminiscences about our experiences in research.

Summing over histories

He told me, of course, of his graduate student days at Princeton and his adventures with his adviser, John Wheeler. Wheeler judged their work on the "absorber theory of radiation" to be too much of a collaboration to qualify as a dissertation for the PhD, and so Richard pursued his interest in Paul Dirac's work on the role of the action S in quantum mechanics. In his book on quantum

mechanics, and even more in his article in the *Physikalische Zeitschrift der Sowjetunion* in 1932, Dirac had carried the idea quite far. He had effectively shown how a quantum mechanical amplitude for the transition from a set of values of the coordinates at one time to another set of values at a later time could be represented as a multiple integral, over the values of the coordinates at closely spaced intermediate times, of $\exp(iS/\hbar)$, where S is the value of the classical action along each sequence of intermediate coordinate values. What Dirac had not done was to state the result in so many words, to point out that this method could be used as the starting point for all quantum mechanics, and to suggest it as a practical way of doing quantum mechanical calculations.

Richard did just those things, I understand, in his 1942 dissertation, and then used the "path integral" or "sum over paths" approach in a great deal of his subsequent research. It was the basis, for example, of his way of arriving at the now standard covariant method of calculation in quantum field theory (which Ernst Stueckelberg reached in a different manner). That method is, of course, always presented in terms of "Feynman diagrams" such as the ones Dick later had painted on his van.

The sum-over-paths formulation is particularly convenient for integrating out one set of coordinates to concentrate on the remaining set. Thus the photon propagator in quantum electrodynamics is obtained[1] by "integrating out" the photon variables, leaving electrons and positrons, both real and virtual, to interact by means of the covariant function $\delta(x^2) + (\pi i x^2)^{-1}$.

In 1963 Feynman and his former student F. L. Vernon Jr, carrying further some research Ugo Fano had earlier done in a different way, showed how in a wide variety of problems of concern to laser physicists, condensed matter physicists and others of a practical bent, one can integrate out variables that are not of interest to throw light on the behavior of the ones that are kept. If initially the density matrix factors into one part depending on the interesting variables and another part depending on the rest, then the subsequent time development of the reduced density matrix for the interesting variables can be expressed in terms of a double path integral in which the coefficient of the initial reduced density matrix is $\exp[i(S - S' + W)/\hbar]$, where S is the action along the path referring to the left-hand side of the density matrix, S' is the action along the path referring to the right-hand side of density matrix, and W is the "influence functional," depending on both paths, that comes from integrating out all the uninteresting variables. Feynman and Vernon worked out a number of cases in detail, and subsequent research by A. O. Caldeira and Anthony Leggett, among others, further clarified some of the issues involved.

Shedding light on quantum mechanics

More recently, in the work of H. Dieter Zeh, of Erich Joos and of Wojciech Żurek and others, this line of research has thrown important light on how quantum mechanics produces decoherence, one of the conditions for the nearly classical behavior of familiar objects. For a planet, or even a dust grain, undergoing collisions with, for example, the photons of the 3-K radiation, the imaginary part of the functional W resulting from the integration over those quanta can yield, in $\exp(iW/\hbar)$, a factor that decreases exponentially with some measure of the separation between the coordinate trajectory on the left side of the density matrix and that on the right. The density matrix can thus be constrained to remain nearly diagonal in the coordinates of the particle, giving rise to decoherence. If in addition the dust grain's inertia is large enough that the grain resists, for the most part, the disturbances of its trajectory caused by the quantum and thermal fluctuations of the background, and also large enough that the quantum spreading of the coordinate is slow, then the behavior of the grain's position operator will be nearly classical.

When an operator comes into correspondence with a nearly classical operator, then the first operator can be measured or observed. Thus work such as that of Feynman and Vernon has led not only to practical applications but also to a better understanding of how quantum mechanics produces the world with which we are familiar.

The path integral approach has proved in numerous situations to be a useful alternative to the conventional formulation of quantum mechanics in terms of operators in Hilbert space. It has many advantages besides the ease of integrating out, under suitable conditions, some of the variables. The path integral method, making use as it does of the action, can usually display in an elegant manner the invariances of the theory and can point the way toward exhibiting those invariances in a perturbation expansion. It is obviously a good approach for deriving the classical limit, and it can also be very helpful in semiclassical approximations, for example, in the description of tunneling. For certain effects, such as tunneling via instantons, it permits calculations that are highly nonperturbative in the usual sense. It is also particularly good for the global study of field configurations in quantum field theory, as it permits a straightforward discussion of topological effects.

Of course the conventional approach is superior for certain purposes, such as exhibiting the unitarity of the S matrix and the fact that probabilities are not negative. Richard would never have contemplated, as he did around 1948, the consistent omission of all closed loops in quantum electrodynamics if he had been thinking in terms of a Hamiltonian formulation, where unitarity, which rules out such an omission, is automatic. (The impossible theory without closed loops could, by the way, realize the remarkable vision of Wheeler, which Richard said Wheeler once awakened him to explain: Not only are positrons electrons going backward in time, but all electrons and positrons represent the same electron going backward and forward, thus explaining why they all have the same absolute value of the electric charge!)

In any case, the path integral formulation remained merely a reformulation of quantum mechanics, equivalent to the usual formulation. I say "merely" because Richard, with his great talent for working out, sometimes in dramatically new ways, the consequences of known laws, was unnecessarily sensitive on the subject of discovering new ones. He wrote, in connection with the discovery of the universal vector and axial vector weak interaction in 1957: "It was the first time, and the only time, in my career that I knew a law of nature that nobody else knew. (Of course, it wasn't true, but finding out later that at least Murray Gell-Mann—and also [E. C. George] Sudarshan and [Robert] Marshak—had worked out the same theory didn't spoil my fun.) . . . It's the only time I ever discovered a new law."[2]

Thus it would have pleased Richard to know (and perhaps he did know, without my being aware of it) that there are now some indications that his PhD dissertation may have involved a really basic advance in physical theory and not just a formal development. The path integral formulation of quantum mechanics may be more fundamental than the conventional one, in that there is a crucial domain where it may apply and the conventional formulation may fail. That domain is quantum cosmology.

Seeking rules for quantum gravity

Of all the fields in fundamental physical theory, the gravitational field is picked out as controlling, in Einsteinian fashion, the structure of space–time. This is true even in a unified description of all the fields and all the particles of nature. Today, in superstring theory, we have the first

JOE MUNROE/CALIFORNIA INSTITUTE OF TECHNOLOGY

respectable candidate for such a theory, apparently finite in perturbation theory and describing, roughly speaking, an infinite set of local fields, one of which is the gravitational field linked to the metric of space–time. If all the other fields are dropped, the theory becomes an Einsteinian theory of gravitation.

Now the failure of the conventional formulation of quantum mechanics, if it occurs, is connected with the quantum mechanical smearing of space–time that is inevitable in any quantum field theory that includes Einsteinian gravitation.

If there is a dominant background metric for space–time, especially a Minkowskian metric, and one is treating the behavior of small quantum fluctuations about the background (for example, the scattering of gravitons by gravitons), then the deep questions about space–time in quantum mechanics do not come to the fore.

Dick played a major part in working out the rules of quantum gravity in that approximation. It so happened that I was peripherally involved in the story of that research. We first discussed it when I visited Caltech during the Christmas vacation of 1954–55 and he was my host. (I was offered a job within a few days—such things would take longer now.) I had been interested in a similar approach, sidestepping the difficult cosmological issues, and when I found that he had made considerable progress I encouraged him to continue, to calculate one-loop effects and to find out whether quantum gravity was really a divergent theory to that order. He was always very suspicious of unrenormalizability as a criterion for rejecting theories, but he did pursue the research on and off. In 1960 he complained to me that he was having trouble. His covariant diagram method was giving results incompatible with unitarity. The imaginary part of the amplitude for a fourth-order process should be related directly to the product of a second-order amplitude and the complex conjugate of a second-order amplitude. That relation was failing.

I suggested that he try the analogous problem in Yang–Mills theory, a much simpler nonlinear gauge theory than Einsteinian gravitation. Richard asked what Yang–Mills theory was. (He must have forgotten, because in 1957 we worked out the coupling of the photon to the charged intermediate boson for the weak interaction and noticed that it was the right coupling for a Yang–Mills theory of those quanta.) Anyway, it didn't take long to teach him the rudiments of Yang–Mills theory, and he threw himself with renewed energy into resolving the contradiction. He found, eventually, that in the Lorentz-covariant formulation of either theory it was necessary to introduce some weird supplementary fields called "ghosts," and they have been used ever since, acquiring more and more importance. He described them at a meeting in Poland (in 1963, I think). Usually they are called "Faddeev–Popov ghosts" after L. D. Faddeev and V. N. Popov, who also studied them.

Thus Feynman was able to report in the 1960s that Einsteinian gravitation was terribly divergent when interacting with electrons, photons or other particles. (The divergences in *pure* quantum gravitation theory turned out to be serious too, but that was shown much later, in the two-loop approximation, by two Caltech graduate students, Marc Goroff and Augusto Sagnotti.)

Those problems may be rectified by unification of all the particles and interactions, as they are in superstring theory. But we must still face up to the issues raised by the fact that the metric is up for quantum mechanical grabs and cannot in general be treated as a simple classical background plus small quantum fluctuations.

Quantum cosmology

Recently there has been great progress in thinking about the cosmological aspects of quantized Einsteinian gravitation. The work of Stephen Hawking and James Hartle, as well as Claudio Teitelboim, Alexander Vilenkin, Jonathan Halliwell and several others, has shown how the path integral method can probably deal with the situation and how it may be possible to generalize the method so as to describe *not only the dynamics of the universe but also its initial boundary condition* in terms of the classical action S. Furthermore, there are now, as I mentioned above, some indications that the conventional formulation of quantum mechanics may not be justifiable except to the extent that a background space–time emerges with small quantum fluctuations. Hartle in particular has emphasized such a possibility.

One crude way to see the argument is to express the wavefunction of the universe (which we assume to be in a pure state) as a path integral over all the fields in nature (for example, the infinity of local fields represented, roughly speaking, by the superstring), reserving the integral over the metric $g_{\mu\nu}$ for last. The total action S can be represented as the Einstein action S_G for pure gravitation plus the actions S_M for all the other, "matter" fields, including their coupling to gravitation. We have, then, crudely,

$$\text{Amplitude} =$$
$$\int \mathscr{D}g_{\mu\nu} \exp \frac{iS_G}{\hbar} \times \int \mathscr{D}(\text{everything else}) \exp \frac{iS_M}{\hbar}$$

For the moment, suppose only $g_{\mu\nu}$ configurations corresponding to a simple topology for space–time are allowed.

Before the integration over $g_{\mu\nu}$ is performed, there is a definite space–time, with the possibility of constructing well-defined space-like surfaces in a definite succession described by a time-like variable. There is an equivalent Hilbert-space formalism; we have unitarity (conservation of positive probability); and we can have conventional causality (it corresponds in the Hilbert-space formulation to the requirement of time ordering of operators in the formula for probabilities).

Now, when the integral over $g_{\mu\nu}$ is done, it is no longer clear that any of that machinery remains, since we are integrating over the structure of space–time and once the integral is performed it is hard to point to space-like surfaces or a succession described by a time-like variable. Of course it may be possible to construct a Hilbert-space formulation, with unitarity and causality, in some new way, perhaps employing a new, external time variable of some kind (what Feynman liked to call a fifth wheel), but it is by no means certain that such a program can be carried out.

At this stage, we may admit the possibility of summing over all topologies of space–time (or of the corresponding space–time with a Euclidean metric). If that is the correct thing to do, then we are immediately transported into the realm of baby universes and wormholes, so beloved of Stephen Hawking and now so fashionable, in which it seems to be demonstrable that the cosmological constant vanishes. In that realm the path integral method appears able to cope, and it remains to be seen to what extent the conventional formulation of quantum mechanics can keep up.

For Richard's sake (and Dirac's too), I would rather like it to turn out that the path integral method is the real foundation of quantum mechanics and thus of physical theory. This is true despite the fact that, having an algebraic turn of mind, I have always personally preferred the operator approach, and despite the added difficulty, in the absence of a Hilbert-space formalism, of interpreting the wavefunction or density matrix of the universe (already a bit difficult to explain in any case, as anyone attending my classes will attest). If notions of transformation theory, unitarity and causality really emerge from the mist only after a fairly clear background metric appears (that metric itself being the result of a quantum mechanical probabilistic process), then we may have a little more explaining to do. Here Dick Feynman's talents and clarity of thought would have been a help.

Turning things around

Richard, as is well known, liked to look at each problem, important or unimportant, in a new way—"turning it around," as he would say. He told how his father, who died when he was young, taught him to do that. This approach went along with Richard's extraordinary efforts to be different, especially from his friends and colleagues.

Of course any of us engaged in creative work, and in fact anyone having a creative idea even in everyday life, has to shake up the usual patterns in some way in order to get out of the rut (or the basin of attraction!) of conventional thinking, dispense with certain accepted but wrong notions, and find a new and better way to formulate some problem. But with Dick, "turning things around" and being different became a passion.

The result was that on certain occasions, in scientific work or in ordinary living, when an imaginative new way of looking at things was needed, he could come up with a remarkably useful innovation. But on many other occa-

sions, when the usual way of doing business had its virtues, he was not the ideal person to consult. Remember his television appearance in which he made fun of the daily habit of brushing one's teeth? (And he didn't even suggest flossing!) Or take his occasional excursions into far-out political choices in the 1950s, during his second marriage. Those certainly set him off from most of his friends. But one day during that time, he called me and sheepishly admitted having voted for a particularly outrageous candidate for statewide office—and then asked me if in the future I would check over such names beforehand and tell him when he was really going off the deep end!

None of the aberrations mentioned here changes the fact that Dick Feynman was a most inspiring person. I have referred to his originality and straightforwardness and to his energy, playfulness and vitality. All of those characteristics showed up in his work and also in the other facets of his life. Indeed, that vitality may be related to the kind of biological (and probably psychological) vitality that enabled him to resist so remarkably and for so long the illness to which he finally succumbed.

When I think of him now, it is usually as he was during that first decade that we were colleagues, when we were both young and everything seemed possible. We phoned each other with good ideas and crazy ones, with serious messages and farcical gags. We yelled at each other in front of the blackboard. We taught stewardesses to say "quark–quark scattering" and "quark–antiquark scattering." We delivered a peacock to the bedroom of our friend Jirayr Zorthian on his birthday, while our wives distracted him. We argued about everything under the Sun.

Later on, we drifted apart to a considerable extent, but I was aware, all the time we were colleagues, that if a really profound question in science came up, there would be fun and profit in discussing it with Dick. Even though on many occasions during the last 20 years, I passed up the opportunity to talk with him in such a case, I knew that I *could* do so, and that made a great difference.

Besides, I did not always pass it up. For example, during the last few months and even weeks of his life, we kept up a running discussion of one of the most basic subjects, the role of "classical objects" in the interpretation of quantum mechanics. We thus resumed a series of conversations on that topic that we had begun a quarter of a century earlier. In between 1963–64 and 1987 those talks about quantum mechanics were rare, but there was at least one remarkable occasion during the last few of those years. Richard sat in on one of my classes on the meaning of quantum mechanics, interrupting from time to time. He did not, however, object to what I was saying; rather, he reinforced the points I was making. The students must have been delighted as they heard the same arguments made by both of us in a kind of counterpoint.

It is hard for me to get used to the fact that now, when I have a deep issue in physics to discuss with someone, Dick Feynman is no longer around.

⋆   ⋆   ⋆

I should like to thank James B. Hartle for many instructive conversations about quantum mechanics and the path integral method in quantum cosmology.

References

1. R. P. Feynman, Phys. Rev. **76**, 749, 769 (1949).
2. R. P. Feynman, *"Surely You're Joking, Mr. Feynman!" Adventures of a Curious Character*, Bantam, New York (1986), p. 229. See also M. Gell-Mann, in *Proc. Int. Mtg. on the History of Scientific Ideas*, M. G. Doncel *et al.*, eds., Bellaterra, Barcelona (1987), p. 474. ∎

PHYSICS TODAY/SEPTEMBER 1984

Peter Kapitza 1894–1984

Pyotr Leonidovich Kapitza, one of the most revered scientists in the Soviet Union and one of the best known Russian physicists in the world, died on 8 April 1984.

Born on 9 July 1894 in Kronstadt, the naval base near Leningrad (then called St. Petersburg), he was the son of a general in the Corps of Engineers. Kapitza received training as an electrical engineer at the Polytechnic Institute of St. Petersburg, and, after graduating in 1918, stayed on as a lecturer. But even before that he commenced research work under A. F. Joffe; he published his first papers in 1916, one of which foreshadowed the Stern–Gerlach experiment.

In 1921 Kapitza came to Cambridge in England for a short visit, which, as it turned out, lasted 13 years; later (in 1966) he described this period of work as his happiest years. Ernest Rutherford, at that time Cavendish Professor, took a great liking to Kapitza, sensing in him a kindred spirit, and offered him a Clerk Maxwell studentship. His first project was a study of α-particle tracks in a strong magnetic field for which he developed a new technique of pulsed magnetic fields; he achieved 32 T in pulses lasting 10 ms.

This experiment was Kapitza's only incursion into nuclear physics. Having built the magnet, he used it as a tool in the area of his major interest, the study of physical properties of matter in strong magnetic fields. Topics of papers published between 1924 and 1932 include electrical conductivity, magnetostriction and the Zeeman effect in such fields. These studies led him on to low-temperature physics, which became his second major area of interest. Again, he entered the new subject by designing the tools for it, in this case a new technique of liquefaction of helium by an adiabatic method. His helium liquefier was for many years the mainstay of cryogenic laboratories.

The research accomplishments of that period gained Kapitza wide recognition. He was elected a Fellow of the Royal Society in 1929 and appointed Research Professor. He also became the Director of the Mond Laboratory in Cambridge—designated for research in magnetic and low-temperature physics. However, he was not to reap the fruits of the plans carefully laid by him for that laboratory.

In 1934, while on his annual trip to Russia to visit his mother, his passport was withdrawn and he was not allowed to return to England. This had a very depressing effect on him and kept him

KAPITZA

away from scientific work for several years, but his spirits were bolstered by frequent letters from Rutherford who urged him to get back to research as the best remedy for his troubles. Heeding this advice, Kapitza accepted the appointment as director of a new Institute for Physical Problems set up by the USSR Academy of Sciences, of which he became a full member in 1939. Kapitza himself designed the building of the Institute, formulated its organization and selected the research team. Rutherford arranged the sale of the equipment of the Mond Laboratory so that Kapitza could continue with his line of research.

Most of the work of the Institute during the ensuing decade was in low-temperature physics, chiefly on the properties of liquid helium. The main achievement in that period was the discovery of the superfluidity of helium below the lambda point. A full theoretical explanation of this phenomenon was given by L. D. Landau, whom Kapitza persuaded to join his Institute, and who received a Nobel Prize in 1962 for this work. Kapitza himself had to wait until 1978 before his own contributions were formally recognized by a Nobel Prize.

In 1946 Kapitza was again in trouble, this time apparently in connection with the Soviet atom bomb project. He was removed from the directorship of the Institute and put under house arrest in his dacha, where he was kept until after Stalin's death. Although in 1941 Kapitza had already spoken publicly about the possibility of the atom bomb, he did not take part in the project. The popular version in the West is that he refused on political or moral grounds; according to Herbert

York, however, it was due to a difference of opinion on technical issues. The version given by his Soviet biographer, Academician A. S. Borovin-Romanov, is that "the method for the production of oxygen proposed by Kapitza was unjustly condemned."

But even during the period of arrest Kapitza managed to carry out valuable research in a small domestic laboratory set up in his dacha, publishing papers on such diverse subjects as the nature of ball lightning, the formation of sea waves by the wind and the stability of a pendulum with a vibrating suspension. At that time he also started work on high-power electronics, a subject to which he later devoted much effort. In 1955 he was reinstated as director of the Institute and resumed the work on low-temperature physics. In his later years Kapitza also developed an interest in plasma physics and in methods of producing high temperatures for thermonuclear reactions.

His interests always extended beyond physics. He held strong views on many subjects: science and technology; education and organization of research; philosophy and politics; international relations and the social impact of science.

Kapitza firmly believed in the unity of science and technology and was himself the best embodiment of that unity, always producing the tools for his fundamental research. He shared with Rutherford a predilection for simple approaches to problems. His favorite quotation was from the Ukrainian philosopher, Skovoroda: "We must be grateful to God that he created the world in such a way that everything simple is true, and everything complicated is untrue."

Kapitza was always conscious of the need to train new generations of scientists and devoted an immense effort to the education of young people and the encouragement of creative talent. A powerful figure in the Academy of Sciences—a member of its presidium—he often criticized its performance and bureaucracy. He was particularly scathing about the lack of debate at sessions of the Academy. In Cambridge he had established what became known as the Kapitza Club, where young physicists gathered to discuss over dinner the developments in science, and he wished to see more opportunities for open discussion in the Soviet Union.

Always outspoken, Kapitza found himself from time to time in conflict with ideological orthodoxy, but his spirits were undaunted. His independence of mind found many expressions, one of them being his refusal in 1973 to join other academicians in condemning Andrei Sakharov.

Kapitza was much concerned about the nuclear arms race and sought ways to stop it. This brought him to Pugwash; he was an active member of the Soviet Pugwash Group and participated in Pugwash Conferences in the USSR, Sweden, France, Finland and Austria. His interest in Pugwash also stemmed from his conviction that scientists are the most likely group to tackle successfully global problems. In the Bernal Lecture, given in the Royal Society in 1976, he said: "The future of civilization depends on whether existing governments are able to provide solutions to global problems.... But, for this, problems must be expressed clearly and convincingly and widely discussed. This can be done mainly by scientists, since they can talk with sufficient authority on the possible solution of global problems for the benefit of mankind. Thus, we should not stand aside from the solution of such problems but realize their connection with our scientific work." One hopes this call will be heeded by the scientific community in the East and West as its tribute to the memory of a great man, big enough to span the ideological divide.

JOSEPH ROTBLAT
University of London

PHYSICS TODAY/JANUARY 1985

Paul Adrien Maurice Dirac 1902–1984

Paul Dirac and Werner Heisenberg in 1933.

Paul Adrien Maurice Dirac, who died on 20 October 1984, was one of the great physicists of the century. Born on 8 August 1902, he obtained an engineering degree at Bristol in 1921, but his outstanding mathematical ability was soon recognized, and after concentrating on mathematics at Bristol for two years he was offered a grant for graduate studies at Cambridge. There he worked under R. H. Fowler; five of his papers written during that time show his already firm command of relativity, quantum theory and statistical mechanics.

His first paper on quantum mechanics was written while he was still a student. He had obtained from Fowler a proof copy of Werner Heisenberg's paper and immediately saw the significance of the new ideas. In particular, he saw the fundamental role of non-commutative algebra and started extending the ideas even before Heisenberg's paper was published. From then on he developed the structure of quantum mechanics in his own way, introducing his own characteristic notation, much of which came into general use. He showed how to combine both Erwin Schrödinger's and Heisenberg's approaches into one common system by using what we now call transformation theory. In this context he introduced the bracket notation and later found it useful to split the brackets in two parts, which he called "bra" and "ket," further pieces of terminology that have come into general use. (I believe he was not aware at the time of the alternative meaning of bra.) The inclusion of continuous variables in the general transformation theory was made possible by the "delta function," another invention of Dirac, which horrified the purists until it was made respectable by Laurent Schwartz.

Dirac took many more important steps in completing the foundations of quantum mechanics, including the first discussion of the emission and absorption of radiation, and of dispersion. He introduced the density matrix, which besides being a convenient tool for quantum statistics is an important concept for the interpretation of quantum mechanics. He saw, independent of Heisenberg and Enrico Fermi, the connection between the symmetry of the wave function and the Pauli principle, leading to Fermi–Dirac statistics.

In 1928 he formulated what is now called the Dirac equation, the relativistic wave equation, which became the keystone of atomic quantum mechanics. Considered as a wave equation for one particle, it is beset by the difficulty of the negative-energy states—but by another stroke of genius he conceived the hole theory, in which all negative-energy states are normally occupied. At first he thought the holes, or empty places in the negative-energy region, were protons, but this hypothesis could not be maintained, and they turned out to be the positrons.

In 1932 Dirac succeeded Sir Joseph Larmor as Lucasian Professor of Mathematics in Cambridge (where, traditionally, theoretical physics is regarded as part of mathematics) and he held this position until his retirement in 1969. In 1971 he became professor of physics at Florida State University.

By the early 1930s, quantum mechanics was essentially complete, and it owed an enormous debt to Dirac, who was recognized with an award of a Nobel prize in 1933.

His work would have secured him a position amongst the great names in physics, even if he had done nothing after 1933, but he continued without slackening his pace. Some of the later work involved applications of quantum mechanics, but most of it was trying new departures to advance into as yet uncharted areas of physics. His suggestion of a magnetic monopole ("one would be surprised if Nature had not made use of [this possibility]") and of the change with time of the gravitational and other constants of physics have attracted much attention, but there is as yet no final verdict on their validity. Many other highly original and ingenious ideas have not made any impact on physics, but it would be rash to assert that we may not learn to make good use of them one day, or that we may not see their significance, as was the case with the brief aside in his book *The Principles of Quantum Mechanics*, which turned out to contain the idea of Feynman's "integral-over-paths" method.

All of Dirac's writings are characterized by brevity and extreme clarity. Often he seems to be saying the obvious—only it was far from obvious before he said it. Where other people would hesitate in following a thought through to the end, letting their prejudices divert them, Dirac would show the direct path.

His unusual way of thinking often showed up in conversation about everyday affairs. There are many "Dirac stories" about—some apocryphal, many genuine. They have led to an image that is far too simple and stereotyped to apply to such an unusual person. He was said to be very silent, but he could be very articulate if he had something to say. He was thought to be concerned only with very abstract things, but he invented a method of isotope separation and, encouraged by Kapitza, set up an experiment to test the method. When, during the war, isotope separation became a very practical problem and a team in Oxford decided to try his method, he made many practical suggestions. He collected puzzles and took a great interest in gardening.

Dirac did not supervise many research students, perhaps because he did not like to discuss unsolved problems or speculate what the answers might be. But any students who asked questions would get clear and patient answers.

He received, of course, numerous honors and distinctions, including the Order of Merit—the highest and most selective British honor for great minds.

RUDOLF PEIERLS
Oxford University

Felix Bloch 1905–1983

Felix Bloch died in Zurich on 10 September 1983 at age 77. He had been working and thinking of physics virtually up to the last day of his life. With his passing went one of the great physicists of the twentieth century.

Bloch was born in Zurich on 23 October 1905. His father urged him to study engineering, and after graduation from secondary school, he entered the Federal Institute of Technology in Zurich with the intention of pursuing such a career. His interests, however, really lay in theoretical directions, and physics and mathematics attracted him much more than did engineering. After a year he transferred to the Institute's division of mathematics and physics. There could hardly have been a better time, for during those years, 1924 to 1927, modern quantum theory emerged in such splendor. In the early part of this climactic period, Louis de Broglie, Werner Heisenberg, P. A. M. Dirac, Erwin Schrödinger, Samuel Goudsmit, George Uhlenbeck, and, a little earlier, Wolfgang Pauli, all made their great contributions to quantum physics.

Peter Debye, with whom Bloch studied at the Institute, suggested that he do his thesis work at Leipzig, where Heisenberg would soon join the faculty. In this way Bloch became Heisenberg's first graduate student. Under Heisenberg's tutelage he attacked the problem of the conductivity of metals and succeeded in his PhD thesis in finding the solution of the quantum theory of metals. This led in turn to the modern quantum theory of solids and eventually to modern radio, television, computers, and the electronic logic behind modern experimentation in virtually all fields of science.

Bloch returned to Zurich as Pauli's assistant in 1928 and worked on the theory of superconductivity. In the next year he became a Lorentz fellow in Holland and worked with both Hendrix Kramers and Adriaan Fokker. During this period he demonstrated that his thesis work could be successfully extended to the theory of electrical resistance, in agreement with experimental results. Moving again to Leipzig in 1930 with Heisenberg, Bloch worked on ferromagnetism and established the nature of the boundaries between do-

BLOCH

mains, which have subsequently been known as "Bloch walls."

Bloch left Leipzig to join Niels Bohr's institute in Copenhagen. One result of his visit was that he became a lifelong friend of Bohr, subsequently visiting Bohr many times in Copenhagen. There he also worked on his famous contribution to the theory of the stopping of charged particles in matter. Again returning to Leipzig in 1932, he became Privat-dozent. When Hitler came to power in 1933, Bloch left Leipzig for his native Zurich. He spent some time as a Rockefeller fellow in Rome, and then, along with many others, left Europe in 1934.

In 1934 Bloch joined the department of physics at Stanford University. Bringing a new kind of physics to Stanford, he started to do neutron studies there. By 1936 he obtained experimental results on the magnetic scattering of neutrons. This familiarity with nuclear physics and his experimental achievements on the magnetic moment of the neutron that he and Luis Alvarez accomplished in 1939 at Berkeley eventually led to his discovery of nuclear induction. For this work, in which he collaborated with William W. Hansen and Martin Packard, Bloch was awarded the Nobel prize in physics for 1952 jointly with Edward Purcell, who had worked separately with Henry Torrey and Robert Pound. The discovery of nuclear induction

became the foundation of new fields in physics, chemistry, biology, physiology and medicine. Indeed, the recent successes of magnetic resonance imaging in diagnostic medicine surely rival the tremendous advance brought about by Wilhelm Roentgen's discovery of x rays.

During the second world war, Bloch worked briefly at Los Alamos and at length in Frederick Terman's Radio Research Laboratory at Harvard on antiradar work. In 1945 Bloch returned to teaching and research at Stanford, where he continued his studies of nuclear induction and, with Leonard Schiff, built a great physics department.

In 1954 Bloch became the first director of CERN, newly formed, in Geneva and formulated and implemented its early scientific policies. Recognizing that his administrative duties prevented him from doing physics, he returned to Stanford a year later. Another sign of the wide appreciation of his accomplishments came in 1965: his presidency of The American Physical Society.

Bloch loved nature, particularly mountains. Fond of skiing, even in his later years he could be seen enjoying the slopes in his old ski suit. He also played the piano well and with tremendous satisfaction. Despite the extraordinary gifts that Bloch gave to the world, he remained modest, but not quietly modest, for he held strong opinions and was usually outspoken in expressing them. No one had any doubt about what Bloch was saying or where he stood on any issue. He enjoyed a good intellectual fight and together with his colleagues at Stanford often produced many sparks.

Felix and Lore Misch were married in 1940 after having met in the previous year at an American Physical Society meeting. Lore also holds a PhD degree in physics and worked in x-ray crystallography.

I was privileged to know Felix Bloch for some 34 years, in which I learned much from him, not only in physics, but also about all the best things in human companionship. I know that I speak for many others as well, and we are all going to miss him more than we know.

ROBERT HOFSTADTER
Stanford University

Jerrold Reinach Zacharias
1905–1986

Jerrold Reinach Zacharias, Institute Professor Emeritus at the Massachusetts Institute of Technology, died on 16 July 1986 at the age of 81. Born in Jacksonville, Florida, he studied at Columbia University and obtained his PhD with Shirley L. Quimby. From 1931 to 1940 he taught at Hunter College while embarking on research in molecular beams with I. I. Rabi at Columbia. The work of that group on hyperfine structure, leading (among other things) to precision measurements of the magnetic moments of the proton and the deuteron, was a classic of atomic physics research in which Zacharias played an important part.

In 1940 Zacharias joined the newly established radiation laboratory at MIT, working on airborne radar. It was here that the first 10-cm radar system was developed. Paralleling British practice, young physicists were given leading roles; they proved themselves very valuable in the field of invention, but did not have the kind of background necessary to design devices that could be mass-produced and used in the field. In 1941 the leaders of the laboratory, in particular Lee DuBridge and Rabi, recognized this need and found in Zacharias the person to deal with it. He had a unique blend of talents as both scientist and technologist—Jerome B. Wiesner has called him the first systems engineer. After spending six months with technical personnel at Bell Laboratories (New York City), Zacharias returned to MIT with the onerous task of convincing his colleagues that engineering design and production was a crucially important part of the job. It became evident at this point that Zacharias was a born leader. He knew how to provide encouragement and get others mobilized, but he was also a perfectionist who could be very demanding. He had the great gift of being able to think a project through from A to Z, and yet change course *en route* if need be. The result was that, by 1943, airborne microwave radar was a reality.

ZACHARIAS

In 1945 Zacharias went to Los Alamos to direct the engineering division of the Manhattan Project; his role was much the same as it had been at the radiation laboratory. His direct involvement there was brief, but its consequences were of profound importance to MIT, for in 1946 he returned to MIT, and brought with him half a dozen outstanding physicists from Los Alamos—chief among them Bruno Rossi and Victor Weisskopf. The result was a transformation of the department in both quality and scope. In addition, resuming his prewar interests, Zacharias established his own molecular beam laboratory. Besides yielding much data of nuclear and atomic interest and providing training in the Zacharias style for a new generation of physicists, the research in this laboratory laid the groundwork for the cesium beam atomic clock—perhaps even more remarkable than radar for its translation of fundamental physics into technological reality.

The postwar years saw Zacharias rise to prominence on the national scene, initially in defense policy. Particularly notable was his leadership in 1950 (together with the late Admiral Forrest Sherman) of Project Hartwell, a study of the protection of surface vessels in the face of modern submarines and nuclear weapons. Also of great significance was his leadership of a summer study in 1952 during which the Distant Early Warning Line was conceived. Then, from 1952 to 1964, during the Eisenhower and Kennedy administrations, he served as a member of the President's Science Advisory Committee.

During the mid 1950s, Zacharias turned his abundant energies and organizing abilities toward problems of national education in science. In 1956 he created the Physical Science Study Committee; its chief product, the PSSC high-school physics course, set a new standard for teaching at this level (see the detailed account in PHYSICS TODAY, September 1986, page 30). From there his interests broadened to education at all levels and in many different fields; indeed, he believed deeply and passionately that education was the key to all the social and political problems that beset a democracy.

During the last two decades of his life, Zacharias became more and more concerned with the potential, for good or for disaster, of nuclear energy, but again he felt that proper education of the public was vital; opinion not rooted in hard facts was worthless. Everyone who worked with him became familiar with his catechism: "What, Why, How, Who, For whom, When, Where, How much . . . ?" And through it all, as Philip Morrison has said, he was a spokesman for values and standards. He was, in the words of another of his associates, "a force of nature"—a very long-range force, one might add. His many honors included the President's Certificate of Merit (1948), the Oersted Medal of the American Association of Physics Teachers (1961), the National Science Teachers Association Citation for distinguished services to science teaching (1969), the Medal of the International Commission on Physics Education (1984), and the I. I. Rabi Award of the Annual Frequency Control Symposium (US Army and IEEE, 1986).

HERMAN FESHBACH
ANTHONY P. FRENCH
ALBERT G. HILL
JOHN G. KING
Massachusetts Institute of Technology
Cambridge, Massachusetts

Frank Oppenheimer
1912–1985

PHYSICS TODAY/NOVEMBER 1985

Frank Oppenheimer died on 3 February 1985 in Sausalito, California. Professionally a physicist, Oppenheimer was also a musician, a rancher and, not least, a teacher and an educator.

Oppenheimer was a graduate of Johns Hopkins; he worked at the Cavendish Laboratory in England and the Istituto di Arceti in Italy and received his PhD at Caltech in 1939. After spending 1939–40 at Stanford University, he went to work with Ernest O. Lawrence on the electromagnetic separation of uranium isotopes at Oak Ridge. He then came to the Los Alamos weapons laboratory in late 1943. There he became right-hand man to Kenneth T. Bainbridge, who was in charge of all preparation for the Trinity test site in the southern New Mexico desert. Oppenheimer had the primary responsibility for the instrumentation of that first nuclear explosion.

Reflections and discussions about the military and historical consequences of nuclear weaponry were never absent from the wartime centers of the Manhattan Project, as Alice Kimball Smith's history makes clear (A. K. Smith, *A Peril and a Hope: The Scientists' Movement in America* 1945–47, Chicago, 1965). At isolated Los Alamos they were muted, though I remember a good many. I was an administrative assistant to Robert Oppenheimer, and I later wrote the wartime history of the laboratory. What was latent at Los Alamos got strong expression after Hiroshima and Nagasaki. Frank Oppenheimer was involved in the beginning of the Association of Los Alamos Scientists, and somewhat later he joined in the national Federation of American Scientists. I think we all soon realized that we were plowing fresh ground, politically. There simply were no informed and intelligent party platforms on the subject, Left, Right or "Center." From then on, for many of us, the cause of nuclear pacificism preempted political energies. Not pacifists from the beginning, those who worked on the Manhattan Project knew what their work had done to two cities, and what upward of a millionfold enhancement of explosive power could mean for the institution of war. If that realization came too late, it has seemed to come far later to the big political world. Though we and others failed in those early efforts toward some system of international controls, the commit-

OPPENHEIMER

ment has remained. With Oppenheimer it surely did.

After the war, Oppenheimer returned to Berkeley for a time and then in 1947 went to the University of Minnesota. There he undertook research that proved to be a landmark in the development of cosmic-ray physics. At that time there was still no firm knowledge of the nature or origin of the rays. The mean value of their upwardly skewed energy distribution was greater, by a factor of a thousand, than the greatest cyclotron energies, and no known astronomical process could produce such energies. Fermi first pointed out that atoms could reach such energies by equipartition with the stars; but whatever the mechanism of their acceleration, the crucial step was to catch the primary rays themselves at high altitudes. Oppenheimer chased balloons across the Minnesota countryside by car and plane and, later, at low altitudes, across the Caribbean—courtesy of the US Navy. His group and that of B. Peters (University of Rochester) collaborated; the one using cloud chambers, the other photo emulsions. In 1948 they shared the discovery that primary rays were a sampling of the whole periodic table (P. Freier, E. J. Lofgren, E. P. Ney, F. Oppenheimer, H. L. Bradt, B. Peters, Phys. Rev. **74**, 213, 1948).

All this was interrupted in 1949, when Oppenheimer was called before the Un-American Activities Committee and asked to denounce as Reds some of his prewar political associates. He had been among those who were later known in intelligence circles by the deftly chosen label "premature antifascists." Having the power to compel testimony, the committee could bring an action for contempt of Congress

against anyone who lacked a legal right of refusal. Oppenheimer was one of the first persons who courageously took the position that he would answer any question about his own past political activities, but he would not talk about those of his friends. In doing so he refused to claim any rights on grounds of self-incrimination. Although the committee brought no action against him, the University of Minnesota, ingloriously intimidated, fired him. And the political climate was such that for a long time no other university administration dared offer him a position.

Oppenheimer's talents stood him in good stead in his next career, that of a Colorado rancher. Already a good carpenter, plumber and mechanic, he and his family became hard-working members of a ranch community high in the Blanco Basin. In 1957 the teacher in their country elementary school left suddenly and Oppenheimer took over. The next year Pagosa Springs High School needed a science teacher and Oppenheimer took the job, promising to get (and getting) teaching credentials.

In 1959 he was invited to help staff a summer institute at the University of Colorado designed to introduce the PSSC physics course to high-school teachers; he was in charge of the laboratory. This led to further work at the University and to his becoming a consultant for the Jefferson County schools. When he was offered a full-time position in the physics department, he was back to teaching and research; his research program in particle physics utilized bubble-chamber films from the accelerator laboratories. For student laboratories Oppenheimer moved in a new direction: In a very large attic space, using standard industrial furniture, he undertook the design of what he called a library of experiments. Much of each experiment was predesigned and hard wired, but the design itself was aimed to give students flexibility in deciding how to use it. Helped by an NSF grant, he and his colleague Malcolm Correll radically transformed the whole year's course in a little more than a year. The stodgy conventional relic of a lab was gone, and in its place was a matrix for a far more investigative way to learn physics.

Oppenheimer soon extended this way of teaching to science teaching for the public at large, for adults and for children. He spent some time looking at science museums here and abroad, took a leave of absence from Colorado and went to San Francisco to create the Exploratorium. For that, of course, he is now famous in the scientific community, to museum specialists and to a remarkably large public.

Exhibits in the Exploratorium are designed to reveal phenomena and extend our perceptions of them. The didactic mode, "explanation," is avoided in favor of asking questions of Nature. In its 15-year history, under Oppenheimer's leadership, the Exploratorium has attracted a wide public, and it has also linked itself with the schools: Teachers and children can use its exhibits in somewhat the way the physics lab in Boulder was intended, a way more persistent and systematic than that of the merely curious visitor. Perhaps in such liaisons between formal and informal education there can be fresh hope for evolving a society in which scientific understanding is not so narrowly and dangerously restricted as in our world today.

One basic mark of the Exploratorium's style is the shop where exhibits are first put together crudely, then tested, revised and tested again. The shop is a highly visible part of the museum and the visitors contribute to the museum's design as well, because their reactions are observed and recorded. What they bring with them, in terms of interest and preconception, determines what they will learn. The staff must be learners themselves: primarily about the exhibit's subject matter and its importance, but also about the visitors' ideas—how they match and mismatch the staff's own. In short, as a staff member one really becomes a teacher. That was Oppenheimer's greatest precept and example. (A special issue of *The Exploratorium* magazine, March 1985, was devoted mostly to Oppenheimer's own talks and writings.)

While contributing significantly to research physics, Oppenheimer's choices were increasingly toward a deepening commitment to education. He was himself a fine and reflective teacher, but his most profound and lasting achievement was surely the creative extension of our resources for science education, on the pattern of the Exploratorium.

DAVID HAWKINS
University of Colorado □

Nicholas C. Christofilos
1916–1972

PHYSICS TODAY/JANUARY 1973

Nicholas C. Christofilos, one of the most original thinkers in physics, died of a heart attack 24 September, at the age of 55. He leaves a legacy of ideas and inventions in the diverse fields of particle accelerators, controlled fusion and military applications. At the time of his death he headed the Astron controlled fusion experiment at the Lawrence Livermore Laboratory.

Born in Boston, Christofilos was taken by his parents to Greece at an early age and remained there through World War II. He studied engineering at the National Technical University in Athens. It was during his employment with a Greek elevator firm, and later in his own firm, that he began the private study of physics that was to reshape his career.

American physicists first learned of Christofilos as the independent co-inventor of the strong focusing principle of magnetic containment for particle accelerators, which he had patented while still in Greece. This principle has resulted in the successful exploration of nuclear phenomena at high energies through a succession of electron accelerators at Cornell, Harvard-MIT, Daresbury, Hamburg and Yerevan; and proton accelerators at Brookhaven, CERN, Serpukhov and the National Accelerator Laboratory. He had also invented independently in 1946 an accelerator similar to the synchrotron. In 1963 he was awarded the Elliott Cresson Medal by the Franklin Institute for this and other outstanding work.

Communication with US scientists about his accelerator ideas led to his first employment as a physicist at the Brookhaven National Laboratory in 1953. There he perfected his Astron concept of controlled fusion, which was first conceived in Greece when the subject was still highly classified. This finally brought him to the University of California Radiation Laboratory at Livermore in 1956, where the Astron project was initiated soon thereafter. He continued to make important contributions to the accelerator field in the development of proton linear accelerators and collective accelerators.

His Astron idea calls for confining a plasma within the magnetic field produced by an intense circulating electron beam (the E-layer). The E-layer

CHRISTOFILOS

electrons would also serve to heat the plasma to thermonuclear ignition temperatures (100 million degrees). Characteristically, his visionary approach first required the development of a new kind of electron accelerator that would by itself have provided a career for lesser men. The result was the Astron linear induction accelerator, which is today among the most powerful electron accelerators in the world. Though deprived of completing the Astron project himself, Christofilos believed that during the last year of his life the Astron facility had been perfected to the point that the physics principles embodied in the concept can now be tested.

Christofilos was intensely proud of his American birth and citizenship. He understood well the military needs of the Nation and conscientiously devoted a significant fraction of his life to improving the US strategic posture. In this, his amazing ability at mental calculations, reminiscent of the late John von Neumann, and his immense reserve knowledge of the work of others placed him in the position of an invaluable technical antagonist. His prolific mind and fresh point of view kept many teams of more conventional scientists busy analyzing the details of this work. His own ideas were urged with a single-minded gusto that at times made him impatient with the "system." The system is much improved by his interaction with it.

Two of his most imaginative defense projects known to the public are Argus and Sanguine. Project Argus has been called the greatest scientific experiment ever conducted. For the first time world-wide measurements were made of geophysical phenomena resulting from a precisely known disturbance. The idea, which Christofilos conceived well before the discovery of the natural radiation belts, involved injecting billions of high-energy electrons into the earth's magnetic field at an altitude of 500 kilometers by means of small nuclear bursts. Experiments carried out in 1958 confirmed the prediction that electrons resulting from decay of the fission products would quickly spread out to form a toroidal electron shell surrounding the earth.

Christofilos was the father of ELF communication which is the basis of the Sanguine system still under consideration at his death. A practical means of exciting Schuman resonance of the earth's ionospheric cavity was not known until he suggested the use of a grounded horizontal antenna near the surface of the earth. Beautifully simple, his idea requires little sophisticated hardware, and yet permits strategic communication to submarines all over the world. Recent experiments have demonstrated the technical feasibility of the concept.

Many of his ideas were controversial and defending them would sometimes seem to demand more ingenuity and unremitting effort than even his great spirit and energy could sustain. Yet he never gave up. To these trying situations he brought a subtle sense of humor that only a few were privileged to know. In recent years, when Astron was his main concern, it was suggested to him that after all Astron was perhaps but a dark horse in the race for controlled fusion. "No," he smiled, "Astron is not a dark horse. It is just a white horse standing in the shadows."

He was devoted to music, especially Beethoven, whose birthday he shared. He had studied concert piano as a child and when he found time he still played, as he did physics: fortissimo. Intensely preoccupied with work, he made few casual friends. Instead, he leaves behind many who admired him greatly and a select group of loyal, close associates who still pursue his dreams. For them, there can never be another Nick.

JOHN S. FOSTER
Department of Defense
T. KENNETH FOWLER
Lawrence Livermore Laboratory
FREDERICK E. MILLS
Brookhaven National Laboratory

Luis W. Alvarez 1911–1988

PHYSICS TODAY/JUNE 1989

Luis W. Alvarez poses with some of the liquid hydrogen bubble chambers he developed. In his hand Alvarez holds the primitive two-inch glass chamber, with which he first explored hydrogen bubble chamber technology.

With the death of Luis W. Alvarez on 1 September 1988, the world lost one of its truly great experimental physicists. Luie, as he was known to all, manifested every one of the talents one associates with experimental physics: inventiveness, quantitative design of experiments, experimental skill and inspired interpretation—and excellent intuition. Luie was not a detailed analyst; although acquainted with theory he did not view the purpose of experimental work to be either to verify or to disprove theoretical prediction. He developed strong convictions as to what was or was not important and what was right and wrong. In making these judgments he tended to disregard the intermediate, and in so doing classified the work of his colleagues as deserving either the highest praise or no attention at all.

Luie's work was divided into several distinct periods, with little overlap. Just this phenomenon—shifting from one type of activity to the next—reveals his habit of sharply dividing activities into those that were worth doing and those that were not, at any given time. Alvarez's work spanned an enormous range of topics and an equally large range of methodologies. Some of it was individually done, "small" science and investigation, yet much became "big" science involving large collaborations. He was always generous in giving credit and public prominence to collaborators and students.

Luie started his career at the University of Chicago as a student of chemistry but quickly shifted to physics. Although his thesis dealt with an optics problem, his most important contribution was based on the adaptation of the then emerging Geiger counter technology to cosmic-ray physics. Under the general direction of Arthur H. Compton, and working largely in Mexico, he carried out an experiment on the "east-west effect" seen in cosmic rays. This work showed for the first time that cosmic rays carry an excess of positive charge.

After taking his degree Luie became a prominent member of Ernest Lawrence's Radiation Laboratory at the University of California, Berkeley. This early period, starting in 1934, was possibly the most concentrated creative epoch in Luie's work with respect to contributions to pure physics. He discovered K capture of atomic electrons as a β-decay process. He measured the characteristics of tritium decay and determined the stability of helium-3. He studied scattering of slow neutrons on many substances including *ortho*- and *para*-hydrogen, and thereby contributed to the knowledge of the spin dependence of nuclear forces. In a classic collaboration with Felix Bloch of Stanford University, Luie combined his experience with slow neutrons with Bloch's expertise in nuclear magnetism to determine the magnetic moment of the neutron. Each one of these papers constituted a major advance in experimental nuclear physics.

Then the war started, and Luie joined the Radiation Laboratory at MIT. Here his inventiveness and "hands on" ability in designing and building equipment became a major force in shaping many of the devices that laboratory contributed to radar. In time Luie became the head of a special section dedicated to original inventions (in contrast to the other sections, which were organized to work on specialized components).

Luie's contributions were many; we will mention only a few prominent ones. He designed the airborne radar called VIXEN, dedicated to the detection of enemy submarines. At that time, the endurance of submarines underwater was low, so they had to spend a fair amount of time on the surface. Luie designed VIXEN so that the intensity of the transmitted radar signals would vary roughly as the cube of the distance between transmitter and target. The consequence was that as the airplane carrying the transmitters and antisubmarine weapons approached the target, the signal as seen by the target (obeying an inverse square law) would decrease, while the signal as seen by the radar receiver (obeying an inverse fourth-power law) would increase. Thus the skipper of the submarine would be fooled into assuming that the attacking aircraft was receding rather than approaching, while the radar operators received an increasing signal-to-noise ratio.

Luie also worked on early-warning

systems and ground-based radars. He was the inventor of the phased-array radar, in which a beam is steered electronically rather than mechanically. Perhaps his most lasting contribution was the invention, development and demonstration of the Ground Controlled Approach radar. Consisting of a ground-based transmitter and receiver that give information on the path of a landing aircraft, GCA enables a ground-based pilot to "talk down" an aircraft to a safe landing under adverse conditions. Luie was involved personally in critical demonstrations of this technology to military "brass" both in the United States and in Great Britain. In the British demonstrations, he helped guide aircraft crippled in combat over Germany, and manned by tired and, frequently, injured crews, to landings theretofore thought impossible. GCA continued to be used throughout the war, and later became the principal tool that made it possible for US aircraft to land at a rate of two per minute during the Berlin blockade. Luie received the highest aviation award given by the US government— the Collier Trophy—for these achievements.

Notwithstanding the strict compartmentalization of information during the war, Luie maintained some communication with the atomic bomb project. When room for invention in the radar work had diminished, Luie joined the nuclear project at the University of Chicago, where he enjoyed the inspiration of working with Enrico Fermi and where he contributed innovations to instrumentation of the early nuclear reactors. However, his work on nuclear weapons did not really come into its own until he and his family moved to Los Alamos. There he worked initially with the explosives division and made seminal contributions to the timing mechanisms that synchronized the detonation of the various lens elements of the implosion device. At the request of Robert Oppenheimer he then pursued yield assessments of the first three nuclear explosions by measuring their shock waves through telemetry. Luie personally directed this activity all the way from design and initial tests to final use, and he participated in the Hiroshima mission himself. The device used in the Hiroshima bombing consisted of a condenser microphone-modulated transmitter, dropped by parachute, that sent a telemetry signal to one of the airplanes participating in the raid. With official approval Alvarez wrote a letter to the well-known Japanese physicist Ryokichi Sagane,

who had worked at Berkeley before World War II, and attached it to the battery case of the shock-wave telemetry package. The letter informed Sagane of the nature of the nuclear weapon and requested that he communicate the relevant facts to the Japanese high command. This letter was recovered by the Japanese and delivered; the telemetry device is still on view in the Hiroshima Memorial Museum. This episode constitutes a truly original means of communicating with the enemy.

Along with many of his colleagues, Luie contemplated what research to pursue upon returning to academic life. In Luie's case this meant giving thought to new concepts of accelerator construction for nuclear physics based on the use of wartime technologies and equipment. Thus his main contributions immediately after the war were in the accelerator arts rather than physics research. The decision to build the Berkeley linear accelerator was based on Luie's realization that the cost of linear accelerators scaled directly with energy, while the cost of existing cyclotrons varied as the cube of the energy. Also contributing to the decision was Luie's expectation that the vast quantities of surplus ground-based radar equipment would provide "free" radiofrequency power sources for a linear accelerator. Thus resulted the first practical linear accelerator, the 32-MeV proton machine at Berkeley, built under Luie's direction. This machine has remained the standard for proton injectors all over the world, notwithstanding that the basic premises on which it was conceived proved incorrect: The invention of phase stability by Edwin McMillan and Vladimir Veksler, applied to the proton synchrotron, changed the scaling law for high-energy proton machines from cubic to linear, and most of the surplus radar gear had to be replaced with more suitable, dedicated equipment.

Luie also invented the tandem electrostatic accelerator, in which negative ions are accelerated from ground potential to a foil maintained at a high positive electrostatic voltage. The foil then strips the negative ions into positive nuclei, which are accelerated back to a target at ground potential. This method not only doubles the available beam voltage but maintains both the ion source and target complex at ground potential. Another of Luie's inventions was the microtron, an accelerator in which the pathlength of successive orbits is increased by one wavelength of the rf power source. (He never reduced this

idea to practice, but others did.)

During the Korean War, Alvarez's accelerator design skills were put to the test by the pressure that developed at Berkeley to contribute to the production of fissionable materials. At that time, Lawrence and Alvarez had concluded that the US was threatened by a cutoff of supplies of uranium, which came largely from overseas. They decided that it would become necessary to "breed" plutonium from either unseparated uranium or uranium depleted in the fissionable isotope U^{235}. Initial plans, later rejected, called for the construction of a fast reactor in the Berkeley hills. These plans were abandoned in favor of breeding plutonium using the neutrons produced by spallation from the impact of high-energy proton or deuterium beams on high-atomic-number targets. After such neutron yields were measured in the synchrocyclotron beam at Berkeley, a prototype accelerator was built under Luie's direction at Livermore, the site Lawrence had selected for the reactor initially planned. The prototype, a deuterium linear accelerator, used the basic principles of the 32-MeV machine but was scaled up vastly in size and average beam current by operating at much lower radiofrequencies. Housed in a 60×60-foot tank, it produced 7-MeV deuterons at an average current of $1/4$ ampere. After its completion it was used for some time for nuclear physics research, but its original purpose disappeared after uranium supplies within the US increased. The plans for a 1-gigawatt production accelerator were abandoned.

With the exception of his participation in proton–proton scattering with the 32-MeV accelerator and some cosmic-ray work using a balloon-borne solenoid, Luie's direct immediate post-war participation in physics research did not match that of his early productive period at Berkeley. This situation was dramatically reversed when he became acquainted with Donald Glaser and his invention, the bubble chamber.

In 1954 the Bevatron came into operation at the Radiation Laboratory in Berkeley, and Luie turned his attention to experiments in particle physics. Lifetime measurements were his first focus of interest: He carried out an emulsion experiment on τ mesons and later a counter experiment designed to measure lifetimes of the various K decay modes.

During the early days of the Bevatron, Luie's contributions to particle physics were threefold. In 1953, right after Glaser announced the invention of the bubble chamber, Luie began research on the construction of a chamber using hydrogen as a working fluid. As soon as he and his team had demonstrated that hydrogen could be used as a working fluid in a laboratory bubble chamber, Luie began construction of a gigantic, 72-inch hydrogen bubble chamber. The important breakthrough here was that he demonstrated that a metal box with glass windows (the so-called dirty chamber) would work just as well as Glaser's small, clean, glass container.

Luie's second major contribution during this period arose from his realization that the old-fashioned projection tables used with cloud chambers would no longer suffice for the large numbers of events that one observes in helium and hydrogen chambers. He spearheaded the construction of automatic scanning and measuring equipment (the so-called Frankensteins) whose output data would be stored on punched cards and then evaluated on large electronic computers.

Luie's third major contribution was the large number of important physics discoveries he and his coworkers made using the tools he had developed. They discovered the Y* (1385 MeV)—the first new resonance—and then in rapid succession the K* (890 MeV)—the first meson resonance—the ω^0 meson and many others. These discoveries represented a major step forward in the physics of particle states, leading to the "eightfold way" and eventually to quarks. In fact, Murray Gell-Mann, the leading figure in those theoretical developments, was in close touch with Luie and his group throughout this period.

The techniques Luie and his group developed were rapidly taken up by many physicists all over the world. During the 1960s large bubble chambers were developed at Brookhaven National Laboratory, at the Rutherford Laboratory in England, at CERN in Switzerland and at the Ecole Polytechnique in Paris. Though these chambers differed in design from Luie's, the confidence that large chambers could be built and the belief that they were essential for particle physics research come directly from Luie's work.

The same is true for the automatic measuring machines and, in particular, the accompanying computer programs that reconstruct the events in space and calculate the kinematical parameters that fit the data. Such programs have been accepted, either outright, with various modifications, or as models, by all physicists engaged in the analysis of bubble chamber pictures. They have also been adapted to reconstruct events in drift chambers and other modern electronic detectors. The Monte Carlo methods Luie's group developed for event simulations and detector efficiency calculations have become a major industry, and are used universally by both experimenters and theorists.

The contributions Luie's group made using the 72-inch bubble chamber continued for well over a decade, and were a major factor in establishing the new spectroscopy of hadrons. In typical Alvarez fashion he observed that the productivity of that chamber was limited by the repetition rate of the Bevatron, rather than by fundamental engineering limits. When the Stanford Linear Accelerator Center developed hadron beams of higher energies and higher repetition rates than were available at the Bevatron, he initiated a re-engineering of the chamber to use that opportunity. He transferred the Bevatron's machinery and much of its operating staff to SLAC, where they received a new lease on life. In taking this action, Luie put the interests of science ahead of institutional and personal interest. The discoveries made with the large hydrogen bubble chambers earned Luie the Nobel Prize.

Paradoxically, Luie's participation in high-energy physics went "full circle." He recognized that Glaser's invention could be exploited only by converting Glaser's small chambers to large, reliable machines, building massive analytical devices and organizing large operating teams: Luie's bubble chamber effort is often cited as the decisive step to "big science." Yet in his later years he expressed a strong dislike of big science and returned to a variety of individual activities including inventions and technical detective work. Much of this work excited the public imagination. Luie searched for hidden burial chambers in Egyptian pyramids using cosmic-ray muons. He studied the events surrounding President Kennedy's assassination by examining the effects of shock waves on Abraham Zapruder's camera and carrying out experiments on the terminal ballistics of bullets in human brains (simulated by melons). He invented lens systems of variable focal length by superimposing appropriately shaped laminae of glass.

His most dramatic application of physics to other fields came with his work bearing on the extinction of species, which he himself considered the most important scientific achievement of his career. (See Alvarez's article in PHYSICS TODAY, July 1987, page 24.) Together with his son Walter, a geologist, Luie discovered an unusually high concentration of an isotope of iridium at the geological boundary between the Cretaceous and Tertiary formations. In association with his colleagues working in radiochemistry at the Lawrence Berkeley Laboratory, Luie determined that this iridium was of extraterrestrial origin and had presumably reached Earth via a meteoric impact. Luie extrapolated that such a massive impact would have raised so much dust into the atmosphere that it would have produced worldwide darkness sufficient to impede biological processes. This he associated with the disappearance of the tiny marine organisms known as foraminifera, as well as dinosaurs and other species, between the Cretaceous and Tertiary periods. While there is little argument about the correctness of the association of the iridium layer with the impact of an extraterrestrial boloid, the details of how this event might be connected to extinctions are still in dispute. Many paleontologists maintain that Alvarez's hypothesis cannot by itself give a satisfactory picture of the Cretaceous–Tertiary extinctions.

Whatever the verdict on this controversy turns out to be, Luie's discovery has induced major changes in thinking about geological and evolutionary processes. Detailed examinations of the soil chemistry of the Cretaceous–Tertiary boundary and other geological strata associated with mass extinctions have been made worldwide. Some studies have correlated apparent periodicities of major extinctions with the frequency of occurrence of meteoric impacts. Luie's findings have triggered interest in the "nuclear winter" phenomenon, which associates danger to life with the dust and soot raised into the atmosphere by nuclear explosives.

Luie's productive life has had its share of controversy, triggered primarily by his attitude that at any one epoch scientific questions have only one best answer and that those who do not accept that verdict are largely wasting their time. Thus when, through his work, the hydrogen bubble chamber became the most productive tool in discovering new hadronic states at high energy, he felt that high-energy physics resources should be almost totally concentrated on bubble chamber work. This caused conflict between Luie and those of his colleagues at the Lawrence Berkeley Laboratory who did not feel that science should be conducted under such a principle of absolute priority. During the period when Lawrence and Alvarez tried to rededicate the resources of the Lawrence Berkeley Laboratory to breeding uranium, not everyone was persuaded that the emergency requiring this action existed. The debate between Luie and the disbelievers in his theory of mass extinctions has at times been acrimonious.

Luie had an amazing knowledge of the entire literature of physics, and he exhibited the opposite of the "not invented here" syndrome: When he felt that others had attacked a problem responsibly, he would not attempt to modify their technique just to prove he might have a better way. For example, when at Los Alamos Luie agreed to assess the yield of nuclear explosions using shock-wave measurements, he ascertained from then-classified reports that Wolfgang K. H. Panofsky had already developed instrumentation he could adapt to that purpose, and he used those devices.

Luie Alvarez, like most academic physicists, immersed himself in military technology during World War II. Unlike many of his academic colleagues he pursued such work intensively (if intermittently) throughout his life. His scientific standards remained high throughout, and he justified his work as the means to peace. He disapproved of SDI. Luie was part of the first wave of US scientists to visit nuclear facilities in the Soviet Union in 1956. He was deeply moved by the occasion and argued that nuclear explosives had made war a thing of the past.

Luie Alvarez was an extraordinarily gifted experimental physicist, an inventor, an investigator and a strong individual who maintained the highest standards of truth in inquiry. The world is poorer without him.

GERSON GOLDHABER
Lawrence Berkeley Laboratory
Berkeley, California

WOLFGANG K. H. PANOFSKY
Stanford Linear Accelerator Center
Palo Alto, California

___For Further Reading___

Several authors represented in this volume have written books related to their Physics Today articles. The individual article may be in the nature of a preview or an overview, or it may concentrate on some particular aspect of the more comprehensive account in the fuller treatment. Examples of such books are Finn Aaserud's *Redirecting Science* (on the Niels Bohr Institute for Theoretical Physics), Spencer Weart's *Scientists in Power* (including Jean Perrin), and Janet Oppenheim's study of spiritualism and psychic research in Victorian England. A. I. Miller's piece on Heisenberg is based on his book, *Imaging in Science*, and R. W. Smith's article on Hubble is at the core of his book on the expanding universe. Mark Walker's controversial article on Heisenberg and Goudsmit is related to his own book on the German quest for nuclear power and to Goudsmit's *Alsos*. A more complete and definitive survey of Heisenberg's role both as a German nationalist and in physics generally will be found in David Cassidy's very recent biography.

Bibliographic information on books and others related to topics discussed here is given below. These books include biographies of Bohr, Dirac, Rabi, and Raman, all people who are featured in this volume. Two of the pieces on Landau correspond to talks given at a Landau Birthday Symposium held in Copenhagen in 1988 and subsequently published in *Landau, the Physicist the Man.* For Alvarez we have not only his autobiography but also a volume of selected works and commentary by students and colleagues.

Also listed are other biographies and some autobiographies of physicists not featured directly in the *Physics Today* articles. These are relevant since the world of physics is essentially one world, and interconnections between people who do physics are inevitable. General books on the history of physics have not been listed, with one exception, namely, the book of selected reprints, including bibliography, edited by S. G. Brush for the American Association of Physics Teachers.

Aaserud, Finn, *Redirecting Science: Niels Bohr, Philanthropy, and the Rise of Nuclear Physics* (Cambridge University Press, New York/Cambridge, 1990).

Abragam, Anatole, *Time Reversal, An Autobiography* (Oxford University Press, New York, 1989).

Alvarez, Luis W., *Alvarez, Adventures of a Physicist* (Basic Books, New York, 1989).

Beyerchen, Alan, *Scientists Under Hitler* (Yale University Press, New Haven, 1977).

Blaadel, Niels, *Harmony and Unity: The Life of Niels Bohr* (Science Tech, Madison, Wisconsin, 1988).

Brown, Laurie M., Max Dresden, and Lillian Hoddeson, eds., *Pions to Quarks: Particle Physics in the 1950s* (Cambridge University Press, New York, 1989).

Brush, Stephen, G., *History of Physics; Selected Reprints* (American Association of Physics Teachers, College Park, Maryland, 1988).

Cassidy, David C., *Uncertainty: The life and science of Werner Heisenberg* (W. F. Freeman and Company, San Francisco, California, 1991).

Dresden, Max, *H. A. Kramers: Between Tradition and Revolution* (Springer-Verlag, New York/Berlin/Heidelberg, 1987).

Fermi, Laura, *Atoms in the Family: My Life with Enrico Fermi* (University of Chicago Press, Chicao, 1954; Reprinted American Institute of Physics, New York, 1987).

Fukushima, Eeichi, *NMR in Biomedicine: The Physical Basis. Key Papers in Physics 2* (American Institute of Physics, New York, 1989).

Goldberg, Stanley and Roger H. Stuewer, eds., *The Michelson Era in American Science 1870–1930* (American Institute of Physics Conference Proceeding 179, New York, 1988).

Goldschmidt, Bertrand, *Atomic Rivals* (Rutgers University Press, New Brunswick, New Jersey, 1990 [Translated from the French edition (1987) by George M. Temmer].

Goudsmit, S. A., *Alsos* (Henry Schuman, Inc., New York, 1947; Reprinted, American Institute of Physics, New York, 1988).

Hafemeister, David, ed., *Physics and Nuclear Arms Today*, Readings from *Physics Today* 4 (American Institute of Physics, New York, 1991).

Hobbie, Russell K., ed., *Medical Physics: Selected Reprints* (American Association of Physics Teachers, College Park, Maryland, 1986).

Hoop, Bernard, ed., *Physical Principles of Physiological Phenomena: Selected Reprints* (American Association of Physics Teachers, College Park, Maryland, 1986).

Jayaraman, A., *Chandrasekhara Venkata Raman* (Allied East–West Press, New Dehli, 1989) (Available from the author, AT&T Bell Laboratories, Murray Hill, New Jersey 07974).

Kursunoglu, B. N. and Wigner, E. P., eds., *Paul Adrien Maurice Dirac: Reminiscences About a Great Physicist* (Cambridge University Press, Cambridge/New York, 1990).

Khyalatnikov, I. M., ed., *Landau, The Physicist and the*

Man, Translated from the Russian by J. B. Sykes (Pergamon Press, New York, 1989).

Kraft, Fritz, "Lise Meitner: Her Life and Times— On the Centenary of the Great Scientist's Birth," *Ange. Chem.*, International Edition in English **17**, 826–842 (1978).

Kragh, Helge S., *Dirac, A Scientific Biography* (Cambridge University Press, Cambridge/New York, 1990).

Laurikainen, K. V., *Beyond the Atom: The Philosophical Thought of Wolfgang Pauli* (Springer-Verlag, New York, 1988).

Lifton, Robert Jay, *The Genocidal Mentality: Nazi Holocaust and Nuclear Threat* (Basic Books, New York, 1990).

Miller, A. I., *Imagery in Scientific Thought: Creating 20th Century Physics* (Berkhauser, Boston, Massachusetts, 1984).

Oppenheim, Janet, *The Other World: Spiritualism and Psychical Research in England, 1850–1914* (Cambridge University Press, New York, 1985).

Pais, Abraham, *Niels Bohr's Times in Physics, Philosophy and Polity* (Clarenden Press, Oxford, 1991).

Pais, Abraham, *Inward Bound: An Autobiography* (Oxford University Press, New York, 1985).

Rigden, John S., *Rabi, Scientist and Citizen* (Basic Books, New York, 1987).

Sime, Ruth Lewin, "Lise Meitner's Escape From Germany," *Am. J. Phys.* **58**, 262–267 (1990).

Smith, Robert W., *The Expanding Universe: Astronomy's "Great Debate" 1990–1931* (Cambridge University Press, Cambridge/New York, 1982).

Trower, Peter, ed., *Discovering Alvarez; Selected Works of Luis W. Alvarez with Commentary by His Students and Colleagues* (University of Chicago Press, Chicago, Illinois, 1988).

Venkataraman, G., *Journey Into Light; Life and Science of C. V. Raman* (Indian Academy of Sciences in cooperation with the Indian National Science Academy, Bangalore, 1988) (Distributed by Oxford University Press).

Wali, Kameshwar, *Biography of S. Chandrasekhar* (University of Chicago Press, Chicago, Illinois, 1991).

Walker, Mark, *German Nationalism and the Quest for Nuclear Power, 1939–1949* (Cambridge University Press, New York, 1989).

Weart, Spencer R., *Nuclear Fear: A History of Images* (Harvard University Press, Cambridge, Massachusetts, 1988).

Weart, Spencer, R., *Scientists in Power* (Harvard University Press, Cambridge, Massachusetts, 1979).

Weart, Spencer R. and Melba Phillips, eds., *History of Physics*, Readings from *Physics Today* **2** (American Institute of Physics, New York, 1985).

Weisskopf, Victor, *The Joy of Insight* (Basic Books, New York, 1991).

Westfall, Catherine, *The First Truly National Laboratory: The Birth of Fermilab* (University Microfilms, Ann Arbor, Michigan, 1988).